Introductory Analysis
Volume 1

Introductory Mathematics for Scientists and Engineers

A Wiley Series, edited by

D. S. JONES, M.B.E., F.R.S.E., F.R.S.

Ivory Professor of Mathematics in the University of Dundee

Already published:
R. C. SMITH and P. SMITH   Mechanics

Forthcoming:
D. S. JONES and D. W. JORDAN   Introductory Analysis Vol. 2 (*in press*)

A. R. MITCHELL   Computational Methods in Partial Differential Equations (*in press*)

H. LIEBECK   Introductory Algebra (*in press*)
J. M. RUSHFORTH   Computers and Computing
C. DIXON   Applied Mathematics
J. D. LAMBERT   Finite Difference Methods in Ordinary Differential Equations

Introductory Analysis
Volume 1

D. S. JONES

*Department of Mathematics, University of
Dundee, Scotland*

and

D. W. JORDAN

*Department of Mathematics, University of
Keele, Staffordshire*

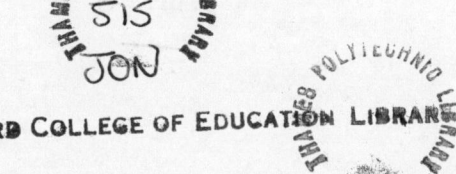

JOHN WILEY & SONS   LONDON NEW YORK SYDNEY TORONTO

Copyright © 1969 John Wiley & Sons Ltd.
All Rights Reserved. No part of this publication
may be reproduced, stored in a retrieval system,
or transmitted, in any form or by any means,
electronic, mechanical photocopying, recording
or otherwise, without the prior written permission of the Copyright owner.

Library of Congress catalog card number 72–93560

SBN 471 44710 2 Cloth bound

SBN 471 44712 9 Paper bound

Printed in Great Britain at the Aberdeen University Press, Aberdeen

This book is dedicated to JIMY

Introductory Mathematics for Scientists and Engineers

Foreword to the Series

The increasing use of high speed digital computers and the growing desire for numerical answers in many disciplines have led to a steady expansion in the numbers of mathematics courses in the first year or two of college and university studies. Many of these courses are intended for students of physics, chemistry, engineering, biology and economics who will regard mathematics as a tool, but a tool with which they must develop some proficiency. This series is designed for such students. However, that does not prevent prospective mathematicians from reading these books with profit.

The authors have, in general, avoided the strict axiomatic approach which is favoured by some pure mathematicians, but there has been no dilution of the standard of mathematical argument. Learning to follow and construct a logical sequence of ideas is one of the important attributes of a course in mathematics.

While the authors' purpose has been to stress mathematical ideas which are central to applications and necessary for subsequent studies, they have attempted, when appropriate, to convey some notion of the connection between a mathematical model and the real world. Exercises have been included which take account of the ready availability of electronic digital computers.

The careful explanation of difficult points and the provision of large numbers of worked examples and exercises should ensure the popularity of the books in this series with students and teachers alike.

D. S. JONES
Department of Mathematics
University of Dundee

Preface

Volumes 1 and 2 of this book together present the coordinate geometry and analysis needed by students of the sciences in the early stages of a university or college course, and by those taking a pre-specialization course for mathematicians. The work is based on lecture courses given at the University of Keele: one to an unselected group of first-year students having eventual specialisms ranging from mathematics to the social sciences but not yet committed to a principal field of study, and another to those already specializing in the physical sciences. These groups had studied calculus at school, but the present book does not assume previous experience of the subjects covered.

Our intention is to stress those mathematical ideas most central in scientific applications, and to emphasize the applicable nature of the subject by providing, as far as possible, physical, graphical or numerical illustration of all principal results. It is important to convey the relation between mathematical and physical language as, for example, with the ideas of density and 'infinite distance', but examples requiring a detailed understanding of scientific topics have been avoided. On the other hand, much attention is given to simple numerical processes, which provide excellent illustrations of the uses of analysis in a field of general importance.

A high standard of mathematical argument is used, mitigated by a leisurely exposition; proofs are generally given in full, but when a proper proof would be too difficult the corresponding result is stated and discussed without proof. Theoretical results are very fully illustrated by textual description and by many worked examples. Some sections are printed in small type. Their omission does not put the reader at a disadvantage in understanding what follows, but they can be read with profit even in a first course.

Volume 1 contains the elements of plane coordinate geometry, the theory of limits, sequences and series (including power series), and elementary integration. The first six chapters are devoted to obtaining the basic concepts in plane geometry and the derivative, together with applications so that students can understand their significance. Advantage has been taken of the fact that many students will also have contact with digital computers at the same time as they are learning this material.

Two very important aspects of mathematics which scientists are constantly meeting are the defining of functions by series and the use of sequential procedures in numerical processes. For this reason there are chapters on sequences (including a discussion of iteration in numerical work) and series before the exponential and logarithm are defined. It is believed that the consequential short delay in the introduction of the exponential is amply repaid by the thorough understanding of analytical and numerical methods which form a basic tool in the scientist's equipment.

The remaining chapters deal with integration. Integration by parts and change of variable are postponed to Volume 2, since most of the commonly needed integrals are obtainable by elementary means. The idea of the integral as a sum is central in scientific problems and a whole chapter is devoted to it. A quasi-Riemann presentation is adopted which is largely descriptive, the main theorem being stated without proof. There is also a more elementary account, so that subsequent sections may be omitted if desired.

Tables giving lists of derivatives, integrals, some useful formulae and the values of certain functions will be found in an Appendix at the end of the book.

The authors are grateful to Miss J. Seivwright and Mrs. M. Brown for their efficiency in typing much of the manuscript.

D. S. Jones
D. W. Jordan

Contents

Preface ix

1 Coordinates
1.1 Coordinates 1
1.2 Distance between points 3
1.3 Trigonometry 7
1.4 The slope of a straight line 11
1.5 The equation of a straight line 20
1.6 Change of origin 25
1.7 Rotation of axes 26
1.8 Polar coordinates 28

2 Conics
2.1 The circle 31
2.2 The parabola 34
2.3 Tangents and normals 39
2.4 The ellipse 44
2.5 The hyperbola 49
2.6 Curves of the second degree 54

3 Functions and graphs
3.1 Functions 64
3.2 The modulus 74
3.3 Polynomials and rational functions . . . 77

4 Limits and continuity
4.1 Limits 84
4.2 One-sided limits 95
4.3 Other limits 99
4.4 Continuous functions 101
4.5 Properties of continuous functions . . . 105
4.6 The bisection method 109
4.7 Flow-charts 112

5 The derivative
5.1 Rate of change 116

5.2	Numerical approximations	124
5.3	Geometrical interpretation	125
5.4	The derivative of a polynomial	130
5.5	Newton's method for roots	135
5.6	Derivatives of products and quotients	138
5.7	The derivative of an implicit function	143
5.8	The chain rule	146
5.9	Trigonometric functions	151
5.10	Small errors and corrections	154

6 Maxima and minima

6.1	Higher derivatives	158
6.2	Maxima and minima—theory	161
6.3	Maxima and minima—applications	168
6.4	Curve sketching	175

7 The mean-value theorem

7.1	Rolle's theorem	181
7.2	The mean-value theorem	184
7.3	Simple error analysis	187
7.4	Generalized mean-value theorem	191
7.5	Indeterminate forms	192
7.6	Taylor's theorem	200
7.7	Numerical work	205
7.8	Interpolation	210

8 Sequences

8.1	Sequences	213
8.2	Properties of sequences	220
8.3	Iteration	227
8.4	Round-off error	238

9 Infinite series

9.1	Infinite series of constant terms	245
9.2	Series of positive terms	251
9.3	Absolute convergence	258
9.4	Power series	267
9.5	Taylor series	277
9.6	Summation notation	282

10 Functions defined by series
- 10.1 The exponential function 285
- 10.2 The natural logarithm 293
- 10.3 The function a^x 299
- 10.4 The general logarithm 304
- 10.5 Trigonometric functions 307
- 10.6 Hyperbolic functions 309
- 10.7 Leibnitz's theorem 315
- 10.8 Inverse functions 318
- 10.9 Inverse hyperbolic functions 328

11 Elementary integration methods
- 11.1 The inverse of differentiation 332
- 11.2 The integrals of a function 333
- 11.3 The symbol for integral 338
- 11.4 A table of basic integrals 340
- 11.5 Functions of $ax + b$ 344
- 11.6 Factorable integrals 351
- 11.7 The integrals of rational functions with a quadratic denominator and related types 356
- 11.8 Integration of more general rational functions by resolution into partial fractions 360

12 The definite integral and some applications
- 12.1 The application of integration to area problems . . 362
- 12.2 Further applications of the definite integral . . . 368
- 12.3 Some important properties of the definite integral . 375
- 12.4 The area analogy for definite integrals . . . 376
- 12.5 The definite integral as a function of its upper limit . 380
- 12.6 Area of sectorial regions in terms of polar coordinates . 381
- 12.7 Volumes of revolution 385
- 12.8 Problems on density 394
- 12.9 The numerical estimation of definite integrals . . 400
- 12.10 Integration of power series 408
- 12.11 Some inequalities involving integrals 410

13 Another view of integration: the integral as a sum
- 13.1 The need for a revised idea of integral . . . 415
- 13.2 Integration as a summation process 416
- 13.3 The Riemann Sum 424

13.4 Some fundamental properties of the definite integral as newly
defined 428
13.5 The integral as a continuous function of its upper limit . 431
13.6 Comparison of the old definition of integration with the new 433
13.7 The indefinite integral 435

Answers 440

Appendix of tables 461

Index 477

1

Coordinates

1.1 Coordinates

The object of analytical geometry is to provide an algebraic method of resolving geometrical problems. The starting point, therefore, is one of identifying points of a page of this book or of a plane by numerical means. First of all, we choose a horizontal straight line, extending indefinitely to the left and to the right, which is called the *x-axis*. A particular point O of the line is chosen and called the *origin*. Then, a unit of length having been selected, the axis is marked off in units starting from O. Those marks to the right of O are identified by attaching the number $+a$ to the point which is a units to the right of O. Similarly, $-a$ is attached to the point a units to the left of O. O, itself, is identified by zero (see figure 1). In this way, by giving a positive or negative decimal number we can specify one, and only one, point of the *x*-axis. Note that a small arrow is placed on the *x*-axis to make sure that you remember which side has the positive numbers.

Draw through O a vertical line, extending indefinitely upwards and downwards; this is called the *y-axis*. Mark it off in units and attach positive numbers to points above O, negative numbers to those below O.

Let P be any given point. Draw a line through P parallel to the *y*-axis to pass through the point marked a on the *x*-axis. Draw a line through P parallel to the *x*-axis to pass through the point marked b on the *y*-axis. Then P will be denoted by (a,b). No other point has the same representation; either a or b or both will be different as you can easily verify.

We say that the *coordinates* of P are (a,b). Correspondingly, the *x*- and *y*-axes are called the *coordinate axes*. Sometimes a is called the *abscissa* of P and b the *ordinate*.

If we are given the coordinates (a,b) of a point we can easily find it by drawing a line parallel to the *y*-axis through the point marked a on the *x*-axis, and a line parallel to the *x*-axis through the point marked b on the *y*-axis. The intersection of these two lines is (a,b).

2   *Introductory Analysis*

The two axes divide the page into four *quadrants*, labelled I, II, III, IV in figure 1 and called the first, second, third and fourth quadrants respectively. Points in the first quadrant have both coordinates positive; this is indicated by the notation $(+,+)$ in figure 1. Similarly, in the third quadrant, both coordinates are negative, and so on.

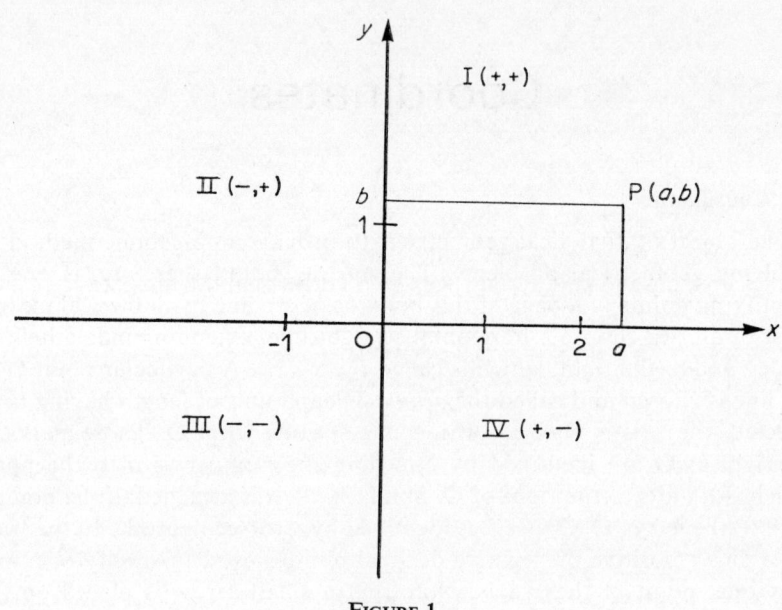

FIGURE 1

Exercises

1. In which quadrant is (i) the abscissa positive, (ii) the ordinate negative, (iii) the abscissa positive and the ordinate negative?

2. Plot the following points (i) (2,1), (ii) (2,−1), (iii) (−1,2), (iv) (−2,2), (v) (−2,−2), (vi) (2·3,1·4), (vii) (2·4,−1·3), (viii) (−1·8,2·1), (ix) (−2,0), (x) (0,−2), (xi) (0,2·1), (xii) (2·1,0).

3. A straight line is drawn through (1,3) and (−1,1). Draw a graph to find where it intersects (i) the y-axis, (ii) the x-axis.

4. A straight line is drawn through (2,1) and (3,2). Draw a graph to find the angle it makes with the positive x-axis.

Coordinates

5(a). A ladder is leaning on top of a vertical wall of height 8 ft 8 in. If the foot of the ladder is a horizontal distance of 5 ft from the wall, find graphically the angle the ladder makes with the horizontal.

(b). In a model ship the mast is 86·7 cm tall and the horizontal support for the sail is 50 cm long. What angle does the slanting side of the sail make with the horizontal?

The system of axes shown in figure 1 is known as *right-handed*. It would be quite possible to use coordinates in which the y-axis was in the opposite direction to that shown in figure 1 so that positive values of y were measured downwards instead of upwards. Such axes form a *left-handed* system. Either right-handed or left-handed systems can be employed so long as one always uses the same type. Most people adopt the right-handed system, and this system will be used throughout this book.

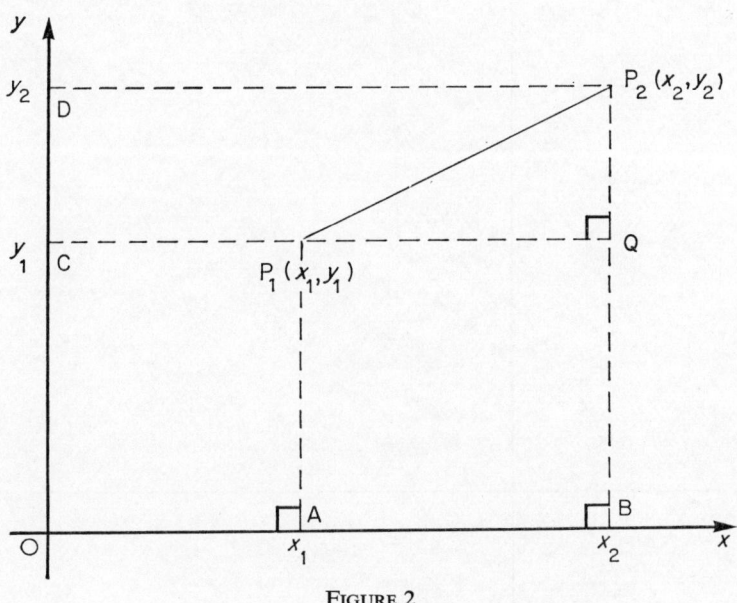

FIGURE 2

1.2 Distance between points

Let $P_1(x_1, y_1)$ and $P_2(x_2, y_2)$ be two given points. Draw parallels to the x- and y-axes through P_1 and P_2 (shown dashed in figure 2). Then A and C give the x- and y-coordinates of P_1; B and D give similar information

for P_2. Clearly the number of units in AB is $x_2 - x_1$. Since P_1Q is the same length as AB, it must have the same number of units, i.e.
$$P_1Q = x_2 - x_1. \qquad (1)$$
Similarly,
$$QP_2 = CD = y_2 - y_1. \qquad (2)$$
From Pythagoras's theorem, applied to triangle P_1QP_2, we find that the distance d between $P_1(x_1,y_1)$ and $P_2(x_2,y_2)$ is given by
$$d^2 = (P_1P_2)^2 = (P_1Q)^2 + (QP_2)^2 = (x_2 - x_1)^2 + (y_2 - y_1)^2$$
from (1) and (2). Since distance is always taken as positive
$$d = \sqrt{[(x_2 - x_1)^2 + (y_2 - y_1)^2]} \qquad (3)$$
where here, and subsequently, the positive value of a square root is used unless otherwise indicated.

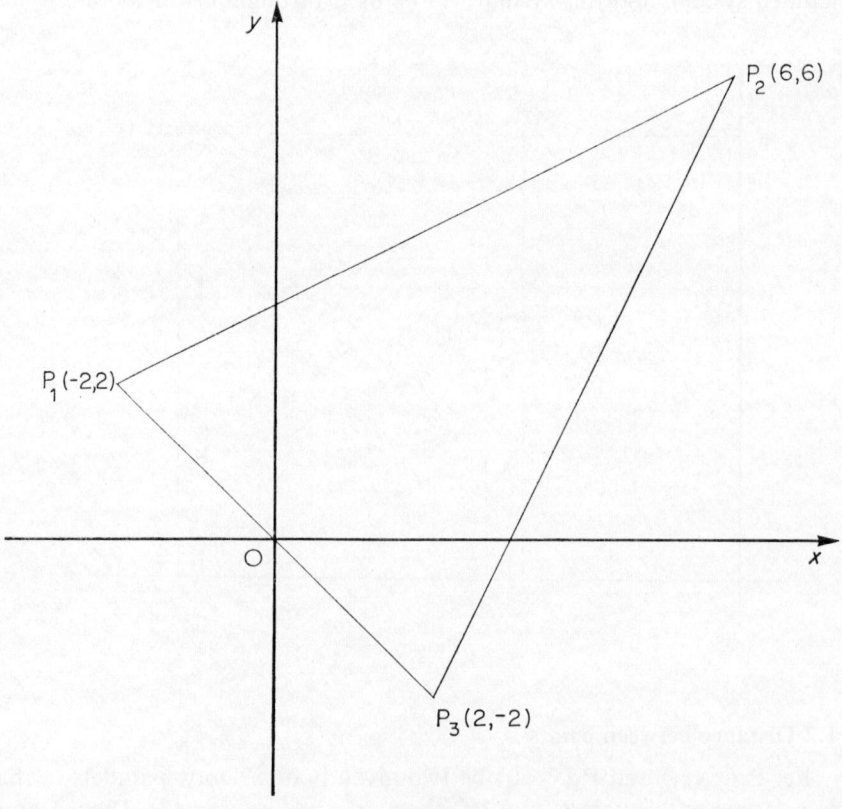

FIGURE 3

Example 1   If P_1 is $(1,2)$ and P_2 is $(2,-1)$, we have
$$x_2 - x_1 = 2 - 1 = 1,$$
$$y_2 - y_1 = -1 - 2 = -3$$
and
$$d = \sqrt{[(1)^2 + (-3)^2]} = \sqrt{10}.$$

Example 2   Show that the points $P_1(-2,2)$, $P_2(6,6)$, $P_3(2,-2)$ are the vertices of an isosceles triangle (*figure 3*).

$$(P_1P_2)^2 = \{6 - (-2)\}^2 + (6 - 2)^2 = 64 + 16 = 80,$$
$$(P_1P_3)^2 = \{2 - (-2)\}^2 + (-2 - 2)^2 = 16 + 16 = 32,$$
$$(P_2P_3)^2 = (6 - 2)^2 + \{6 - (-2)\}^2 = 16 + 64 = 80.$$

Since $P_1P_3 = P_2P_3$, the triangle is isosceles.

Example 3   Show that the points $P_1(1,2)$, $P_2(2,5)$, $P_3(3,8)$ lie on a straight line (*figure 4*).

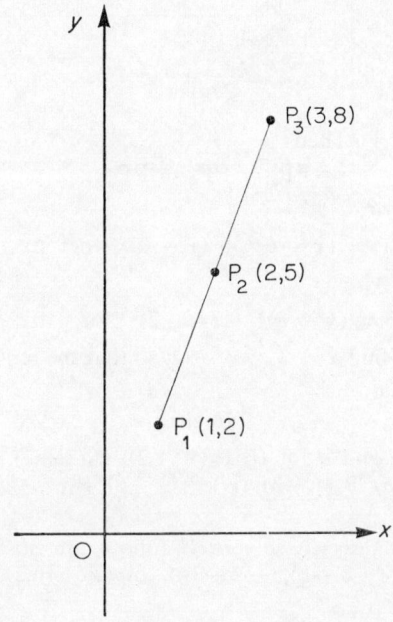

FIGURE 4

$$(P_1P_2)^2 = (2 - 1)^2 + (5 - 2)^2 = 1 + 9 = 10,$$
$$(P_1P_3)^2 = (3 - 1)^2 + (8 - 2)^2 = 4 + 36 = 40,$$
$$(P_2P_3)^2 = (3 - 2)^2 + (8 - 5)^2 = 1 + 9 = 10.$$

Hence $P_1P_2 = \sqrt{10}$, $P_1P_3 = 2\sqrt{10}$ $P_2P_3 = \sqrt{10}$. Since $P_1P_2 + P_2P_3 = P_1P_3$ the points lie on a straight line.

Example 4 Find the point which is equidistant from $P_1(3,3)$, $P_2(6,2)$, $P_3(8,-2)$ (*figure 5*).

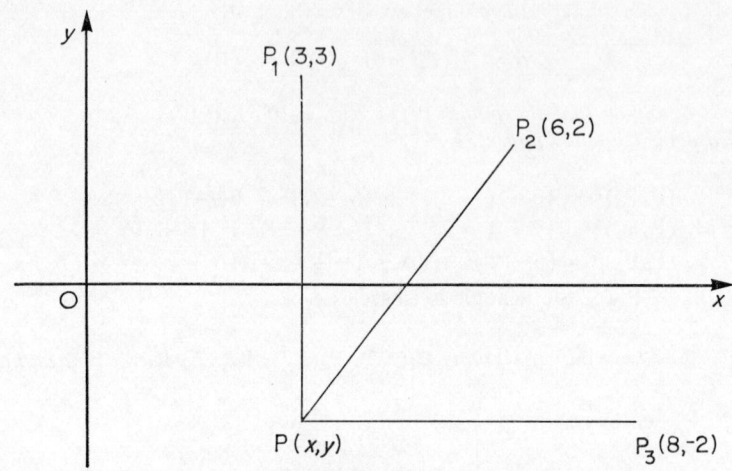

FIGURE 5

Let the point be $P(x,y)$. Then
$$PP_1 = PP_2 = PP_3.$$
Since $(PP_1)^2 = (PP_2)^2$,
$$(x-3)^2 + (y-3)^2 = (x-6)^2 + (y-2)^2 \quad \text{or} \quad 6x - 2y - 22 = 0. \quad (4)$$
Since $(PP_1)^2 = (PP_3)^2$
$$(x-3)^2 + (y-3)^2 = (x-8)^2 + (y+2)^2 \quad \text{or} \quad 10x - 10y - 50 = 0. \quad (5)$$
Solving (4) and (5) we find $x = 3$, $y = -2$, so that the required point is $(3,-2)$.

Exercises

6. Find Q, P_1Q, QP_2 and d for (i) $P_1(-1,2)$, $P_2(2,-2)$, (ii) $P_1(-6,4)$, $P_2(2,-11)$, (iii) $P_1(-2,-5)$, $P_2(1,-3)$, (iv) $P_1(3,-6)$, $P_2(3,-2)$.

7. Show that the triangles, whose vertices follow, are isosceles: (i) (4,8), (−10, 3), (−7,−2), (ii) (6,0), (−8,−8), (−2,−14), (iii) (3,5), (6,2), (7,6).

8. Show that the triangle with vertices (9,4), (2,1), (5,−6) is right-angled.

9. Show that the following points lie on straight lines: (i) (2,4), (5,−2), (0,8), (ii) (0,1), (−4,9), (3,−5), (iii) (0,0), (−3,−6), (2,4).

10. Find the point which is equidistant from (i) (4,4), (7,3), (9,−1), (ii) (0,0), (3,−1), (5,−5).

1.3 Trigonometry

The purpose of this section is to review briefly results of trigonometry which the reader is likely to have met previously. Those who are familiar with the results should proceed straight to the next section.

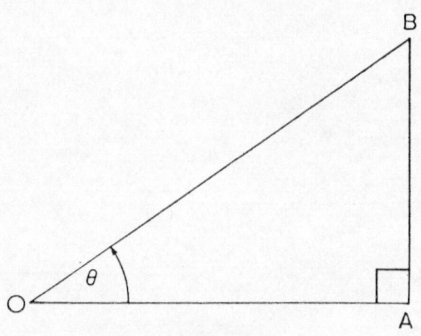

FIGURE 6

In the right-angled triangle OAB the angle $\angle$AOB is acute and denoted by θ in figure 6. Then trigonometric quantities are defined by

$$\sin \theta = \frac{1}{\operatorname{cosec} \theta} = \frac{AB}{OB}, \qquad (6)$$

$$\cos \theta = \frac{1}{\sec \theta} = \frac{OA}{OB}, \qquad (7)$$

$$\tan \theta = \frac{1}{\cot \theta} = \frac{AB}{OA} = \frac{\sin \theta}{\cos \theta}, \qquad (8)$$

From Pythagoras's theorem
$$(OA)^2 + (AB)^2 = (OB)^2$$
or
$$\left(\frac{OA}{OB}\right)^2 + \left(\frac{AB}{OB}\right)^2 = 1.$$

It follows from (6) and (7) that
$$\cos^2 \theta + \sin^2 \theta = 1. \qquad (9)$$

Tables of sin, cos and tan have been prepared for all positive acute angles. An extract of values will be found in Table 2 (Appendix).

When θ is obtuse we continue to use (6), (7) and (8) but in a slightly different way. Let O be the origin and OA the positive x-axis of a right-handed system of axes. Let B be (x,y). Then, in figure 1.6, OA $= x$, AB $= y$, OB $= \sqrt{(x^2 + y^2)}$ and

$$\sin \theta = \frac{y}{\sqrt{(x^2 + y^2)}}, \quad \cos \theta = \frac{x}{\sqrt{(x^2 + y^2)}}, \quad \tan \theta = \frac{y}{x}. \qquad (10)$$

Introductory Analysis

Now we agree to use (10) whatever quadrant B lies in. Thus, when B is in the second quadrant as in figure 7, the sin is positive, the cos is negative and the tan is negative. Figure 8 shows the signs in the various quadrants. Some people find the mnemonic 'all, sin, tan, cos' a convenient way of remembering which are positive in the various quadrants.

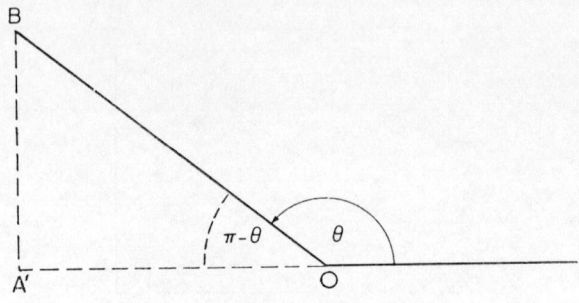

FIGURE 7

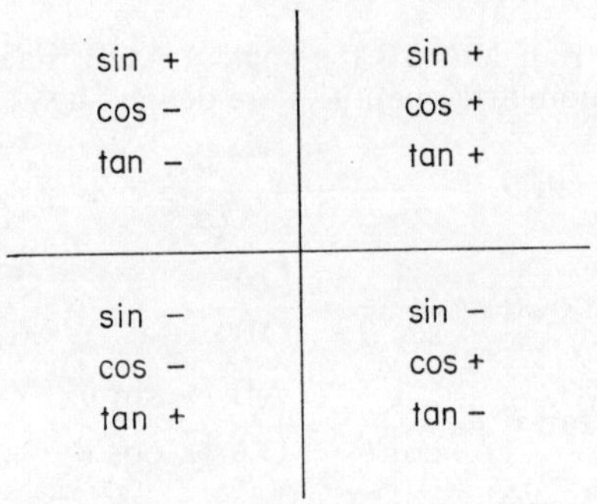

sin +	sin +
cos −	cos +
tan −	tan +
sin −	sin −
cos −	cos +
tan +	tan −

FIGURE 8

In triangle A'OB of figure 7, $\pi - \theta$ is an acute angle. Therefore

$$\sin(\pi - \theta) = \frac{A'B}{OB} = \frac{y}{\sqrt{(x^2 + y^2)}} = \sin\theta, \tag{11}$$

$$\cos(\pi - \theta) = \frac{OA'}{OB} = \frac{-x}{\sqrt{(x^2 + y^2)}} = -\cos\theta \tag{12}$$

because the x-coordinate of B is negative so that the length of OA' is $-x$. Dividing (11) by (12) we obtain

$$\tan(\pi - \theta) = -\tan\theta. \tag{13}$$

The reader should draw figures and verify that

$$\sin(\tfrac{1}{2}\pi - \theta) = \cos\theta, \qquad \cos(\tfrac{1}{2}\pi - \theta) = \sin\theta,$$
$$\tan(\tfrac{1}{2}\pi - \theta) = \cot\theta, \tag{14}$$

$$\sin(\tfrac{1}{2}\pi + \theta) = \cos\theta, \qquad \cos(\tfrac{1}{2}\pi + \theta) = -\sin\theta,$$
$$\tan(\tfrac{1}{2}\pi + \theta) = -\cot\theta, \qquad (15)$$
$$\sin(\pi + \theta) = -\sin\theta, \qquad \cos(\pi + \theta) = -\cos\theta,$$
$$\tan(\pi + \theta) = \tan\theta. \qquad (16)$$

Formulae (11)–(16) have been written in a form which is suitable when θ is measured in radians (for a reminder of *radian measure* see Example 8 of Chapter 4). If θ is expressed in terms of degrees, π should be replaced by 180°. However, in the calculus, it turns out to be most convenient to use angles in radian measure.

By convention, positive angles are drawn anticlockwise with right-hand systems of axes as illustrated in figure 7.

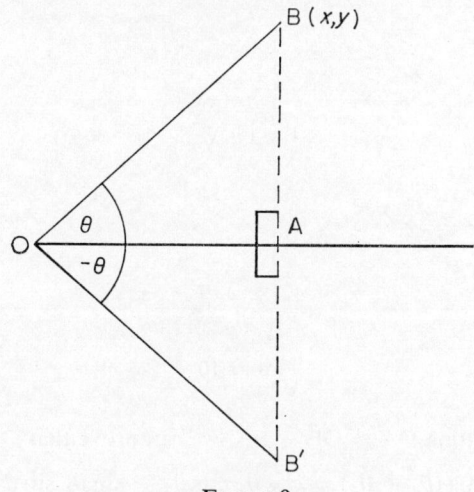

FIGURE 9

Negative angles can be handled by taking B′ the reflection of B in the x-axis (figure 9). Then

$$\sin(-\theta) = \frac{-y}{\sqrt{(x^2 + y^2)}} = -\sin\theta, \qquad (17)$$

$$\cos(-\theta) = \frac{x}{\sqrt{(x^2 + y^2)}} = \cos\theta, \qquad (18)$$

$$\tan(-\theta) = -\tan\theta. \qquad (19)$$

More complicated formulae involving angles are also often useful. For example, in figure 10,

$$\sin(\theta_1 + \theta_2) = \frac{AB}{OB}.$$

Draw BD perpendicular to OC, DE perpendicular to OA and DF perpendicular to AB. Then

$$AB = AF + FB = DE + BD \cos \angle FBD = OD \sin \theta_1 + BD \cos \angle FBD$$
$$= OB \cos \theta_2 \sin \theta_1 + OB \sin \theta_2 \cos \theta_1$$

since $\angle FBD = 90° - \angle BCD = 90° - \angle ACO = \angle COA = \theta_1$. Hence

$$\sin(\theta_1 + \theta_2) = \sin \theta_1 \cos \theta_2 + \cos \theta_1 \sin \theta_2. \tag{20}$$

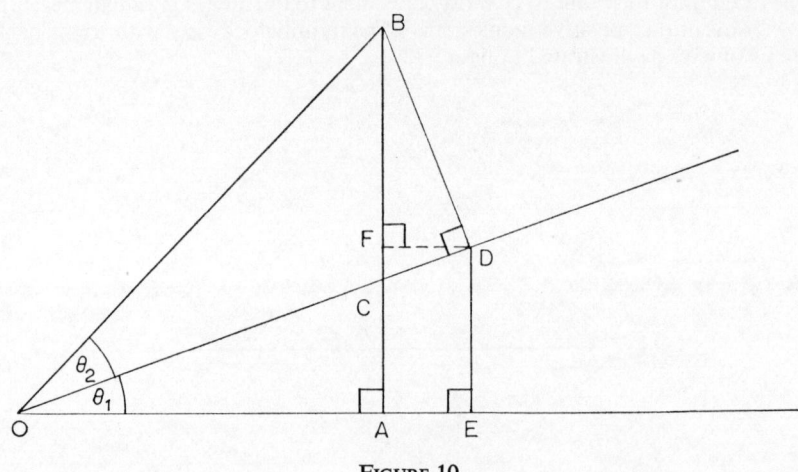

FIGURE 10

Similarly, by putting $OA = OE - EA$, we can prove that

$$\cos(\theta_1 + \theta_2) = \cos \theta_1 \cos \theta_2 - \sin \theta_1 \sin \theta_2. \tag{21}$$

These formulae obtained when the angles in figure 10 are acute are also valid when the angles are not acute; the proof is left to the reader.

Special cases arise when $\theta_1 = \theta_2 = \theta$; then

$$\sin 2\theta = 2 \sin \theta \cos \theta, \tag{22}$$

$$\cos 2\theta = \cos^2 \theta - \sin^2 \theta = 2 \cos^2 \theta - 1 = 1 - 2 \sin^2 \theta \tag{23}$$

when (9) is used to replace firstly $\sin^2 \theta$ by $1 - \cos^2 \theta$ and then $\cos^2 \theta$ by $1 - \sin^2 \theta$.

If the lengths of the sides of a triangle are a, b, c as shown in figure 11, $CD = b \sin A$ from triangle ADC and $CD = a \sin B$ from triangle DBC. Hence

$$b \sin A = a \sin B$$

or

$$\frac{\sin A}{a} = \frac{\sin B}{b}.$$

By drawing the perpendicular from B to AC a similar relation involving C can be obtained and so

$$\frac{\sin A}{a} = \frac{\sin B}{b} = \frac{\sin C}{c}. \tag{24}$$

Furthermore
$$(CB)^2 = (DB)^2 + (CD)^2 = (AB - AD)^2 + (AC)^2 - (AD)^2.$$
Now, $AB = c$, $AD = b \cos A$ so that
$$(AB - AD)^2 = c^2 - 2bc \cos A + b^2 \cos^2 A$$

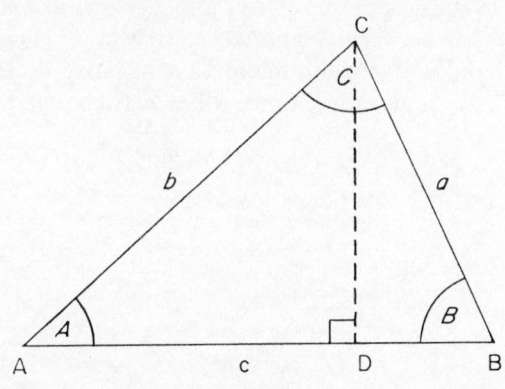

FIGURE 11

and, since $CB = a$ and $AC = b$,
$$a^2 = b^2 + c^2 - 2bc \cos A. \tag{25}$$
Finally,
$$\text{area ABC} = \text{area ADC} + \text{area DBC}$$
$$= \tfrac{1}{2}AD.DC + \tfrac{1}{2}DB.DC$$
$$= \tfrac{1}{2}(AD + DB).DC = \tfrac{1}{2}AB.DC.$$
Hence the area of the triangle is $\tfrac{1}{2}bc \sin A$.

1.4 The slope of a straight line

Consider a straight line P_1P_2 which is parallel to neither the x-axis nor the y-axis (figure 12). Let θ be the angle between P_1P_2 and the *positive* direction of the x-axis. Then the *slope m* of the line is defined by
$$m = \tan \theta.$$
From the triangle P_1QP_2
$$m = \frac{QP_2}{P_1Q} = \frac{y_2 - y_1}{x_2 - x_1}$$
which enables the calculation of the slope from the positions of two points on the line.

12   *Introductory Analysis*

Example 5 *Let P_1 and P_2 be $(-4,-5)$ and $(2,1)$, respectively. Then*

$$m = \frac{1-(-5)}{2-(-4)} = \frac{6}{6} = 1.$$

The angle θ corresponding to this slope is 45° since $\tan 45° = 1$.

It should be noted that the slope of a line of the sort shown in figure 12 is positive. This is because both $y_2 - y_1$ and $x_2 - x_1$ are positive. To put it another way, θ lies between 0° and 90° so that $\tan \theta$ is positive.

On the other hand, the slope in figure 13 is negative because $y_2 - y_1$ is positive but $x_2 - x_1$ is negative. Since θ lies between 90° and 180°, $\tan \theta$ is negative.

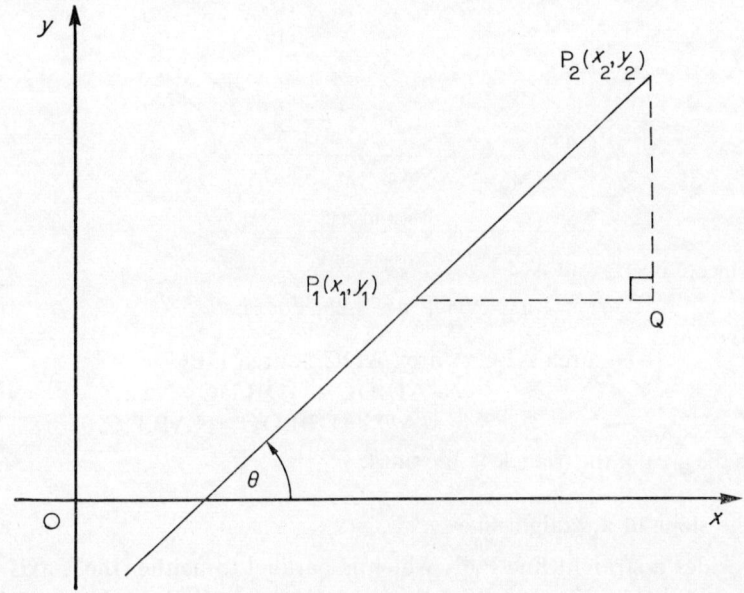

FIGURE 12

From figure 12 it can be seen that as θ becomes smaller and smaller P_1P_2 becomes more and more parallel to the x-axis. However, figure 13 shows that P_1P_2 becomes parallel to the x-axis, as θ approaches 180° or π radians. Thus, given a line parallel to the x-axis, we cannot say whether θ is 0° or 180° unless some rule is provided as to how it was obtained; the slope m of the line is 0 whatever the precise identification attached to θ.

The slope of a line parallel to the y-axis cannot be specified. If, in figure 12, θ steadily increases m increases and becomes very large when θ is near 90°, e.g. $m = 57 \cdot 29$ when $\theta = 89°$. Therefore, m becomes infinite as θ reaches 90° and, from this point of view, the slope of a line parallel to the y-axis might be regarded as positively infinite. However, if we allow θ to fall to 90° in figure 13, m becomes more and more negative—for $\theta = 91°$,

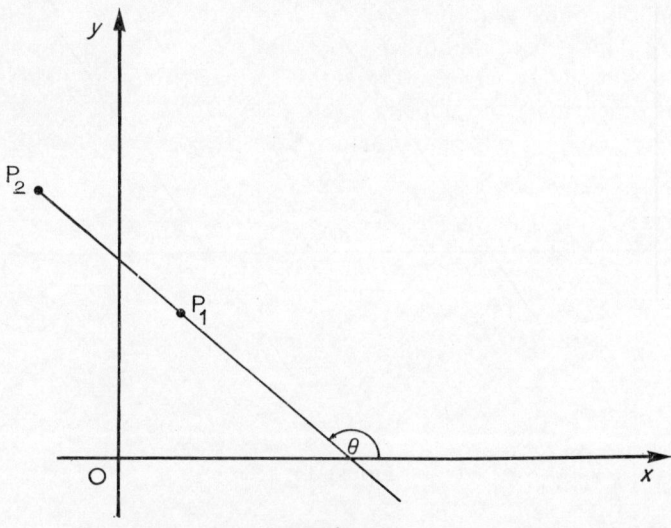

FIGURE 13

$m = -57 \cdot 29$. Thus, this point of view would give the slope of a line parallel to the y-axis as negatively infinite. Hence we cannot say precisely what the slope of such a line is, although we can say what an approach from either side would give.

If two lines are parallel (figure 14) $\theta_1 = \theta_2$ and $m_1 = m_2$. Conversely, if $m_1 = m_2$ then
$$\tan \theta_1 = \tan \theta_2$$
and the only solution of this with both θ_1 and θ_2 between 0° and 180° is $\theta_1 = \theta_2$. Hence, two lines are parallel if, and only if, their slopes are the same.

If two lines are perpendicular (figure 15) $\theta_2 = 90° + \theta_1$. Therefore
$$m_2 = \tan \theta_2 = \tan (90° + \theta_1) = -\cot \theta_1 = -\frac{1}{\tan \theta_1} = -\frac{1}{m_1}$$
i.e.
$$1 + m_1 m_2 = 0.$$

14 *Introductory Analysis*

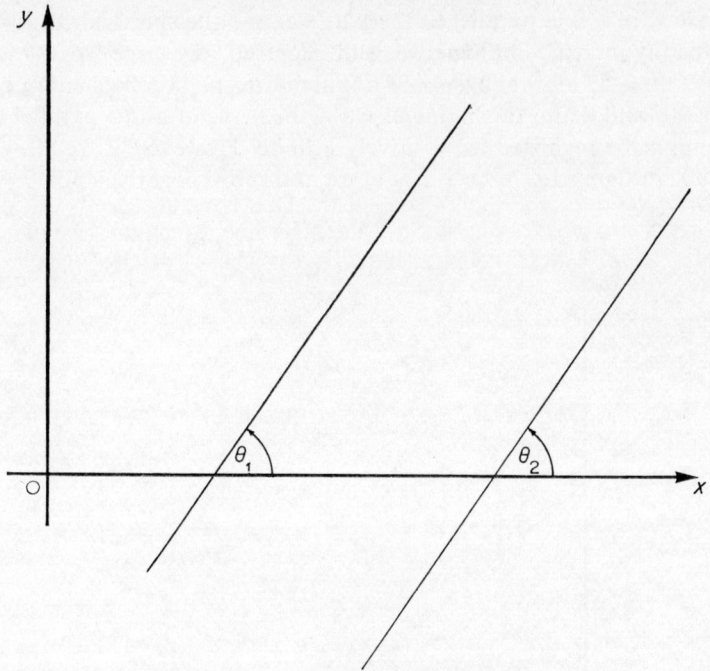

FIGURE 14

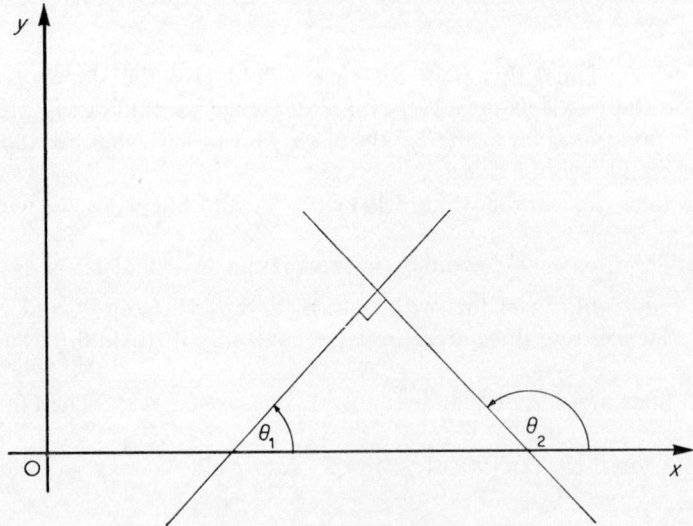

FIGURE 15

Conversely, if $1 + m_1 m_2 = 0$, $-\cot \theta_1 = \tan \theta_2$ of which the only solutions with θ_1 and θ_2 both between 0° and 180° are $\theta_1 = 90° + \theta_2$ and $\theta_2 = 90° + \theta_1$. In either case, the two lines are perpendicular. Thus, *two lines are perpendicular if, and only if,* $1 + m_1 m_2 = 0$.

This formula does not apply if one of the lines is parallel to the *x*-axis; the slope is then zero whereas the slope of the other line, which must be parallel to the *y*-axis, is infinite. The reader should examine whether the formula $m_2 = -1/m_1$ would still be valid.

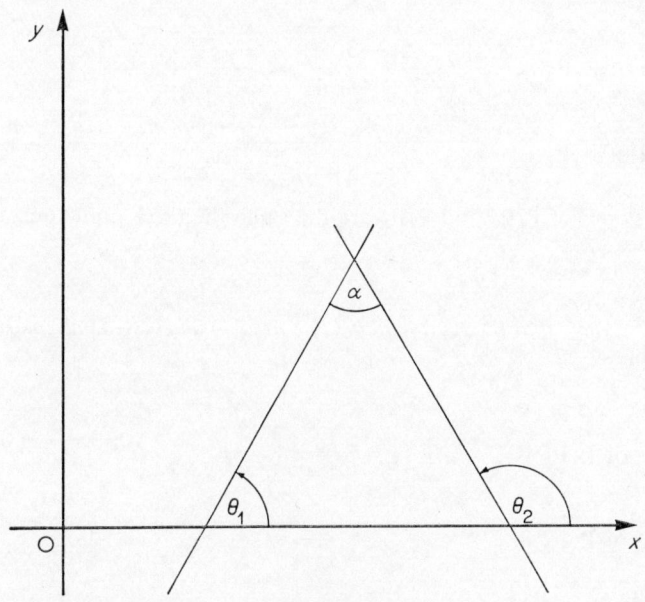

FIGURE 16

The angle α between two lines (figure 16) can be calculated from $\theta_2 = \theta_1 + \alpha$ or $\alpha = \theta_2 - \theta_1$. Therefore,

$$\sin \alpha = \sin (\theta_2 - \theta_1) = \sin \theta_2 \cos \theta_1 - \cos \theta_2 \sin \theta_1$$

from (20), (17) and (18). Similarly,

$$\cos \alpha = \cos (\theta_2 - \theta_1) = \cos \theta_2 \cos \theta_1 + \sin \theta_2 \sin \theta_1.$$

Hence,

$$\tan \alpha = \frac{\sin \theta_2 \cos \theta_1 - \cos \theta_2 \sin \theta_1}{\cos \theta_2 \cos \theta_1 + \sin \theta_2 \sin \theta_1} = \frac{\tan \theta_2 - \tan \theta_1}{1 + \tan \theta_1 \tan \theta_2}$$

on dividing numerator and denominator by $\cos \theta_1 \cos \theta_2$. Consequently,

$$\tan \alpha = \frac{m_2 - m_1}{1 + m_1 m_2}. \tag{26}$$

Observe that, for perpendicular lines, $1 + m_1 m_2 = 0$ and $m_2 \neq m_1$ (why not?) so that $\tan \alpha$ is infinite as it should be.

Example 6 *Show that the three points* $P_1(-4,3)$, $P_2(-1,5)$ *and* $P_3(2,7)$ *lie on the same straight line.*

The slope of $P_1 P_2$ is $\quad \dfrac{5 - 3}{-1 - (-4)} = \dfrac{2}{3}.$

The slope of $P_1 P_3$ is $\quad \dfrac{7 - 3}{2 - (-4)} = \dfrac{2}{3}.$

Since the slopes of $P_1 P_2$ and $P_1 P_3$ are the same the three points must lie on a straight line.

Example 7 *Show that the lines joining* $(1,1)$ *to* $(5,7)$ *and* $(4,-1)$ *are perpendicular* (*figure 17*).

The slope of $P_1 P_2$ is $\quad \dfrac{7 - 1}{5 - 1} = \dfrac{3}{2}.$

The slope of $P_1 P_3$ is $\quad \dfrac{-1 - 1}{4 - 1} = \dfrac{-2}{3}.$

Since $\dfrac{(-2)}{(3)} \dfrac{(3)}{(2)} + 1 = 0$ the lines are perpendicular.

Example 8 *Find the angle* $\angle P_1 P_2 P_3$ *in figure 17.*

The slope of $P_3 P_2$ is $(7 + 1)/(5 - 4) = 8$ and the slope of $P_1 P_2$ is $\frac{3}{2}$ from Example 7. It follows from (26) that

$$\tan \angle P_1 P_2 P_3 = \frac{8 - \frac{3}{2}}{1 + 8 \cdot \frac{3}{2}} = \frac{1}{2}$$

so that $\angle P_1 P_2 P_3$ lies between $26°$ and $27°$.

Coordinates

In fact, according to the Table 2 (Appendix), the values of the tangent lie on a straight line when plotted against angle, for the angles 25°, 26°, 27° and 28°; the line rises by 0·022 for each degree. To get a rise of 0·5 − 0·488 = 0·012 we need 0·012/0·022 of a degree or 0·55° approximately. Hence, $\angle P_1P_2P_3$ is about 26·55°.

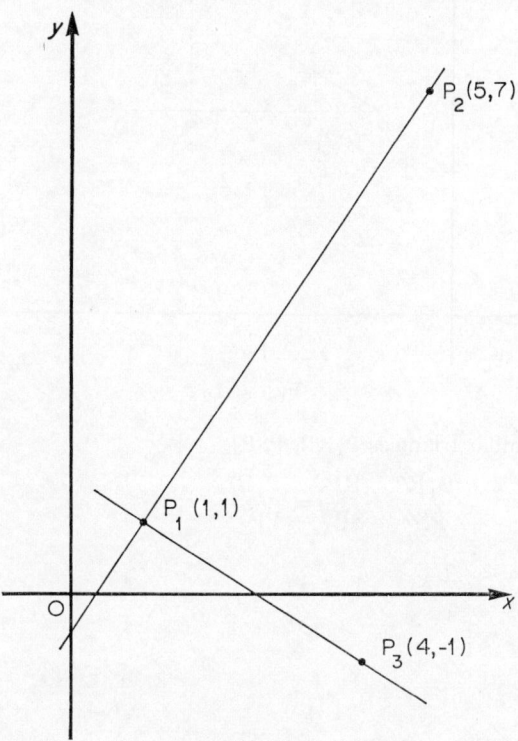

FIGURE 17

This is an example of *interpolation* in which an intermediate value is found from the values given in a table. As another example, suppose that tan 47·3° is required. The values for 46°, 47°, 48°, 49° do not lie strictly on a straight line but are not far off. Therefore we approximate the curve of tan x between 47° and 48° by the straight line joining tan 47° and tan 48°. In other words, we *approximate the curve by its chord;* this seems reasonable over a short distance when values are not changing too rapidly. Thus, 0·3° gives a rise of

$$0.3 \times (1.111 - 1.072) = 0.012.$$

Hence this method of interpolation gives tan 47·3° = 1·084. For more sophisticated interpolation the reader should refer to J. M. Rushforth's book in this series.

Example 9 *Find the point* P *on the line* P_1P_2 *such that* $P_1P/PP_2 = r$, *i.e. find the point which divides a line segment in a given ratio (figure 18).*

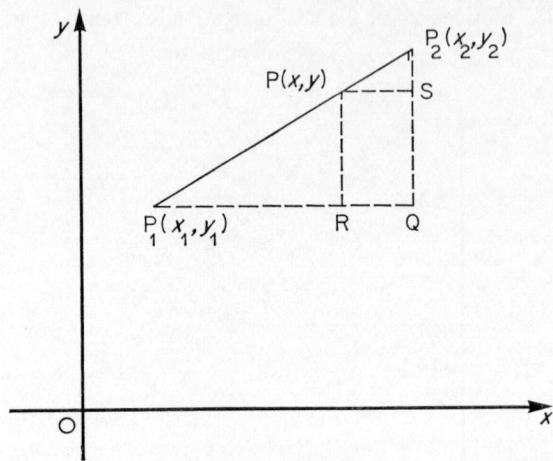

FIGURE 18

From the similar triangles P_1RP, PSP_2

$$\frac{P_1R}{PS} = \frac{RP}{SP_2} = \frac{P_1P}{PP_2} = r.$$

Therefore,

$$\frac{x - x_1}{x_2 - x} = \frac{y - y_1}{y_2 - y} = r. \tag{27}$$

Hence,

$$x = \frac{x_1 + rx_2}{1 + r}, \qquad y = \frac{y_1 + ry_2}{1 + r}. \tag{28}$$

When P lies beyond P_2 on P_1P_2 produced, PP_2 is taken as negative and in this case r lies between $-\infty$ and -1. However, since $x_2 - x$ and $y_2 - y$ are both negative when P is in this position, (27) remains valid and hence (28) is still true. Similarly, when P lies to the left of P_1 on P_2P_1 produced, P_1P is taken as negative and r lies between -1 and 0; again (28) continues to hold. Thus, all possible positions of P on P_1P_2 are covered by formula (28).

As a specific illustration, let P_1 be $(-1,2)$ and P_2 be $(3,6)$. Then, if $r = \frac{1}{3}$, (28) gives

$$x = \frac{-1 + \frac{1}{3} \cdot 3}{1 + \frac{1}{3}} = 0, \qquad y = \frac{2 + \frac{1}{3} \cdot 6}{1 + \frac{1}{3}} = 3.$$

If, however, $r = -3$

$$x = \frac{-1 - 3 \cdot 3}{1 - 3} = 5, \qquad y = \frac{2 - 3 \cdot 6}{1 - 3} = 8$$

whereas, if $r = -\frac{1}{3}$,

$$x = \frac{-1 - \frac{1}{3} \cdot 3}{1 - \frac{1}{3}} = -3, \qquad y = \frac{2 - \frac{1}{3} \cdot 6}{1 - \frac{1}{3}} = 0.$$

The relative positions of these points are shown in figure 19.

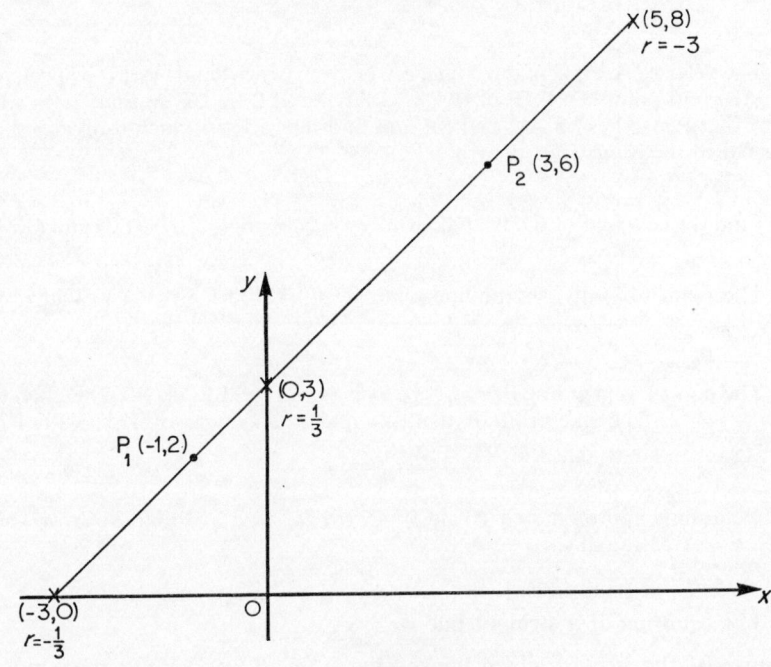

FIGURE 19

Exercises

11. Find the slopes of the lines joining the following pairs of points (i) $P_1(2,3)$, $P_2(3,4)$, (ii) $P_1(1,2)$, $P_2(-1,3)$, (iii) $P_1(1,2)$, $P_2(3,1)$, (iv) $P_1(-7,5)$, $P_2(3,5)$, (v) $P_1(x,0)$, $P_2(0,y)$.

12. Which of the following sets of points lie on a straight line? (i) $P_1(1,3)$, $P_2(2,5)$, $P_3(-3,-5)$, (ii) $P_1(1,-2)$, $P_2(-2,7)$, $P_3(-1,3)$, (iii) $P_1(-2,1)$, $P_2(0,2)$, $P_3(2,3)$, (iv) $P_1(6,0)$, $P_2(6,-6)$, $P_3(-2,2)$.

13. Find which of $P_1P_2P_3P_4$ in the following determine a parallelogram and which a rectangle (i) $P_1(-2,0)$, $P_2(-1,3)$, $P_3(-4,4)$, $P_4(-5,1)$, (ii) $P_1(1,1)$, $P_2(2,3)$, $P_3(0,5)$, $P_4(-1,2)$, (iii) $P_1(4,0)$, $P_2(2,1)$, $P_3(1,-1)$, $P_4(3,-2)$, (iv) $P_1(1,1)$, $P_2(-1,0)$, $P_3(0,1)$, $P_4(2,0)$.

14. Find the angles between the three lines which join $(1,-7)$, $(-3,-5)$ and $(-1,-1)$.

15. Find the point P on the line P_1P_2 such that $P_1P/PP_2 = r$ for (i) $P_1(1,2)$, $P_2(2,1)$, $r = \frac{1}{2}$, (ii) $P_1(-3,-2)$, $P_2(1,-1)$, $r = 3$, (iii) $P_1(2,-2)$, $P_2(-1,1)$, $r = -2$, (iv) $P_1(-1,2)$, $P_2(-2,0)$, $r = -\frac{1}{4}$.

16. The vertices A, B, C of a triangle are (x_1,y_1), (x_2,y_2) and (x_3,y_3) respectively. Find the mid-points D, E, F of BC, CA, AB. Find P on CF so that $CP = \frac{2}{3}CF$. Show that P also lies on AD and BE and find the ratio in which it divides them. (P is called the *centroid* of the triangle.)

17. Find the centroid of the triangle whose vertices are $(3,2)$, $(-1,3)$ and $(2,-1)$.

18. The point $(4,8)$ divides the line joining $P_1(3,11)$ to $P_2(x_2,y_2)$ so that $r = \frac{1}{4}$. Find P_2.

19. The point P on the line P_1P_2 where P_1 is $(4,-1)$ and P_2 is $(-2,2)$ is such that $P_1P/PP_2 = 2$. The line from P to $(x,2)$ makes an angle of 135° with P_1P_2. Find x.

20. Determine approximately (i) sin 30·4°, (ii) cos 64·2°, (iii) tan 81·7°, (iv) sin θ when $\theta = 1\cdot25$ radians.

1.5 The equation of a straight line

Suppose that we are given two points $P_1(x_1,y_1)$ and $P_2(x_2,y_2)$ which lie on a straight line. Then we want to know what relation (if any) exists between the coordinates x and y of P (x,y) in order that P lies on P_1P_2.

If $x_1 = x_2$ then P_1P_2 is parallel to the y-axis (figure 20) and all points on it must have the same x-coordinate. Hence, P is on P_1P_2 if and only if $x = x_1$. In this case there is no relation between x and y. Provided that $x = x_1$, different points on the line are obtained for different y and, as y runs from $-\infty$ to ∞, the whole line is traced.

If $x_1 \neq x_2$, i.e. x_1 is not the same as x_2, then figure 18 applies. Then it is clear that P will lie on the line if and only if the slope of P_1P is the same as the slope of P_1P_2, i.e. we must have

$$\frac{y - y_1}{x - x_1} = m$$

or

$$(y - y_1) = m(x - x_1). \tag{29}$$

Equation (29) is one form of the equation of a line. In it x_1, y_1 and m are fixed, but x can take any negative, positive or zero value. To each value of x corresponds one value of y such that (29) is satisfied. As x varies continuously from the value x_1 to the value x_2 the corresponding point (x,y) traces a portion of the straight line.

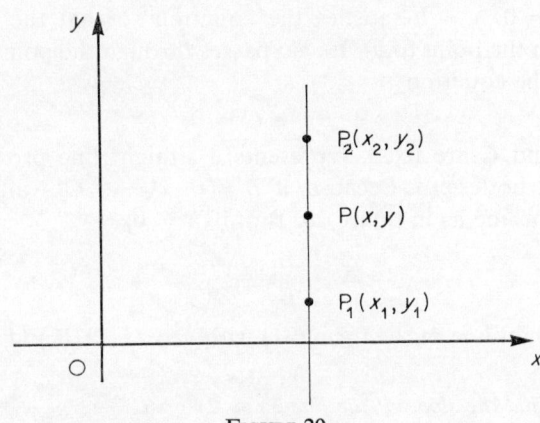

FIGURE 20

Another way of writing (29) is to use the formula $m = (y_2 - y_1)/(x_2 - x_1)$. Then (29) becomes

$$\frac{y - y_1}{x - x_1} = \frac{y_2 - y_1}{x_2 - x_1} \tag{30}$$

when $x_1 \neq x_2$.

This form is especially appropriate when we are given two points P_1 and P_2 on the line.

Example 10 *Find the equation of the line passing through* P_1 *and* P_2 *for* (i) $P_1(-3,3)$, $P_2(2,1)$ *and* (ii) $P_1(a,0)$, $P_2(0,b)$, $(a \neq 0, b \neq 0)$.

(i) From (30) the equation of the line is

$$\frac{y + 3}{x + 3} = \frac{1 + 3}{2 + 3} = \frac{4}{5}$$

or

$$4x - 5y - 3 = 0.$$

(ii) From (30) the equation of the line is

$$\frac{y - 0}{x - a} = \frac{b - 0}{0 - a} = -\frac{b}{a}$$

or
$$\frac{x}{a} + \frac{y}{b} = 1.$$

If we put $b = y_1 - mx$ in (29) the equation becomes
$$y = mx + b. \tag{31}$$
Note that $x = 0$, $y = b$ satisfies the equation so that the straight line passes through the point $(0,b)$. It also passes through the point $(-b/m, 0)$.

In general the equation
$$Ax + By + C = 0 \tag{32}$$
where A, B and C are fixed, represents a straight line provided that A and B are not both zero. Because, if $B = 0$, $x = -C/A$ and we have a vertical straight line as in figure 20. But, if $B \neq 0$,
$$y = -\frac{A}{B}x - \frac{C}{A}$$
which is a straight line of the form (31) with $m = -A/B$ and $b = -C/A$.

Example 11 Find the slope of the line $3x + 2y = 6$.

Solving for y, we have
$$y = -\frac{3}{2}x + 3$$
and comparison with (31) gives $m = -\frac{3}{2}$, $b = 3$. Hence the slope is $-\frac{3}{2}$.

Example 12 *The perpendicular distance* ON *from the origin to a straight line is p, and* ON *makes an angle* α *with the positive x-axis. Find the equation of the line (figure 21).*

The angle α is supposed to be between $0°$ and $360°$. Let the straight line pass through $(a,0)$ and $(0,b)$; from Example 10 its equation is
$$\frac{x}{a} + \frac{y}{b} = 1.$$
From the right-angled triangle OAN (7) gives
$$p = a \cos \alpha.$$
Similarly, from OBN and observing that $\angle \text{BON} = \tfrac{1}{2}\pi - \alpha$, we have
$$p = b \cos(\tfrac{1}{2}\pi - \alpha) = b \sin \alpha$$
from (14). Substitute for a and b in the equation of the line; then
$$\frac{x \cos \alpha}{p} + \frac{y \sin \alpha}{p} = 1$$

or
$$x \cos \alpha + y \sin \alpha = p.$$

This is the required equation.

Because of (9),
$$\frac{p^2}{a^2} + \frac{p^2}{b^2} = 1$$

which shows that
$$p = \sqrt{\left(\frac{a^2 b^2}{a^2 + b^2}\right)}. \tag{33}$$

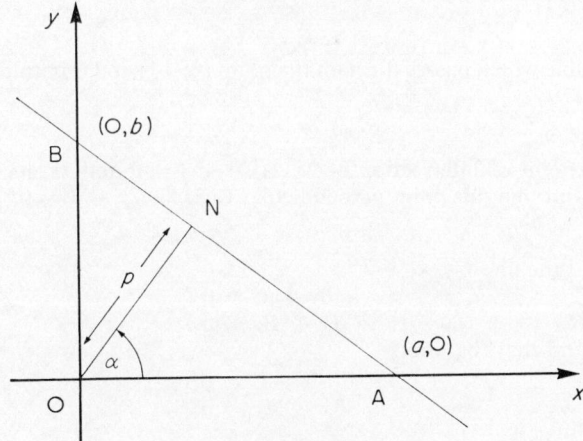

FIGURE 21

Example 13 Find the perpendicular distance of the origin from the line $x - 2y + 6 = 0$.

Write the equation in the form
$$-\frac{x}{6} + \frac{y}{3} = 1$$

so that comparison with Example 12 gives $a = -6$, $b = 3$. Hence, from (33),
$$p = \sqrt{\left(\frac{36 \cdot 9}{36 + 9}\right)} = \frac{6}{\sqrt{5}}.$$

Sometimes OA is known as the *projection* of AB on Ox; OB is the projection of AB on Oy.

Exercises

21. Find the equations of the lines satisfying the conditions (i) through (1,2), $m = 5$, (ii) through $(0,-1)$, $m = -1$, (iii) through (2,1), $m = -\frac{1}{3}$, (iv) through $(0,-2)$, $m = \frac{1}{4}$.

22. Find the equations of the lines joining the following pairs of points (i) (0,0), $(-2,1)$, (ii) (0,0), (0,1), (iii) (2,1), $(-3,-2)$, (iv) $(4,-2)$, $(-3,-2)$, (v) $(2,-1)$, $(-3,2)$.

23. Find the slope of (i) $y = 4x + 3$, (ii) $3x + 4y = 7$, (iii) $4x + 3y + 7 = 0$, (iv) $x = 5y - 8$.

24. Find the line which passes through the point $(3,-1)$ and is parallel to the line $4x - 3y = 12$.

25. Find the point of intersection of $2x + 7y - 3 = 0$ and $7x + y + 13 = 0$. Find the line through this point perpendicular to $2x + 7y - 3 = 0$.

26. Prove that the lines
$$Ax + By + C = 0,$$
$$Ax + By = 0$$
are parallel, and that the lines
$$Ax + By + C = 0,$$
$$Bx - Ay = 0$$
are perpendicular.

27. Show that the perpendicular distance of (x_1, y_1) from the line
$$x \cos \alpha + y \sin \alpha = p$$
is $x_1 \cos \alpha + y_1 \sin \alpha - p$ when (x_1, y_1) is on the opposite side of the line to the origin. What is the distance when (x_1, y_1) is on the same side as the origin?

28. Use Exercise 27 to determine the distance from (i) $8x + 6y = 1$ to $(3,-1)$, (ii) $12x + 5y = 2$ to $(-6,1)$.

29. The speed of a falling stone when plotted against the time is a straight line. Its speeds at times 0 and 10 are 2 and 100 respectively. When was its speed zero and what will its speed be at time 50?

30. The graph of the Centigrade temperature plotted against the Fahrenheit is a straight line. The Centigrade temperatures of 0 and 100 are the Fahrenheit

temperatures 32 and 212 respectively. Verify the following scheme for calculating the Fahrenheit temperature from the Centigrade: add 40 to C, multiply the result by $\frac{9}{5}$ and then subtract 40.

1.6 Change of origin

It will be seen quite often in future chapters that the equations being considered can be made much simpler by choosing the origin and axes carefully. We shall now see what happens when the origin and axes are changed, dealing with the two possibilities separately.

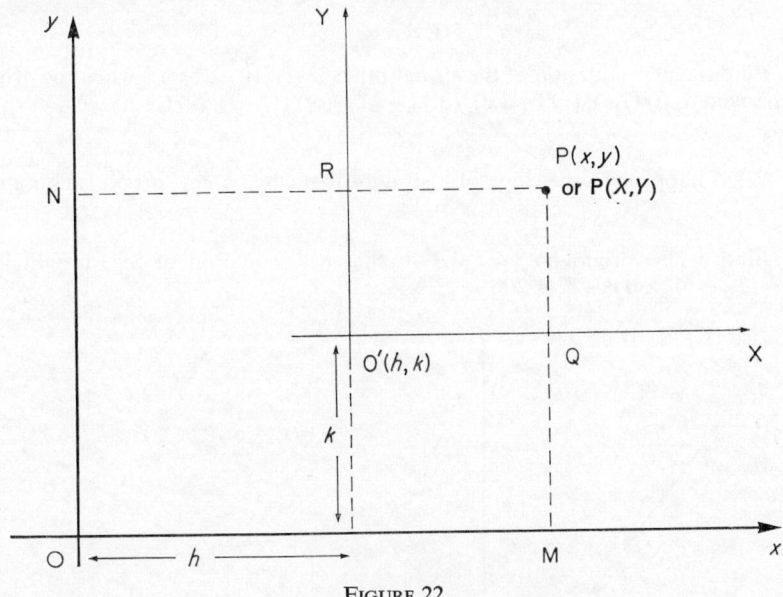

FIGURE 22

Let us suppose that we are given an origin O and axes Ox, Oy (figure 22). Let O' be the point (h,k) in these axes. Draw straight lines O'X, O'Y parallel to Ox, Oy, respectively. Then O' can be taken as a new origin and O'X, O'Y as new axes. Let a point P have coordinates (x,y) in the old axes and coordinates (X, Y) in the new axes. Then PM = y, PN = x, PQ = Y, PR = X. Therefore,

$$x = PN = PR + RN = X + h,$$
$$y = PM = PQ + QM = Y + k.$$

By means of these equations we can go from coordinates in the old system to those in the new, or vice versa.

26   *Introductory Analysis*

Example 14 *The equation of a straight line is* $5x + 3y - 2 = 0$. *The point* $(1,2)$ *is chosen as origin. What is the new equation of the line?*

If X and Y are coordinates with the new origin
$$x = X + 1, \qquad y = Y + 2$$
since $h = 1$ and $k = 2$. Hence the equation of the line becomes
$$5(X + 1) + 3(Y + 2) - 2 = 0$$
or
$$5X + 3Y + 9 = 0.$$

Exercises

31. Find the new equation of the straight line $3x - 4y + 2 = 0$ when the origin is changed to (i) $(2,3)$, (ii) $(-2,3)$, (iii) $(-\frac{2}{3},0)$, (iv) $(1,-2)$, (v) $(-3,-\frac{7}{4})$.

32. What happens to the slope of a straight line when a new origin is chosen?

33. Find a new origin on $y = -1$ so that new equation of the straight line $4x - 2y - 6 = 0$ is $4X = 2Y$.

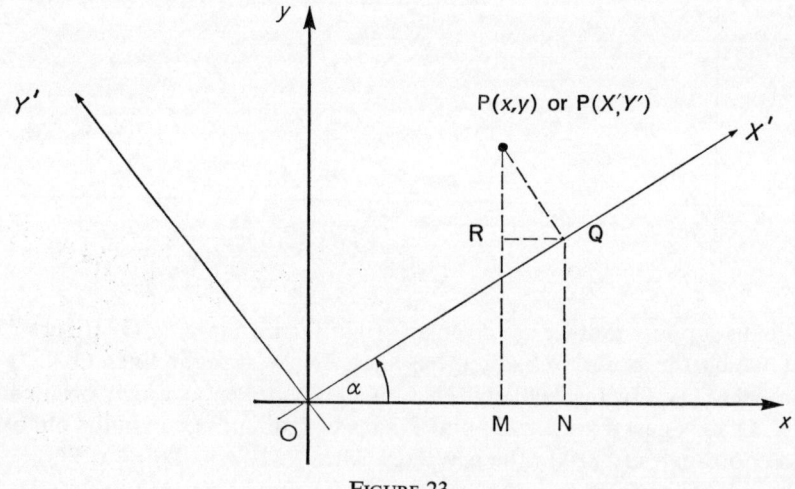

FIGURE 23

1.7 Rotation of axes

Now suppose that the origin is not changed but that the directions of the axes are altered. Let Ox and Oy be the given axes and let OX' and OY'

be new axes (figure 23) so that both systems have the same origin O. Let $\angle xOX' = \alpha$ so that $\angle yOY'$ is also α. Then we can get from Oxy to $OX'Y'$ by rotating through an angle α, so this is called a *rotation of axes*. Let a point P have coordinates (x,y) in the old axes and coordinates (X',Y') in the new axes. Then PM $= y$, OM $= x$, PQ $= Y'$, OQ $= X'$. Draw QN, QR perpendicular to Ox and PM respectively. Then

$$ON = OQ \cos \alpha = X' \cos \alpha,$$
$$MN = RQ = PQ \sin \alpha = Y' \sin \alpha$$

since $\angle QPR = \alpha$. Therefore,

$$x = OM = ON - MN = X' \cos \alpha - Y' \sin \alpha,$$
$$y = MP = MR + RP = X' \sin \alpha + Y' \cos \alpha.$$

Example 15 *The axes $OX'Y'$ are obtained from Oxy by a rotation through $60°$. Find the new coordinates of $(2,\sqrt{3})$.*

Since $\alpha = 60°$, $\cos \alpha = \frac{1}{2}$ and $\sin \alpha = \sqrt{\frac{3}{2}}$. Therefore,

$$2 = \tfrac{1}{2}(X' - \sqrt{3}Y'),$$
$$\sqrt{3} = \tfrac{1}{2}(\sqrt{3}X' + Y').$$

The solution of these equations is $X' = \frac{5}{2}$, $Y' = -\sqrt{\frac{3}{2}}$ so the point is $(\frac{5}{2}, -\sqrt{\frac{3}{2}})$ in the new coordinate system.

Exercises

34. Prove that
$$X' = x \cos \alpha + y \sin \alpha,$$
$$Y' = y \cos \alpha - x \sin \alpha.$$
Also verify the formulae for the rotation of axes when α is larger than $90°$.

35. Make a rotation of axes so that the line $3x - 4y = 0$ becomes $Y' = 0$. Deduce the perpendicular distance of the point $x = h$, $y = k$ from this line.

36. Make a rotation of axes so that the Y'-axis is parallel to $4x - 3y + 3 = 0$ and deduce the perpendicular distance of $x = 2$, $y = 2$ from this line.

37. Find a coordinate system (X_1, Y_1) so that the straight lines $x + 2y - 5 = 0$, $2x + y - 4 = 0$ have equations of the form
$$Y_1 - mX_1 = 0, \qquad Y_1 + mX_1 = 0.$$

Explain how you would do the same if the equations of the lines were

$$Ax + By + C = 0, \qquad A_1 x + B_1 y + C_1 = 0.$$

Interpret your result geometrically. (Hint: use both a rotation and a change of origin.)

38. Show that the area of a triangle with vertices $(0,0)$, $(a,0)$ and (b,c) is $\tfrac{1}{2}ac$. Use this result and a change of coordinates to show that the area of a triangle with vertices (x_1,y_1), (x_2,y_2) and (x_3,y_3) is

$$\tfrac{1}{2}(x_1 y_2 - x_2 y_1 + x_2 y_3 - x_3 y_2 + x_3 y_1 - x_1 y_3).$$

1·8 Polar coordinates

Instead of determining the position of a point by means of its rectangular coordinates it is sometimes convenient to use its distance from the origin as one of the coordinates. The location of the point is then fixed by giving the angle made with the x-axis by its direction from the origin. Thus, in figure 24, r is the distance of P from O and θ is the angle between OP and Ox.

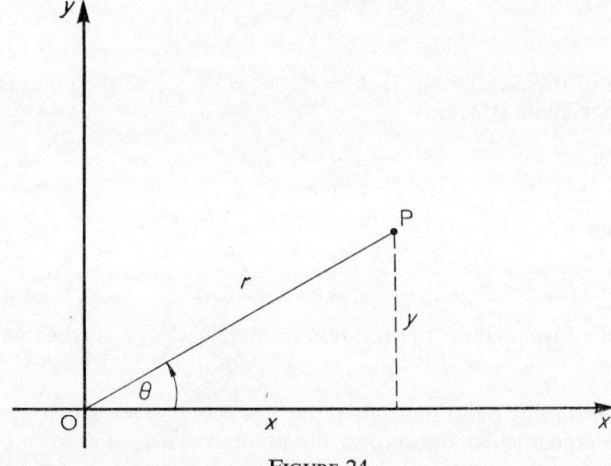

FIGURE 24

The coordinates r and θ are called *polar coordinates*. Sometimes in this system O is called the *pole* and Ox the *polar axis*.

The polar coordinates of P are written (r,θ). The distance r is always taken to be positive. The angle θ is always positive when measured in a counter-clockwise direction from Ox as in the diagram; it is negative when measured in a clockwise direction.

Coordinates

The relation between rectangular and polar coordinates is given by
$$x = r \cos \theta, \qquad y = r \sin \theta,$$
$$r = \sqrt{(x^2 + y^2)}, \qquad \tan \theta = \frac{y}{x}.$$

Later (see section 10.8) we shall write $\theta = \tan^{-1}(y/x)$.

Example 16 Find the distance between the points P_1 and P_2 with polar coordinates (r_1, θ_1) and (r_2, θ_2) respectively (figure 25).

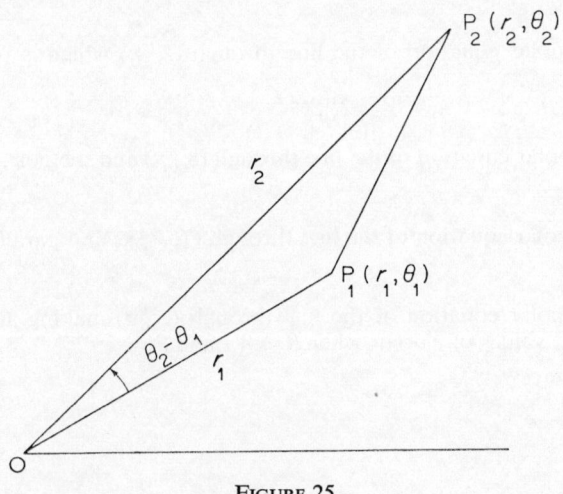

FIGURE 25

In the triangle OP_1P_2 the two sides OP_1, OP_2 and the included angle $\angle P_1OP_2$ are known. Hence, from (25),
$$P_1P_2 = \sqrt{[r_1^2 + r_2^2 - 2r_1r_2 \cos(\theta_2 - \theta_1)]}.$$

Exercises

39. Obtain the formula of Example 16 by substituting the polar coordinate form for the rectangular coordinates in (3).

40. Plot the points whose polar coordinates are $(1, \frac{1}{4}\pi)$, $(3, 0)$, $(2, \frac{5}{6}\pi)$, $(1, \frac{4}{3}\pi)$, $(3, 2\pi)$, $(1, -\frac{2}{3}\pi)$, $(5, \frac{1}{2}\pi)$, $(4, -\frac{1}{2}\pi)$, $(4, \frac{3}{2}\pi)$, $(3, \frac{11}{6}\pi)$, $(0, \frac{1}{3}\pi)$, $(0, \frac{4}{3}\pi)$.

41. Find the distance between the points $(3, \frac{3}{4}\pi)$ and $(4, \frac{1}{4}\pi)$.

42. Find the area of the triangle whose vertices are $(0,0)$, (r_1,θ_1) and (r_2,θ_2).

43. Show that the straight line $Ax + By + C = 0$ becomes, in polar coordinates,
$$r \cos(\theta - \beta) + C/\sqrt{(A^2 + B^2)} = 0$$
where
$$\cos \beta = A/\sqrt{(A^2 + B^2)}, \qquad \sin \beta = B/\sqrt{(A^2 + B^2)}.$$

44. Find the polar equation of the line through $(\sqrt{2}, \tfrac{1}{4}\pi)$ which is perpendicular to Ox.

45. Find the polar equation of the line through $(2, \tfrac{2}{3}\pi)$ which is perpendicular to Ox.

46. Find the polar equation of the line through $(8, \tfrac{5}{6}\pi)$ and the pole.

47. Find the polar equation of the line through $(1, -\tfrac{1}{6}\pi)$ and parallel to Ox.

48. Find the polar equation of the line through $(2, \tfrac{1}{6}\pi)$ making an angle $5\pi/6$ with Ox. What values of r occur when $\theta = \tfrac{1}{3}\pi$ and $\theta = \tfrac{4}{3}\pi$?

2
Conics

2.1 The circle

A *circle* is composed of all those points which are at the same distance r from a given point C(h,k). The point C is known as the *centre* of the circle and r is called the *radius* (figure 26).

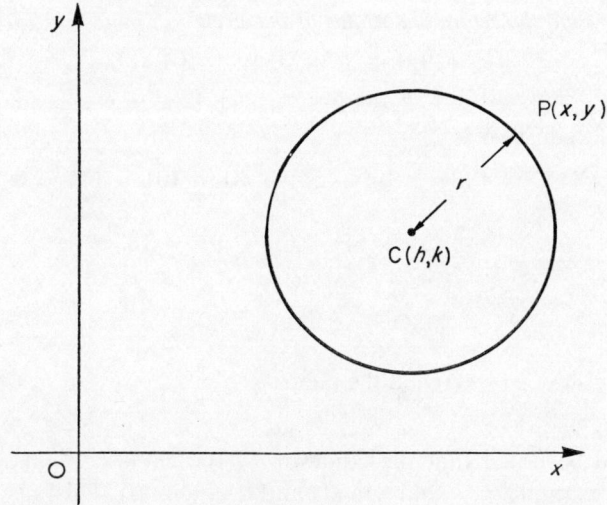

FIGURE 26

Now, if P(x,y) is any point on the circle, we must have CP = r and so (CP)^2 = r^2. It follows from section 1.2 that

$$(x - h)^2 + (y - k)^2 = r^2, \tag{1}$$

which is the equation of a circle. If the centre is at the origin the equation reduces to

$$x^2 + y^2 = r^2.$$

By squaring out the brackets in (1) we obtain
$$x^2 + y^2 + 2gx + 2fy + c = 0 \qquad (2)$$
where $g = -h, f = -k, c = h^2 + k^2 - r^2$.

Conversely, suppose that (2) is given and we enquire whether it represents a circle. Write it as
$$(x^2 + 2gx + g^2) + (y^2 + 2fy + f^2) + c - g^2 - f^2 = 0$$
or
$$(x + g)^2 + (y + f)^2 = g^2 + f^2 - c. \qquad (3)$$
The left-hand side cannot be negative. Therefore, *if $g^2 + f^2 - c$ is negative*, there are no real x and y which satisfy (2).

However, if $g^2 + f^2 - c$ is positive, (3) is of the same form as (1) and is a circle with centre $(-g, -f)$ and radius $\sqrt{(g^2 + f^2 - c)}$.

If $g^2 + f^2 - c = 0$, the only solution of (3) is $x = -g, y = -f$. In this case (2) represents the single point $(-g, -f)$.

Example 1 Find the centre and radius of the circle
$$9x^2 + 9y^2 + 36x - 54y + 101 = 0.$$

Writing the equation as
$$9x^2 + 36x + 36 + 9y^2 - 54y + 81 + 101 - 117 = 0$$
or
$$(3x + 6)^2 + (3y - 9)^2 = 16$$
we have
$$(x + 2)^2 + (y - 3)^2 = \frac{16}{9}$$
so that the centre is $(-2, 3)$ and the radius is $\frac{4}{3}$.

It should be noted that the equation of the circle contains three constants; for example, h, k and r in (1) and g, f, c in (2). This means that, in general, a circle can be found that will satisfy three conditions, e.g. a circle can be determined to pass through three given points provided that they do not lie on a straight line.

Example 2 Find the circle which passes through $(3,2)$, $(-4,1)$ and $(2,-2)$.

Let the equation of the circle be
$$x^2 + y^2 + 2gx + 2fy + c = 0.$$

Then, since the points lie on it,
$$9 + 4 + 6g + 4f + c = 0,$$
$$16 + 1 - 8g + 2f + c = 0,$$
$$4 + 4 + 4g - 4f + c = 0.$$
The solution of these equations is $g = \frac{7}{18}, f = -\frac{13}{18}, c = -\frac{112}{9}$. Hence the equation of the circle is
$$9x^2 + 9y^2 + 7x - 13y - 112 = 0.$$

Exercises

1. Find the circles whose centres and radii are (i) C(3,4), $r = 2$; (ii) C(−4,3), $r = 5$; (iii) C(−6,−4), $r = 1$.

2. Find the circle whose centre is (3,−2) and passes through the point (−1,2).

3. Find the circle whose centre is (0,−3) and passes through the origin.

4. What do the following equations represent? Give the centre and radius of each circle. (i) $x^2 + y^2 - 2x = 3$, (ii) $x^2 + y^2 + 3x - 2y + 1 = 0$, (iii) $4x^2 + 4y^2 + 3x + 9y + 1 = 0$, (iv) $x^2 + y^2 + 24x - 10y + 169 = 0$, (v) $x^2 + y^2 + x - y + 1 = 0$, (vi) $x^2 + y^2 + x - y - 1 = 0$.

5. Find the equation of the circle which passes through the three points (6,2), (7,1) and (8,−2).

6. Find the equation of the circle which passes through the vertices of the triangle formed by the straight lines $7y - x = 17, 4x - y = 13$ and $x + 2y = 1$.

7. The lines joining P(x,y) to (3,2) and (−5,−4) are perpendicular. Find the equation satisfied by (x,y).

8. A straight line through the origin intersects $x^2 + y^2 = r^2$ in P_1 and P_2. If P is any other point on the circle show that PP_1 and PP_2 are perpendicular.

9. The sum of the squares of the distances of P from the lines $2x + 3y - 4 = 0$ and $3x - 2y + 2 = 0$ is 2. Show that P lies on the circle
$$13x^2 + 13y^2 - 4x - 32y = 6.$$

10. The point P(x,y) moves so that the sum of the squares of its distances from (2,1) and (−1,2) is 3. Find its locus.*

*A *locus* is the curve which consists of all those points P which satisfy the given conditions.

11. The point P(x,y) moves so that its distance from (2,3) is 5 times its distance from $(-1,0)$. Find its locus.

12. The point P(x,y) moves so that its distance from (x_1,y_1) is k times its distance from (x_2,y_2). Show that P describes a circle unless $k = 1$ when P describes a straight line. (The circles obtained by giving k different values are called *coaxial circles*.)

13. The point P(x,y) moves so that the lines joining it to $(-a,0)$ and $(b,0)$ always make the angle θ with each other. Show that P describes a circle.

14. The point P(x,y) is given by
$$x = \frac{3\xi}{\xi^2 + \eta^2}, \qquad y = \frac{3\eta}{\xi^2 + \eta^2}.$$
If (ξ,η) moves on the straight line $2\xi - \eta = 8$ find the curve described by P.

15. Is the point (11,5) inside or outside the circle $x^2 + y^2 - 4x - 2y = 95$?

16. What is the equation of a circle with centre (0,0) and radius a in polar coordinates?

2.2 The parabola

A *parabola* is composed of all those points P(x,y) which are such that P is the same distance from the point F as it is from a straight line L which does not pass through F. F is called the *focus* of the parabola and the line L is known as the *directrix*.

Suppose that F is the point $(a,0)$ and L is the line $x = -a$ (figure 27). Then for a parabola
$$PF = PQ$$
where PQ is the perpendicular to L. Hence,
$$\sqrt{[(x-a)^2 + y^2]} = x + a.$$
Squaring both sides we obtain
$$x^2 - 2ax + a^2 + y^2 = x^2 + 2ax + a^2$$
or
$$y^2 = 4ax. \tag{4}$$
This is the equation of the parabola shown in figure 27.

Conics

We notice that the curve has exactly the same shape above the x-axis as below. Now two points are said to be *symmetrical about a line* if that line is the perpendicular bisector of the line joining the two points. Thus the parabola is symmetrical about the x-axis because (4) is unchanged if y is replaced by $-y$.

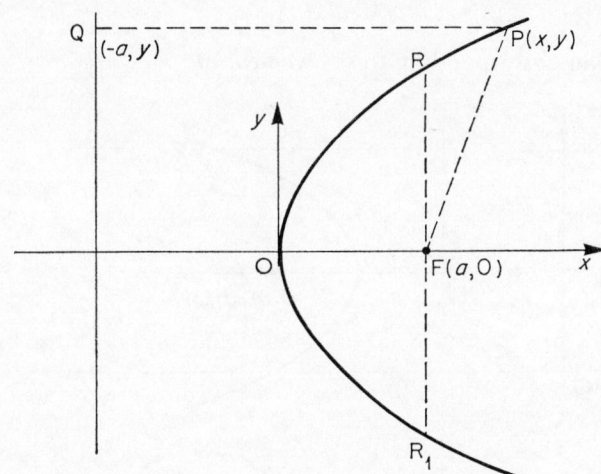

FIGURE 27

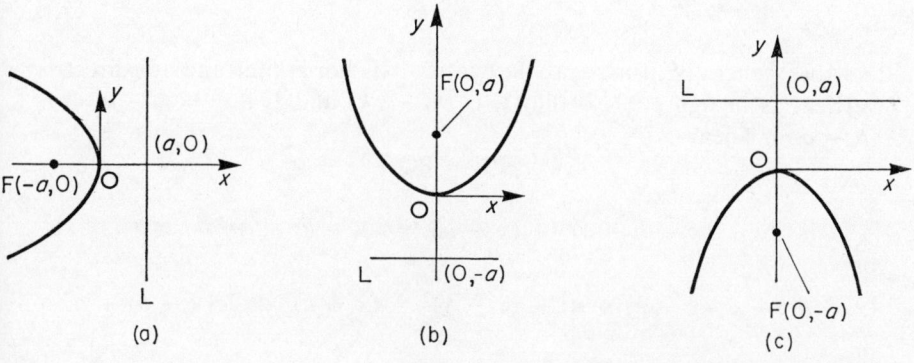

FIGURE 28

The point O at which the curve cuts its axis of symmetry is called the *vertex* of the parabola. The chord RR_1 through the focus and parallel to the directrix is called the *latus rectum*. When $x = a$ (4) gives $y^2 = 4a^2$ so that $y = \pm 2a$. Therefore, R is the point $(a, 2a)$ and R_1 is $(a, -2a)$. Thus the latus rectum is of length $4a$.

1—3

Different parabolas can be obtained by making different choices of F and L. For example, in figure 28(a) with F($-a$,0) and L as $x = a$, the parabola is

$$y^2 = -4ax.$$

Similarly, figures 28(b) and (c) show the parabolas

$$x^2 = \pm 4ay,$$

the upper sign corresponding to (b) when $a > 0$.

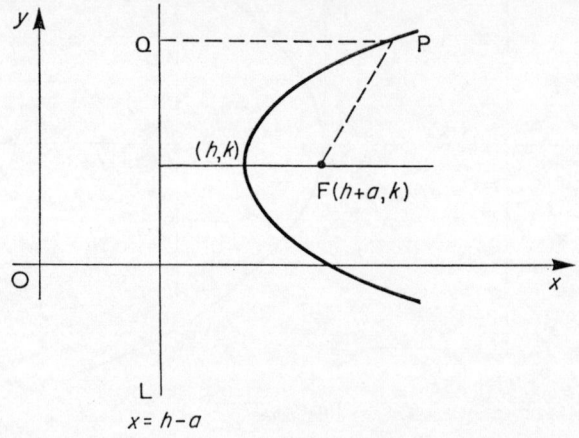

FIGURE 29

More generally, if a parabola has its axis horizontal and its directrix vertical as in figure 29 so that F is $(h + a, k)$ and L is $x = h - a$, Q is $(h - a, y)$. Then

$$PF = PQ$$

gives

$$(x - h - a)^2 + (y - k)^2 = (x - h + a)^2$$

or

$$(x - h)^2 - 2a(x - h) + a^2 + (y - k)^2 = (x - h)^2 + 2a(x - h) + a^2$$

or

$$(y - k)^2 = 4a(x - h). \tag{5}$$

This is the equation of the parabola; its axis is $y = k$ and its vertex is (h,k). It can easily be converted into the form (4) by a change of origin (section 1.6).

On the other hand, if we are given an equation

$$y^2 = bx + cy + d \tag{6}$$

we write it as

$$(y - \tfrac{1}{2}c)^2 = bx + d + \tfrac{1}{4}c^2$$
$$= b\{x + (d + \tfrac{1}{4}c^2)/b\}.$$

Comparison with (5) shows that we have a parabola with latus rectum b if b is positive (if b is negative the parabola faces as in figure 28(a) and the length of the latus rectum is $-b$). The vertex is $[-(d + \tfrac{1}{4}c^2)/b, \tfrac{1}{2}c]$, the focus is $[\tfrac{1}{4}b - (d + \tfrac{1}{4}c^2)/b, \tfrac{1}{2}c]$ and the directrix is

$$x = -\tfrac{1}{4}b - (d + \tfrac{1}{4}c^2)/b.$$

Example 3 Find the focus, directrix and latus rectum of (i) $y^2 = \tfrac{2}{3}x$, (ii) $y^2 = 2x + 2y - 7$.

(i) Comparing with (4) we see that $a = \tfrac{1}{6}$ so that the focus is $(\tfrac{1}{6}, 0)$, the directrix $x = -\tfrac{1}{6}$ and the latus rectum is $\tfrac{2}{3}$.

(ii) Writing the equation as

$$(y - 1)^2 = 2x - 6 = 2(x - 3)$$

we see that $a = \tfrac{1}{2}$, $h = 3$ and $k = 1$ in (5). Hence the focus is $(3\tfrac{1}{2}, 1)$, the directrix $x = 2\tfrac{1}{2}$ and the length of the latus rectum is 2.

The equation of the parabola contains three constants a,h,k in (5) or b,c,d in (6). Therefore, in general, a parabola with a horizontal axis can be drawn through three given points provided that they do not lie on a straight line and that no two of them lie on the same horizontal line.

Exercises

17. Find the vertex, focus and directrix of the parabolas (i) $y^2 = 8x$, (ii) $x^2 = 5y$, (iii) $4y^2 = -3x$, (iv) $y^2 = 6y + 4x - 7$, (v) $y^2 + 4y + 3x + 1 = 0$, (vi) $x^2 = 4x + 2y + 8$.

18. Find the focus F and vertex V of the parabola $x^2 = 12y$. The point P moves so that its distance from V is twice its distance from F. Determine the path traced by P.

19. Find the equations of the parabolas with (i) focus (2,0) and directrix $x = -2$, (ii) focus $(0,-3)$ and directrix $y = 3$, (iii) focus (2,3) and directrix $x = 1$, (iv) focus (3,0) and directrix $y = 0$.

20. Find the parabolas with (i) focus (2,0) and vertex (1,0), (ii) focus (3,2) and vertex (2,2).

21. The strongest simple masonry arch is parabolic. Such an arch has its bases at (4,0) and (−4,0), and its highest point is (0,1); find its equation.

22. Find the height of the arch in question 21 at (2,0).

23. Find the equation of the parabola, with horizontal axis, which passes through the points (−1,1), (11/3,−1) and (−3,3).

24. Find the equation of the parabola, with vertical axis, which passes through the points (1,11/2), (2,8) and (−2,8).

25(a). When an object is thrown horizontally with a speed v ft/sec it describes a parabola
$$d = 16x^2/v^2$$
in which x is the horizontal distance travelled from the point of projection and d is the vertical distance fallen. If a ball is thrown horizontally with speed 100 ft/sec at a height 9 ft, how far has it travelled horizontally when it strikes the ground?

(b). If distances are measured in metres and speeds in metres per second, the number 16 in the formula in (a) is replaced by 4·9. If a stone is thrown horizontally with speed 7 m/sec from a height of 0·9 m, how far will it go horizontally before hitting the ground?

26(a). Water issues horizontally from a hosepipe with speed 12 ft/sec. How high must it be held for the water to strike a flower bed 6 ft away?

(b). Find the height if the water issues with speed 4 m/sec and hits a flower bed 2 m away.

27(a). A plane is flying horizontally at 800 ft/sec and is 2,500 ft high. How soon before it passes over its target must it release a bomb and what will be the horizontal distance travelled by the bomb?

(b). Repeat (a) with the aeroplane flying at 210 m/sec and 250 m high.

28(a). The cable of a suspension bridge hangs in a parabola when the load is uniformly distributed. The supporting towers are 50 ft high and 400 ft apart, and the lowest point of the cable is 10 ft above the roadway. How high is the cable above the roadway 100 ft from a tower?

(b). Repeat (a) but suppose that the given dimensions were metres instead of feet, i.e. the towers are 50 m high, etc.

29. A comet describes a parabola with the Sun as focus. The nearest distance of the comet to the Sun is 20 million kilometres. How far away is it at the end of the latus rectum?

2.3 Tangents and normals

Let P and P_1 be two points on a curve (see figure 30). Draw the straight line PP_1. Take a point P_2 between P and P_1 and draw the line PP_2. Choose P_3 between P and P_2 and continue the process so that the chosen

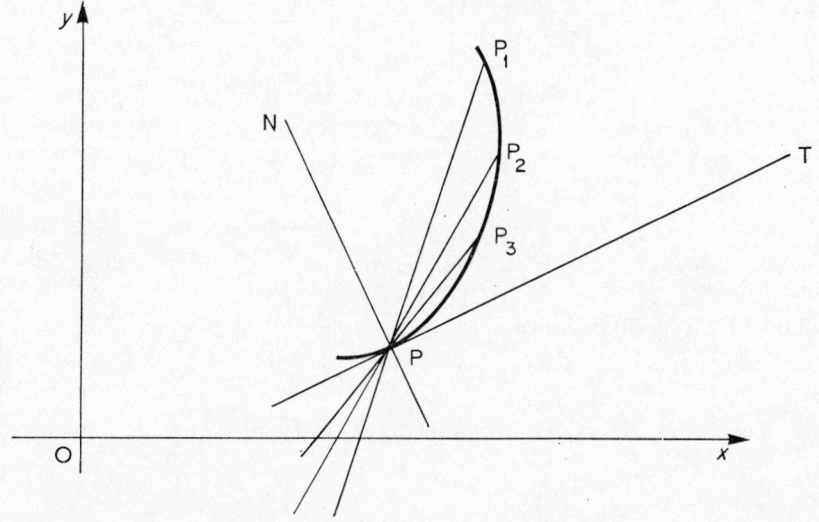

FIGURE 30

point gets closer and closer to P. Finally, as the point coincides with P the line joining them becomes PT. The line PT is called the *tangent* to the curve at P and is said to *touch* the curve at P. P is also called the *point of contact* of the tangent.

The straight line PN through P perpendicular to the tangent at P is called the *normal* to the curve.

40 Introductory Analysis

Example 4 Find the tangent to the circle $x^2 + y^2 = a^2$ at the point $P(x_0, y_0)$ which is on the circle.

Choose another point $P_1(x_1, y_1)$ on the circle (figure 31). Since both P and P_1 lie on the circle

$$x_0^2 + y_0^2 = a^2, \tag{7}$$

$$x_1^2 + y_1^2 = a^2. \tag{8}$$

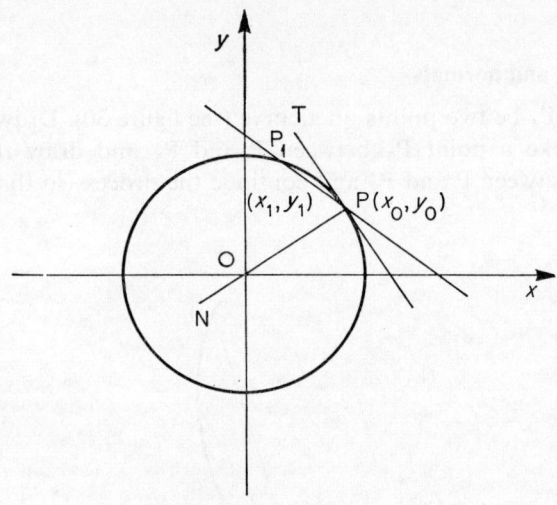

FIGURE 31

Subtract (7) from (8) and then

$$x_1^2 - x_0^2 + y_1^2 - y_0^2 = 0$$

or

$$(x_1 - x_0)(x_1 + x_0) + (y_1 - y_0)(y_1 + y_0) = 0$$

or

$$\frac{y_1 - y_0}{x_1 - x_0} = -\frac{x_1 + x_0}{y_1 + y_0}$$

Therefore the slope of PP_1 which is $(y_1 - y_0)/(x_1 - x_0)$ can be taken as $-(x_1 + x_0)/(y_1 + y_0)$. As P_1 approaches P, x_1 becomes x_0 and y_1 becomes y_0. Hence the slope of PP_1 becomes $-2x_0/2y_0$ or $-x_0/y_0$. But, when P_1 coincides with P, PP_1 becomes the tangent PT and thus the slope of PT is $-x_0/y_0$.

Since PT passes through P the equation of the tangent is (see equation (29) of Chapter 1)

$$y - y_0 = -\frac{x_0}{y_0}(x - x_0)$$

or
$$yy_0 - y_0^2 = -xx_0 + x_0^2$$
or
$$xx_0 + yy_0 = a^2 \tag{9}$$
on using (7).

The normal PN is perpendicular to PT and so its slope is y_0/x_0 (section 1.4). Since PN passes through P its equation is

$$y - y_0 = \frac{y_0}{x_0}(x - x_0)$$

or
$$x_0 y = y_0 x.$$

This shows that the normal passes through the origin, i.e. a *radius of a circle is perpendicular to the tangent at its end*.

Example 5 Find the tangent to the circle $x^2 + y^2 + 2gx + 2fy + c = 0$ at the point $P(x_0, y_0)$ on the circle.

Let $P_1(x_1, y_1)$ also be on the circle. Then
$$\begin{aligned} x_0^2 + y_0^2 + 2gx_0 + 2fy_0 + c &= 0, \\ x_1^2 + y_1^2 + 2gx_1 + 2fy_1 + c &= 0. \end{aligned} \tag{10}$$

By subtraction
$$x_1^2 - x_0^2 + 2g(x_1 - x_0) + y_1^2 - y_0^2 + 2f(y_1 - y_0) = 0$$
whence
$$\frac{y_1 - y_0}{x_1 - x_0} = -\frac{x_1 + x_0 + 2g}{y_1 + y_0 + 2f}.$$

Thus the right-hand side is another expression for the slope of PP_1. As P_1 approaches P, x_1 and y_1 become x_0 and y_0, respectively, so that the slope of PP_1 becomes $-\dfrac{2x_0 + 2g}{2y_0 + 2f}$. Hence the equation of the tangent PT is

$$y - y_0 = -\frac{x_0 + g}{y_0 + f}(x - x_0)$$

or
$$yy_0 + yf - y_0^2 - y_0 f = -xx_0 - xg + x_0^2 + gx_0$$
or
$$xx_0 + yy_0 + g(x + x_0) + f(y + y_0) + c = 0 \tag{11}$$
on account of (10).

Exercises

30. Find the slope of the tangent and normal to the following circles at the points given (i) $x^2 + y^2 = 1$, $(1/\sqrt{2}, 1/\sqrt{2})$; (ii) $x^2 + y^2 = 25$, $(-4,-3)$; (iii) $x^2 + y^2 + 3x - 4y - 7 = 0$, $(2,1)$.

31. Find the equations of the tangent and normal to the following circles at the given points (i) $x^2 + y^2 = 25$, $(4,3)$; (ii) $x^2 + y^2 = 4$, $(2,0)$; (iii) $2x^2 + 2y^2 - 4x + 5y - 23 = 0$, $(4,1)$.

32. Obtain (11) from (9) by making a change of origin.

33. A circle touches $y - 18x + 52 = 0$ at $(3,2)$ and also passes through $(1,1)$. Show that its equation is
$$7x^2 + 7y^2 - 24x - 29y + 39 = 0.$$

34. Show that there are two circles which pass through $(-4,3)$, have their centres on $3y + x = 0$ and touch $2y = x$. Find their equations and the points of contact with the last line.

35. Prove that the length of the tangent from (X,Y) to the circle $x^2 + y^2 = a^2$ is $\sqrt{(X^2 + Y^2 - a^2)}$. Hence, show that the length of the tangent from (X,Y) to $x^2 + y^2 + 2gx + 2fy + c = 0$ is $\sqrt{(X^2 + Y^2 + 2gX + 2fY + c)}$.

36. Find the length of the tangent from $(2,1)$ to $x^2 + y^2 + 4x - 3y + 6 = 0$.

37. Show that there is a point from which the lengths of the tangents to
$$x^2 + y^2 + 3x - 2y + 2 = 0, \qquad x^2 + y^2 + 2x - 2y = 0,$$
$$3x^2 + 3y^2 + 3x - 4y - 12 = 0$$
are all equal. Find the equation of the circle which passes through the points of contact.

Example 6 Find the tangent to the parabola $y^2 = 4ax$ at the point $P(x_0, y_0)$ on the parabola.

Select another point $P_1(x_1, y_1)$ also on the parabola (figure 32). Then
$$y_0^2 = 4ax_0, \qquad y_1^2 = 4ax_1.$$

By subtraction,
$$y_1^2 - y_0^2 = 4a(x_1 - x_0)$$
or
$$\frac{y_1 - y_0}{x_1 - x_0} = \frac{4a}{y_1 + y_0}.$$

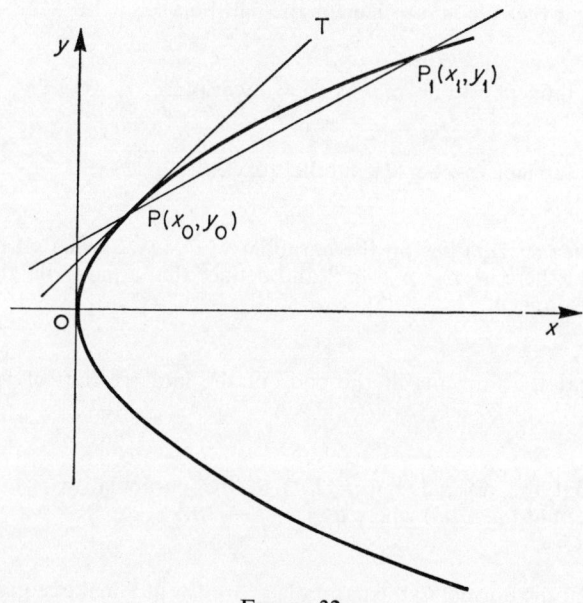

FIGURE 32

Therefore the slope of PP_1 as P_1 comes into coincidence with P becomes $4a/2y_0$ or $2a/y_0$. Hence the equation of the tangent PT is
$$y - y_0 = \frac{2a}{y_0}(x - x_0)$$
or
$$yy_0 - y_0^2 = 2a(x - x_0)$$
or
$$yy_0 = 2a(x + x_0)$$
since $y_0^2 = 4ax_0$.

The slope of the normal is $-y_0/2a$ and its equation is
$$2a(y - y_0) + y_0(x - x_0) = 0.$$

The reader should note that in each of Examples 4, 5 and 6 the equation of the tangent is obtained correctly if we adopt the following

Rule: To find the equation of the tangent at the point of contact $P(x_0,y_0)$ replace, in the equation of the curve, x^2 by xx_0, y^2 by yy_0, x by $\frac{1}{2}(x + x_0)$, y by $\frac{1}{2}(y + y_0)$ and xy by $\frac{1}{2}(xy_0 + x_0y)$.

Exercises

38. Verify that the rule is satisfied for the parabola $x^2 = bx + cy + d$.

39. Find the tangent and normal to $x^2 = 8y$ at (4,2).

40. Find the tangent to $y^2 = 2x$ parallel to $x - 2y = 3$.

41. The point $(at^2, 2at)$ lies on the parabola $y^2 = 4ax$. Show that the equation of the tangent there is $yt = x + at^2$, and deduce that a parabola always lies on one side of its tangent.

42. Show that the tangents at the ends of the latus rectum of $y^2 = 4ax$ are perpendicular.

43. Show that the normal at $(at^2, 2at)$ to the parabola $y^2 = 4ax$ meets the parabola again at $(au^2, 2au)$ where $u = -t - 2/t$.

44. Show that the normal to the parabola $y^2 = 4ax$ at P makes equal angles with SP, S being the focus, and the x-axis. Deduce the behaviour of rays of light originating at the focus of a parabolic mirror. (Assume that, when a ray of light strikes the surface of a mirror, a reflected ray is produced making an equal angle with, and on the opposite side of, the normal.)

45. The angle between the lines joining the points P and Q of the parabola $y^2 = 4ax$ to the vertex is a right angle. Show that the mid-point of PQ lies on the parabola $y^2 = 2ax - 8a^2$.

2.4 The ellipse

An *ellipse* is composed of all those points $P(x,y)$ which are such that the sum of the distances PF, PF' from two fixed points F,F' is constant. F and F' are called the *foci* of the ellipse (figure 33).

Suppose that F is $(c,0)$ and that F' is $(-c,0)$. Then, for an ellipse,

$$\sqrt{[(x - c)^2 + y^2]} + \sqrt{[(x + c)^2 + y^2]} = \text{constant} = 2a \text{ (say)}.$$

Hence,
$$\sqrt{[(x-c)^2 + y^2]} = 2a - \sqrt{[(x+c)^2 + y^2]}$$
so that squaring both sides gives
$$(x-c)^2 + y^2 = 4a^2 + (x+c)^2 + y^2 - 4a\sqrt{[(x+c)^2 + y^2]}$$
or
$$cx + a^2 = a\sqrt{[(x+c)^2 + y^2]}.$$

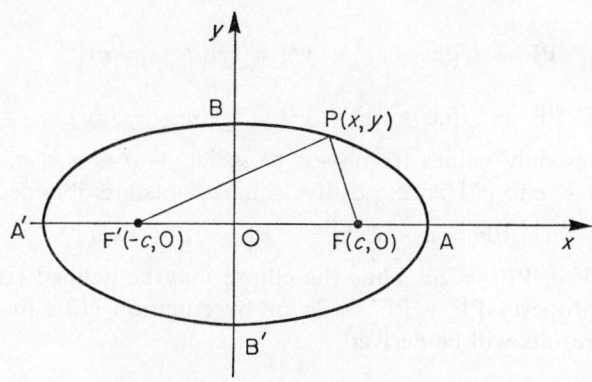

FIGURE 33

Squaring again we obtain
$$c^2x^2 + 2a^2cx + a^4 = a^2(x^2 + 2cx + c^2) + y^2a^2$$
or
$$(a^2 - c^2)x^2 + a^2y^2 = a^2(a^2 - c^2). \tag{12}$$

One point which satisfies this is $(a,0)$ which is the point A, so that clearly $a^2 - c^2$ is positive. Write
$$b = \sqrt{(a^2 - c^2)}.$$
Then (12) can be expressed as
$$\frac{x^2}{a^2} + \frac{y^2}{b^2} = 1. \tag{13}$$

Thus the curve is symmetrical about the x- and y-axes and intersects the axes at the points $(\pm a, 0)$ and $(0, \pm b)$. O is called the *centre* of the ellipse. A′A is known as the *major axis*—its length is $2a$. B′B is known as the *minor axis*—its length is $2b$.

Note that x can never be less than $-a$ or greater than a, while y always lies between $-b$ and b.

The number e defined by $c = ae$ or $b = a\sqrt{(1 - e^2)}$ is called the *eccentricity* of the ellipse. Since c is less than a the eccentricity is less than unity. If $e = 0$ then $b = a$ and (13) becomes the equation of a circle of radius a. Thus the circle can be regarded as a special kind of ellipse in which the eccentricity is zero.

It has been shown that, if $PF + PF' = 2a$, equation (12) must be satisfied. Conversely, if x and y satisfy (12) with* $0 < c < a$, then

$$y^2 = (a^2 - c^2)\frac{a^2 - x^2}{a^2}$$

so that
$$PF = \sqrt{[(x - c)^2 + y^2]} = \sqrt{[(a - cx/a)^2]}$$
and
$$PF' = \sqrt{[(x + c)^2 + y^2]} = \sqrt{[(a + cx/a)^2]}.$$

Since x takes only values from $-a$ to a, i.e. $-a \leqslant x \leqslant a$, cx/a takes values from $-c$ to c. Hence, positive square roots are obtained by taking

$$PF = a - cx/a, \qquad PF' = a + cx/a$$

and then $PF + PF' = 2a$. Thus the ellipse may be defined either by the geometric property $PF + PF' = 2a$ or by equation (12); in either case equivalent results will be derived.

Example 7 A straight line QPR, *with* QP $= b$ *and* PR $= a$, *is drawn with* Q *and* R *on the x and y-axes respectively. Find the locus of* P.

Draw MP,NP parallel to the x- and y-axes respectively (figure 34) so that MP $= x$, NP $= y$. Let $\angle$NQP $= \theta$; then $\angle$MPR $= \theta$. From triangle NQP, $y = b \sin \theta$, and from triangle MPR, $x = a \cos \theta$. Hence,

$$\frac{x^2}{a^2} + \frac{y^2}{b^2} = \cos^2 \theta + \sin^2 \theta = 1.$$

Therefore, P lies on an ellipse and, by allowing Q to occupy all permissible positions on the positive and negative x-axis while R may be on the positive or negative y-axis, we see that the locus of P consists of the whole ellipse.

The tangent at the point $P(x_0, y_0)$ on the ellipse (13) is derived in the same way as for the circle and parabola. Let $P_1(x_1, y_1)$ be another point on the ellipse (figure 35). Then,

$$\frac{x_0^2}{a^2} + \frac{y_0^2}{b^2} = 1, \qquad \frac{x_1^2}{a^2} + \frac{y_1^2}{b^2} = 1. \tag{14}$$

---

* $c < a$ means 'c is less than a'; $a > c$ means 'a is greater than c'; $c \leqslant a$ means 'c is less than or equal to a'; $a \geqslant c$ means 'a is greater than or equal to c'.

Hence, $(x_0^2 - x_1^2)/a^2 = (y_1^2 - y_0^2)/b^2$ or

$$\frac{y_1 - y_0}{x_1 - x_0} = -\frac{x_1 + x_0}{y_1 + y_0} \cdot \frac{b^2}{a^2}.$$

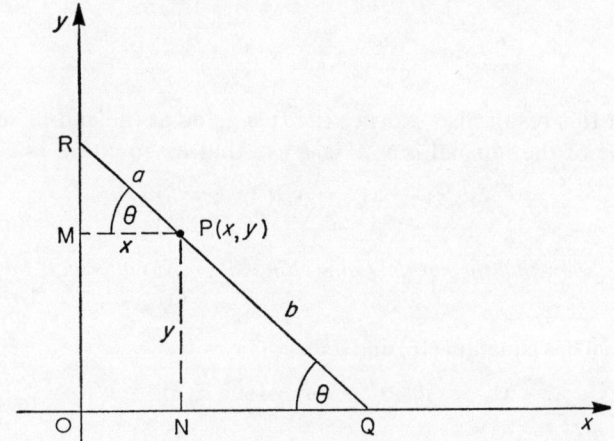

FIGURE 34

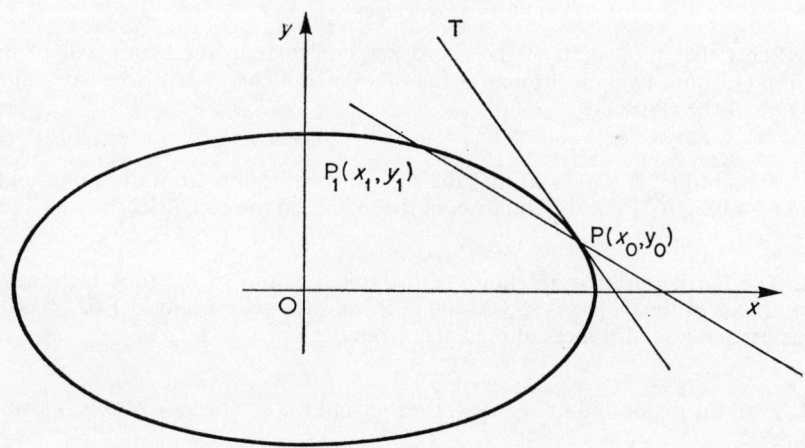

FIGURE 35

This gives the slope of PP_1 and, as P_1 approaches P, it becomes $-x_0 b^2/y_0 a^2$, which is the slope of the tangent PT. Hence the equation of the tangent is

$$y - y_0 = -\frac{x_0 b^2}{y_0 a^2}(x - x_0)$$

or

$$\frac{yy_0}{b^2} - \frac{y_0^2}{b^2} = -\frac{xx_0}{a^2} + \frac{x_0^2}{a^2}.$$

Using (14) we obtain for the equation of PT

$$\frac{xx_0}{a^2} + \frac{yy_0}{b^2} = 1. \tag{15}$$

Notice that this result also satisfies the *rule* given at the end of section 2.3. The slope of the normal is $y_0 a^2/x_0 b^2$ so that its equation is

$$x_0 b^2 (y - y_0) = y_0 a^2 (x - x_0). \tag{12}$$

Example 8 Show that the normal to the ellipse at (x_0, y_0) intersects the x-axis at $(e^2 x_0, 0)$.

The normal has equation (16) and intersects $y = 0$ at

$$x = x_0 - x_0 b^2/a^2 = x_0 - x_0(1 - e^2) = e^2 x_0$$

since $b^2 = a^2(1 - e^2)$.

Exercises

46. Show that the length of the chord perpendicular to the major axis of the ellipse (13) and passing through a focus is $2b^2/a$. (This chord is called a *latus rectum* of the ellipse).

47. A point $P(x,y)$ moves so that the sum of its distances from the points $(4,0)$ and $(-4,0)$ is 10. Find the equation of the locus and the eccentricity.

48. The Earth's orbit is an ellipse with the Sun as one of the foci. If the semi-major axis of the ellipse is 93,000,000 miles and the eccentricity is 1/62, find the greatest and least distances of the Earth from the Sun.

49. Find the major and minor axes, the foci and the eccentricity of the following ellipses:
(i) $(x^2/169) + (y^2/144) = 1$; (ii) $(x^2/25) + (y^2/16) = 1$; (iii) $(x^2/10) + (y^2/5) = 1$.

50. A straight line QR, 7 units long, contains a point P 5 units from R. The line moves so that Q is always on the x-axis and R on the y-axis. Find the locus of P.

51. An ellipse has foci $(\pm \sqrt{2}, 0)$ and passes through $(2, \sqrt{3})$. Find its equation.

Conics

52. A point P moves so that its distance from F(ae,0), where $e < 1$, is e times its distance from the line $x = a/e$. Show that the locus of P is the ellipse (13) with $b^2 = a^2(1 - e^2)$. (The line $x = a/e$ is called the *directrix* corresponding to the focus F. There is also a directrix $x = -a/e$ corresponding to the focus F'.)

53. Find the equation of the ellipse with focus (4,0), directrix $x = 9$ and eccentricity $\frac{2}{3}$.

54. Determine the locus of a point P so that the product of the slopes of the lines joining P to $(1,-2)$ and $(-1,2)$ is -3.

55. Find the tangent and normal to the ellipse $5x^2 + y^2 = 5$ at $(-\frac{2}{3},\frac{5}{3})$.

56. Show that the tangents to the ellipse $x^2/a^2 + y^2/b^2 = 1$ which have slope m are
$$y = mx \pm \sqrt{(a^2m^2 + b^2)}.$$

57. A ray of light issuing from a focus of an ellipse is reflected by the ellipse according to the law of question 44. Show that the reflected ray passes through the other focus. (The phenomenon occurs in some whispering galleries.)

58. The tangent at P to an ellipse intersects at T the directrix corresponding to the focus F. Show that FP and FT are perpendicular.

59. A straight rod touches an elliptical cam $3x^2 + 5y^2 = 15$. If the rod makes an angle θ with the major axis, find its perpendicular distance from the centre of the ellipse.

60. If p and p' are the lengths of the perpendiculars from the foci to a tangent to the ellipse $x^2/a^2 + y^2/b^2 = 1$, show that $pp' = b^2$.

61. Show that, in polar coordinates, with the pole at a focus, the equation of an ellipse can be written as
$$\frac{b^2}{ar} = 1 + e \cos \theta.$$

2.5 The hyperbola

One branch of a *hyperbola* consists of those points P(x,y) which are such that the difference of the distances PF', PF from the fixed points F(c,0) and $F'(-c,0)$ is a constant $2a$, where $a < c$. F and F' are known as the *foci* of the hyperbola (figure 36).

On a hyperbola
$$PF' - PF = 2a$$
so that
$$\sqrt{[(x + c)^2 + y^2]} - \sqrt{[(x - c)^2 + y^2]} = 2a. \qquad (17)$$

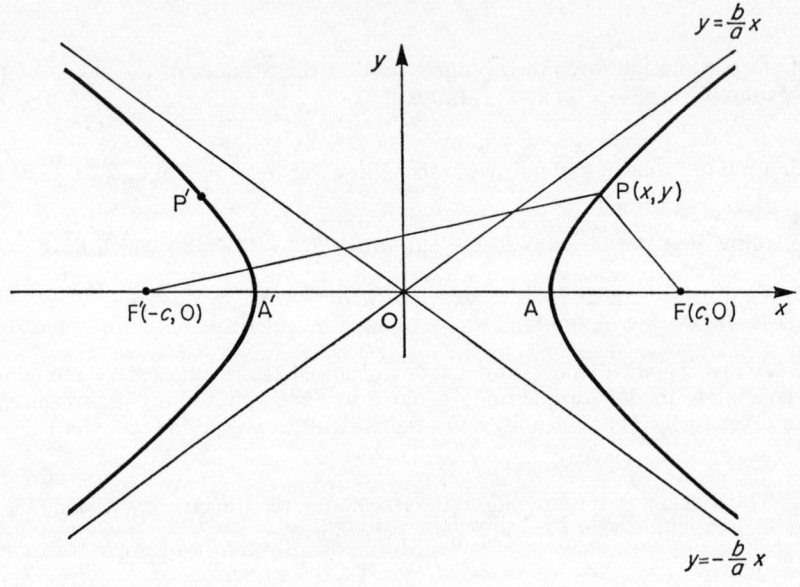

FIGURE 36

Hence
$$(x + c)^2 + y^2 = 4a^2 + (x - c)^2 + y^2 + 4a\sqrt{[(x - c)^2 + y^2]}$$
or
$$cx - a^2 = a\sqrt{[(x - c)^2 + y^2]}.$$
Squaring both sides we obtain
$$c^2x^2 - 2a^2cx + a^4 = a^2(x^2 - 2cx + c^2) + a^2y^2$$
or
$$(c^2 - a^2)x^2 - a^2y^2 = a^2(c^2 - a^2).$$
Since $c > a$ we write $c^2 - a^2 = b^2$ and the equation becomes
$$\frac{x^2}{a^2} - \frac{y^2}{b^2} = 1. \qquad (18)$$

The curve is symmetrical about the x- and y-axes and intersects the x-axis at $(\pm a, 0)$. It does not intersect the y-axis (figure 36) since $-y^2 = b^2$ has no real solutions. In fact, it is only the portion lying to the right of the

Conics

y-axis and passing through A which satisfies $PF' - PF = 2a$. The branch to the left of the y-axis is such that $P'F - P'F' = 2a$; for this, $2a$ in (17) must be replaced by $-2a$ but there is no alteration to (18). The two branches constitute the hyperbola.

That (18) gives both branches can be confirmed as follows. Let (x,y) satisfy (18); then $x^2/a^2 \geqslant 1$ so that either $x \geqslant a$ and (x,y) is on the right part of the diagram or $x \leqslant -a$ and (x,y) is on the left part. Now, with $c^2 = a^2 + b^2$,

$$PF = \sqrt{[(x-c)^2 + y^2]} = \sqrt{[(a - cx/a)^2]},$$
$$PF' = \sqrt{[(x+c)^2 + y^2]} = \sqrt{[(a + cx/a)^2]}.$$

If $x \geqslant a$, positive square roots are obtained by taking

$$PF = cx/a - a, \qquad PF' = a + cx/a$$

since $c > a$; thus, when P is to the right of $x = a$, $PF' - PF = 2a$. If $x \leqslant -a$, positive square roots are given by

$$PF = a - cx/a, \qquad PF' = -(a + cx/a)$$

and so, when P is to the left of $x = -a$, $PF - PF' = 2a$. Consequently, any point which fulfils the geometrical conditions must satisfy (18) and, conversely, any point which satisfies (18) also satisfies the geometrical conditions.

Observe that, since $y^2 = b^2(x^2 - a^2)/a^2$, y^2 gets larger and larger as x^2 increases. Thus, as x goes to positive infinity one value of y goes to positive infinity and the other to negative infinity. Therefore the right branch of the hyperbola goes off to infinity as shown in figure 36. Similar considerations apply to the left branch. Indeed, as x^2 becomes very large, a^2 can be neglected in comparison with it and $y^2 = b^2 x^2/a^2$ approximately. Therefore, at infinity, the hyperbola approximates the straight lines $y = \pm bx/a$. In other words, the perpendicular distance of a point on the hyperbola from one of the straight lines becomes smaller and smaller as the point goes off to infinity. The lines $y = \pm bx/a$ are called the *asymptotes* of the hyperbola and their intersection is called the *centre*.

The *eccentricity* e is defined by $c = ae$ or $b^2 = a^2(e^2 - 1)$. It is clear that $e > 1$. When $e = \sqrt{2}$, so that $a = b$, the curve is sometimes known as a *rectangular hyperbola*.

Example 9 *A point moves in the region $-\frac{2}{3}x < y < \frac{2}{3}x$ so that the product of its distances from the lines $2x - 3y = 0$ and $2x + 3y = 0$ is $1/13$. Find its locus.*

The distance of $P(x,y)$ from $2x - 3y = 0$ is $(2x - 3y)/\sqrt{13}$ and from $2x + 3y = 0$ is $(2x + 3y)/\sqrt{13}$ (see section 1.5). Hence

$$(2x - 3y)(2x + 3y)/13 = 1/13$$

or
$$4x^2 - 9y^2 = 1$$
which is a hyperbola with the given lines as asymptotes. Thus the point describes the branch of this hyperbola which lies in $x > 0$.

To find the tangent to the hyperbola at the point (x_0, y_0) choose another point (x_1, y_1) on the hyperbola. Then

$$\frac{x_0^2}{a^2} - \frac{y_0^2}{b^2} = 1, \qquad \frac{x_1^2}{a^2} - \frac{y_1^2}{b^2} = 1. \tag{18}$$

Therefore, by subtraction, $(x_1^2 - x_0^2)/a^2 = (y_1^2 - y_0^2)/b^2$ or

$$\frac{y_1 - y_0}{x_1 - x_0} = \frac{x_1 + x_0}{y_1 + y_0} \cdot \frac{b^2}{a^2}.$$

As (x_1, y_1) approaches (x_0, y_0) this slope becomes $x_0 b^2 / y_0 a^2$ which is therefore the slope of the tangent. Hence the equation of the tangent is

$$y - y_0 = \frac{x_0 b^2}{y_0 a^2}(x - x_0)$$

or

$$\frac{y y_0}{b^2} - \frac{y_0^2}{b^2} = \frac{x x_0}{a^2} - \frac{x_0^2}{a^2}.$$

From (18) we obtain

$$\frac{x x_0}{a^2} - \frac{y y_0}{b^2} = 1 \tag{19}$$

as the equation of the tangent. Once again the *rule* of section 2.3 is complied with.

The slope of the normal is $-y_0 a^2 / x_0 b^2$ and so its equation is

$$x_0 b^2 (y - y_0) + y_0 a^2 (x - x_0) = 0. \tag{20}$$

Exercises

62. Show that the length of the chord perpendicular to the line joining the foci of the hyperbola (18) and passing through a focus is $2b^2/a$. (This chord is called a *latus rectum* of the hyperbola.)

63. A point $P(x, y)$ moves so that difference between its distances from $(4, 0)$ and $(-4, 0)$ is 6. Find the equation of the locus and the eccentricity.

64. Find the foci, eccentricity, asymptotes and centre of the hyperbolae:
(i) $(x^2/25) - (y^2/9) = 1$; (ii) $x^2 - 2y^2 = 10$; (iii) $2x^2 - y^2 = 10$.

Conics

65. A point moves in $-\frac{3}{4}x < y < \frac{3}{4}x$ so that the product of its distances from the lines $3x - 4y = 0$ and $3x + 4y = 0$ is $\frac{1}{3}$. Find its locus.

66. A hyperbola has foci $(\pm 5, 0)$ and passes through $(\sqrt{10}, -\sqrt{20})$. Find its equation.

67. A hyperbola has asymptotes $y = \pm 4x$ and passes through $(1,2)$. Find its equation.

68. A point P moves so that its distance from $F(ae, 0)$, where $e > 1$, is e times its distance from the line $x = a/e$. Show that the locus of P is the hyperbola (18) with $b^2 = a^2(e^2 - 1)$. (The line $x = a/e$ is called the *directrix* corresponding to the focus F. There is also a directrix corresponding to the focus F'.)

69. Find the equation of the hyperbola with focus $(3,0)$, directrix $x = \frac{1}{3}$ and eccentricity 3.

70. In a radar navigation system a master station sends out a signal. There are two slave stations each of which sends out its own signal as soon as it receives the master signal. A receiver measures the time that elapses between the arrival of a master signal and of a slave signal. Show that, from these time intervals, the position of the receiver is determined by the intersection of two hyperbolae.

71. Show that the equation

$$\frac{x^2}{4 - \lambda} + \frac{y^2}{3 - \lambda} = 1$$

represents (i) an ellipse if λ is a constant less than 3, (ii) a hyperbola if λ is greater than 3 but less than 4, (iii) no real locus if λ exceeds 4. Show also that all the ellipses in (i) and all the hyperbolae in (ii) have foci $(\pm 1, 0)$. (In this case the ellipses and hyperbolae are examples of *confocal conics*, i.e. conics with the same focus.)

72. Find the equations of the tangents to the hyperbola $3x^2 - 2y^2 = 10$ at the points where it is intersected by the line $3x - 2y = 4$.

73. Show that the tangents to the hyperbola $x^2/a^2 - y^2/b^2 = 1$ which have slope m are
$$y = mx \pm \sqrt{(a^2 m^2 - b^2)}.$$
Deduce that no tangent passes through the origin.

74. The normal to the hyperbola $x^2/a^2 - y^2/b^2 = 1$ at P intersects the x-axis at Q. If OT is the perpendicular distance of the origin O from the tangent at P, prove that OT . PQ = b^2.

75. A tangent is drawn to the ellipse $4x^2 + 9y^2 = 36$ and a tangent is drawn to the hyperbola $3x^2 - 2y^2 = 6$. If the two tangents are perpendicular show that they intersect on the circle $x^2 + y^2 = 6$.

76. Prove that the ellipse $x^2/25 + y^2/9 = 1$ and the hyperbola $x^2 - y^2 = 8$ intersect at right angles.*

77. Show that the parabolae $y^2 = 8x$ and $x^2 = 4y - 12$ touch at (2,4).

78. Prepare programs so that a graph plotter will draw (i) an ellipse, (ii) a hyperbola, (iii) a parabola. Arrange also to plot a tangent and a normal.

79. Consider the possibility of answering Exercises 8, 11, 23, 28, 33, 44, 52, 57, 67 and 74 by means of a computer.

2.6 Curves of the second degree

All the curves discussed in this chapter so far have equations which are particular cases of the general equation of the second degree

$$Ax^2 + Bxy + Cy^2 + Dx + Ey + F = 0. \qquad (21)$$

For instance, (21) gives a circle when $A = C$ and $B = 0$. An ellipse is obtained if $A = 1/a^2$, $C = 1/b^2$, $F = -1$, $B = D = E = 0$ and a hyperbola if $A = 1/a^2$, $C = -1/b^2$, $F = -1$, $B = D = E = 0$. The equation supplies a straight line if $A = B = C = 0$ but is then no longer of the second degree.

We now enquire what curves (21) can represent when one at least of A,B,C is non-zero. The method is to choose new coordinates so as to simplify (21) as much as possible and make it as close to one of our known equations as we can.

First, we take (h,k) as a new origin and put (section 1.6)

$$x = X + h, \qquad y = Y + k.$$

Then (21) becomes

$$AX^2 + BXY + CY^2 + (2Ah + Bk + D)X + (Bh + 2Ck + E)Y +$$
$$Ah^2 + Bhk + Ck^2 + Dh + Ek + F = 0. \qquad (22)$$

---

* The *angle between two curves* which intersect at a point P is defined to be the angle between their tangents at P. Two curves which have the same tangent at P are said to *touch* at P.

Conics

Now we try to choose h and k so that the coefficients of X and Y disappear, i.e. so that
$$\left.\begin{array}{c} 2Ah + Bk + D = 0, \\ Bh + 2Ck + E = 0. \end{array}\right\} \quad (23)$$
These equations determine h and k uniquely unless $2A/B = B/2C$, i.e. $B^2 = 4AC$. It will therefore be necessary to consider the cases $B^2 \neq 4AC$ and $B^2 = 4AC$ separately.

(i) $B^2 \neq 4AC$.

In this case choose h and k so that (23) are satisfied and then (22) becomes
$$AX^2 + BXY + CY^2 + F' = 0 \quad (24)$$
where
$$\begin{aligned} F' &= Ah^2 + Bhk + Ck^2 + Dh + Ek + F \\ &= \tfrac{1}{2}Dh + \tfrac{1}{2}Ek + F \\ &= \frac{1}{4AC - B^2}(4ACF + BDE - AE^2 - CD^2 - FB^2) \end{aligned} \quad (25)$$
on using (23). Note that (21) is in the form (24) when $D = E = 0$; when this happens (23) give $h = 0$, $k = 0$ and no change of origin is necessary.

Now rotate the axes through an angle α (section 1.7) so that
$$X = X' \cos \alpha - Y' \sin \alpha, \qquad Y = X' \sin \alpha + Y' \cos \alpha.$$
Then (24) becomes
$$(A \cos^2 \alpha + B \cos \alpha \sin \alpha + C \sin^2 \alpha)X'^2 +$$
$$[B(\cos^2 \alpha - \sin^2 \alpha) + 2(C - A) \sin \alpha \cos \alpha]X'Y' +$$
$$(A \sin^2 \alpha - B \sin \alpha \cos \alpha + C \cos^2 \alpha)Y'^2 + F' = 0. \quad (26)$$
But $\cos^2 \alpha - \sin^2 \alpha = \cos 2\alpha$ and $2\sin \alpha \cos \alpha = \sin 2\alpha$ so that the coefficient of $X'Y'$ is $B \cos 2\alpha + (C - A) \sin 2\alpha$. Therefore we can make the coefficient of $X'Y'$ vanish by choosing α so that
$$\tan 2\alpha = \frac{B}{A - C}. \quad (27)$$
With this choice (26) goes over to
$$A'X'^2 + C'Y'^2 + F' = 0 \quad (28)$$
where
$$A' = A \cos^2 \alpha + B \cos \alpha \sin \alpha + C \sin^2 \alpha,$$
$$C' = A \sin^2 \alpha - B \sin \alpha \cos \alpha + C \cos^2 \alpha.$$
However, α can be chosen to satisfy (27) and so that
$$\cos^2 \alpha = \tfrac{1}{2}(1 + \cos 2\alpha) = \tfrac{1}{2} + \tfrac{1}{2}(A - C)/\sqrt{[(A - C)^2 + B^2]},$$
$$\sin^2 \alpha = \tfrac{1}{2}(1 - \cos 2\alpha) = \tfrac{1}{2} - \tfrac{1}{2}(A - C)/\sqrt{[(A - C)^2 + B^2]},$$
$$\sin \alpha \cos \alpha = \tfrac{1}{2} \sin 2\alpha = \tfrac{1}{2}B/\sqrt{[(A - C)^2 + B^2]},$$

and so
$$A' = \tfrac{1}{2}(A + C) + \tfrac{1}{2}\sqrt{[(A - C)^2 + B^2]}, \tag{29}$$
$$C' = \tfrac{1}{2}(A + C) - \tfrac{1}{2}\sqrt{[(A - C)^2 + B^2]}. \tag{30}$$
Note that, as a consequence,
$$A' + C' = A + C, \tag{31}$$
$$4A'C' = 4AC - B^2 \tag{32}$$
which provide useful checks on any numerical work.

Observe that neither A' nor C' can be zero. For, if either A' or C' were zero, (32) shows that $B^2 = 4AC$ contrary to the assumption made in the derivation of (24).

There are now several possibilities to consider.

(a) $F' = 0$. In this case (28) reduces to
$$A'X'^2 + C'Y'^2 = 0.$$
If A' and C' have the same sign, i.e. $A'C' > 0$, the only real solution is $X' = 0$, $Y' = 0$; in other words, the curve consists only of a single point.

If A' and C' have opposite signs, i.e. $A'C' < 0$, suppose A' is not negative. Then the equation can be written as
$$[X'\sqrt{A'} + Y'\sqrt{(-C')}][X'\sqrt{A'} - Y'\sqrt{(-C')}] = 0$$
so that the curve consists of the two straight lines
$$X'\sqrt{A'} = -Y'\sqrt{(-C')}, \quad X'\sqrt{A'} = Y'\sqrt{(-C')}$$
passing through $X' = 0$, $Y' = 0$. If C' is positive, instead of A', a similar conclusion may be drawn.

(b) $F' \neq 0$. If A', C', F' all have the same sign (28) has no real solution and there is no real locus.

If A', C' have the same sign but F' has the opposite sign the equation can be written as
$$\left(-\frac{A'}{F'}\right)X'^2 + \left(-\frac{C'}{F'}\right)Y'^2 = 1 \tag{33}$$
and represents an ellipse since A'/F' and C'/F' are both negative. If $A' = C'$ the ellipse degenerates into a circle; from (29) and (30), this requires $(A - C)^2 + B^2 = 0$ which necessitates $A = C$ and $B = 0$.

If A' and C' have opposite signs, i.e. $A'C' < 0$, then the form (33) demonstrates that the locus is a hyperbola. In this case the asymptotes are $Y' = \pm X'\sqrt{(-A'/C')}$.

This completes the discussion of the possible cases when $B^2 \neq 4AC$. It will be observed that $X' = 0$, $Y' = 0$ plays a special role for all the real loci; either it is the only point on the curve, or it is the intersection of the two lines, or it is the centre of the ellipse, or it is the centre of the hyperbola

Conics

depending upon which curve is represented by the equation. In other words, when $B^2 \neq 4AC$, the point $x = h$, $y = k$ as given by (23) can be regarded as the *centre* of the locus represented by (21).

We turn now to the cases $B^2 = 4AC$.

(ii) $B^2 = 4AC$.

With the origin at $x = 0$, $y = 0$ rotate the axes through the angle α given by (27). Then (21) goes over to

$$A'X_1^2 + C'Y_1^2 + D'X_1 + E'Y_1 + F = 0$$

where $D' = D \cos \alpha + E \sin \alpha$, $E' = E \cos \alpha - D \sin \alpha$.

From (32), $A'C' = 0$ so that one at least of A' and C' is zero. But both A' and C' cannot be zero. For, if $A' = C' = 0$, (29) and (30) show that $A + C = 0$ and $(A - C)^2 + B^2 = 0$ which imply $B = 0 = A = C$ contrary to the original assumption that one at least of A, B, C is non-zero.

It is convenient to assume that $A' \neq 0$; if this is not so, it can always be arranged by a rotation of the axes through a right angle.

Thus the equation to be considered is

$$A'X_1^2 + D'X_1 + E'Y_1 + F = 0$$

with $A' \neq 0$.

(a) $E' = 0$. The equation is then $A'X_1^2 + D'X_1 + F = 0$ which has either two real and unequal roots representing two lines parallel to the Y_1-axis, or two real equal roots representing a line parallel to the Y_1-axis twice, or two non-real roots when there is no real locus. All these possibilities are regarded as *degenerate* cases of the parabola.

(b) $E' \neq 0$. The equation can now be written as

$$Y_1 = -\frac{A'}{E'}\left(X_1 + \frac{D'}{2A'}\right)^2 - \frac{F}{E'} + \frac{D'^2}{4A'E'}.$$

Choose $(-D'/2A', (-4A'F + D'^2)/4A'E')$ as a new origin so that

$$X_1 = \xi - D'/2A', \qquad Y_1 = \eta + (-4A'F + D'^2)/4A'E',$$

and the equation becomes

$$\xi^2 = 4a\eta$$

with $4a = -E'/A'$. In this case the curve is a parabola. This concludes the discussion of (ii).

The findings of this section may be summarized as follows: the equation

$$Ax^2 + Bxy + Cy^2 + Dx + Ey + F = 0$$

represents

(I) A *parabola*, including the degenerate cases of parallel lines, if $B^2 - 4AC = 0$ (see (ii)).

(II) An *ellipse*, or possibly no real locus, if $B^2 - 4AC < 0$. The ellipse is a *circle* if $B = 0$, $A = C$ (see (i)(b)). The ellipse degenerates to a point if $F' = 0$ (see (i)(a)).

(III) A *hyperbola* if $B^2 - 4AC > 0$ and $F' \neq 0$ (see (i)(b)).

(IV) A *pair of intersecting straight lines* if $B^2 - 4AC > 0$ and $F' = 0$ (see (i)(a)).

In these conditions F' is given by (25). For (II), (III) and (IV) the centre is (h,k), satisfying (23).

This demonstrates the way in which the parabola, ellipse and hyperbola are connected through the equation of the second degree. They are also related by the *focus-directrix property*, namely, that if P moves so that its distance PF from the focus F is e times its distance PL from a fixed line, i.e.

$$PF = e \cdot PL,$$

the curve is a *parabola* if $e = 1$ (section 2.2), an *ellipse* with eccentricity e if $e < 1$ (question 52), a *hyperbola* of eccentricity e if $e > 1$ (question 68).

The curve of the second degree is known as a *conic section* or, more briefly, as a *conic*.

Figures demonstrating how conic sections are derived from a cone can be found in Chapter 7 of Volume II. (See Figs. 94, 95, 96). The ellipse and hyperbola are often classified as *central conics*, because each possesses a centre.

Example 10 Determine the nature of the curve

$$9x^2 + 24xy + 16y^2 - 136x + 152y + 136 = 0.$$

Here $B^2 - 4AC = 24^2 - 4 \cdot 9 \cdot 16 = 0$ so that (I) applies and the curve is a parabola or one of its degeneracies.

The angle α of (27) is given by

$$\tan 2\alpha = 24/(9 - 16) = -24/7.$$

Therefore $\cos^2 \alpha = \frac{1}{2} + \frac{1}{2}(9 - 16)/\sqrt{[(9 - 16)^2 + 24^2]} = \frac{1}{2} - \frac{1}{2} \cdot \frac{7}{25} = \frac{9}{25}$ whence $\cos \alpha = \frac{3}{5}$ and then $\sin \alpha = \frac{4}{5}$ in order that $\sin 2\alpha$ be positive as required by the choice leading to (29) when $B > 0$. The substitution

$$x = \tfrac{1}{5}(3X_1 - 4Y_1), \qquad y = \tfrac{1}{5}(4X_1 + 3Y_1)$$

gives

$$25X_1^2 + 40X_1 + 200Y_1 + 136 = 0$$

or

$$25(X_1 + \tfrac{4}{5})^2 + 200(Y_1 + \tfrac{3}{5}) = 0.$$

Consequently the curve is a parabola with vertex $X_1 = -\frac{4}{5}$, $Y_1 = -\frac{3}{5}$. i.e. $x = 0$, $y = -1$, and axis $X_1 + \frac{4}{5} = 0$ or $3x + 4y + 4 = 0$. The parabola is sketched in figure 37.

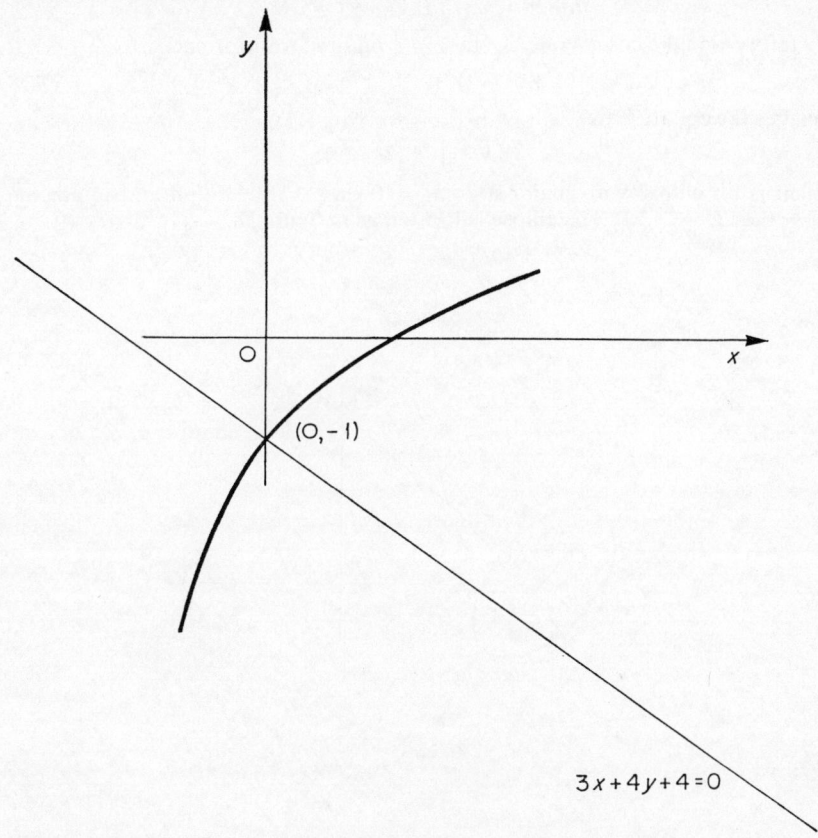

FIGURE 37

Example 11 *Determine the nature of the conic*
$$11x^2 + 9y^2 + 2\sqrt{3}xy + 44x + 4\sqrt{3}y + 40 = 0.$$

Since $B^2 - 4AC = 9^2 - 4 \cdot 11 \cdot 2\sqrt{3} < 0$, (II) is relevant and the curve is an ellipse or one of its degeneracies. The centre (h,k) satisfies
$$22h + 2\sqrt{3}k + 44 = 0,$$
$$2\sqrt{3}h + 18k + 4\sqrt{3} = 0$$
which lead to $h = -2$, $k = 0$. The substitution
$$x = X - 2, \qquad y = Y$$

gives
$$11X^2 + 9Y^2 + 2\sqrt{3}XY = 4.$$
Now
$$\tan 2\alpha = 2\sqrt{3}/(11-9) = \sqrt{3}.$$
Therefore we take $\cos \alpha = \frac{1}{2}\sqrt{3}$, $\sin \alpha = \frac{1}{2}$ and the transformation
$$X = \tfrac{1}{2}(\sqrt{3}X' - Y'), \qquad Y = \tfrac{1}{2}(X' + \sqrt{3}Y')$$
changes the equation to
$$12X'^2 + 8Y'^2 = 4$$
which is an ellipse with major axis $X' = 0$ or $\sqrt{3}X + Y = 0$ and minor axis $Y' = 0$ or $X = \sqrt{3}Y$. The ellipse is located as in figure 38.

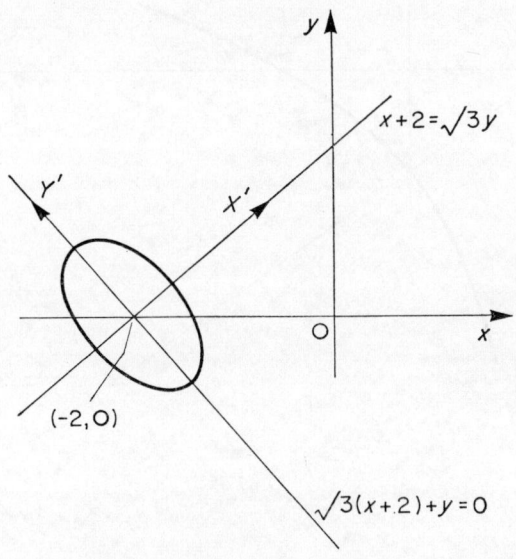

FIGURE 38

Example 12 Determine the nature of the conic
$$x^2 - 2xy + y^2 + x - y - 2 = 0.$$

In this case $B^2 - 4AC = 2^2 - 4 = 0$ so that, according to (I), the curve is a parabola or one of its degeneracies. In fact, the equation can be written
$$(x - y)^2 + (x - y - 2) = 0$$
or
$$(x - y + 2)(x - y - 1) = 0$$
showing that the locus consists of the parallel straight lines $x - y + 2 = 0$ and $x - y = 1$.

Example 13 Determine the nature of the conic
$$9x^2 + 12xy + 6y^2 + 10 = 0.$$

Here $B^2 - 4AC = 12^2 - 4.9.6 < 0$ so that, from (II), the curve is an ellipse or one of its degeneracies. In fact, the equation can be expressed as
$$(3x + 2y)^2 + 2y^2 + 10 = 0$$
which cannot be satisfied by any real x and y. The locus is not real.

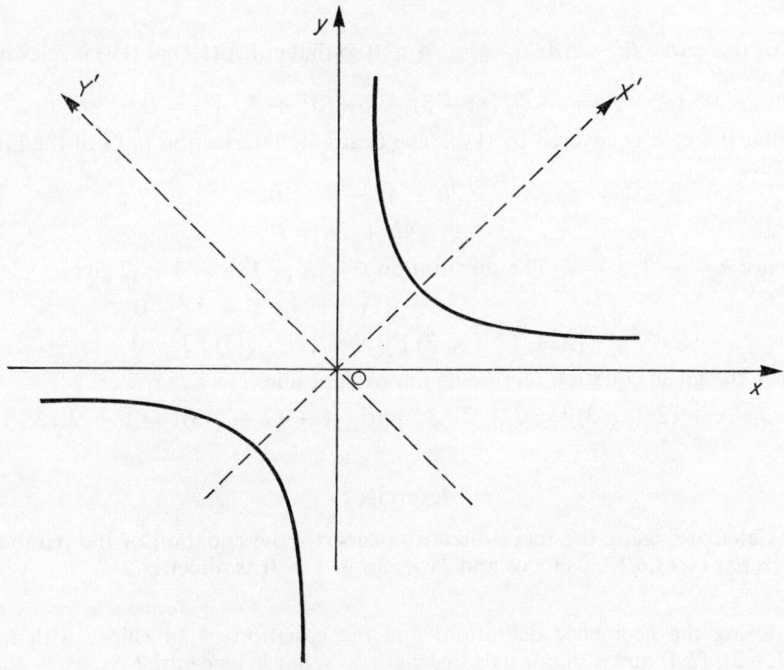

FIGURE 39

Example 14 Determine the nature of the conic
$$xy = a^2.$$

Since $B^2 - 4AC = 1 > 0$, (III) or (IV) is applicable. From (25)
$$F' = (-1)(-1) \neq 0$$
so that the curve is a hyperbola. There are no x or y terms and so the centre is the origin. Also $\tan 2\alpha$ is infinite and so we can take $2\alpha = \tfrac{1}{2}\pi$. The rotation of axes is then
$$x = (X' - Y')/\sqrt{2}, \qquad y = (X' + Y')/\sqrt{2}$$

leading to
$$X'^2 - Y'^2 = 2a^2.$$
Thus the hyperbola has foci $X' = \pm 2a$, $Y' = 0$ or $x = \pm \sqrt{2}a$, $y = \pm \sqrt{2}a$, eccentricity $\sqrt{2}$ and asymptotes $X' = \pm Y'$ or $x = 0$ and $y = 0$. The curve is sketched in figure 39.

Example 15   Determine the nature of the conic
$$x^2 + 4xy + y^2 - 6x - 3 = 0.$$

For this curve $B^2 - 4AC = 4^2 - 4 > 0$ so that either (III) or (IV) is relevant. But
$$F' = -\tfrac{1}{12}[4(-3) - (-6)^2 + 3 \cdot 4^2] = 0$$
so that the case is covered by (IV). The centre or intersection (h,k) of the lines satisfies
$$2h + 4k - 6 = 0,$$
$$4h + 2k = 0$$
whence $h = -1$, $k = 2$. The substitution $x = X - 1$, $y = Y + 2$ gives
$$X^2 + 4XY + Y^2 = 0$$
or
$$[X + (2 + \sqrt{3})Y][X + (2 - \sqrt{3})Y] = 0.$$
Hence the given equation represents the straight lines
$$x + (2 + \sqrt{3})y = 3 + 2\sqrt{3} \quad \text{and} \quad x + (2 - \sqrt{3})y = 3 - 2\sqrt{3}.$$

Exercises

80. Calculate, using the focus-directrix property, the equation of the parabola which has $(4/13, 6/13)$ as focus and $2x + 3y + 1 = 0$ as directrix.

81. Using the geometric definition, find the equation of an ellipse with foci $(-1,-2)$, $(2,4)$ and a major axis of length 8. What is its centre?

82. Using the geometric definition, find the hyperbola produced by a point moving so that the difference of its distances from $(-1,-2)$ and $(2,4)$ is 4. What is its centre?

83. Using the focus-directrix property, derive the equation of the ellipse which has focus $(1 + 2\sqrt{2}, -2\sqrt{2})$, corresponding directrix $x - y = 1 + 9\sqrt{2}$ and eccentricity $\tfrac{2}{3}$.

84. Using the focus-directrix property, derive the equation of the hyperbola which has focus $(9/5, -17/5)$, corresponding directrix $9x - 12y = 17$ and eccentricity 3.

85. Determine what type of conic each of the following equations represents and locate the centre if one exists:
   (i) $3x^2 + 5xy - 2y^2 - x + 5y - 2 = 0$.
   (ii) $3x^2 + 5xy - 2y^2 - x + 5y + 2 = 0$,
   (iii) $9x^2 - 12xy + 4y^2 + x + 8 = 0$,
   (iv) $10xy - 2y^2 - 2x + 2y + 1 = 0$,
   (v) $41x^2 - 84xy + 76y^2 - 84x + 152y = 92$,
   (vi) $x^2 + y^2 + x - y = 0$.

86. By changing the origin and rotating the axes as necessary, reduce the following equations to standard form and identify the conic:
   (i) $194x^2 + 120xy + 313y^2 + 776x + 240y + 607 = 0$,
   (ii) $9x^2 + 6xy + y^2 = 4$,
   (iii) $xy + 2x + y = 0$,
   (iv) $x^2 + 4xy + 4y^2 - 6x + 3y = 0$,
   (v) $x^2 + y^2 - 2x + 4y + 9 = 0$.

87. If the equation $AX^2 + BXY + CY^2 + DX + EY + F = 0$ represents a pair of straight lines prove that $AX^2 + BXY + CY^2 = 0$ is a pair of straight lines, and that the angle between the first pair is the same as that between the second pair. Find the angle between the lines
$$4y^2 - 48xy - 25x^2 - 52x - 104y = 0.$$

88. If $a \neq 0$ show that
$$ab(x^2 - y^2) + (a^2 - b^2)xy - a^2bx + ab^2y = 0$$
represents a pair of perpendicular straight lines. Prove that they intersect on the circle $x^2 + y^2 - ax = 0$.

89. Show that the tangent to
$$Ax^2 + Bxy + Cy^2 + Dx + Ey + F = 0$$
at (x_0, y_0) has equation
$$Axx_0 + \tfrac{1}{2}B(xy_0 + x_0y) + Cyy_0 + \tfrac{1}{2}D(x + x_0) + \tfrac{1}{2}E(y + y_0) + F = 0.$$
This agrees with the *rule* of section 2.3. In general the rule does not apply to curves of higher degree (see section 5.7).

90. The tangent at the point P of the hyperbola $xy = a^2$ intersects the asymptotes at Q and R. Find the locus of the mid-point of QR and show that the area of the triangle OQR is $2a^2$.

91. Make a program to graph any curve of the second degree. Arrange for the program to display the foci, directrices, centre, axes and asymptotes where relevant. Allow also for the possibility of drawing a tangent and normal at any point of the conic.

3

Functions and Graphs

3.1 Functions

We have already encountered in the two preceding chapters several instances of the equations of curves. For example, the equation of a straight line is
$$y = mx + b.$$
Here x and y are the coordinates of any point on the line and can take values from $-\infty$ to $+\infty$. On the other hand, m and b are fixed for any particular line. Therefore, x and y are called *variables* of the equation whereas m and b are called *constants*. For different straight lines m and b have different constant values.

Another example is provided by the ellipse with equation
$$\frac{x^2}{a^2} + \frac{y^2}{b^2} = 1, \tag{1}$$
Again x and y are variables while a and b are constants. In this case, however, x is limited to values from $-a$ to a and y to values from $-b$ to b because no real values of x and y satisfy (1) outside these values.

Again, in the equation of the hyperbola
$$\frac{x^2}{a^2} - \frac{y^2}{b^2} = 1,$$
x and y are variables whereas a and b are constants. Here x takes only values from $-\infty$ to $-a$ and from a to $+\infty$ while y can take values from $-\infty$ to $+\infty$.

In general, we can regard each of these equations as supplying a rule to determine y when x is known. Whenever we are provided with a rule, relating the variables x and y, which determines y when x is one of a certain set of values, y is said to be a *function* of x. Thus the word function is a shorthand way of saying that there is a rule; different functions correspond to different rules.

Functions and Graphs

The function tells us what y corresponds to a given x. The rule that specifies this does not have to be an equation though it may be. We might, for example, have a table (as in Appendix A) in which the values of x are listed with the corresponding values of y alongside. Or the information might be given graphically and then a line drawn through a given x parallel to the y-axis will intersect the graph in the corresponding y. Yet again, the rule might be stored in a computer; then, when the given x is fed into the input, the corresponding y is printed out by the machine.

The variable x is often referred to as the *independent variable* and y is often called the *dependent variable*. This terminology indicates that free choice of the value of x is available but that, once this value has been assigned, the corresponding y is calculated from it.

That y is a function of x is usually denoted by writing
$$y = f(x).$$
The notation does not mean that y is f times x. It just means that y is a function of x and is read 'y equals a function f of x'. Sometimes x is, called the *argument* of the function f. If, in the same problem, other functions of x occur then other letters are used to denote them. Thus $g(x)$, $h(x)$, $F(x)$, $\phi(x)$ are all used to denote functions of x.

The set of values of x under consideration is called the *domain*. The corresponding set of y is called the *range*. If a is in the domain then $f(a)$ denotes the value assumed by $f(x)$ when $x = a$.

For example, if $f(x) = x^2 + 4x - 5$,
$$f(1) = (1)^2 + 4.1 - 5 = 0,$$
$$f(0) = -5,$$
$$f(-2) = (-2)^2 + 4(-2) - 5 = -9,$$
$$f(a) = a^2 + 4a - 5,$$
$$f(a + 1) = (a + 1)^2 + 4(a + 1) - 5 = a^2 + 6a.$$

For the straight line $y = mx + b$, the domain contains all real numbers from $-\infty$ to ∞ and the range likewise. For the parabola $y^2 = x$ the domain contains only those $x \geqslant 0$ while the range contains all real numbers. If $y = \sqrt{(1 - x^2)}$ the domain is $-1 \leqslant x \leqslant 1$ and the range is $0 \leqslant y \leqslant 1$.

As another illustration consider $y = 1/(x - 1)$. A definite value of y is obtained for every real x except $x = 1$. In this case the domain consists of all real x except $x = 1$.

If to each value of x in the domain there corresponds one and only one value of y, then y is said to be a *single-valued* function of x. For example, $y = 3x$ and $y = x^2 + 2$ are single-valued functions of x.

If more than one value of y corresponds to each value of x in the domain, then y is said to be a *multiple-valued* function of x. For example, $y^2 = x$ is satisfied by both $y = \sqrt{x}$ and $y = -\sqrt{x}$ and therefore defines y as a double-valued function of x; in this case, real values of y are obtained only if x is not negative.

On the other hand, if we specified definitely that y must be positive, we should be obliged to take $y = \sqrt{x}$ and then y would be a single-valued function of x. In general, the notation $y = f(x)$ will be used only when y is a single-valued function of x.

The reader should be careful to observe that the statement that y is a single-valued function of x does not imply that x is a single-valued function of y. This can be seen easily from the counter-example $y = x^2$ which gives y as a single-valued function of x but defines x as a double-valued function of y.

Exercises

1. If $f(x) = 3x^3 - 2x^2 - 5x + 4$, find the values of $f(1), f(2), f(0), f(-1), f(-4)$.

2. If $f(x) = (x + 1)(x + 5)/(x - 1)^2$ find the values of $f(2), f(0), f(-1), f(-1\cdot 1), f(-2), f(x + 1)$.

3. If $f(x) = x^4 - 13x^2 + 36$ find $f(2), f(-2), f(3), f(-3)$.

4. If $f(x) = ax^2 + bx + c$ find $f(x + 1), f(x - 1), f(x + h) - f(x), f(1/x)$.

5. If $f(x) = x^2 - x$ prove that $f(x + 1) = f(-x)$.

6. If $y = f(x)$ and $f(x) = (3x + 2)/(5x - 3)$ prove that $x = f(y)$.

7. If $f(x) = 3^x$ prove that $f(m) \cdot f(n) = f(m + n)$, $f(m)/f(n) = f(m - n)$, $f(x + 2) - f(x + 1) = 6f(x), f(x + 3)/f(x - 1) = f(4)$.

8. Given the following functions, find how many values of y correspond to each value of x and determine the domain for x so that x and y are both real:
(i) $y = x^3$, (ii) $(y - 2)^2 = x$, (iii) $y = \tan^2 x$, (iv) $y^m + x^m = a^m$, (v) $x^2 + y^2 = 6$, (vi) $x^2 + y^2 - 2x = 0$, (vii) $y = x/(x + 1)$, (viii) $y^2(4 - x^2) = 1$, (ix) $y^2(2 - x) = x$.

9(a). A rectangular plot of land requires 500 ft of fence to enclose it. If one side is x ft find the area y as a function of x. What are the possible values of x?

(b). Repeat (a) if 600 m of fence are required.

A function which is unaltered when x is changed to $-x$ is said to be an *even* function of x. For example, $x^6 + 2x^2 + 3$ and $\cos x$ are even functions because $(-x)^6 + 2(-x)^2 + 3 = x^6 + 2x^2 + 3$ and $\cos(-x) = \cos x$. In general, $f(x)$ is an even function if $f(x) = f(-x)$ and vice versa.

If $y = f(x)$ and f is even, the graph of y against x will be symmetrical about the y-axis because the same values of y are obtained for x and $-x$.

A function which changes sign when x is changed to $-x$ is called an *odd* function of x. Thus x, $\tan x$ and $x^3 + 2x$ are all odd functions because $(-x) = -(x)$, $\tan(-x) = -\tan x$ and $(-x)^3 + 2(-x) = -(x^3 + 2x)$. In general, $f(x)$ is an odd function if $f(-x) = -f(x)$ and vice versa.

An odd function of x must vanish at $x = 0$. For, if

$$f(-x) = -f(x)$$

then, when $x = 0$, $f(0) = -f(0)$ whence $f(0) = 0$.

If $y = f(x)$ and f is odd, the sign of y is altered when the sign of x is changed. In this case, the graph can be regarded as symmetrical about the origin because, if P is a point on the graph, so is P′ where P′ is the point on PO produced such that PO = OP′.

Functions such as $x^3 + x^2 + 1$, in which some terms change sign and some do not when x is replaced by $-x$, are neither even nor odd.

Recognition that a function is odd or even is a help in the drawing of graphs of elementary functions since the resulting symmetry reduces the number of points which have to be calculated.

Example 1 *In* $y = x^2$, f *is even and there is symmetry about the* y-*axis (figure 40). The curve must pass through the origin and also through* $(1,1)$ *since* $(0,0)$ *and* $(1,1)$ *both satisfy* $y = x^2$. *The equation defines* y *as a single-valued function of* x.

The equation $y = x^3$ also defines y as a single-valued function of x and gives a curve which passes through $(0,0)$ and $(1,1)$. However, x^3 is an odd function and consequently there is symmetry about the origin. If x is between 0 and 1, x^3 is less than x^2 and therefore the graph is nearer the x-axis than that of $y = x^2$. But, if $x > 1$, x^3 is greater than x^2 and the graph is nearer the y-axis than that of $y = x^2$; in fact, the curve rises quite steeply as x increases beyond 1. The two curves are compared in figure 40.

More generally, the curve $y = x^n$ is similar to that of $y = x^2$ when n is an even integer; the graph is similar to that of $y = x^3$ when n is an odd integer. All the graphs pass through the origin and (1,1). The greater the value of n, the flatter the curve is near the origin, i.e. when x is between 0 and 1 the graph

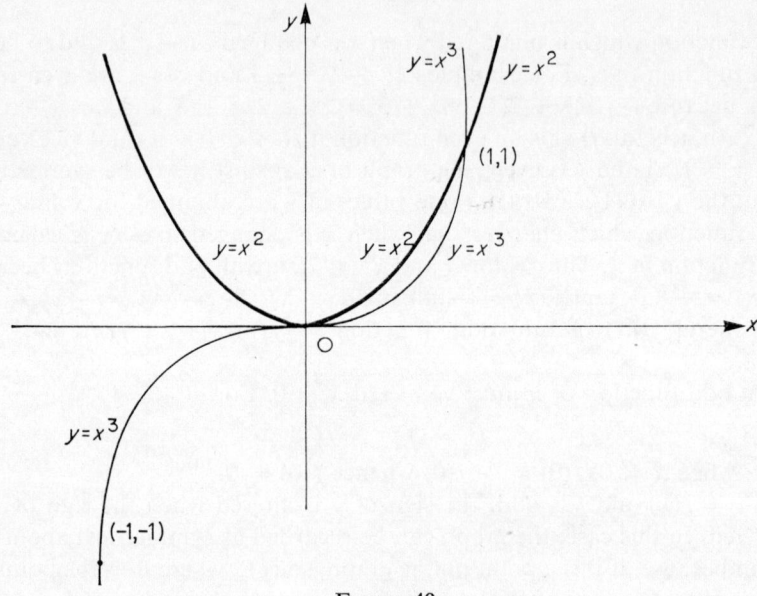

FIGURE 40

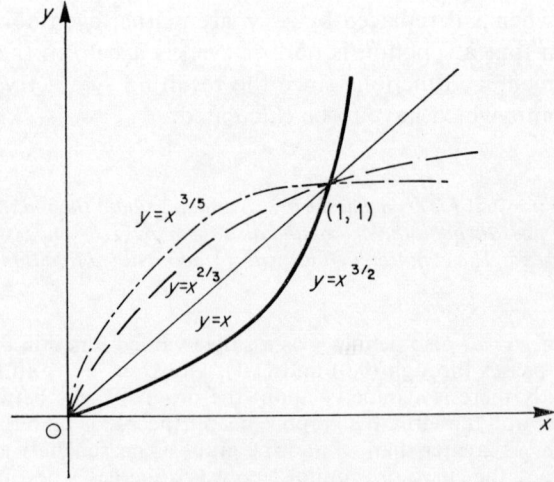

FIGURE 41

of $y = x^n$ is below that of $y = x^m$ when $n > m$. On the other hand, the greater the value of n, the steeper the curve is beyond (1,1), i.e. when $x > 1$ the graph of $y = x^n$ is above that of $y = x^m$ when $n > m$.

The graph of $y = x^{\frac{3}{2}}$ is shown in figure 41. The curve passes through (0,0) and (1,1). When x is between 0 and 1, $x^{\frac{3}{2}}$ is greater than x^2 but less than x so that the curve lies between $y = x^2$ and $y = x$ in this region. For $x > 1$, $x^{\frac{3}{2}}$ is greater than x and less than x^2. On the other hand, the graph of $y = x^{\frac{3}{2}}$ is above $y = x$ for x between 0 and 1 but below $y = x$ when $x > 1$.

Thus the graphs of $y = x^a$ for differing a provide a family of curves in which, when x is between 0 and 1, $y = x^a$ is above $y = x^b$ when $a < b$ and, when $x > 1$, is below.

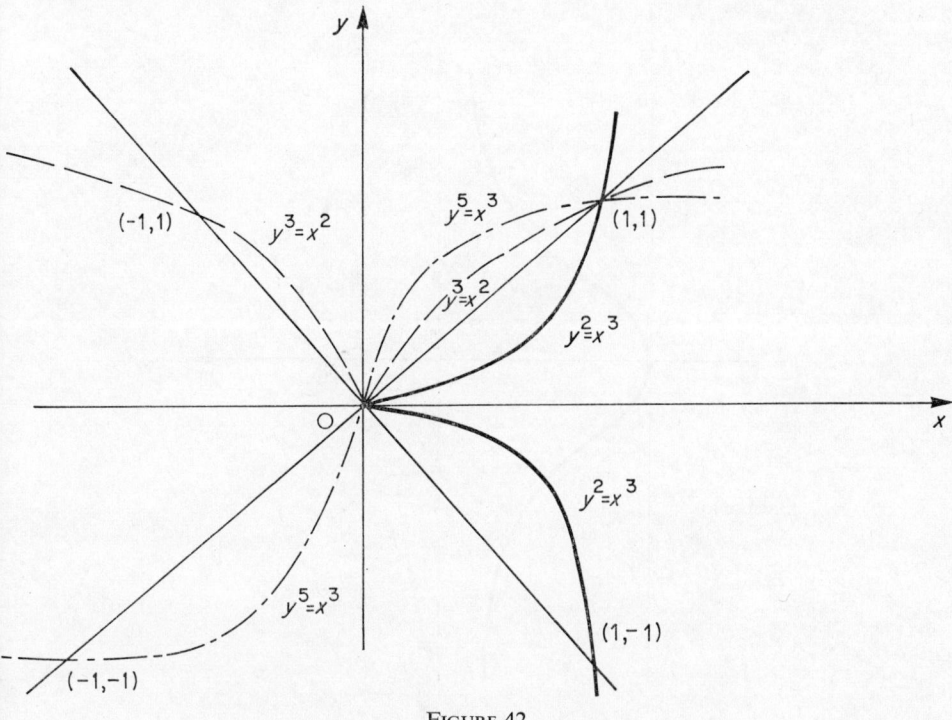

FIGURE 42

Example 2 *The equation $y^2 = x^3$ defines y as a double-valued function of x with domain $x \geqslant 0$. The two values of y are given by $y = x^{\frac{3}{2}}$ and $y = -x^{\frac{3}{2}}$. From Example 1 we see that the graph is as shown in figure 42.*

The equation $y^3 = x^2$ defines y as a single-valued function of x with range $y \geqslant 0$. Since changing x to $-x$ does not alter y the curve is symmetrical about the y-axis and so lies in the first and second quadrants.

The equation $y^5 = x^3$ defines y as a single-valued function of x with y positive when x is positive and with y negative when x is negative. From Example 1 we can draw the graph as in figure 42.

Example 3 The equation $x = y^2/(1+y^2)$ defines x as a single-valued function of y. (Written as $y^2 = x/(1-x)$ it expresses y as a double-valued function of x.)

When $y \neq 0$, y^2 is positive and it follows that x is positive unless $y = 0$ when $x = 0$. Further, $1 + y^2 > y^2$ and so $x < 1$; hence the curve does not leave the

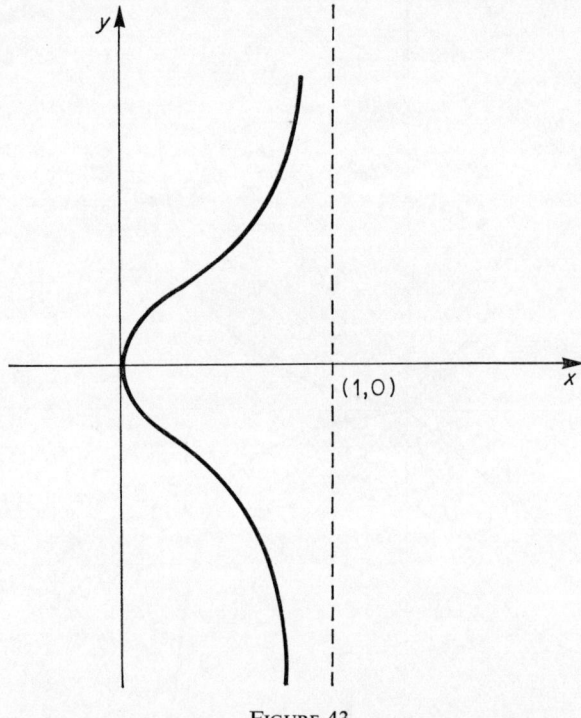

FIGURE 43

region bounded by $x = 0$ and $x = 1$. Next, note that the curve is symmetrical about the x-axis since replacing y by $-y$ does not alter x. Also

$$x - 1 = \frac{-1}{1 + y^2}$$

and the right-hand side gets smaller and smaller as y gets larger and larger, and can be made as small as we like by taking y large enough. This shows that $x = 1$ is an asymptote of the curve (see section 2.5).

Functions and Graphs 71

Example 4 The equation $y^2 = 1/x$ defines y as a double-valued function of x, with domain $x > 0$.

The curve is symmetrical about the x-axis since both $y = 1/\sqrt{x}$ and $y = -1/\sqrt{x}$ satisfy the equation. By taking x large enough y can be made as small as we like so that $y = 0$ is an asymptote (figure 44). On the other hand, the form $x = 1/y^2$ shows that x can be made as small as we like by taking y large enough. Therefore $x = 0$ is also an asymptote.

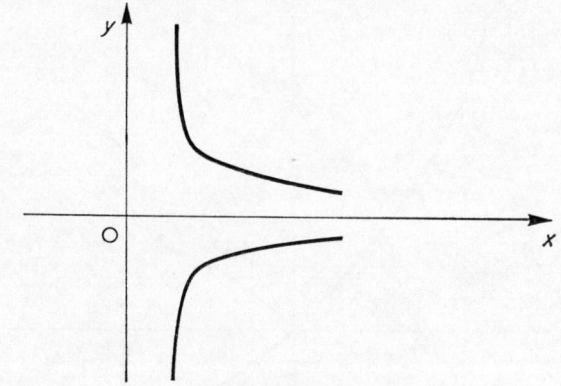

FIGURE 44

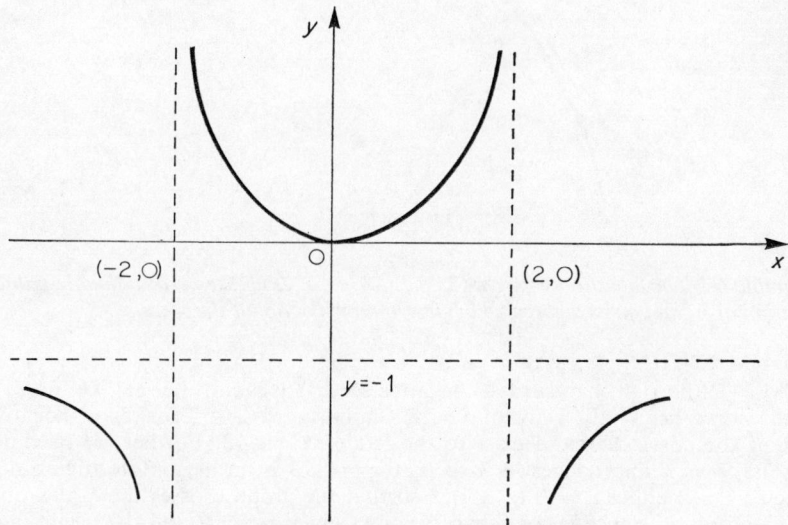

FIGURE 45

72   *Introductory Analysis*

Example 5   *The equation $y = x^2/(4 - x^2)$ gives y as a single-valued even function of x.*

The graph is therefore symmetrical about the y-axis; it also passes through the origin. If $x^2 < 4$, then $y \geqslant 0$ and, as x^2 approaches 4, y becomes larger and larger. Thus the lines $x = \pm 2$ are asymptotes (figure 45). If $x^2 > 4$ then $y < 0$ and, as x^2 approaches 4 from this region, y becomes more and more negative. Again, as x^2 becomes large, we see from $y + 1 = 4/(4 - x^2)$ that $y + 1$ becomes as small as we like if we choose x^2 large. Hence $y = -1$ is also an asymptote.

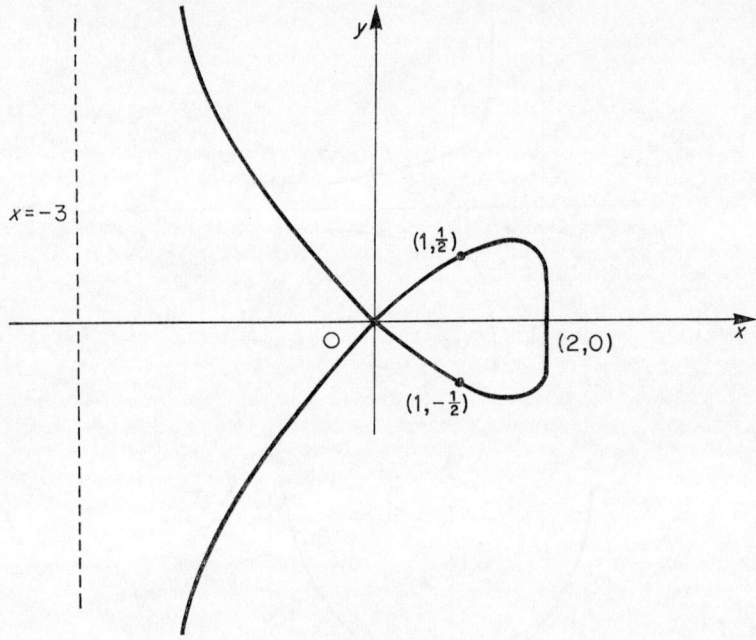

FIGURE 46

Example 6   *The equation $y^2 = x^2(2 - x)/(3 + x)$ expresses y as a double-valued function of x, and gives a curve which is symmetrical about the x-axis.*

If $x > 2$ the right-hand side is negative so that y cannot be real, i.e. no part of the curve lies to the right of $x = 2$. Similarly, $x < -3$ makes y^2 negative so that the curve does not exist to the left of $x = -3$. Further, y^2 becomes very large as x approaches -3 so that $x = -3$ is an asymptote (figure 46). When $y^2 = 0$ either $x = 0$ or $x = 2$ so that the graph crosses the x-axis only at these points. Hence it consists of a loop between $x = 2$ and the origin, and approaches the asymptote $x = -3$ from above and from below.

All the functions of the preceding examples have produced curves having smooth graphs. Not all functions, however, give smooth graphs and, indeed, some functions define only isolated points (see question 12). As an illustration consider

Example 7 At one time the charges, in pence, of the British General Post Office for parcels were

Not over	Charge	Not over	Charge	Not over	Charge
2 lb	33	6 lb	42	14 lb	66
3 lb	36	8 lb	48	18 lb	78
4 lb	39	10 lb	54	22 lb (max.)	90

If y is the cost of a parcel in pence and x its weight in pounds, the graph of y as a function of x is shown in figure 47. Notice that y is not defined if x exceeds 22. The graph consists of a series of portions of horizontal straight lines. Thus, for $6 < x \leqslant 8$, y is always 48, but observe that when $x = 6$, $y = 42$. Consequently,

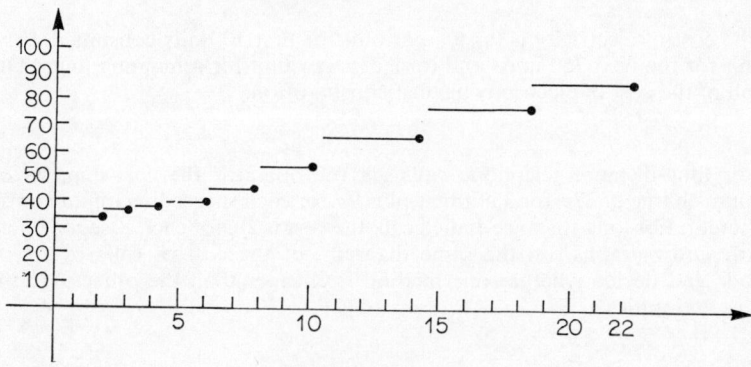

FIGURE 47

the function suddenly jumps from one value to another without taking any of the intermediate values. Such a jump is called a *discontinuity*. In figure 47 there are discontinuities at $x = 2, 3, 4, 6, 8, 10, 14$ and 18. It may also be considered that there is a discontinuity at $x = 0$ since there is no cost if there is no parcel.

A function, such as that of figure 47, whose graph consists of a number of horizontal lines is called a *step function*.

Exercises

10. Draw graphs of the following functions: (i) $y = x^2$, (ii) $y = -x^2$, (iii) $y^2 = \frac{1}{9}x^3$, (iv) $y^2 = -x^3$, (v) $x^2 + y^2 = 9$, (vi) $x^2 - y^2 = 9$, (vii) $y^2 = 1/x$, (viii) $y = x/(4 + x^2)$, (ix) $y = x/(4 - x^2)$, (x) $y = x^2/(4 - x^2)$, (xi) $y^2 =$

$x^3/(2-x)$, (xii) $y = \frac{1}{4}x - 3/x$, (xiii) $y = 1/y$, (xiv) $y^2 = x^2(2+x)/(3-x)$, (xv) $y = 1/(x^2-1)$, (xvi) $y^2 = 1/(x^2-1)$, (xvii) $y^2 = 1 + 2/x$, (xviii) $y^2 = x^2(2+x)/(x-3)$.

11. According to the British Highway Code the distance a good driver takes to stop on a dry road is composed of two parts—the *thinking distance*, the distance travelled before the driver can get the brakes applied, and the *braking distance*, the distance travelled with the brakes applied. The total stopping distance is 75 ft at a speed of 30 m.p.h. and 175 ft at 50 m.p.h. Assume that the total stopping distance y ft is related to the speed v m.p.h. by $y = bv + cv^2$. Find the total stopping distance at 20, 40, 60 and 80 m.p.h. Deduce that the driver's reaction time is 15/22 sec and that the brakes produce a deceleration of about $\frac{2}{3}g$, where g is the acceleration due to gravity. Find the thinking distance at the four previous speeds.

12. Draw graphs showing the cost of purchasing (i) oranges, (ii) chocolates, (iii) soap powder.

13. The cost of electricity is (a) $6p$. per unit for first 70 units consumed, (b) $2p$. per unit for the next 250 units and (c) $1 \cdot 25p$. per unit for remaining units. Draw a graph of the cost of electricity against consumption.

14. For long-distance telephone calls via the operator the cost consists of a minimum charge of $27p$. for 3 minutes plus $9p$. for each succeeding minute or fraction thereof. For long-distance dialled calls the cost is $2p$. for each 20 sec or fraction thereof. Draw graphs, on the same diagram, of the cost of calls by the two methods and decide whether one method is cheaper than the other. Are there any discontinuities?

15. Make a computer program to answer one of the preceding questions.

3.2 The modulus

In certain parts of mathematics it is desirable to be sure that one is not handling negative numbers. This is facilitated by the introduction of a special entity. The *modulus* of the real number a is written $|a|$ and defined by

$$|a| = a, \text{ if } a \text{ is positive or zero}$$
$$= -a, \text{ if } a \text{ is negative.}$$

For example, $|2| = 2$, $|-2| = -(-2) = 2$, $|4-1| = |1-4| = 3$. Sometimes the modulus is also called the *absolute value*.

Functions and Graphs

The same notation is used for functions. Thus

$$|f(x)| = f(x), \text{ if } f(x) \text{ is positive or zero}$$
$$= -f(x), \text{ if } f(x) \text{ negative.}$$

The function $|x|$ has a geometrical interpretation which is often useful. If $(x,0)$ is a point of the x-axis then $|x|$ is the distance of that point from the origin.

Thus, if x is less than 2 units from the origin we can say $|x| < 2$. That is, $|x| < 2$ says the same as $-2 < x < 2$.

More generally, if x is less than 2 units from the number a we can write $|x - a| < 2$. If $x \geqslant a$, $|x - a| = x - a$ so that $|x - a| < 2$ requires $x - a < 2$ or $x < 2 + a$. On the other hand, if $x < a$, $|x - a| = a - x$ and then we must have $a - 2 < x$ to satisfy $|x - a| < 2$. Thus $|x - a| < 2$ is the same as $a - 2 < x < a + 2$.

Similarly, $|x - a| \leqslant 3$ is another way of writing $a - 3 \leqslant x \leqslant a + 3$.

If a and b are any two numbers it is clear that

$$|a + b| = |b + a|, \quad |a - b| = |b - a|.$$

For instance, $a - b$ and $b - a$ differ only in sign so that when we take the modulus we must obtain the same positive number in both cases.

Also

$$|ab| = ab \text{ if } a \text{ and } b \text{ have the same sign}$$
$$= -ab \text{ if } a \text{ and } b \text{ have opposite signs.}$$

If a and b are both positive $a = |a|$, $b = |b|$ and $ab = |a| \, |b|$, whereas if a and b are both negative $a = -|a|$, $b = -|b|$ and $ab = |a| \, |b|$. Again, if a is positive and b is negative, $a = |a|$ and $b = -|b|$ so that $-ab = |a| \, |b|$. Thus, in all cases,

$$|ab| = |a| \cdot |b|. \tag{2}$$

It may be shown similarly that, if $b \neq 0$,

$$\left|\frac{a}{b}\right| = \frac{|a|}{|b|}.$$

Further, since $(-a - b)^2 = (-1)^2(a + b)^2 = (a + b)^2$,

$$|a + b|^2 = (a + b)^2 = a^2 + 2ab + b^2. \tag{3}$$

But $ab \leqslant |ab|$ and so

$$|a + b|^2 \leqslant a^2 + 2|ab| + b^2$$
$$\leqslant |a^2| + 2|a| \, |b| + |b|^2$$

from (2). Hence,

$$|a + b|^2 \leqslant (|a| + |b|)^2$$

so that, by taking the positive square root of both sides, we obtain

$$|a + b| \leqslant |a| + |b|. \tag{4}$$

Similarly, by using $ab \geq -|ab|$ in (3), we find
$$|a+b|^2 \geq (|a|-|b|)^2.$$
The positive square root of the right-hand side is $|a|-|b|$ if $|a| \geq |b|$ and $|b|-|a|$ if $|b| > |a|$. Both possibilities are covered by $||a|-|b||$ so that
$$|a+b| \geq ||a|-|b||. \tag{5}$$

By changing the sign of b throughout (4) and (5) we obtain
$$|a-b| \leq |a|+|b|; \qquad |a-b| \geq ||a|-|b||. \tag{6}$$

Note, in addition, that since the square of a real number is not negative
$$(|a|-|b|)^2 \geq 0 \tag{7}$$
or
$$|a|^2 - 2|a||b| + |b|^2 \geq 0$$
or
$$\tfrac{1}{2}(|a|^2 + |b|^2) \geq |a||b|.$$
If we put $|a|^2 = |x|$ and $|b|^2 = |y|$ this can be written as
$$\tfrac{1}{2}(|x|+|y|) \geq \sqrt{(|x||y|)}. \tag{8}$$
Since the inequality (7) is an equality only when $|a|=|b|$, (8) is an equality only when $|x|=|y|$. Often (8) is described as saying that *the arithmetic mean of two moduli is not less than their geometric mean.*

The relation (8) may be interpreted another way. Since $\sqrt{(|x||y|)} < \tfrac{1}{2}(|x|+|y|)$ if $|x| \neq |y|$ it follows, if the sum $|x|+|y|$ is given, that the product $|x||y|$ is greatest when $|x|=|y|$ because then equality holds in (8). Alternatively, if the product $|x||y|$ is given the sum $|x|+|y|$ is least when $|x|=|y|$.

Consider now the product $|x||y||z|$. If $|x|$ and $|y|$ are not equal they can both be replaced by $\tfrac{1}{2}(|x|+|y|)$ without altering the sum $|x|+|y|+|z|$ but the product will be increased. Hence, as long as the product contains two unequal factors it can be increased without altering the sum of the factors; therefore the product will be greatest when all the factors are equal. Then the value of the three factors is $\tfrac{1}{3}(|x|+|y|+|z|)$ and the greatest value of the product is $[\tfrac{1}{3}(|x|+|y|+|z|)]^3$. Consequently,
$$[\tfrac{1}{3}(|x|+|y|+|z|)]^3 \geq |x||y||z|$$
or
$$\tfrac{1}{3}(|x|+|y|+|z|) \geq (|x||y||z|)^{\tfrac{1}{3}}.$$
Equality holds only if $|x|=|y|=|z|$.

Example 8 *For what values of x is $x^2 + 4x - 21 \geq 0$?*

The relation can be written as $(x+7)(x-3) \geq 0$ which can be satisfied only if both $x+7$ and $x-3$ have the same sign or are zero. If both are positive we must have $x > -7$ and $x > 3$; these two inequalities are satisfied when $x > 3$. If $x + 7$ and $x - 3$ are both negative then $x < -7$ and $x < 3$ which require $x < -7$. Since $x = -7$ and $x = 3$ certainly satisfy the relation we see that the required values of x are $x \geqslant 3$ and $x \leqslant -7$.

The relation $x^2 + 4x - 21 > 0$ is satisfied by $x > 3$ and $x < -7$.

Exercises

16. Describe, without using the modulus sign, the values of x which satisfy (i) $|x| < 3$, (ii) $|x| \geqslant 3$, (iii) $|x - 2| < 3$, (iv) $|x - 2| \geqslant 3$, (v) $\frac{1}{2} < |x + 3| < 1$, (vi) $\frac{1}{2} < |x + 3| \leqslant 1$, (vii) $|x + 3| < 2$ and $|x + 2| \geqslant 1$.

17. Express the following without modulus signs (i) if $|x - 3| < 2$ then $|f(x)| < 5$, (ii) if $|x - 3| \geqslant 2$ then $|f(x) - 2| < 4$, (iii) if $|x| \leqslant 1$ then $\frac{1}{2} < |f(x) + 2| < 1$.

18. Draw a graph of $y = |1 - x^2|$ for $|x| \leqslant 2$.

19. Prove that $x^2 + y^2 + z^2 \geqslant |xy + yz + zx|$.

20. Prove that $|x| + (1/|x|) \geqslant 2$. When does the equals sign hold?

21. If $a^2 + b^2 + c^2 = 1$ and $x^2 + y^2 + z^2 = 1$ prove that $ax + by + cz \leqslant 1$.

22. Prove that $(x + y + z)^3 > 27(y + z - x)(z + x - y)(x + y - z)$ when each of the factors on the right-hand side is positive.

23. If $a, b, c, \ldots, k$ are n positive numbers show that

$$\frac{1}{n}(a + b + c + \ldots + k) \geqslant (abc \ldots k)^{1/n}.$$

3.3 Polynomials and rational functions

A function which is the sum of a finite number of terms of the type ax^n, where a is a constant and n is zero or a positive integer, is called a *polynomial* in x. Thus

$$3x^2 + 2x - 2, \quad -x, \quad (x^2 - 2)^4, \quad 5, \quad 7x + 2$$

are all polynomials in x. Similarly, $\frac{1}{2}ft^2$ is a polynomial in t but $\sqrt{(2as)}$ is not a polynomial in s. In general, a polynomial can be written as $ax^n + bx^{n-1}$

$+ \ldots + k$ where $a,b,\ldots,k$ are constants (some of which may be zero); when the highest power of x occurring is x^n we sometimes speak of a *polynomial of degree n* in x and call the constants $a,b,\ldots,k$ the *coefficients*.

Both $x^3 + x$ and $x^2 - x + 2$ are polynomials and
$$(x^3 + x)(x^2 - x + 2) = x^3(x^2 - x + 2) + x(x^2 - x + 2)$$
$$= x^5 - x^4 + 3x^3 - x^2 + 2x$$
which is another polynomial. In the same way, we can show that, if P and Q are polynomials in x, PQ is also a polynomial in x; if P is of degree n and Q of degree m then PQ will be of degree $n + m$. In particular, P^r is a polynomial of degree rn when r is a positive integer.

Let now $p_0, p_1, \ldots, p_n$ be numbers. Then
$$P(x) = p_0 x^n + p_1 x^{n-1} + \ldots + p_n$$
is a polynomial of degree n. We want to show that we can always write
$$P(x) = (x - a)Q(x) + P(a) \tag{9}$$
where Q is a polynomial of degree $n - 1$. Suppose that
$Q(x) = q_0 x^{n-1} + q_1 x^{n-2} + \ldots + q_{n-1}$; then
$(x - a)Q(x) = q_0 x^n + (q_1 - aq_0)x^{n-1} +$
$\qquad (q_2 - aq_1)x^{n-2} + \ldots + (q_{n-1} - aq_{n-2})x - aq_{n-1}$.

Choose $q_0 = p_0$, then choose q_1 so that $q_1 - aq_0 = p_1$, then q_2 so that $q_2 - aq_1 = p_2$ and so on until finally q_{n-1} is chosen so that $q_{n-1} - aq_{n-2} = p_{n-1}$. Then $P(x) - (x - a)Q(x)$ has no terms involving x and must be a constant. The value of the constant can be determined by giving x any convenient value; putting $x = a$ we see that the constant must be $P(a)$ since $Q(a)$ is finite. Hence (9) has been demonstrated.

If $P(a) = 0$, (9) becomes
$$P(x) = (x - a)Q(x) \tag{10}$$
from which we deduce that *a polynomial in x which vanishes when $x = a$ is divisible by $x - a$.*

Let now $P(x)$ vanish when x is equal to each of the n different values $a_1, a_2, \ldots, a_n$. From (10)
$$P(x) = (x - a_1)Q(x). \tag{11}$$
Putting $x = a_2$ we obtain
$$0 = P(a_2) = (a_2 - a_1)Q(a_2).$$
But $a_2 \neq a_1$ and so $Q(a_2) = 0$. Therefore $Q(x)$ is divisible by $x - a_2$ and
$$Q(x) = (x - a_2)R(x) \tag{12}$$
where $R(x)$ is a polynomial of degree $n - 2$.

Now $x = a_3$ in (11) gives $Q(a_3) = 0$ and then $x = a_3$ in (12) gives $R(a_3) = 0$; thus R is divisible by $x - a_3$ and may be written as a product

of $x - a_3$ and a polynomial of degree $n - 3$. Proceeding in this way we find
$$P(x) = p_0(x - a_1)(x - a_2) \ldots (x - a_n).$$
Suppose now that $P(x)$ also vanishes when $x = a_{n+1}$ where a_{n+1} is different from each of $a_1, a_2, \ldots, a_n$. Then
$$p_0(a_{n+1} - a_1)(a_{n+1} - a_2) \ldots (a_{n+1} - a_n) = 0$$
and therefore $p_0 = 0$ since none of the other factors is zero. Thus, if P vanishes for more than n different values of x, $p_0 = 0$ and P reduces to
$$p_1 x^{n-1} + \ldots + p_n.$$
But now this expression vanishes for more than n values of x, and therefore $p_1 = 0$.

Similarly, we may show that each of $p_2, p_3, \ldots, p_n$ must be zero. *Hence, if a polynomial of degree n vanishes for more than n different values of x it must vanish for every value of x.*

A corollary to this result is that, *if two polynomials of degree n take the same values for more than n different values of x, their coefficients are the same.* For, let the polynomials be
$$p_0 x^n + p_1 x^{n-1} + \ldots + p_n, \quad q_0 x^n + q_1 x^{n-1} + \ldots + q_n.$$
Then $(p_0 - q_0)x^n + (p_1 - q_1)x^{n-1} + \ldots + p_n - q_n$ is a polynomial of degree n which vanishes for more than n different values of x; therefore, by the preceding result,
$$p_0 - q_0 = 0, \quad p_1 - q_1 = 0, \ldots, p_n - q_n = 0$$
which show that the coefficients of the two polynomials are the same.

Exercises

24. Show that $x^3 - 2x + 4$ is divisible by $x + 2$.

25. Find the remainder when $3x^5 + 11x^4 + 90x^2 - 19x + 53$ is divided by $x + 5$.

26. Find the connection between a and b in order that $2x^4 - 7x^3 + ax + b$ is divisible by $x - 3$.

27. If $(x + 1)(x + 2) = a + b(2x + 1) + c(3x^2 + 3x + 1) + 4dx^3$ for $x = 1, 2, 3, 4$ find a, b, c and d.

28. Under what conditions is $ax^3 + bx^2 + cx + d$ a perfect cube?

29. If $x^5 - 5ax + 4b$ is divisible by $(x - c)^2$, show that $a^5 = b^4$.

30. Find a polynomial of degree $(n - 1)$ which takes the value y_1 at $x = x_1$, the value y_2 at $x = x_2,...$, the value y_n at $x = x_n$. No two of $x_1, x_2,...,x_n$ are the same. (Hint: Construct $P_k(x)$ so that $P_k(x_i) = 0$ $(i \neq k)$, $P_k(x_k) = 1$ and consider $y_k P_k(x)$.) (This is *Lagrange's interpolation formula*.)

Although the product of two polynomials is a polynomial, the ratio of two polynomials is not in general a polynomial. Thus $1/(x + 1)$ cannot be expressed as a polynomial $P(x)$. If it could we should have
$$1 = (x + 1)P(x) = (x + 1)(p_0 x^n + p_1 x^{n-1} + ... + p_n)$$
say. From the preceding theory, the coefficients on both sides must be the same, i.e.
$$p_0 = 0, \quad p_1 + p_0 = 0,..., \quad p_n + p_{n-1} = 0, \quad p_n = 1$$
which cannot be satisfied since they require p_n to be both 0 and 1.

A function $P(x)/Q(x)$, where P and Q are polynomials, is called a *rational function* of x. Only under special circumstances will a rational function be a polynomial.

If y is given as a function of x by means of an equation of the form
$$R_0(x) y^m + R_1(x) y^{m-1} + ... + R_m(x) = 0$$
where m is a positive integer and $R_0, R_1,..., R_m$ are rational functions, y is said to be an *algebraic function* of x.

Functions of x which are not algebraic functions are called *transcendental functions* of x.

A rational function can often be expressed more conveniently as a sum of fractions with simpler denominators. The technique of doing this is known as the method of *partial fractions*. To see how the method works consider
$$\frac{1}{x + 2} + \frac{2}{x - 3} = \frac{x - 3 + 2(x + 2)}{(x + 2)(x - 3)} = \frac{3x + 1}{x^2 - x - 6}.$$
Now, let us attempt to go in the opposite direction, i.e. find constants A and B such that
$$\frac{3x + 1}{x^2 - x - 6} = \frac{A}{x + 2} + \frac{B}{x - 3}$$
assuming that we do not know the previous result. Multiplying both sides by $x^2 - x - 6$ we have
$$3x + 1 = A(x - 3) + B(x + 2). \tag{13}$$

Since this equation is identically true, we equate coefficients of x and then
$$A + B = 3, \qquad -3A + 2B = 1$$
whence $A = 1$, $B = 2$.

The same result is achieved by first putting $x = -2$ and then $x = 3$ in (13).

The general method of dealing with the rational function $P(x)/Q(x)$ consists of several parts:

(i) Check that the degree of P is less than the degree of Q. If this is not so, carry out a division until the remainder term is of this form and then work with the remainder. The remainder can always be put into the required form. For example,
$$\frac{x^3}{x+2} = x^2 - 2x + 4 - \frac{8}{x+2}$$

(ii) If the factor $ax + b$ occurs n times in Q, put into the partial fractions the expression
$$\frac{A_1}{ax+b} + \frac{A_2}{(ax+b)^2} + \ldots + \frac{A_n}{(ax+b)^n}$$
where $A_1, A_2, \ldots, A_n$ are constants to be determined.

(iii) If $ax^2 + bx + c$ does not have real factors and occurs m times in Q, put into the partial fractions the expression
$$\frac{A_1 x + B_1}{ax^2 + bx + c} + \frac{A_2 x + B_2}{(ax^2 + bx + c)^2} + \ldots + \frac{A_m x + B_m}{(ax^2 + bx + c)^m}$$
where $A_1, A_2, \ldots, A_m, B_1, B_2, \ldots, B_m$ are constants to be determined.

(iv) Write the rational function as equal to the sum of all the partial fractions from (ii) and (iii). Then multiply both sides by $Q(x)$ and equate coefficients of like powers of x to determine the constants, though sometimes this is more quickly done by choosing particular values for x.

Example 9 Express
$$\frac{2x+3}{x^3 - x^2 - x + 1}$$
in partial fractions.

Since $x^3 - x^2 - x + 1 = (x-1)^2(x+1)$, (ii) tells us that
$$\frac{2x+3}{x^3 - x^2 - x + 1} = \frac{A}{x-1} + \frac{B}{(x-1)^2} + \frac{C}{x+1}$$
and
$$2x + 3 = A(x-1)(x+1) + B(x+1) + C(x-1)^2.$$

82 *Introductory Analysis*

Before equating coefficients we note that putting $x = 1$ gives $B = 2$ and that putting $x = -1$ gives $C = \frac{1}{4}$. And now equating the constant terms on both sides we have
$$3 = -A + \tfrac{5}{2} + \tfrac{1}{4}$$
or $A = -\tfrac{1}{4}$. Hence
$$\frac{2x + 3}{x^3 - x^2 - x + 1} = \frac{-\tfrac{1}{4}}{x - 1} + \frac{\tfrac{5}{2}}{(x - 1)^2} + \frac{\tfrac{1}{4}}{x + 1}.$$

Example 10 Express
$$\frac{1}{x^2(x^2 + x + 1)}$$
in partial fractions.

By (ii) and (iii)
$$\frac{1}{x^2(x^2 + x + 1)} = \frac{A}{x} + \frac{B}{x^2} + \frac{Cx + D}{x^2 + x + 1}$$
whence
$$1 = Ax(x^2 + x + 1) + B(x^2 + x + 1) + x^2(Cx + D).$$
By putting $x = 0$ we see that $B = 1$. Then equating coefficients of x^3, x^2 and x on both sides we have
$$A + C = 0,$$
$$A + 1 + D = 0,$$
$$A + 1 = 0.$$
Hence $A = -1$, $D = 0$ and $C = 1$. Therefore,
$$\frac{1}{x^2(x^2 + x + 1)} = -\frac{1}{x} + \frac{1}{x^2} + \frac{x}{x^2 + x + 1}.$$

Example 11 *Resolve into partial fractions*
$$\frac{x^4}{(x^2 + 1)^2}.$$

Since the numerator is not of lower degree than the denominator write first
$$\frac{x^4}{(x^2 + 1)^2} = 1 - \frac{2x^2 + 1}{(x^2 + 1)^2}.$$
From (iii),
$$\frac{2x^2 + 1}{(x^2 + 1)^2} = \frac{Ax + B}{x^2 + 1} + \frac{Cx + D}{(x^2 + 1)^2}$$

whence
$$2x^2 + 1 = (Ax + B)(x^2 + 1) + Cx + D.$$
Equating the coefficients of x^3, x^2, x and x^0 on both sides gives
$$A = 0, \quad B = 2, \quad C = 0, \quad B + D = 1.$$
Hence
$$\frac{x^4}{(x^2 + 1)^2} = 1 - \frac{2}{x^2 + 1} + \frac{1}{(x^2 + 1)^2}.$$

Exercises

31. Resolve into partial fractions:

   (i) $\dfrac{x^2 - 10x + 13}{(x - 1)(x^2 - 5x + 6)}$, (ii) $\dfrac{7x - 1}{6x^2 - 5x + 1}$,

   (iii) $\dfrac{2x^3 + x^2 - x - 3}{x(x - 1)(2x + 3)}$, (iv) $\dfrac{3x^3 - 8x^2 + 10}{(x - 1)^4}$,

   (v) $\dfrac{2x^2 - 11x + 5}{(x - 3)(x^2 + 2x - 5)}$, (vi) $\dfrac{2x^2 - x + 1}{(x - 1)^3}$, (vii) $\dfrac{x^2}{x^4 - a^4}$,

   (viii) $\dfrac{2x^2 + 3}{(x^2 + 1)^2}$, (ix) $\dfrac{3x^4 + 9x^3 + 16x^2 + 9x + 13}{(x - 1)^2(x^2 + 2x + 2)^2}$.

32. See if you can devise a computer program that will resolve a rational function into partial fractions, assuming that any real zeros of the denominator are integers between -100 and 100.

4

Limits and Continuity

4.1 Limits

Suppose that $y = x^2 + 2x + 3$. Then, as x takes the values 1·1, 1·01, 1·001, y takes the values 6·41, 6·0401, 6·004001, respectively. This suggests that, as x gets closer and closer to 1, y gets closer and closer to 6. Indeed, we can see by actual substitution that y is 6 when x is 1. However, it proves to be important to distinguish between a statement such as '$y = 6$ when $x = 1$' and a statement such as 'y approaches 6 as x approaches 1'.

Consider, for example, $y = (x^2 - 1)/(x - 1)$. In this case y is not defined for $x = 1$ because both numerator and denominator vanish there. Nevertheless, when x has the values 1·2, 1·02, 1·002, y takes the values 2·2, 2·02, 2·002 which indicate that it might be reasonable to make the statement 'y approaches 2 as x approaches 1'.

Now we want to turn this idea into a usable mathematical tool. Our first thought might be that it would be sufficient to require that y approached a number b as x took in succession values which steadily approached a number a. However, it is clear that we do not want to approach a different number if x follows a different sequence of values; for example, when $y = (x^2 - 1)/(x - 1)$ and x has the values 1·1, 1·01, 1·001, y takes the values 2·1, 2·01, 2·001 which approach 2. These values are different from the set of values obtained above but both sets approach 2. Therefore, we shall want to say that y is close to b for *all* values of x sufficiently close to a. On the other hand, to avoid the difficulty that y may not be defined for $x = a$ we must exclude the possibility that x takes the value a. Accordingly, we frame

DEFINITION 4.1 *If $f(x)$ is defined for $0 < |x - a| < c$ and $|f(x) - b|$ can be made less than any given positive quantity (however small) by taking x sufficiently close to a, and remains less for all values of x which are still nearer to a (other than $x = a$), then $f(x)$ is said to approach the limit b as x approaches a.*

When the definition holds we write
$$\lim_{x \to a} f(x) = b$$
which should be read as 'the limit, as x approaches a, of $f(x)$ is b'. An alternative notation is to write '$f(x) \to b$ as $x \to a$', which can be read as '$f(x)$ tends to b as x tends to a'.

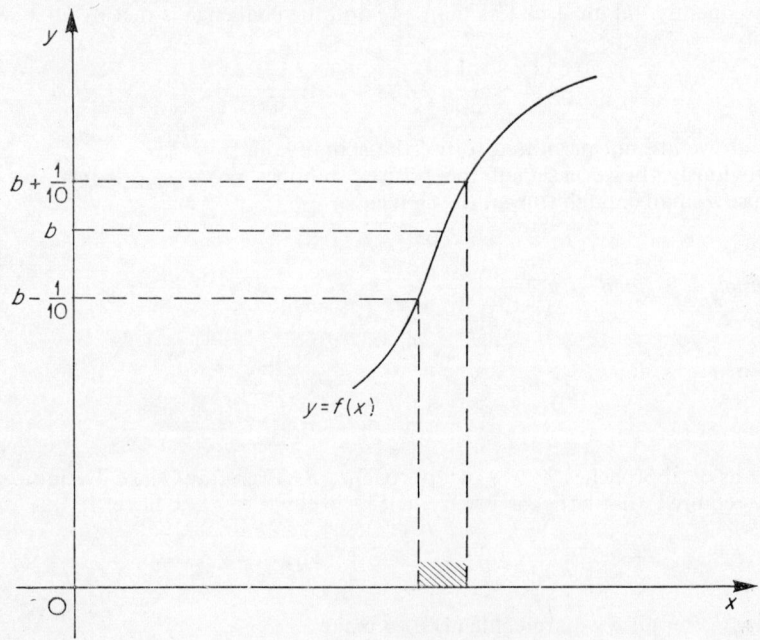

FIGURE 48

The definition represents a challenge to the reader because he is asked to show that, no matter how small a positive number is given in advance, he can find an interval of x's near a which make $|f(x) - b|$ less than this positive number. For instance, in figure 48 the values of x for which $|f(x) - b|$ does not exceed $\frac{1}{10}$ are shown shaded.

Example 1 Find $\lim_{x \to 2} (3x + 4)$.

Put $x = 2 + h$; then $3x + 4 = 10 + 3h$. As x approaches 2, h approaches zero and so $3x + 4$ approaches 10. Thus
$$\lim_{x \to 2} (3x + 4) = 10.$$

To see how the challenge of the definition is met specifically imagine that we are asked to make $3x + 4$ differ from 10 by less than $\frac{1}{100}$ without taking $x = 2$. Then we must choose x so that

$$|3x + 4 - 10| < \frac{1}{100}$$

or

$$|3h| < \frac{1}{100}.$$

Consequently, $|h|$ must be less than $\frac{1}{300}$ and the challenge is met by all x which satisfy

$$2 - \frac{1}{300} < x < 2 + \frac{1}{300}$$

though we are not permitted to use the actual value $x = 2$.

Obviously, however small the positive number given in advance, we can choose h small enough to beat the challenge.

Example 2   Find $\lim\limits_{x \to 4} \dfrac{x^2 - 16}{x - 4}$.

When $x \neq 4$,

$$\frac{x^2 - 16}{x - 4} = x + 4$$

and, as x approaches 4, $x + 4$ approaches 8. Therefore, since Definition 4.1 only requires what happens when x is *nearly* equal to 4, we have

$$\lim_{x \to 4} \frac{x^2 - 16}{x - 4} = 8.$$

Again, suppose we are challenged to make

$$\left| \frac{x^2 - 16}{x - 4} - 8 \right| < \frac{1}{1000} \qquad (1)$$

for $|x - 4|$ sufficiently small but with $x \neq 4$. If we put $x = 4 + h$, (1) becomes

$$|h| < \frac{1}{1000}$$

so that we can meet the challenge by taking $3 \cdot 999 < x < 4 \cdot 001$.

Example 3   Find $\lim\limits_{x \to 1} \dfrac{\sqrt{(2 - x)} - \sqrt{x}}{3 - 3x}$.

If 1 is substituted for x the numerator and denominator are both zero, so the expression is not defined for $x = 1$. However, Definition 4.1 only requires values of x which are different from unity. Now, multiply the numerator and denominator by $\sqrt{(2-x)} + \sqrt{x}$ to obtain

$$\frac{(2-x)-x}{(3-3x)[\sqrt{(2-x)}+\sqrt{x}]} = \frac{2(1-x)}{3(1-x)[\sqrt{(2-x)}+\sqrt{x}]}$$

$$= \frac{2}{3[\sqrt{(2-x)}+\sqrt{x}]}$$

when $x \neq 1$. As x approaches 1, this approaches the value $2/3(1+1)$ or $\frac{1}{3}$. Thus

$$\lim_{x \to 1} \frac{\sqrt{(2-x)} - \sqrt{x}}{3-3x} = \frac{1}{3}$$

even though the expression on the left-hand side is not defined when $x = 1$.

Example 4 Find $\lim_{x \to 0} (a_0 x^n + a_1 x^{n-1} + \ldots + a_n)$ where n is a positive integer.

It is evident that the terms which involve x can be made as small as desired by taking x sufficiently small. Therefore

$$\lim_{x \to 0} (a_0 x^n + a_1 x^{n-1} + \ldots + a_n) = a_n.$$

More generally, consider the rational function

$$\frac{a_0 x^n + a_1 x^{n-1} + \ldots + a_n}{b_0 x^m + b_1 x^{m-1} + \ldots + b_m}.$$

Again the terms containing x can be made as small as we please by taking x small enough. Therefore, if $b_m \neq 0$,

$$\lim_{x \to 0} \frac{a_0 x^n + a_1 x^{n-1} + \ldots + a_n}{b_0 x^m + b_1 x^{m-1} + \ldots + b_m} = \frac{a_n}{b_m}.$$

When $b_m = 0$, further discussion is necessary and will be found in section 7.5.

The calculation of limits is often simplified by the following theorem which gives rules for the addition, multiplication and division of limits.

THEOREM 4.1 *If* $\lim_{x \to a} f(x) = b$ *and* $\lim_{x \to a} g(x) = c$, *then*

(i) $\lim_{x \to a} kf(x) = kb$, *for any fixed number* k,

(ii) $\lim_{x \to a} [f(x) \pm g(x)] = \lim_{x \to a} f(x) \pm \lim_{x \to a} g(x) = b \pm c$,

(iii) $\lim_{x \to a} [f(x)g(x)] = \lim_{x \to a} f(x) \cdot \lim_{x \to a} g(x) = bc$,

(iv) $\lim_{x \to a} \frac{f(x)}{g(x)} = \frac{\lim_{x \to a} f(x)}{\lim_{x \to a} g(x)} = \frac{b}{c}$ *if* $c \neq 0$,

(v) $\lim_{x \to a} [f(x)]^{1/n} = [\lim_{x \to a} f(x)]^{1/n} = b^{1/n}$ provided that $[f(x)]^{1/n}$ and $b^{1/n}$ are real.

Note. The theorem starts by assuming that limits of both f and g exist. It is always important to check that this is true before applying the theorem.

Proof. From Definition 4.1, $\lim_{x \to a} f(x) = b$ implies that $f(x) = b + \beta$ where $\beta \to 0$ as $x \to a$. Similarly, $g(x) = c + \gamma$ where $\gamma \to 0$ as $x \to a$.

(i) Therefore, $kf(x) = kb + k\beta$ and, since $k\beta \to 0$ as $x \to a$, $kf(x)$ approaches the limit kb.

(ii) In this case
$$f(x) \pm g(x) = b \pm c + (\beta \pm \gamma)$$
and, since $\beta \pm \gamma \to 0$ as $x \to a$, $f \pm g$ approaches the limit $b \pm c$.

Obviously, this result remains valid for the addition of any *finite* number of limits.

(iii) Here
$$f(x)g(x) = (b + \beta)(c + \gamma)$$
and
$$f(x)g(x) - bc = b\gamma + c\beta + \beta\gamma.$$
The right-hand side tends to zero as β and γ tend to zero, and so $f(x)g(x)$ approaches the limit bc as $x \to a$.

Again, this result is true for any *finite* number of factors.

(iv) Now,
$$\frac{f(x)}{g(x)} - \frac{b}{c} = \frac{b + \beta}{c + \gamma} - \frac{b}{c} = \frac{c\beta - b\gamma}{c(c + \gamma)}.$$

So long as $c \neq 0$, the right-hand side tends to zero as β and γ tend to zero. Hence f/g tends to the limit b/c.

(v) If $\lim_{x \to a} [f(x)]^{1/n} = b^{1/n}$, there must be values of x as close to a as we please, such that $|[f(x)]^{1/n} - b^{1/n}|$ exceeds any given small quantity. If the given small quantity be denoted by ε, it follows that there are x, as close to a as we like, such that either $[f(x)]^{1/n} < b^{1/n} + \varepsilon$ or $[f(x)]^{1/n} < b^{1/n} - \varepsilon$. On raising these inequalities to the nth power we see that $f(x)$ must differ from b by more than a small quantity for these x. But no such x exist since $\lim_{x \to a} f(x) = b$. Hence, our starting hypothesis must be wrong.
Consequently $\lim_{x \to a} [f(x)]^{1/n} = b^{1/n}$.

Example 5   Prove that
$$\lim_{x \to a}(p_0 + p_1 x + p_2 x^2 + \ldots + p_n x^n) = p_0 + p_1 a + p_2 a^2 + \ldots + p_n a^n.$$

Since $\lim_{x \to a} x = a$, Theorem 4.1 (iii) shows that
$$\lim_{x \to a} x^2 = \lim_{x \to a} x . \lim_{x \to a} x = a^2.$$
If $\lim_{x \to a} x^{m-1} = a^{m-1}$, then Theorem 4.1 (iii) shows that
$$\lim_{x \to a} x^m = \lim_{x \to a} x . \lim_{x \to a} x^{m-1} = a . a^{m-1} = a^m.$$
Hence, by induction* on m, $\lim_{x \to a} x^m = a^m$. By Theorem 4.1 (i), $\lim_{x \to a} p_m x^m = p_m a^m$
and the result follows from Theorem 4.1 (ii).

An immediate consequence of Theorem 4.1 (iv) is that
$$\lim_{x \to a} \frac{p_0 + p_1 x + \ldots + p_n x^n}{q_0 + q_1 x + \ldots + q_m x^m} = \frac{p_0 + p_1 a + \ldots + p_n a^n}{q_0 + q_1 a + \ldots + q_m a^m}$$
provided that the denominator on the right-hand side is non-zero.

Example 6
$$\lim_{x \to 1} \frac{x^3 - 4x^2 + 7x - 4}{x^2 - 4x + 3} = \lim_{x \to 1} \frac{(x-1)(x^2 - 3x + 4)}{(x-1)(x-3)}$$
$$= \lim_{x \to 1} \frac{x^2 - 3x + 4}{x - 3} = \frac{1 - 3 + 4}{1 - 3} = -1$$
from Example 5.

Example 7   Investigate $\lim_{x \to a} \dfrac{x^\nu - a^\nu}{x - a}$.

(a) Suppose that ν is a positive integer n. Then
$$(x - a)(x^{n-1} + ax^{n-2} + a^2 x^{n-3} + \ldots + a^{n-1}) = x^n - a^n$$
so that
$$\frac{x^n - a^n}{x - a} = x^{n-1} + ax^{n-2} + \ldots + a^{n-1}$$

---

* It has been demonstrated that assuming the result for $m - 1$ leads to the result for m. In other words, if it is true for $m - 1$ then it is true for m. But it has been proved for $m = 2$; therefore it holds for $m = 3$; it is therefore true for $m = 4$; and so on. Thus the result is true for all integer values of m.

when $x \neq a$. It follows from Example 5 that
$$\lim_{x \to a} \frac{x^n - a^n}{x - a} = na^{n-1}.$$

(b) Suppose that $v = p/q$, where p and q are positive integers. Put $x = y^q$ and $a = b^q$. Then $x^v = x^{p/q} = (y^q)^{p/q} = y^p$ and similarly $a^v = b^p$. On account of Theorem 4.1 (v), $y \to b$ as $x \to a$ and so
$$\lim_{x \to a} \frac{x^{p/q} - a^{p/q}}{x - a} = \lim_{y \to b} \frac{y^p - b^p}{y^q - b^q} = \lim_{y \to b} \frac{y^p - b^p}{y - b} \cdot \frac{y - b}{y^q - b^q} = \frac{pb^{p-1}}{qb^{q-1}}$$
by (a) and Theorem 4.1 (iv) so long as $b \neq 0$. Since $b^{p-q} = (b^q)^{p/(q-1)} = a^{(p/q)-1}$ we have
$$\lim_{x \to a} \frac{x^{p/q} - a^{p/q}}{x - a} = \frac{p}{q} a^{(p/q)-1}$$
provided that $a \neq 0$.

(c) Suppose that $v = -p/q$, where p and q are positive integers. Put $v = -\mu$ so that $\mu = p/q$. Then
$$\lim_{x \to a} \frac{x^{-\mu} - a^{-\mu}}{x - a} = \lim_{x \to a} \frac{(a^\mu - x^\mu)/x^\mu a^\mu}{x - a}$$
$$= \lim_{x \to a} -\frac{1}{x^\mu a^\mu} \cdot \frac{x^\mu - a^\mu}{x - a}$$
$$= -\frac{1}{a^{2\mu}} \cdot \mu a^{\mu-1} = -\mu a^{-\mu-1}$$
by (b) and Theorem 4.1 (iv).

All three cases can be included in a single formula, namely
$$\lim_{x \to a} \frac{x^v - a^v}{x - a} = va^{v-1}$$
which has been established for all rational values of v except for the case when $a = 0$ and v is not a positive integer.

A theorem which is used frequently is

THEOREM 4.2 *If $g(x) \leq f(x) \leq h(x)$ for $0 < |x - a| < c$ and if $\lim_{x \to a} g(x) = \lim_{x \to a} h(x) = b$, then $\lim_{x \to a} f(x) = b$.*

Proof. By hypothesis, given a positive number ε we can find positive numbers δ_1 and δ_2 such that
$$|g(x) - b| < \varepsilon \quad \text{for} \quad 0 < |x - a| < \delta_1,$$
$$|h(x) - b| < \varepsilon \quad \text{for} \quad 0 < |x - a| < \delta_2.$$
Let δ be the smallest of c, δ_1 and δ_2. Then
$$b - \varepsilon < g(x) \leq f(x) \leq h(x) < b + \varepsilon \quad \text{for} \quad 0 < |x - a| < \delta$$

and so
$$|f(x) - b| < \varepsilon \quad \text{when} \quad 0 < |x - a| < \delta.$$
This demonstrates that $\lim_{x \to a} f(x) = b$.

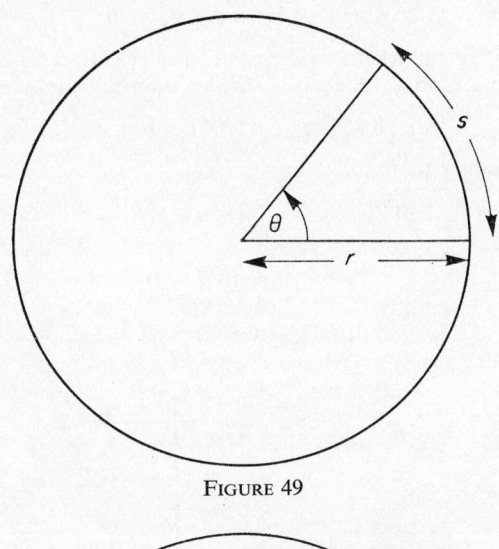

FIGURE 49

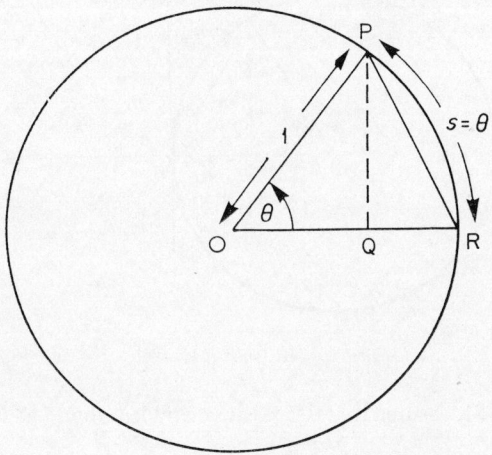

FIGURE 50

Example 8 The radian measure of an angle is defined by figure 49
$$\theta = s/r$$
where s is the length of arc which the angle intercepts on a circle of radius r when the centre of the circle is at the vertex of the angle. In figure 50 ROP is an

acute angle with O the centre of the circle of unit radius. The arc PR is then the radian measure of the angle θ. Obviously the length of the straight line PR is less than the arc PR. Therefore

$$PQ^2 + QR^2 < \theta^2$$

or
$$\sin^2 \theta + (1 - \cos \theta)^2 < \theta^2$$

and in this form it is true whether θ be positive or negative. Since both terms on the left-hand side are positive, each is smaller than the sum, i.e.

$$\sin^2 \theta < \theta^2, \quad (1 - \cos \theta^2) < \theta^2,$$

from which we deduce that

$$|\sin \theta| < |\theta|, \quad |1 - \cos \theta| < |\theta|.$$

Hence (cf. Theorem 4.2)

$$\lim_{\theta \to 0} \sin \theta = 0, \tag{2}$$

$$\lim_{\theta \to 0} (1 - \cos \theta) = 0. \tag{3}$$

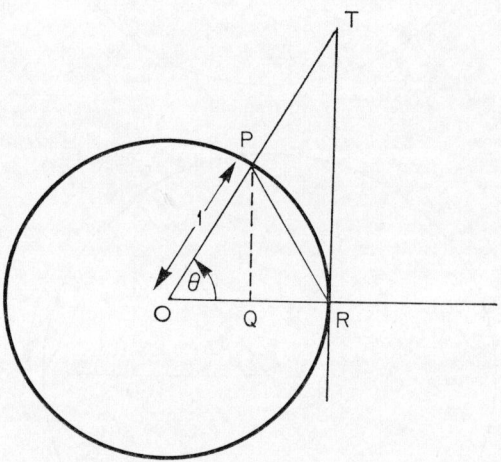

FIGURE 51

Now, in figure 51, assume that θ is positive and less than $\tfrac{1}{2}\pi$. Then it is obvious that

area of $\triangle$ROP < area of sector ROP < area of $\triangle$ROT.

Since OR = 1 and PQ = $\sin \theta$,

area $\triangle$ROP = $\tfrac{1}{2} \sin \theta$.

Since OR = 1 and RT = $\tan \theta$,

area $\triangle$ROT = $\tfrac{1}{2} \tan \theta$.

Limits and Continuity

The area of sector ROP is $\frac{1}{2}OR^2 \cdot \theta = \frac{1}{2}\theta$ since OR = 1. Hence

$$\tfrac{1}{2}\sin\theta < \tfrac{1}{2}\theta < \tfrac{1}{2}\tan\theta$$

or, dividing throughout the inequality by the positive quantity $\frac{1}{2}\sin\theta$,

$$1 < \frac{\theta}{\sin\theta} < \frac{1}{\cos\theta}$$

when $0 < \theta < \frac{1}{2}\pi$.

Suppose now $0 > \theta > -\frac{1}{2}\pi$. Put $\theta = -\phi$ so that $0 < \phi < \frac{1}{2}\pi$. Then

$$1 < \frac{\phi}{\sin\phi} < \frac{1}{\cos\phi}$$

or

$$1 < \frac{-\theta}{\sin(-\theta)} < \frac{1}{\cos(-\theta)}.$$

But $\sin(-\theta) = -\sin\theta$ and $\cos(-\theta) = \cos\theta$ (equations (1.17) and (1.18)) and so we have established

$$1 < \frac{\theta}{\sin\theta} < \frac{1}{\cos\theta} \quad \text{when} \quad 0 < |\theta| < \tfrac{1}{2}\pi.$$

Take reciprocals, and therefore reverse the inequality signs, to obtain

$$1 > \frac{\sin\theta}{\theta} > \cos\theta \quad \text{when} \quad 0 < |\theta| < \tfrac{1}{2}\pi. \tag{4}$$

Applying (3) and Theorem 4.2 we obtain

$$\lim_{\theta \to 0} \frac{\sin\theta}{\theta} = 1. \tag{5}$$

It must be remembered in this formula that radian measure is used for the angle θ. Also

$$\lim_{x \to 0} \frac{\sin 2x}{x} = \lim_{2x \to 0} 2\frac{\sin 2x}{2x}$$

$$= 2 \lim_{\theta \to 0} \frac{\sin\theta}{\theta} = 2$$

where Theorem 4.1 (i) has been used. Similarly

$$\lim_{x \to 0} \frac{\sin ax}{x} = \lim_{ax \to 0} a \cdot \frac{\sin ax}{ax} = a \cdot 1 = a,$$

$$\lim_{x \to 0} \frac{\sin ax}{\sin bx} = \lim_{x \to 0} \frac{(\sin ax)/ax}{(\sin bx)/bx} \cdot \frac{a}{b} = \frac{1}{1} \cdot \frac{a}{b} = \frac{a}{b}$$

on using Theorem 4.1 (iv).

Further important results can be deduced from (5). Thus
$$\lim_{x \to 0} \frac{\tan x}{x} = \lim_{x \to 0} \frac{\sin x}{x} \cdot \frac{1}{\cos x} = 1 \cdot 1 = 1$$
by Theorem 4.1 (iii). Also
$$\lim_{x \to 0} \frac{1 - \cos x}{x^2} = \lim_{x \to 0} \frac{1 - \cos^2 x}{x^2(1 + \cos x)} = \lim_{x \to 0} \frac{\sin^2 x}{x^2} \cdot \frac{1}{1 + \cos x} =$$
$$\left(\lim_{x \to 0} \frac{\sin x}{x}\right)^2 \lim_{x \to 0} \frac{1}{1 + \cos x} = \tfrac{1}{2}$$
by Theorem 4.1 (iii). Consequently
$$\lim_{x \to 0} \frac{1 - \cos x}{x} = \lim_{x \to 0} \frac{1 - \cos x}{x^2} \cdot x = 0.$$

Exercises

1. Evaluate the following:

   (i) $\lim\limits_{x \to 2} (x^2 - 3x)$,   (ii) $\lim\limits_{x \to -1} (x^3 + 3x + 2)$,

   (iii) $\lim\limits_{x \to -3} \dfrac{(x + 2)^3}{(3x - 1)^2}$,   (iv) $\lim\limits_{x \to 3} \dfrac{x^2 - 9}{x - 3}$,   (v) $\lim\limits_{x \to 2} \dfrac{x - 2}{x^2 - 4}$,

   (vi) $\lim\limits_{x \to -1} \dfrac{x^3 + 1}{x + 1}$,   (vii) $\lim\limits_{x \to 2} \dfrac{x^3 - 8}{x^2 - 4}$,   (viii) $\lim\limits_{x \to 2} \dfrac{x^2 - 4}{x^2 - 5x + 6}$,

   (ix) $\lim\limits_{x \to 1} \dfrac{x - 1}{\sqrt{(x^2 - 1)}}$,   (x) $\lim\limits_{x \to 0} \dfrac{ax + b}{cx + d}$,   (xi) $\lim\limits_{x \to 0} \dfrac{4x^2 - 3x + 2}{3x^2 + 5x + 7}$,

   (xii) $\lim\limits_{x \to 0} \dfrac{x^2 - p_1 x + p_2}{q_1 x + q_2}$,   (xiii) $\lim\limits_{x \to 0} \dfrac{x^2 + a^2}{x^3 + b^3}$,

   (xiv) $\lim\limits_{x \to 0} \dfrac{x^2(2x - 1)}{(x + 1)^2(3 + x)}$,   (xv) $\lim\limits_{h \to 0} \dfrac{(x + h)^3 - x^3}{h}$,

   (xvi) $\lim\limits_{x \to 2} \dfrac{\sqrt{(5x - 8)} - \sqrt{x}}{x - 2}$,   (xvii) $\lim\limits_{x \to 0} \dfrac{x}{\sqrt{(1 + x)} - \sqrt{(1 - x)}}$,

   (xviii) $\lim\limits_{x \to 2a} \dfrac{\sqrt{(6a - x)} - \sqrt{(x + 2a)}}{4x - 8a}$,   (xix) $\lim\limits_{x \to 1} \dfrac{x^6 - 1}{x - 1}$,

   (xx) $\lim\limits_{x \to a} \dfrac{x^7 - a^7}{x - a}$,   (xxi) $\lim\limits_{x \to a} \dfrac{\sqrt{x} - \sqrt{a}}{x - a}$,   (xxii) $\lim\limits_{x \to a} \dfrac{x^7 - a^7}{x^5 - a^5}$,

   (xxiii) $\lim\limits_{x \to 0} \dfrac{1 - \cos 2x}{x^2}$,   (xxiv) $\lim\limits_{\theta \to 0} \dfrac{\tan a\theta}{\theta}$,   (xxv) $\lim\limits_{\theta \to 0} \dfrac{\tan a\theta}{\tan b\theta}$,

(xxvi) $\lim\limits_{\theta \to 0} \dfrac{\cos a\theta - \cos b\theta}{\theta^2}$, (xxvii) $\lim\limits_{x \to 0} \dfrac{1 - \cos ax}{1 - \cos bx}$,

(xxviii) $\lim\limits_{\theta \to \frac{1}{2}\pi} (\sec \theta - \tan \theta)$, (xxix) $\lim\limits_{x \to 0} \dfrac{\sin ax}{\tan bx}$,

(xxx) $\lim\limits_{x \to 0} x \sin\left(\dfrac{1}{x}\right)$, (xxxi) $\lim\limits_{x \to \frac{1}{2}\pi} \dfrac{\cos 3x}{\cos 5x}$.

2. Show that if x is measured in degrees $\lim\limits_{x \to 0} \dfrac{\sin x}{x} = \dfrac{\pi}{180}$. (This explains why radians are used in the calculus.)

3. Find a δ so that, if $0 < |x - 2| < \delta$, $|x^2 + 3x - 10|$ will be less than (a) $\tfrac{1}{100}$, (b) $\tfrac{1}{1000}$, (c) ε, where ε is any positive number less than $\tfrac{1}{2}$.

4. If $f(x) \geqslant 0$ for $0 < |x - a| < c$ prove that $\lim\limits_{x \to a} f(x) \geqslant 0$. Deduce that, if $g(x) \leqslant f_1(x) \leqslant h(x)$, $\lim\limits_{x \to a} g(x) \leqslant \lim\limits_{x \to a} f_1(x) \leqslant \lim\limits_{x \to a} h(x)$.

We have not actually shown that our definition of limit is unambiguous. In other words, we have not settled whether it is possible that $\lim\limits_{x \to a} f(x) = b$ and $\lim\limits_{x \to a} f(x) = b'$ with $b \neq b'$. Suppose it were possible; then, since $|b - b'| > 0$, there must be x close to a, say $0 < |x - a| < \delta_1$, such that
$$|f(x) - b| < \tfrac{1}{2}|b - b'| \qquad (6)$$
and also x, say $0 < |x - a| < \delta_2$, such that
$$|f(x) - b'| < \tfrac{1}{2}|b - b'|. \qquad (7)$$
If δ is the smaller of δ_1 and δ_2, we have, for $0 < |x - a| < \delta$,
$$|b - b'| = |b - f(x) + f(x) - b'| \leqslant |b - f(x)| + |f(x) - b'|$$
from (3.4). From (6) and (7), this leads to $|b - b'| < |b - b'|$ which is a contradiction. Therefore, our hypothesis that $b \neq b'$ must be wrong. Thus, $f(x)$ cannot have two different limits at a point.

4.2 One-sided limits

In the preceding section it has not made any difference to the calculation of a limit as $x \to a$ whether x be greater or less than a. In fact, our definition of the limit requires $|f(x) - b|$ to be small when $|x - a|$ is sufficiently small so that $f(x) - b$ must be small both when x is a little greater than a and when x is a little less than a. However, some functions cannot meet this condition. Consider, for example, the function $H(x)$ (sometimes called the *Heaviside unit function*) defined by (figure 52)

$$H(x) = 1 \qquad (x \geqslant 0)$$
$$ = 0 \qquad (x < 0).$$

If x is a little greater than 0, $H(x) = 1$, whereas, if x is slightly less than 0, $H(x) = 0$. There is now no b which will satisfy our definition of limit

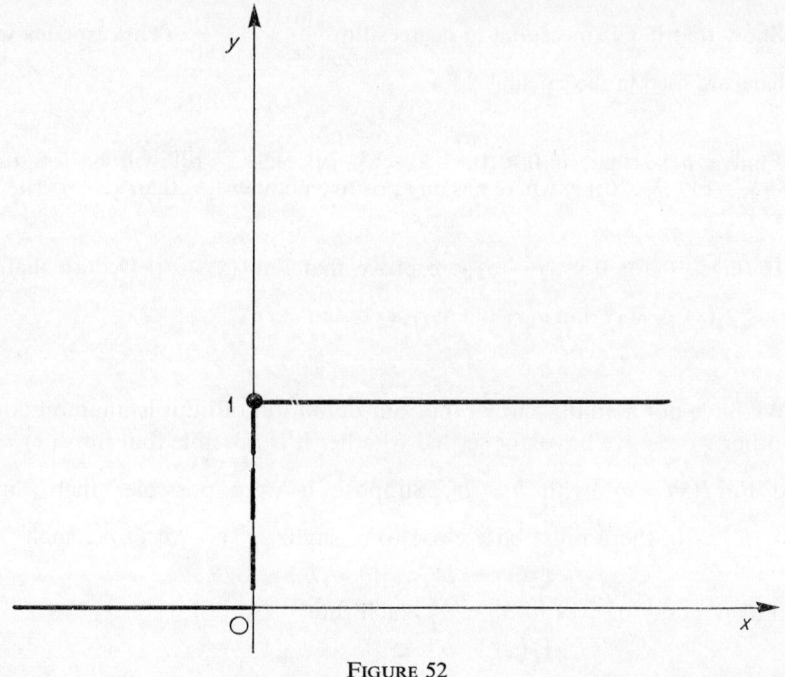

FIGURE 52

because, when x is near 0 some values of $H(x)$ are 0 and some are 1 so that they cannot *all* be close to one number b. Consequently $\lim_{x \to 0} H(x)$ does not exist.

However, if we restricted x to positive values only, a limit would exist. When x is a little greater than 0, $H(x)$ is always 1 and so we write

$$\lim_{x \to 0^+} H(x) = 1.$$

The notation $x \to 0^+$ means that x tends to 0 from greater values, i.e. from values to the right of O; for this reason this type of limit is called a *right limit*.

Limits and Continuity

Similarly, by restricting x to values less than 0, i.e. to the left of O, we obtain the *left limit*

$$\lim_{x \to 0^-} H(x) = 0.$$

These ideas can be generalized to other functions by saying that the right limit

$$\lim_{x \to a^+} f(x) = b$$

when the conditions of Definition 4.1 are satisfied with x confined to $x < a$, and that the left limit

$$\lim_{x \to a^-} f(x) = b$$

when the conditions of Definition 4.1 are satisfied with $x < a$ only.

The Heaviside unit function above demonstrates that both right and left limits can exist without $\lim_{x \to a} f(x)$ existing. However, if both right and left limits exist and are equal, i.e. $\lim_{x \to a^+} f(x) = \lim_{x \to a^-} f(x)$, then $\lim_{x \to a} f(x)$ exists and is the same as the other two limits. The converse statement is also obviously true. Note, though, that the existence of the right limit does not imply the existence of the left limit or conversely.

Example 9 *The function* $\sqrt{(4 - x^2)}$ *is defined for* $-2 \leqslant x \leqslant 2$.

If $-2 < a < 2$, $\lim_{x \to a} \sqrt{(4 - x^2)} = \sqrt{(4 - a^2)}$ but, if $a = -2$, the limit as $x \to -2$ does not exist because the function has not been defined for $x < -2$. If we let x tend to -2 from the right we obtain $\lim_{x \to -2^+} \sqrt{(4 - x^2)} = 0$; the left limit does not exist because the function has not been given for $x < -2$.

Similarly, $\lim_{x \to 2^-} \sqrt{(4 - x^2)} = 0$ but the right limit does not exist at $x = 2$.

One of the most important one-sided limits is associated with infinity. We shall say that x tends to *positive infinity*, and write $x \to +\infty$, if it eventually becomes and remains greater than any given positive number, no matter how large. Notice that, although we can say that x is large, we have not said that it is close to infinity because we have not given any rule for measuring distance from infinity.

Similarly we shall say that x tends to *negative infinity*, and write $x \to -\infty$, if it eventually becomes and remains less than any given negative number. Obviously, if $x \to +\infty$ then $-x \to -\infty$ and, if $x \to -\infty$ then $-x \to +\infty$.

The symbols $+\infty$ and $-\infty$ are introduced to indicate a certain type of behaviour, and are not to be regarded as new numbers. In terms of these symbols we say
$$\lim_{x \to +\infty} f(x) = b$$
if $|f(x) - b|$ is less than any given positive quantity for all sufficiently large positive x.

Since $1/x$ can be made as small as we like by making x large,
$$\lim_{x \to +\infty} \frac{1}{x} = 0.$$
Similarly, $\lim_{x \to -\infty} f(x) = b$ if $|f(x) - b|$ is less than any given positive quantity for all sufficiently negative x. Thus
$$\lim_{x \to -\infty} \frac{1}{x} = 0.$$
As $x \to +\infty$, $1/x$ approaches 0 through positive values and so
$$\lim_{x \to +\infty} f(x) = \lim_{y \to 0^+} f\left(\frac{1}{y}\right).$$
Similarly,
$$\lim_{x \to -\infty} f(x) = \lim_{y \to 0^-} f\left(\frac{1}{y}\right).$$
These two results are often helpful in evaluating limits.

By methods of proof which are very similar to those of Theorems 4.1 and 4.2 we can show

THEOREM 4.3 *Theorems 4.1 and 4.2 are valid when a is $+\infty$ (or $-\infty$) provided that, in Theorem 4.2, $g(x) \leq f(x) \leq h(x)$ for all sufficiently large x (or sufficiently negative x).*

Example 10 On dividing numerator and denominator by x^2, we have
$$\lim_{x \to +\infty} \frac{p_0 + p_1 x + p_2 x^2}{q_0 + q_1 x + q_2 x^2} = \lim_{x \to +\infty} \frac{p_2 + (p_1/x) + (p_0/x^2)}{q_2 + (q_1/x) + (q_0/x^2)}.$$

The terms with x or x^2 in the denominator can be made as small as we please by taking x sufficiently large. Consequently, the numerator tends to p_2 and the denominator tends to q_2 as $x \to +\infty$. Therefore
$$\lim_{x \to +\infty} \frac{p_0 + p_1 x + p_2 x^2}{q_0 + q_1 x + q_2 x^2} = \frac{p_2}{q_2}$$
provided, of course, that $q_2 \neq 0$.

It may be checked easily that the same limiting value is obtained as $x \to -\infty$.

Limits and Continuity

Example 11 *On putting* $x = 1/y$ *we find*

$$\lim_{x \to +\infty} x \sin \frac{1}{x} = \lim_{y \to 0+} \frac{\sin y}{y} = 1$$

from (5).

Exercises

5. Find the limit as $x \to +\infty$ of the functions given in question 1(viii), (ix), (x), (xi), (xiii), (xiv). Are any of these limits changed if $x \to -\infty$?

6. Determine the limit as $x \to +\infty$ of (i) $\sqrt{(2+x)} - \sqrt{x}$, (ii) $\sqrt{(x^2 + 6x + 4)} - x$.

7. Do the following exist? If so, what are they?
   (i) $\lim_{x \to 0^-} |x|$, (ii) $\lim_{x \to 0^+} |x|$, (iii) $\lim_{x \to 0} |x|$, (iv) $\lim_{x \to 0^+} [x + H(x)]$,
   (v) $\lim_{x \to 0} [x + H(x)]$, (vi) $\lim_{x \to 0} xH(x)$, (vii) $\lim_{x \to 1^-} \sqrt{(1 - x^2)}$,
   (viii) $\lim_{x \to 1^-} \sqrt{(x^2 - 1)}$, (ix) $\lim_{x \to -\infty} (1 - \cos x)$, (x) $\lim_{x \to +\infty} \frac{1 - \cos x}{x}$.

4.3 Other limits

There are certain cases when limits do not exist in the sense of the preceding section in which it is convenient to use the limit notation. For example, if $f(x)$ becomes steadily larger and larger as x approaches a, and eventually exceeds any number we care to name, it is often appropriate to write $f(x) \to +\infty$ in the same notation that we have used in section 4.2 for x tending to positive infinity. In fact, we shall write

$$\lim_{x \to a} f(x) = +\infty$$

if $f(x)$ can be made greater than any given positive number, however large, by taking x sufficiently close to a and remains greater for all values of x, except $x = a$, which are still nearer to a. Similarly we write

$$\lim_{x \to a} f(x) = -\infty$$

if $f(x)$ can be made less than any given negative number, no matter how negative.

One-sided infinite limits $\lim_{x \to a^+} f(x) = \pm\infty$, $\lim_{x \to a^-} f(x) = \pm\infty$ can be similarly defined by restricting x to one side or the other of a.

1—5

For example, $1/x$ can be made as large as desired by choosing x positive and sufficiently small. Thus

$$\lim_{x \to 0^+} \frac{1}{x} = +\infty.$$

Similarly

$$\lim_{x \to 0^-} \frac{1}{x} = -\infty,$$

and

$$\lim_{x \to 0} \frac{1}{|x|} = +\infty.$$

The behaviour as $x \to +\infty$ or $x \to -\infty$ can be dealt with in the same way. For example,

$$\lim_{x \to +\infty} f(x) = +\infty$$

if $f(x)$ is larger than any given positive number for all sufficiently large x. It is evident that

$$\lim_{x \to +\infty} x^2 = +\infty, \quad \lim_{x \to -\infty} x^2 = +\infty, \quad \lim_{x \to +\infty} (-x) = -\infty,$$

$$\lim_{x \to -\infty} x^3 = -\infty.$$

The reader must not apply Theorem 4.1 to the limits of this section. This is because no meaning has been assigned to expressions such as $+\infty/+\infty$ or $+\infty - \infty$. As an instance of the difficulties that arise note that $\lim_{x \to +\infty} (x^2 + 1) = +\infty$ and $\lim_{x \to +\infty} (-x^2) = -\infty$ but $\lim_{x \to +\infty} [(x^2 + 1) - x^2] = 1$; on the other hand $\lim_{x \to +\infty} (-x^2 - 3) = -\infty$ but $\lim_{x \to +\infty} [(x^2 + 1) - x^2 - 3] = -2$, which shows that Theorem 4.1 (ii) certainly cannot be asserted when it would involve adding $+\infty$ and $-\infty$.

Exercises

8. Find each point where the denominator in the following functions vanishes. If such a point is $x = a$ consider the limits as $x \to a^+$ and $x \to a^-$:

(i) $\dfrac{3}{x}$, (ii) $\dfrac{x+2}{(x-1)(x+3)}$, (iii) $\dfrac{x^2 - ax + b}{x+q}$,

(iv) $\dfrac{(x+1)(x+2)}{(x-3)^2}$, (v) $\dfrac{(2x-1)^3}{x(x+2)^2}$.

9. Examine the behaviour of the functions in question 7 as $x \to +\infty$ and $x \to -\infty$.

10. Evaluate

(i) $\lim\limits_{x \to 1^+} \dfrac{\sqrt{(x-1)}}{x^2 - 1}$, (ii) $\lim\limits_{x \to +\infty} \dfrac{p_0 x^n + p_1 x^{n-1} + \ldots + p_n}{p_0 x^m + q_1 x^{m-1} + \ldots + q_m}$ with $p_0 \neq 0$, (iii) $\lim\limits_{x \to 1^-} \dfrac{x-1}{x^2 - 1}$.

11. Does $\lim\limits_{x \to 0^+} (-1)^x$ exist? (Hint: consider x taking the values $\tfrac{2}{3}, \tfrac{1}{3}, \tfrac{2}{5}, \tfrac{1}{5}, \tfrac{2}{7}, \tfrac{1}{7}, \ldots$.)

4.4 Continuous functions

The functions considered so far in this chapter and in preceding chapters have exhibited a number of different characteristics. Some of these properties are not desirable for future developments and it will be necessary to classify from time to time types of function to which processes can be applied. As a start let us see whether we can categorize functions for which it is possible to draw graphs without breaks in them.

If y is a function of x given by $y = f(x)$ this will require that a small change in the value of x produces only a small change in the value of y. And we must certainly exclude infinite values of y because the graph of $y = 1/x$ shows that breaks can occur at such values. Combining together these two requirements we formulate

DEFINITION 4.2 *A function $f(x)$ is said to be continuous at $x = a$ if*
 (i) *$f(a)$ is defined as a finite number,*
 (ii) $\lim\limits_{x \to a} f(x)$ *is a finite number,*
 (iii) $\lim\limits_{x \to a} f(x) = f(a)$.

The use of the term continuous is to indicate the unbroken nature of the graph.

Observe that (iii) is not a consequence of (i) and (ii) because in the definition of a limit the value at $x = a$ is deliberately excluded.

Example 12   Consider the behaviour of $y = x^2$ at $x = a$.

In this case (i) is certainly satisfied because $f(a) = a^2$. Also
$$f(x) - f(a) = x^2 - a^2 = (x-a)(x+a)$$
which tends to zero as $x \to a$. Hence
$$\lim\limits_{x \to a} f(x) = f(a)$$

which shows that (ii) and (iii) are complied with. Therefore the function x^2 is continuous at $x = a$. Since a was chosen arbitrarily, it follows that x^2 is continuous for all finite x.

A function which is continuous at every point of an interval is said to be *continuous on that interval*. Thus, if $f(x)$ is continuous for every x which satisfies $a < x < b$, $f(x)$ is continuous on $a < x < b$. For example, x^2 is continuous on $-5 < x < 38$.

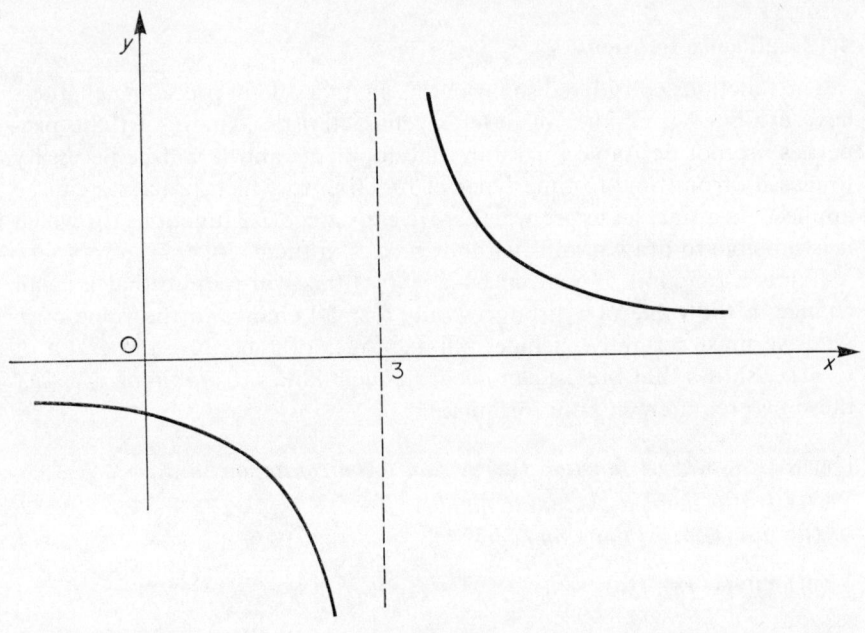

FIGURE 53

A slight modification of this definition is necessary if we want to include the end-point of an interval. This is because $f(x)$ might not be defined beyond the end-point so that condition (ii) would be violated. To avoid this difficulty, we say $f(x)$ is *continuous on* $a < x \leqslant b$ if $f(x)$ is continuous on $a < x < b$ and $\lim_{x \to b^-} f(x) = f(b)$ with $f(b)$ a finite number. Similarly, $\lim_{x \to a^+} f(x) = f(a)$ with $f(a)$ finite and $f(x)$ continuous on $a < x < b$ would be described as $f(x)$ *continuous on* $a \leqslant x < b$.

If, both $\lim_{x \to a^+} f(x) = f(a)$ and $\lim_{x \to b^-} f(x) = f(b)$, and $f(x)$ is continuous on $a < x < b$ then $f(x)$ is continuous on $a \leqslant x \leqslant b$.

For instance, $\sqrt{(1-x)}$ is continuous for any finite x less than 1 and
$$\lim_{x \to 1^-} \sqrt{(1-x)} = 0 = \sqrt{(1-1)}.$$
Hence $\sqrt{(1-x)}$ is continuous for any finite $x \leqslant 1$, but is not defined for $x > 1$. In contrast $1/\sqrt{(1-x)}$ is continuous only for finite $x < 1$ because it does not have a finite value at $x = 1$. Similarly, $\sqrt{(1+x)}/\sqrt{(1-x)}$ is continuous on $-1 \leqslant x < 1$ and $\sqrt{(1-x^2)}$ is continuous on $-1 \leqslant x \leqslant 1$.

If one of the conditions (i), (ii) or (iii) of Definition 4.2 fails at $x = a$, $f(x)$ is said to be *discontinuous* at that point. One type of discontinuity has already occurred in Chapter 3, figure 47 where there are discontinuities at $x = 2, 3, 4, 6, 8, 10$; at these values the function changes abruptly so that $\lim_{x \to a^+} \neq \lim_{x \to a^-}$ and condition (ii) is not met.

The function $1/(x-3)$ illustrates another kind of discontinuity (figure 53). This function is discontinuous at $x = 3$ because $f(3)$ is not defined as a finite number and $\lim_{x \to 3} f(x)$ does not exist. So neither (i) nor (ii) of Definition 4.2 holds. This is an example of an *infinite discontinuity*. The reader should note that $1/(x-3)$ is continuous for any finite $x > 3$ or any finite $x < 3$.

The function $\tan x$ supplies another example of a function with an infinite discontinuity. But, in this case, since $\tan x$ is infinite when $x = \pm \frac{1}{2}\pi, \pm \frac{3}{2}\pi, \pm \frac{5}{2}\pi,\ldots$ we have a function which is discontinuous at an infinite number of isolated points; it is continuous for $-\frac{1}{2}\pi < x < \frac{1}{2}\pi$, $\frac{1}{2}\pi < x < \frac{3}{2}\pi$, etc., but not for $0 < x < \pi$ or any interval which includes one or more of the discontinuities.

Example 13 *The function $f(x) = (x^2 - 9)/(x - 3)$ is not defined at $x = 3$ because both numerator and denominator vanish there. Thus $f(x)$ does not meet condition (i) and so must be discontinuous at $x = 3$.*

This type of discontinuity is called *removable* because it can be removed by adding the further definition $f(3) = 6$. Since $\lim_{x \to 3} f(x) = 6$, a function is obtained which is continuous for all finite x. Actually, $y = (x^2 - 9)/(x - 3)$ consists of the straight line $y = x + 3$ with a 'hole' at $x = 3$ (figure 54); filling this 'hole' removes the discontinuity.

The function $f(x) = (x^2 - 4)/(x - 2)$, $f(2) = 6$ is discontinuous at $x = 2$ because $\lim_{x \to 2} f(x) = 4$. In this case the 'hole' has not been filled. However, the discontinuity can be removed by redefining $f(2) = 4$.

Since a small change of x produces only a small change in a continuous function, the graph of a continuous function consists of a curve without any breaks in it. This fact is very important in modern numerical methods because it enables us to draw curves from a knowledge of the positions of a relatively small number of points. For, once we have plotted these points, we can complete the graph by drawing an unbroken curve through them.

Indeed, we can go on to determine the value of the function at an intermediate value of x from the graph. Naturally, such a value will be an approximation since the graph will not have been drawn precisely

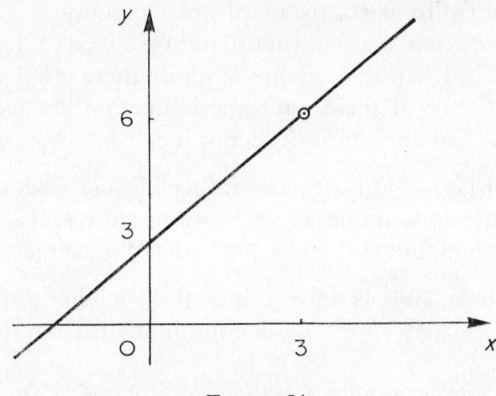

FIGURE 54

accurately, but the closer together the calculated points are the more exact the approximation is likely to be. This idea of joining exactly calculated points by a suitable continuous curve which is used to predict intermediate values is in widespread use and we have already had an example of it in the interpolation of section 1.4 where the continuous curve drawn between two points was, in fact, a straight line.

Despite the valuable properties of continuous functions it must not be supposed that they are the only functions likely to be encountered. Functions with finite discontinuities are met in everyday life and occur in almost every financial transaction because there is a minimal unit of currency. Other examples are the flight of a ball suddenly terminated by a good catch, the motion of a pedestrian stopping abruptly in front of a shop window and the collision of a car with a solid brick wall.

Much more discontinuous functions have been devised by the mathematicians. A favourite example is the function defined by

$f(x) = 1$ when x is of the form p/q, where p is an integer and q is a positive integer,

$ = 0$ for all other values of x.

Limits and Continuity

Certainly, $f(x)$ has a finite value for every x in the interval $0 \leqslant x \leqslant 1$. However, if $x = a$ is a point of this interval and $f(a) = 0$, there are points x as close to a as we please where $f(x) = 1$. Thus, $|f(x) - f(a)|$ is not small when x is close to a. Similarly, if $x = a$ is a point where $f(a) = 1$ there are nearby points where $f(x) = 0$ and $|f(x) - f(a)|$ cannot be made arbitrarily small. In other words, this function is discontinuous at every point of the interval. It consists of the x-axis except for the points of the form $x = p/q$, together those points of $y = 1$ which are of the form $x = p/q$. Thus the graph is two parallel lines, each of which is full of 'holes', and cannot be drawn.

4.5 Properties of continuous functions

An immediate consequence of Definition 4.2 and Theorem 4.1 is

THEOREM 4.4 *If $f(x)$ and $g(x)$ are continuous at $x = a$, then* (i) $f(x) \pm g(x)$ *and* (ii) $f(x)g(x)$ *are continuous at $x = a$. If in addition, $g(a) \neq 0$ then $f(x)/g(x)$ is also continuous at $x = a$.*

If m is a positive integer or zero, x^m is continuous at $x = a$ for any finite a. Therefore, by Theorem 4.4 (i) so is $ax^m + bx^{m+1}$ and, by a further application, so is $ax^m + bx^{m+1} + cx^{m+2}$. Proceeding in this way, we see

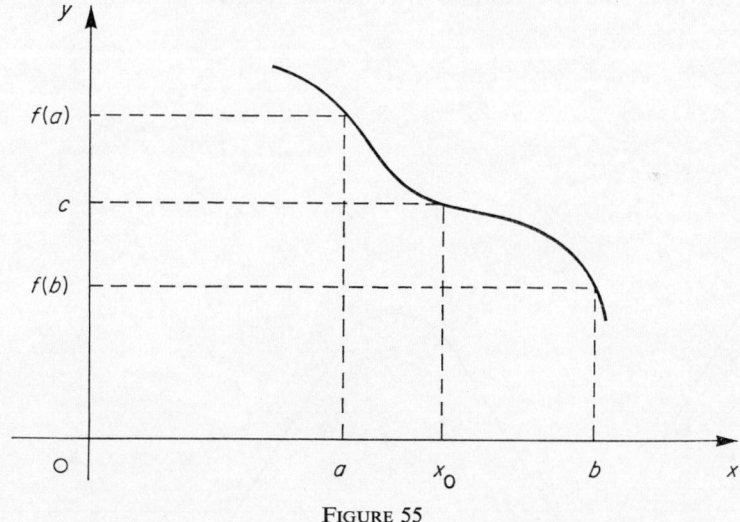

FIGURE 55

that *a polynomial is continuous on any finite interval*. It follows from Theorem 4.4 that a rational function of x is continuous for every finite x except those values at which the denominator is zero.

In sketching the graph of a continuous function any two points are joined by an unbroken curve. Consequently, in passing from one value to another a continuous function must pass through every intermediate value at least once. To put it another way, if $f(x)$ is continuous on $a \leqslant x \leqslant b$ and if $f(a) \neq f(b)$ then, for any number c between $f(a)$ and $f(b)$, there is at least one value of x, say $x = x_0$, for which $f(x_0) = c$. See figure 55 for a typical example. A rigorous proof of this statement is beyond the scope of this book.

If $f(a)$ and $f(b)$ have opposite signs, it follows that there is at least one x_0 where $f(x_0) = 0$. This is illustrated in figure 56 where there are, in fact, three points between a and b where $f(x) = 0$.

This property is not necessarily valid for a function which is not continuous as can be seen in figure 53 where a function changes sign, as x passes through 3, without vanishing.

Again, we have a property which is very important in numerical work. Consider the polynomial $p_0 x^n + p_1 x^{n-1} + \ldots + p_n$. Suppose that it is positive when $x = a$ and negative when $x = b$. Then we know that the equation
$$p_0 x^n + p_1 x^{n-1} + \ldots + p_n = 0$$
has at least one root (the number must be *odd*) between $x = a$ and $x = b$. By subdividing the interval into m equal parts and examining the signs at

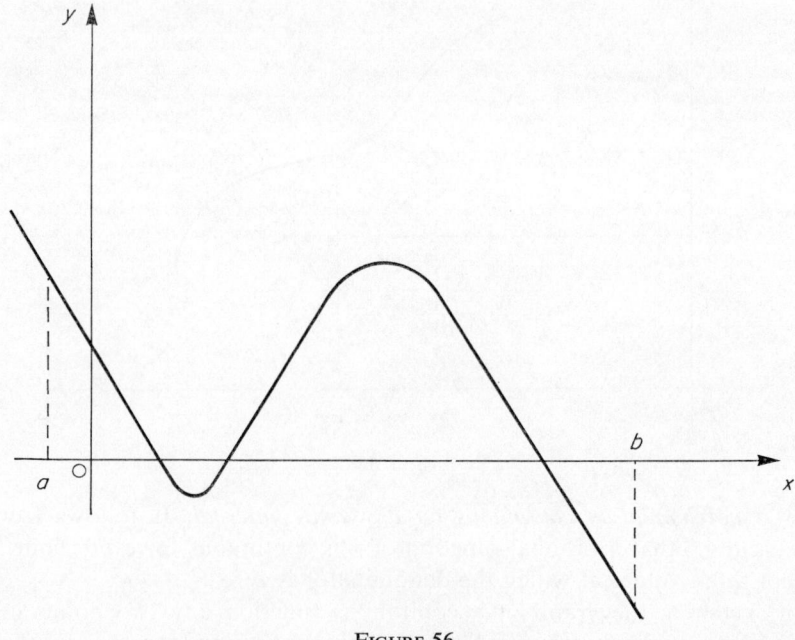

FIGURE 56

the dividing points we can locate the roots more accurately—the larger m the more exact will be the determination. In principle, the same method will work for solving
$$f(x) = 0$$
when f is continuous; the process has been refined and systematized so that it can be done automatically by a digital computer (see section 4.6).

Another property of a function continuous on $a \leqslant x \leqslant b$ is that *it has a least value m and a greatest value M*. This is evident from any graph (see,

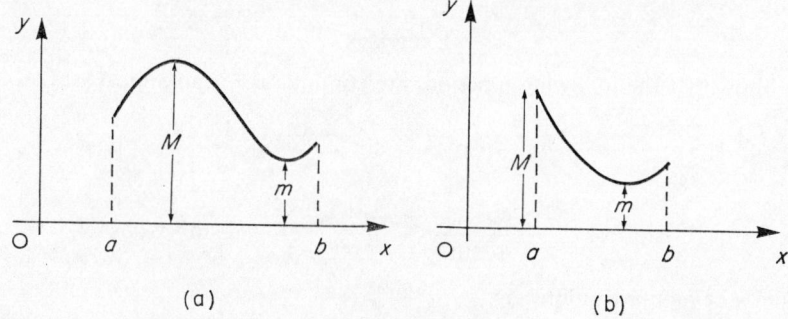

FIGURE 57

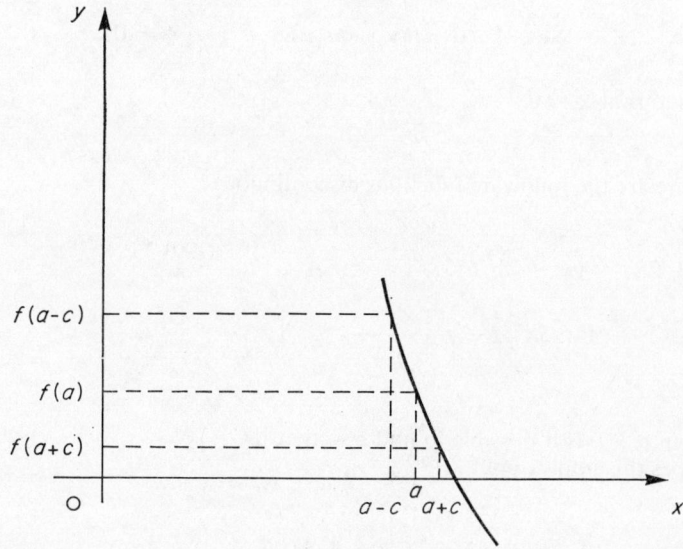

FIGURE 58

for example, figure 57(a) and (b)) but is rather difficult to prove rigorously. Again, figure 53 shows that this need not be true for a discontinuous function; for, on the interval $2 \leqslant x \leqslant 4$ the function has neither a greatest value nor a least value.

Finally, if $x = a$ is a point inside an interval of continuity of f and $f(a) > 0$, there is a $c > 0$ such that $f(x) > 0$ for $a - c < x < a + c$. For, since $\lim_{x \to a} f(x) = f(a)$ we can find points near to a such that $|f(x) - f(a)| < \tfrac{1}{2}f(a)$ which requires $f(x) > \tfrac{1}{2}f(a) > 0$. See figure 58.

Exercises

12. Show that the following functions are continuous for all finite x:

$$2 + 3x; \quad a + bx - cx^2; \quad \sin x; \quad \frac{1}{x^2 + 1}; \quad \cos^2 x;$$

$$x^4; \quad \sin^4 x; \quad \frac{\cos x}{3 + \sin x}; \quad \frac{x^2}{x^2 + 1}; \quad \sin^m x \cos^n x,$$

m and n being positive integers.

13. Show that the function defined as

$$\frac{\tan x}{x} \quad (0 < |x| < \tfrac{1}{2}\pi), \qquad 1 \quad (x = 0)$$

is continuous at $x = 0$.

14. Where are the following functions discontinuous?

$$\frac{x}{3x + 2}; \quad \frac{x^2}{(x-2)^3}; \quad \frac{x^2 + 8x - 48}{x - 4}; \quad \cot x; \quad \frac{1}{x^4 - 6x^2 + 8};$$

$$\sec 2x; \quad \tan 3x; \quad \frac{3 + \cos x}{1 + \sin x}; \quad |x|.$$

15. Given $\varepsilon > 0$ is it possible to find c so that $||x| - 0| < \varepsilon$ for $-c < x < c$? What does this imply about $|x|$?

16. What value must be assigned to $(x^3 + 64)/(x + 4)$ when $x = -4$ to make it continuous?

Limits and Continuity

17. Are the following discontinuous at the given point:

   (i) $2^{1/x}$ at $x = 0$;   (ii) $x^{\frac{1}{3}}$ at $x = 0$;   (iii) $\dfrac{x - a}{(\sqrt{x} - \sqrt{a})^2}$ at $x = a$?

18. Prove that the equation $x^3 + 7x^2 - 4x - 1 = 0$ has a root between 0 and 1.

19. Prove that the equation $x^5 + 5x^4 - 20x^2 - 19x - 2 = 0$ has roots between 2 and 3, and between -4 and -5.

20. Find the maximum and minimum of $1 - |x|$ for $-1 \leqslant x \leqslant 1$.

21. Does x^3 have a maximum or minimum for (i) $0 < x < 1$, (ii) $0 < x \leqslant 1$, (iii) $0 \leqslant x \leqslant 1$?

22. If $f(x)$ is continuous on $a \leqslant x \leqslant b$, if c lies between a and b, and $f(c) < 0$ show that there is a $d > 0$ such that $f(x) < 0$ for $c - d < x < c + d$.

23. If $\lim_{h \to 0} (1/h)[f(a + h) - f(a)]$ exists prove that $f(x)$ is continuous at $x = a$. (Hint: The existence of the limit implies that $f(a + h) - f(a) \to 0$ as $h \to 0$.)

24. If $f(x)$ is continuous at $x = a$ prove that $|f(x)|$ is continuous there. (Hint: Equation (3.6).)

4.6 The bisection method

It was seen in the preceding section that, if $f(a)$ and $f(b)$ have opposite signs, there is at least one x_0 between a and b where $f(x_0) = 0$ when f is continuous. This result can be used to provide a simple and effective method of determining the root of an equation which cannot be handled analytically. To see how the method is constructed let us consider an example.

Example 14   Suppose we wish to compute a real root of the equation
$$x^3 + 3x^2 + 3x - 2 = 0.$$
Denote $x^3 + 3x^2 + 3x - 2$ by $f(x)$. Then note that $f(0) = -2$ and $f(1) = 5$. Hence there is a root between $x = 0$ and $x = 1$.

A systematic procedure for locating the root more exactly is as follows. Evaluate $f(x)$ for the half-way value $x = \tfrac{1}{2}(0 + 1) = 0.5$; the result is $f(0.5) = 0.375$.

Since this is positive and $f(0)$ is negative the root lies between $x = 0$ and $x = 0.5$. Choose the half-way point again, namely $x = 0.25$, and find $f(0.25) = -1.047$. Thus the root lies between $x = 0.25$ and $x = 0.5$. Selecting the half-way point again and continuing in this way we obtain the sequence of values

$$x = 0.375 \quad 0.438 \quad 0.469 \quad 0.454 \quad 0.446 \quad 0.442 \quad 0.444 \quad 0.443$$
$$f(x) = -0.4 \quad -0.026 \quad 0.167 \quad 0.088 \quad 0.023 \quad -0.002 \quad 0.011 \quad 0.004$$

where we have worked to three decimal places. The table shows that the root lies between $x = 0.442$ and $x = 0.443$ and our next approximation is

$$\tfrac{1}{2}(0.442 + 0.443) = 0.442_5.$$

Placing the figure 5 at a lower level than the other figures indicates that, while we are sure of the first three decimal places 442, we are not certain of the fourth. The figure 5 itself shows that, at this stage of approximation, our method does not tell us whether the root is closer to 0.442 or to 0.443.

The method of finding a root illustrated in Example 14 is known as the *bisection method*. It is one of the methods of finding a root by a systematic and repetitive procedure which gives numbers which approximate the root more and more closely as the procedure is continued. Such procedures are called *iterative* and are well adapted for use in digital computers. Later on (see section 8.3) we shall examine iterative processes with a view to determining whether they do give closer and closer approximations and if they do, the accuracy that can be expected from a calculation.

For the moment we want to crystallize the method of Example 14 into a definite rule for finding a root of the equation $f(x) = 0$. A rule which lays down certain operations which will give the solution to a problem is often known as an *algorithm*. So now we aim to provide an algorithm for finding a root by the bisection method.

Suppose that $f(x)$ is continuous for $a \leqslant x \leqslant b$ and that $f(a)$ and $f(b)$ have opposite signs, i.e. $f(a).f(b) < 0$. Then we know that $f(x) = 0$ has at least one root between $x = a$ and $x = b$.

ALGORITHM FOR THE BISECTION METHOD *Define $a_0 = a$, $b_0 = b$ and then form the numbers $a_1, b_1, a_2, b_2, \ldots$ successively by the following procedure. Put*

$$c = \tfrac{1}{2}(a_{r-1} + b_{r-1})$$

and calculate $f(c)$. If $f(c) = 0$ then $x = c$ is the required root. If $f(c) \neq 0$ then either (i) $f(c)f(a_{r-1}) > 0$ *and then we define $a_r = c$, $b_r = b_{r-1}$, or* (ii) $f(c)f(a_{r-1}) < 0$ *and then we define $a_r = a_{r-1}$, $b_r = c$. Stop the process when $|a_r - b_r| \leqslant \varepsilon$, where ε is some assigned number.*

Let us first check that the algorithm does follow the route of Example 14. In this example $a_0 = 0$, $b_0 = 1$. Then $c = 0.5$ and $f(c) = 0.375$;

therefore $f(c).f(a_0) < 0$ and (ii) occurs. Hence we take $a_1 = 0$, $b_1 = 0.5$. On the next round $c = \frac{1}{2}(0 + 0.5) = 0.25$ and $f(0.25) = -1.047$ so that $f(c).f(a_1) > 0$, i.e. (i) occurs; therefore we take $a_2 = 0.25$, $b_2 = 0.5$. The next repetition gives $c = 0.375$, $f(c) = -0.4$ and (i) occurs; therefore $a_3 = 0.375$, $b_3 = 0.5$. Proceeding in this way we reach $a_9 = 0.442$, $b_9 = 0.444$ and $a_{10} = 0.442$, $b_{10} = 0.443$. Now $|a_{10} - b_{10}| = 0.001$ and since we are working to three decimal places we cannot guarantee getting a_r and b_r any closer. In other words we have chosen $\varepsilon = 0.001$ in this example by restricting the calculation to three decimal places. In this case the process stops when $r = 10$.

Exercise

25. Use the bisection method to solve
    (i) $8x^3 - 4x - 5 = 0$ to three decimal places,
    (ii) $x = 1/(1 + x^2)$ to two decimal places,
    (iii) $2x = \tan x$ in radians to two decimal places.

It will now be proved that the algorithm does, in fact, lead to a solution of the problem of solving $f(x) = 0$.

THEOREM 4.5 *Under the conditions of the algorithm*
  (i) $b_r - a_r = (1/2^r)(b - a)$,
  (ii) $|x_0 - \frac{1}{2}(a_r + b_r)| < \frac{1}{2}(b_r - a_r)$,
  (iii) $|x_0 - \frac{1}{2}(a_r + b_r)| < (1/2^{r+1})(b - a)$,
where x_0 is a root of $f(x) = 0$.

Proof. If (i) of the algorithm applies
$$b_r - a_r = b_{r-1} - c = b_{r-1} - \tfrac{1}{2}(a_{r-1} + b_{r-1}) = \tfrac{1}{2}(b_{r-1} - a_{r-1}).$$
If (ii) applies
$$b_r - a_r = c - a_{r-1} = \tfrac{1}{2}(a_{r-1} + b_{r-1}) - a_{r-1} = \tfrac{1}{2}(b_{r-1} - a_{r-1}).$$
Thus, in either case,
$$b_r - a_r = \tfrac{1}{2}(b_{r-1} - a_{r-1}).$$
Changing r to $r - 1$ we have
$$b_{r-1} - a_{r-1} = \tfrac{1}{2}(b_{r-2} - a_{r-2})$$
so that
$$b_r - a_r = (1/2^2)(b_{r-2} - a_{r-2}).$$
Proceeding in this way we obtain
$$b_r - a_r = (1/2^r)(b_0 - a_0) = (1/2^r)(b - a)$$
and this proves the first part of the theorem.

To prove (ii) observe that
$$x_0 - \tfrac{1}{2}(a_r + b_r) = \tfrac{1}{2}(x_0 - a_r) + \tfrac{1}{2}(x_0 - b_r).$$
Now $x_0 - a_r$ is positive and $x_0 - b_r$ is negative so that the right-hand side must be less than $\tfrac{1}{2}(x_0 - a_r)$ and greater than $\tfrac{1}{2}(x_0 - b_r)$. However, $x_0 < b_r$ so that $x_0 - a_r < b_r - a_r$, and $x_0 > a_r$ so that $x_0 - b_r > a_r - b_r$. Thus the right-hand side is smaller than $\tfrac{1}{2}(b_r - a_r)$ and larger than $\tfrac{1}{2}(a_r - b_r)$, i.e.
$$|x_0 - \tfrac{1}{2}(a_r + b_r)| < \tfrac{1}{2}(b_r - a_r)$$
which proves (ii). Part (iii) follows at once from (i) and (ii).

Theorem 4.5(i) tells us that the successive intervals which contain the root become smaller and smaller so that the root can be located with any desired degree of accuracy. In fact Theorem 4.5 (ii) and (iii) indicate how accurately the root has been found. If the process is stopped when $b_r - a_r \leqslant \varepsilon$ the error in $\tfrac{1}{2}(a_r + b_r)$ as an approximation to x_0 does not exceed $\tfrac{1}{2}\varepsilon$. Part (iii) demonstrates how large r must be to give a specified degree of accuracy. Suppose we want the error in the approximation not to exceed $\tfrac{1}{2}\varepsilon$; then we choose r so that
$$(1/2^{r+1})(b - a) \leqslant \tfrac{1}{2}\varepsilon$$
or such that $2^r \geqslant (b - a)/\varepsilon$. Then we can be sure that the repetitions will certainly cease by this value of r; of course, in a particular example the process *may* terminate for a smaller value of r. If $b - a = 1$ and $\varepsilon = 0 \cdot 001$ we can be confident that ten repetitions will suffice since $2^{10} = 1024$.

These conclusions and Theorem 4.5 itself assume that, if x is a given number, $f(x)$ can be calculated exactly. In general, however, when $f(x)$ is computed numerically either by hand or by machine only a certain accuracy will be possible because there is a limit to the number of decimal places which can be handled practically. As a result an approximation to $f(x)$ is calculated, instead of $f(x)$ itself. The effect of this phenomenon, called *round-off error*, on practical calculations will be examined later (see section 8.4).

4.7 Flow-charts

The operations required by an algorithm are sometimes difficult to sort out and so it is often helpful to have available a pictorial representation. Such a diagram is called a *flow-chart*. A flow-chart for the algorithm of the preceding section is shown in figure 59 (for simplicity it has been assumed that at no stage of the calculation does $f(c)$ vanish). It will be seen that the flow-chart should be read from the top, following the arrows from box to box.

Certain conventions are observed concerning the boxes. *Circles* containing START and STOP are used to show where the beginning and end of the flow-chart are. There is only one START box but there can be any number of STOP boxes.

Each *rectangular box* specifies a certain operation which is to be carried out before the box is left. Often the operation requires the evaluation of a certain formula and then a convenient shorthand is used. Thus

$$c := \tfrac{1}{2}(a_r + b_r)$$

means 'calculate c from the expression on the right of $:=$' or 'calculate $\tfrac{1}{2}(a_r + b_r)$ and call the result c'. Only one symbol should appear on the

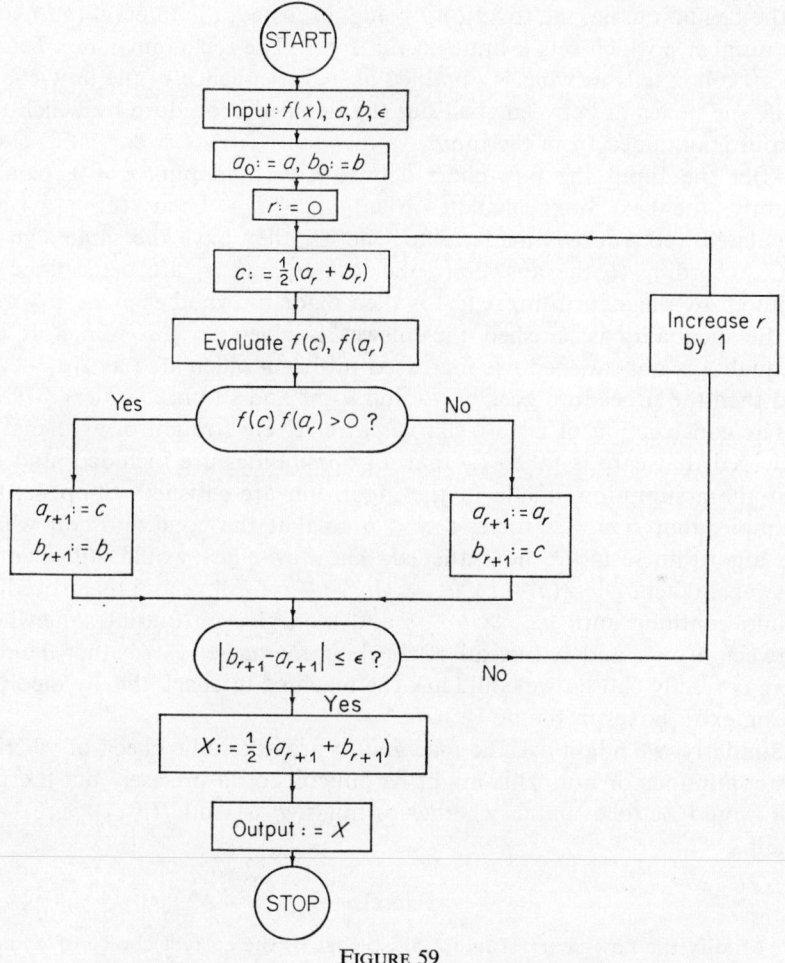

FIGURE 59

left of $:=$. All symbols on the right of $:=$ must have been defined at an earlier stage in the flow-chart.

Each *oval box* contains a question. There is a separate line leaving the box for each possible answer to the question in the box. Therefore the line followed on leaving the box depends on the answer to the question in the box. In contrast there is only one exit from a rectangular box.

Sometimes rectangular and oval boxes are called *assertion* and *test boxes*, respectively, to indicate their different purposes.

Considering the flow-chart of figure 59 in more detail we see that the rectangular box after the START box gives the information that is available at the beginning, i.e. the function f being discussed, the interval (a,b), and the number ε which sets a limit on the error. The rectangular box before the STOP box tells us what is obtained as a consequence of the flow-chart, while the boxes in between spell out the detailed procedure by which the output is obtained from the input.

After the Input the flow-chart defines a_0, b_0 and puts $r = 0$. Consequently, the next box calculates c as $\frac{1}{2}(a_0 + b_0)$. Then $f(c)$, $f(a_0)$ are calculated and a test is made as to whether they have the same sign or not. According to the answer to the test, a_1 and b_1 are determined as required by the algorithm. A test is then made on whether $|b_1 - a_1| \leqslant \varepsilon$. If the inequality is satisfied the answer is given as $\frac{1}{2}(a_1 + b_1)$. If the inequality is not satisfied r is increased by 1, c is calculated as $\frac{1}{2}(a_1 + b_1)$ and then the procedure goes on to find a_2, b_2 and so on.

The construction of a flow-chart is a relatively straightforward matter provided that care is taken (i) that all possibilities are included, and (ii) that the assumptions made in the algorithm are satisfied. Suppose, for example, that $f(x) = 1$ for $a \leqslant x \leqslant b$ so that the conditions on which the algorithm rests are not satisfied. The flow-chart would still give an answer; in fact $a_1 = \frac{1}{2}(a + b)$, $a_2 = \frac{1}{2}[b + \frac{1}{2}(a + b)]$,... and the procedure would continue until $a_{r+1} \geqslant b - \varepsilon$ and then give an output somewhere between $b - \frac{1}{2}\varepsilon$ and b. Instead of checking for ourselves whether a and b were correctly chosen we could ask the machine to check this by incorporating extra boxes in the flow-chart.

Similarly, we might ask the machine to provide some check of whether f is continuous or not. This might be difficult to do precisely but a crude test would be to examine whether or not $f(a_{r+1})$ and $f(b_{r+1})$ were both small.

Exercise

26. Modify the flow-chart of figure 59 so that (i) the correct choice of a and b is checked and (ii) $f(x)$ is computed only once per iteration instead of twice.

The use of flow-charts is not restricted to mathematics. As an illustration we give a flow-chart in figure 60 of how the location of an airport might be decided when the position proposed by the government is challenged in the High Court. For simplicity, not all possible courses of action are shown and it is assumed that the Minister can always command a majority in the House of Commons although not necessarily in the House of Lords.

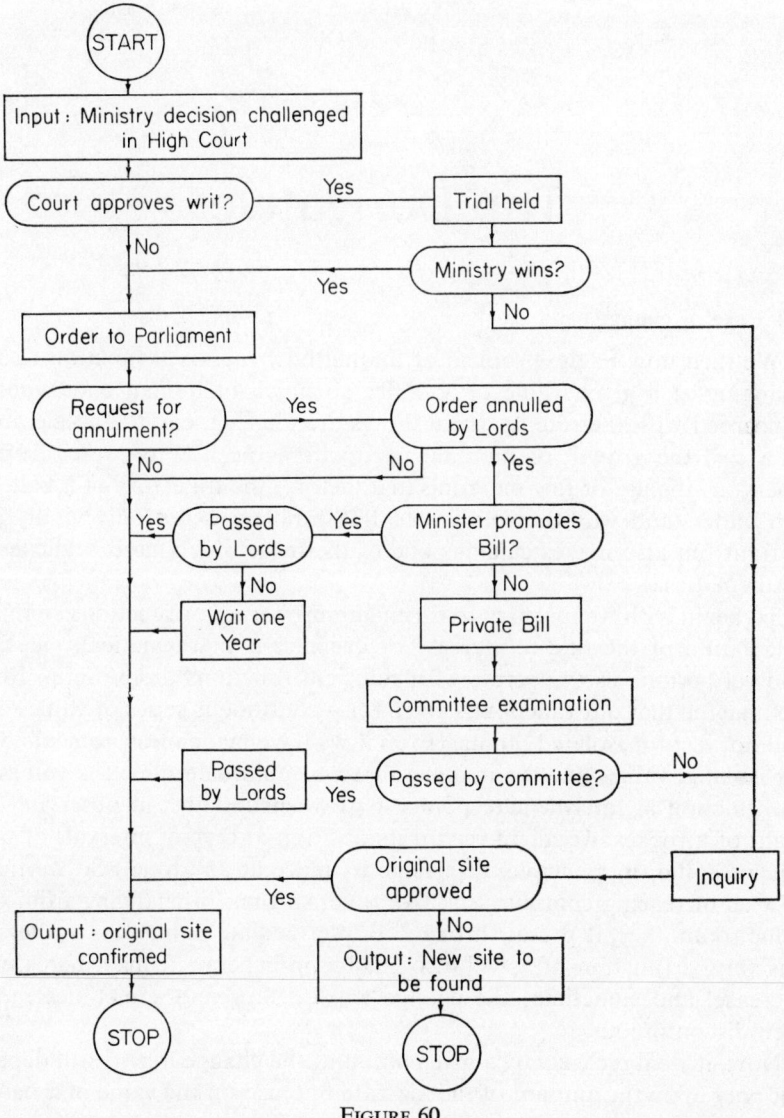

FIGURE 60

27. Draw a flow-chart showing the courses in mathematics at your institution which could be taken by a student from the time of entering until the time of leaving.

5

The Derivative

5.1 Rate of change

We turn now to the problem of finding how rapidly a function of x is changing at a given value of x. Since so much of human endeavour is concerned with the rate at which things are altering, e.g. the acceleration of a car, the growth of national production, the flow of water from a reservoir, the use of raw materials in a factory, the trajectory of a ball, we can understand why the solution to this problem is not only vitally important but also makes calculus one of the most widely used branches of mathematics.

To begin with we must try to formulate properties of functions so that a calculation of the rate of increase or decrease is mathematically feasible and yet permits us to operate with sufficient functions to be useful. First, we imagine that our function is given for a continuous series of values of x and not just for isolated points. (Even if we have made measurements only for isolated values of x we imagine that we could determine the values of the function at intermediate points if we wished. Thus, in observing the flight of a rocket, we might record its position at certain intervals of time but we could, in principle, determine its place at any time.) So we think of x as increasing continuously, like time, so that, in changing from one value to another, it passes through all intermediate points. As x alters in this way, a function of x will also change in general; it may increase or decrease, and sometimes the change will be slow and sometimes rapid, even discontinuous.

Now, if we increase x by a given amount, the change in $f(x)$ will depend not only upon the amount of the increase but also on the value of x before the increase—this is obvious from an inspection of the graphs in the preceding chapters. Indeed, there is no reason why a rocket should travel the same distance in the first second after firing as in the tenth second. We would be surprised if it did because at first it is at rest and then it

The Derivative

begins to move, commencing slowly and gradually getting faster and faster. Then, perhaps, the fuel will be cut off and the rocket will coast along steadily. Another period of burning fuel, boosting the rate at which the rocket is going, may follow. To return through the atmosphere, retro-rockets may be fired to slow down the motion. Thus the distance travelled in 1 second may be large or small, may increase or decrease rapidly but it will rarely be constant for every second of the flight.

Suppose we knew that the rocket had covered 1 kilometre in 1 second. Then we might be tempted to say that it was travelling at a speed of 1 kilometre per second. But this would be unwise because, if the second were the first second of flight, we know that at the beginning the rocket

TABLE 1

t	s	Average speed
1	1	1
$\frac{1}{2}$	$\frac{1}{4}$	$\frac{1}{2}$
$\frac{1}{4}$	$\frac{1}{16}$	$\frac{1}{4}$

would be hardly moving at all; in fact, the rocket might take $\frac{1}{2}$ second to achieve the first $\frac{1}{4}$ kilometre. So, instead, if the rocket travels a distance s in the time t measured from some fixed time we say that s/t is the *average speed* for that particular time interval. Possible values of the average speeds corresponding to different time intervals at the start of flight are shown in table 1. It is clear that the average speed varies considerably with the time interval and is therefore not a suitable measure of the speed at a precise time. However, the distance is a continuous function of time and so we might expect that, as the time interval became smaller and smaller, the average speed would tend to a definite limiting value. This limit can be defined as the *speed* at the beginning of the interval.

So, when we say that the speed of a rocket at the end of the first second of flight is 2 kilometres per second, we do not mean that it will travel 2 kilometres in the next second; it may go much further. What we mean is that in 1/10 of a second it will cover about 0·2 kilometre or 200 metres, and in 1/1000 second the distance will be quite close to 2 metres because in 1/10 of a second the speed will not change too much, and in 1/1000 of a second it will alter even less.

Now argue in the same way about any function of x, replacing the time of flight by x and the distance covered by $f(x)$. Then the change in $f(x)$ divided by the change in x is called the *average rate of change*; now find the limit of the average rate of change as the change in x tends to zero.

This limit is regarded as the rate of change at the given value of x. In other words, if x is altered to $x + h$, $f(x)$ changes to $f(x + h)$ and the average rate of change is

$$\frac{f(x + h) - f(x)}{x + h - x}$$

and the rate of change of f at x is

$$\lim_{h \to 0} \frac{f(x + h) - f(x)}{h}. \tag{1}$$

The argument given so far does not guarantee that the limit will exist. However, at each point x where the limit does exist, the function $f(x)$ is said to possess a *derivative* (or to be *differentiable*) and the limit is called the derivative or differential coefficient at x.

The process of finding the derivative of a function is one of the central problems of the calculus and will occupy us for several chapters. The problem of providing a rule which ensures that a function has a derivative is too difficult to solve and will be left on one side. However, we note from Exercise 22 of Chapter 4 that the existence of the derivative of f at a point implies that f must be continuous at that point. Unfortunately, the converse statement is not true, i.e. it is not true that a continuous function must possess a derivative. Even if failure were permitted at a few isolated points it is not true because the mathematicians can construct continuous functions which do not have a derivative at any point.

So, although the requirement of continuity is necessary, it is not sufficient to guarantee the existence of a derivative. In order to make progress this difficulty will be avoided in this book and all continuous functions encountered in it will have derivatives at most points.

Example 1 Suppose that $f(x) = x^2$ and we wish to find its rate of change or derivative at $x = 5$.

Now
$$f(5 + h) = (5 + h)^2 = 25 + 10h + h^2$$
so that
$$f(5 + h) - f(5) = 10h + h^2.$$
Therefore, according to (1), the rate of change is
$$\lim_{h \to 0} \frac{10h + h^2}{h} = \lim_{h \to 0} (10 + h) = 10.$$

This means that, when x has the value 5, x^2 is increasing at the rate of 10 units per unit increase of x, i.e. if x increases by the small quantity ε then x^2 increases by 10ε very nearly, the accuracy being better the smaller ε.

The Derivative

In general, when $f(x) = x^2$,
$$f(x + h) - f(x) = x^2 + 2hx + h^2 - x^2$$
so that
$$\lim_{h \to 0} \frac{f(x + h) - f(x)}{h} = \lim_{h \to 0} (2x + h) = 2x.$$

Thus the rate of change of x^2 is $2x$; when $x = 5$, $2x = 10$ in agreement with the calculation above. Consequently, the derivative of x^2 exists for every finite x and is $2x$.

Again, this means that, if x increases by the small amount ε, x^2 increases by approximately $2x\varepsilon$. The smaller ε, the closer the increase in x^2 is to $2x\varepsilon$.

These results are capable of geometric interpretation. Suppose that we have a square of side x. Then its area is x^2. Therefore if the side of the square is increased slightly to $x + \varepsilon$ the above result tells us that the area will be increased by $2x\varepsilon$ approximately.

Another geometrical illustration is obtained by considering a circle of radius x. Its area is πx^2. Therefore, if the radius is increased by the small amount ε, the area increases by $2\pi x\varepsilon$ approximately. We can see this another way: the small increase in radius adds a very narrow rim all round the circle. The circumferences of the inner and outer edges of the rim are both effectively $2\pi x$ (the outer is actually a little more); therefore, the approximate increase in area is $2\pi x \times$ increase in radius $= 2\pi x\varepsilon$.

The reader should convince himself that the geometrical interpretations are unaltered whether ε be positive or negative.

Exercises

1. Find the average rate of change in x^3 when x increases from (i) 5 to 6, (ii) 5 to 5·1, (iii) 5 to 5·01, (iv) 5 to $5 + h$. Do these average rates tend to a limit? What happens when x increases from $5 - h$ to 5?

2. Find the derivative of x^3.

3. The side of a square is increasing at the rate of 1 cm/sec. Find the rate of increase of (i) the area, (ii) the perimeter at the instant when the side is (a) 3 cm, (b) 5 cm, (c) 10 cm.

4. The side of a cube is increasing at the rate of 1 in./sec. Find the rate of increase of (i) the volume, (ii) a diagonal when the side is (a) 3 in., (b) 1 ft.

5. The radius of a sphere is increasing by 1 m/sec. Find the rate of increase of (i) the volume, (ii) the surface area at the instant when the radius is 3 m. (The volume and surface area of a sphere of radius x are $\frac{4}{3}\pi x^3$ and $4\pi x^2$ respectively.)

6. The area of a circle is increasing at the rate of 3 ft²/sec. At what rate is the circumference increasing at the instant when the radius is 9 ft?

7. The volume of a cube is increasing at the rate of 12 cm³/sec. Find the rate at which (i) the surface area, (ii) the side is increasing at the instant when the side is 2 cm.

8. The surface area of a sphere is increasing at the rate of 3 in.²/min. At what rate is (i) the volume, (ii) the radius increasing at the instant when the radius is 3 in.?

9. Water is poured into a cone, with axis vertical, at the rate of 16 m³/min. If the radius of the water's surface is the same as the depth, find the rate at which the depth is increasing at the instant when it is 3 m.

10. The side of a cube is the same length as the radius of a sphere and both are increasing at the same rate. Which has the faster increasing (i) volume, (ii) surface area and by how much?

The derivative is so important that special notations are used for it. One notation is to denote the derivative of $f(x)$ by $f'(x)$ (read as f-dashed x or f-prime x), i.e.

$$f'(x) = \lim_{h \to 0} \frac{f(x+h) - f(x)}{h}. \tag{2}$$

Thus, from Example 1, if $f(x) = x^2$, $f'(x) = 2x$.

Another notation arises in the following way. Let y be a continuous function of x; then a small change in x will cause a small change in y. Suppose that we denote a point near x by $x + \delta x$, where the *symbol* δx is used to indicate that we have made a small deviation from x; it does *not* mean that we have multiplied x by δ. In the same way we can use $y + \delta y$ to signify that the new value of y is slightly different from y. If $y = f(x)$ we must have

$$y + \delta y = f(x + \delta x).$$

By subtraction

$$\delta y = f(x + \delta x) - f(x),$$

a formula which gives the small change in y due to a small change in x.

As long as δx is not zero we can divide by it and obtain

$$\frac{\delta y}{\delta x} = \frac{f(x + \delta x) - f(x)}{\delta x}.$$

The Derivative

If, now, we let δx approach zero, while x is held fixed, we see that the right-hand side gives exactly the same as (1), because h could be replaced by δx in the limiting process. Hence the derivative can also be written as

$$\lim_{\delta x \to 0} \frac{\delta y}{\delta x}.$$

An abbreviation employed for this limit is the symbol $\dfrac{\mathrm{d}y}{\mathrm{d}x}$ or $\mathrm{d}y/\mathrm{d}x$ (read either as '$\mathrm{d}y$ by $\mathrm{d}x$' or as 'the derivative of y with respect to x'), i.e.

$$\frac{\mathrm{d}y}{\mathrm{d}x} = \lim_{\delta x \to 0} \frac{\delta y}{\delta x}.$$

It is extremely important to remember that $\mathrm{d}y/\mathrm{d}x$ is not a fraction formed by dividing the quantity $\mathrm{d}y$ by the quantity $\mathrm{d}x$; on no account must the d's be cancelled. It is a *symbol* to represent the limiting value of the fraction $\delta y/\delta x$. Indeed, $\mathrm{d}y/\mathrm{d}x$ must be regarded as a combination of y and the symbol $\mathrm{d}/\mathrm{d}x$; the combination denotes that a certain operation must be performed on y, in the same way that $\sin y$ indicates the result of carrying out a certain operation on y, namely the determination of the sine of the angle y.

The purpose of these symbols is to show quite clearly the function y whose derivative is being calculated and the variable x whose change gives rise to the derivative. Thus it is best to interpret $\mathrm{d}y/\mathrm{d}x$ as $(\mathrm{d}/\mathrm{d}x)(y)$ or $(\mathrm{d}/\mathrm{d}x)y$. With this interpretation we can write the derivative of $f(x)$ as

$$\frac{\mathrm{d}}{\mathrm{d}x} f(x)$$

if we wish.

The notation is very flexible. For example, if the volume of a sphere of radius r is V, the derivative $\mathrm{d}V/\mathrm{d}r$ tells us the rate at which the volume is altering with changes in r. Similarly, if s is the distance travelled at time t, the derivative $\mathrm{d}s/\mathrm{d}t$ gives the rate of change of distance with time at time t, i.e. it gives the speed. Again, if F is the amount of fuel in the tank of a car at time t, the derivative $\mathrm{d}F/\mathrm{d}t$ supplies the rate at which fuel is being consumed by the car at time t.

To summarize, the derivative of x^n may be denoted by any of the following notations

$$(x^n)' \text{ or } \frac{\mathrm{d}}{\mathrm{d}x}(x^n) \text{ or } \frac{\mathrm{d}}{\mathrm{d}x}x^n.$$

and the derivative of $\cos x$ by

$$(\cos x)' \text{ or } \frac{\mathrm{d}}{\mathrm{d}x}(\cos x) \text{ or } \frac{\mathrm{d}}{\mathrm{d}x}\cos x.$$

Whichever notation is used the procedure for calculating the derivative is the same and involves four operations: (i) make a small non-zero change in x, (ii) calculate the resulting small change in y, (iii) form the ratio of (ii) to (i), and (iv) find the limit of this ratio as the small change in x tends to zero.

Example 2   Find $f'(x)$ when $f(x) = 1/x$.

If x is altered to $x + h$, f becomes $1/(x + h)$. Hence the change in f is

$$f(x + h) - f(x) = \frac{1}{x + h} - \frac{1}{x} = \frac{-h}{x(x + h)}.$$

Dividing this by h we have

$$\frac{f(x + h) - f(x)}{h} = \frac{-1}{x(x + h)}$$

and now, from (2),

$$f'(x) = \lim_{h \to 0} \frac{-1}{x(x + h)} = \frac{-1}{x^2}.$$

Alternatively, this could be written as

$$\frac{d}{dx}\frac{1}{x} = -\frac{1}{x^2}.$$

It is clear that the derivative does not exist at $x = 0$.

Example 3   Determine dy/dx when $x > 0$ if $y = \sqrt{x}$.

When x is increased to $x + \delta x$, y alters to $y + \delta y$ where

$$y + \delta y = \sqrt{(x + \delta x)}$$

so that

$$\delta y = \sqrt{(x + \delta x)} - \sqrt{x}$$

and

$$\frac{\delta y}{\delta x} = \frac{\sqrt{(x + \delta x)} - \sqrt{x}}{\delta x}.$$

In order to eliminate δx from the denominator before allowing $\delta x \to 0$, we multiply both the numerator and denominator by the non-zero quantity $\sqrt{(x + \delta x)} + \sqrt{x}$. Thus

$$\frac{\delta y}{\delta x} = \frac{\sqrt{(x + \delta x)} - \sqrt{x}}{\delta x} \cdot \frac{\sqrt{(x + \delta x)} + \sqrt{x}}{\sqrt{(x + \delta x)} + \sqrt{x}}$$

$$= \frac{(x + \delta x) - x}{\delta x[\sqrt{(x + \delta x)} + \sqrt{x}]} = \frac{1}{\sqrt{(x + \delta x)} + \sqrt{x}}.$$

Letting δx approach zero we have

$$\frac{dy}{dx} = \lim_{\delta x \to 0} \frac{\delta y}{\delta x} = \frac{1}{2\sqrt{x}}.$$

Example 4 If $y = (4x + 3)/(2x + 1)$ *find* dy/dx.

When x is increased to $x + \delta x$, y changes to $y + \delta y$ where

$$y + \delta y = \frac{4(x + \delta x) + 3}{2(x + \delta x) + 1}$$

and so

$$\delta y = \frac{4(x + \delta x) + 3}{2(x + \delta x) + 1} - \frac{4x + 3}{2x + 1}$$

$$= \frac{[4(x + \delta x) + 3](2x + 1) - (4x + 3)[2(x + \delta x) + 1]}{[2(x + \delta x) + 1](2x + 1)}$$

$$= \frac{-2\delta x}{[2(x + \delta x) + 1](2x + 1)}.$$

Therefore

$$\frac{\delta y}{\delta x} = \frac{-2}{[2(x + \delta x) + 1](2x + 1)}$$

and, proceeding to the limit as δx approaches zero, we obtain

$$\frac{dy}{dx} = \lim_{\delta x \to 0} \frac{\delta y}{\delta x} = \frac{-2}{(2x + 1)^2}.$$

This might also be written as

$$\frac{d}{dx} \frac{1}{2x + 1} = \frac{-2}{(2x + 1)^2}.$$

Notice that this derivative does not exist at $x = -\frac{1}{2}$.

Exercises

11. Find $f'(x)$ when $f(x)$ is (i) x^2, (ii) $3x + 4$, (iii) $ax + b$, (iv) x^3, (v) $1/x^3$, (vi) $3x^2 - 2x + 1$, (vii) $ax^2 + bx + c$, (viii) $(4x + 3)/(2x + 1)$.

12. Find dy/dx when y is (i) $2x/(x + 2)$, (ii) $(5x + 3)/(x - 2)$, (iii) $(ax + b)/(cx + d)$, (iv) $x/(x^2 - 1)$, (v) $2\sqrt{x}$, (vi) $3/\sqrt{x}$, (vii) $a + b/x$, (viii) $\sqrt{(ax + b)}$, (ix) $\sqrt{(x^2 + 2)}$.

13. If $y = x^2 - 2x + 3$ find where dy/dx vanishes.

14. If $f(x)$ is $(3x + 2)/(3 - x)$ show that $f'(x)$ is always positive.

5.2 Numerical approximations

We have already seen on page 119 how a knowledge of the rate of change or derivative of a function enables us to make an estimate of the change in the function when x is altered slightly. We now consider this useful attribute of the derivative in rather more detail.

By definition
$$f'(x) = \lim_{h \to 0} \frac{f(x+h) - f(x)}{h}.$$
Now, this tells us that the difference
$$f'(x) - \frac{f(x+h) - f(x)}{h}$$
approaches zero as $h \to 0$. Let us denote this difference by ε, i.e.
$$f'(x) - \frac{f(x+h) - f(x)}{h} = \varepsilon. \tag{3}$$
Then, we are saying that
$$\lim_{h \to 0} \varepsilon = 0. \tag{4}$$
But (3) can also be written as
$$f(x+h) = f(x) + h[f'(x) + \varepsilon].$$
The smaller h the smaller ε is, on account of (4), and therefore the more negligible it is compared with $f'(x)$, provided that $f'(x)$ is non-zero. Thus, we are led to the conclusion: *if $f'(x) \neq 0$ an approximation to $f(x+h)$ is supplied by*
$$f(x+h) = f(x) + hf'(x), \tag{5}$$
the approximation being better the smaller h.

As an illustration suppose that $f(x) = x^2$ so that $f'(x) = 2x$. Taking $x = 3$ we have
$$f(3) = 9, \quad f'(3) = 6.$$
According to (5) an approximation to $f(3 \cdot 01)$ would be obtained with $h = 0 \cdot 01$ so that
$$f(3 \cdot 01) = 9 + 0 \cdot 01 \times 6 = 9 \cdot 06 \text{ approximately}.$$
To see how much this is in error, we remark that the exact value is
$$f(3 \cdot 01) = (3 \cdot 01)^2 = 9 \cdot 0601$$
so that the approximation is correct to the third decimal place. If the value of $f(3 \cdot 1)$ is required so that $h = 0 \cdot 1$ we would expect the error to be greater; on the other hand, for $f(3 \cdot 001)$ and $h = 0 \cdot 001$ the error would have been much smaller. The reader should confirm these conjectures.

Example 5 If $f(x) = 1/x$, estimate approximately the change in f when x increases from 8 to 8·001.

In this case, $f'(x) = -1/x^2$ from Example 2, so that
$$f(8) = \tfrac{1}{8} = 0·125, \qquad f'(8) = -\tfrac{1}{64} = -0·015625.$$
Since $h = 0·001$, (5) gives
$$f(8·001) = 0·125 - 0·001 \times 0·015625 = 0·124984375$$
approximately. Since the true value is 0·124984377 to 9 decimal places the error in the approximation is less than 0·000000002.

Example 6 Find $\sqrt{122}$.

Since $\sqrt{121} = 11$ this can be written as $11\sqrt{(122/121)}$ or $11\sqrt{(1 + \tfrac{1}{121})}$. Now $\sqrt{1} = 1$ and we use our approximate method with $f(x) = \sqrt{x}$ and $h = \tfrac{1}{121}$. From Example 3, $f'(x) = 1/2\sqrt{x}$ so that $f(1) = 1$ and $f'(1) = \tfrac{1}{2}$. Hence
$$\sqrt{(1 + \tfrac{1}{121})} = 1 + \tfrac{1}{242}$$
approximately, and
$$\sqrt{122} = 11(1 + \tfrac{1}{242}) = 11 + \tfrac{1}{22} = 11·04545$$
approximately.

The correct value is 11·04536..... .

Exercises

15. Find the difference between $f(x + h)$ calculated exactly and calculated approximately by (5) when $f(x)$ is (i) $x^2 - x$, (ii) $x^3 + 2x$, (iii) x^4, (iv) $1/x$.

16. Calculate approximately by (5) (i) $\sqrt{9·05}$, (ii) $1/\sqrt{101}$, (iii) $(12·02)^2$, (iv) $1/(12·02)^2$, (v) $\sqrt{[(3·02)^2 + 16]}$.

17. Find approximately (i) $(7x - 4)/(5x + 10)$ when $x = 18·06$, (ii) $1/(x^2 - 1)$ when $x = 8·98$.

18. Formula (3) suggests that a numerical approximation to the derivative is $(1/h)[f(x + h) - f(x)]$. Write a computer program to calculate this quantity and compare the values obtained for a series of values of h, starting at $h = 0·1$ and going down to a very small value of h, with *the exact derivative* at $x = 1$ when $f(x) = x^2$ and $f(x) = x^3$.

5.3 Geometrical interpretation

The method of obtaining the tangent PT at a point P of a curve has already been described in section 2.3. Points $P_1, P_2, \ldots$ getting steadily closer and closer to P are chosen (figure 61); the chords $PP_1, PP_2, \ldots$, then approach closer and closer to the tangent PT. Suppose that the equation of the curve is $y = f(x)$; then if the abscissa of P is x its ordinate is $f(x)$. If $x + \delta x$ is the abscissa of a nearby point (say P_i) its ordinate is $f(x + \delta x)$. The slope of the straight line PP_i is then given by

$$m_{\text{chord}} = \frac{f(x + \delta x) - f(x)}{x + \delta x - x}.$$

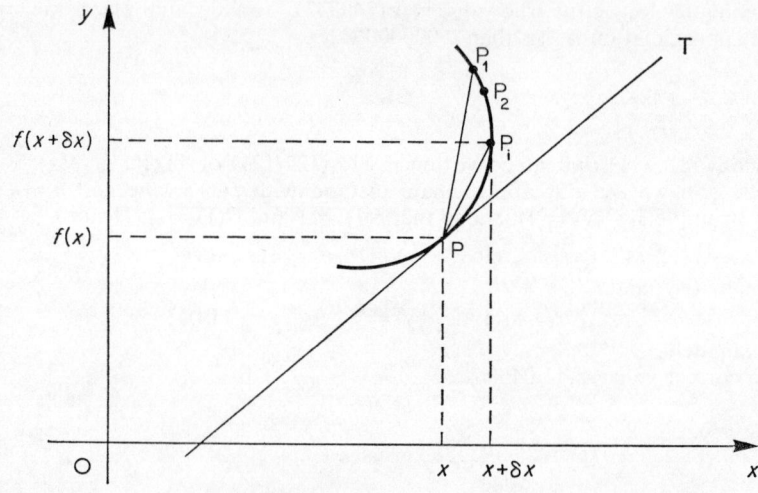

FIGURE 61

By letting $\delta x \to 0$ we can make the chord coincide with PT and so the slope m of PT is given by

$$m = \lim_{P_i \to P} m_{\text{chord}} = \lim_{\delta x \to 0} \frac{f(x + \delta x) - f(x)}{\delta x} + f'(x).$$

Thus the derivative of f gives the slope of the tangent to $y = f(x)$. Equally well, we could say that the slope is dy/dx.

The fact that the same limit is obtained whether δx is positive or negative means that the tangent is the same whether P_i approaches P from the left or the right.

Of course, it may happen that the limit does not exist for δx both positive and negative but that one-sided limits (section 4.2) exist for positive δx and negative δx separately. In that event, as R approaches P the chord PR tends to the tangent PT (figure 62), but as S approaches P the chord PS becomes the tangent PT_1. So the curve is continuous at P but its slope is not; in other words the function is continuous but its rate of change is discontinuous.

The Derivative

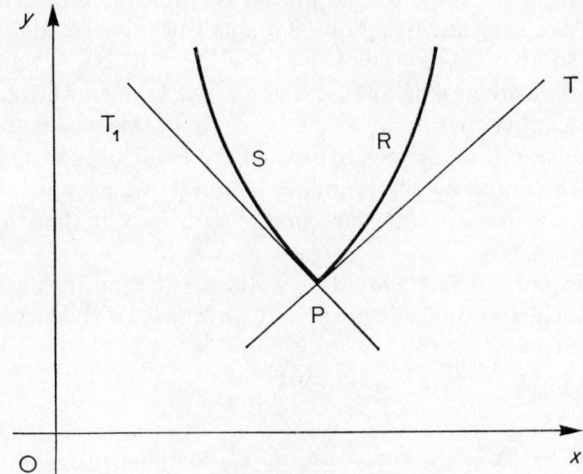

FIGURE 62

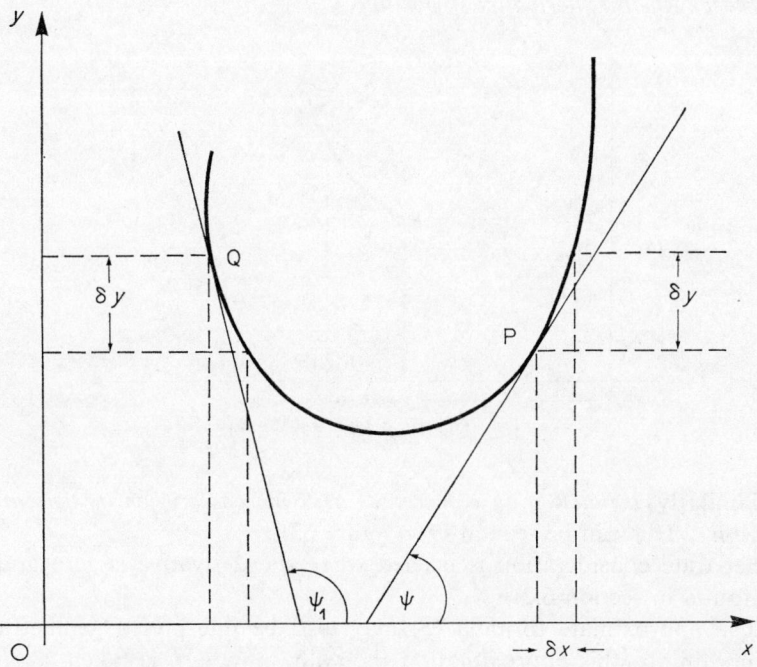

FIGURE 63

It is important to recognize the precise relationship between dy/dx and the slope of the tangent. In figure 63 a small positive δx at P produces a positive δy so that dy/dx is positive; therefore $dy/dx = \tan \psi$, because $\tan \psi$ is positive when ψ is an acute angle. At Q, however, a positive δx produces a negative δy so that dy/dx is negative; this means that $dy/dx = \tan \psi_1$, because $\tan \psi_1$ is negative when ψ_1 is an obtuse angle. We can combine both results by saying $dy/dx = \tan \psi$, where *ψ is the angle the tangent makes with the positive direction of the x-axis*, in the sense shown in figure 63.

Another important fact should be noticed. At P an increase of x gives an increase of y and dy/dx is positive. Conversely, if dy/dx is positive we recall that

$$\frac{\delta y}{\delta x} = \frac{dy}{dx} + \varepsilon$$

where $\varepsilon \to 0$ as $\delta x \to 0$. Therefore, if δx is sufficiently small, ε will be negligible compared with dy/dx and $dy/dx + \varepsilon$ will be positive. Hence, for such δx, $\delta y/\delta x$ will be positive; therefore, if δx is positive, so is δy, i.e. y increases. It has thus been shown that *a function of x increases as x increases when its derivative is positive*.

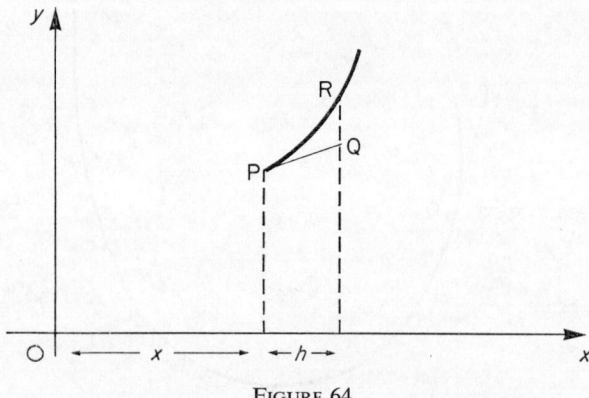

FIGURE 64

Similarly, *a function of x decreases as x increases when its derivative is negative*. This can be seen at Q in figure 63.

Separate consideration is needed when the derivative is zero and will be found in section 6.2.

The approximate formula (5) may also be interpreted geometrically. According to this approximation the value at $x + h$ is taken as $f(x) + hf'(x)$. Since $f'(x)$ is the slope of the tangent at P, this approximation locates

The Derivative

the point at Q. The correct position is R so that the error is given by the length QR. Clearly this error diminishes as $h \to 0$ because both Q and R approach P (See figure 64).

The geometrical significance of the derivative enables one to estimate the rate of change of data given graphically. All that is necessary is to draw the tangent as accurately as the graph will permit and measure its slope; this will give the rate of change at the point where the tangent has been drawn. In this way one can determine, for example, the rate at which a bath is filling with water at any time or the rate at which one's money is increasing or decreasing.

Exercises

19. Find the slope of the curves (i) $y = x^2$, (ii) $4ay = x^2$, (iii) $y = 1/x^3$, (iv) $y = ax^2 + bx + c$.

20. Show that $y = x^2 - 2x$ increases, as x increases, when $x > 1$. What happens for $x < 1$? Check by drawing a graph.

21. Show that $(3x + 1)/(2 - 5x)$ decreases as x increases except in passing through $x = 5/2$. Check by drawing a graph.

22. The distance s that a body has moved in time t is given by $s = a + ut + \frac{1}{2}gt^2$ where a, u and g are constants. Find the speed v at time t where $v = ds/dt$. How is the speed represented on a graph of s against t?

23. The position of a body at time t is given by the table below. Draw a smooth curve through these points and use it to estimate the speed at (i) $t = 1$, (ii) $t = 6$, (iii) $t = 2$.

$t =$	0	1	2	3	4	5	6	7	8	9	10	11
$s =$	22.2	26.5	27.8	26.7	23.8	19.7	15	10.3	6.2	3.3	2.2	3.5

What is the *average* speed over the whole interval?

24. During a certain period the price of oil at time t is $\frac{1}{900}(8100 + 60t - t^2)$ in a certain currency. Find the rate at which the price is changing at $t = 15$. Does the price increase or decrease during the period from $t = 0$ to $t = 15$, and what is the *average* rate of change for this period?

25. The total cost to a manufacturer of making x plastic toys is $404 - 40x + x^2$. At what rate are his costs altering when x is (i) 10, (ii) 30, (iii) 20?

26. Does x^2 increase more rapidly than x as x increases over positive values?

27. Each of $(1/h)[f(x + h) - f(x)]$, $(1/h)[f(x) - f(x - h)]$, $(1/2h)[f(x + h) - f(x - h)]$ represents an approximation to $f'(x)$. Interpret these geometrically and indicate why the last of the three might be expected to be the most accurate. Given the following values

$$x = 1{\cdot}05 \qquad 1{\cdot}10 \qquad 1{\cdot}15$$
$$f(x) = 1{\cdot}0247 \qquad 1{\cdot}0488 \qquad 1{\cdot}0724$$

estimate $f'(1{\cdot}10)$ by each of the formulae. Compare with the exact result given that $f(x) = x^{\frac{1}{2}}$.

Exercises 18 and 27 demonstrate that straightforward approximations can lead to quite inaccurate numerical values for derivatives and that even 4 decimal place accuracy in f does not necessarily give 2 place accuracy in f'.

5.4 The derivative of a polynomial

The object of this section and section 5.6 is to provide a number of rules which will simplify the calculation of a derivative.

THEOREM 5.1 *The derivative of a constant is zero.*

Proof. If y has the constant value c for all values of x in an interval (figure 65), then
$$y = c, \qquad y + \delta y = c$$
so that $\delta y = 0$ and $\delta y / \delta x = 0$.
Hence
$$\frac{dy}{dx} = \lim_{\delta x \to 0} \frac{\delta y}{\delta x} = 0.$$

Geometrically, $y = c$ is a straight line parallel to the x-axis, so that its slope is zero and therefore $dy/dx = 0$ by the preceding section.

THEOREM 5.2 *The derivative of x^n, where n is any positive integer, is nx^{n-1}.*

Proof. If $f(x) = x^n$, then $f(x + h) = (x + h)^n$ and
$$\frac{f(x + h) - f(x)}{h} = \frac{(x + h)^n - x^n}{h}.$$

The Derivative

Now, if we denote x by a and $x + h$ by y, this is $(y^n - a^n)/(y - a)$ and, as $h \to 0$, $y \to a$. But, in Example 7(a) of Chapter 4 it has been shown that

$$\lim_{y \to a} \frac{y^n - a^n}{y - a} = na^{n-1}.$$

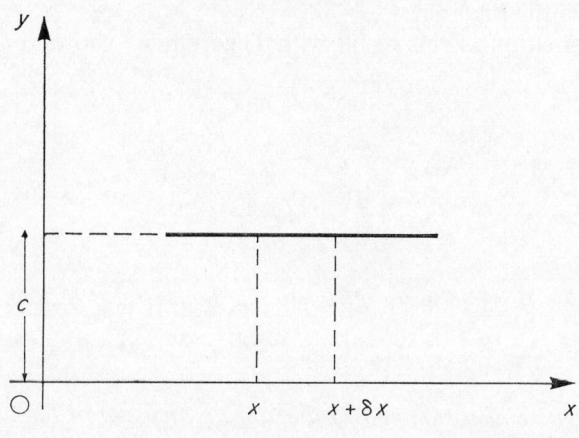

FIGURE 65

Hence, turning a back into x, we obtain

$$\lim_{h \to 0} \frac{1}{h}[f(x + h) - f(x)] = nx^{n-1}$$

i.e.
$$f'(x) = nx^{n-1}$$

the required result.

One conclusion from this theorem is that, if $y = x$, $dy/dx = 1$, which is consistent with the fact that $y = x$ is a straight line and so has constant slope.

THEOREM 5.3 *If c is a constant and y has a derivative, then the derivative of cy is $c.dy/dx$.*

Proof. If $u = cy$, then changing x to $x + \delta x$ gives
$$u + \delta u = c(y + \delta y).$$
Hence
$$\delta u = c\delta y$$
and
$$\frac{\delta u}{\delta x} = c\frac{\delta y}{\delta x}.$$

Since y possesses a derivative $\lim_{\delta x \to 0} \dfrac{\delta y}{\delta x} = \dfrac{dy}{dx}$. Therefore, by Theorem 4.1(i),

$$\frac{du}{dx} = \lim_{\delta x \to 0} \frac{\delta u}{\delta x} = \lim_{\delta x \to 0} c\frac{\delta y}{\delta x} = c \lim_{\delta x \to 0} \frac{\delta y}{\delta x} = c\frac{dy}{dx}$$

which proves the theorem.

The combination of this result with Theorem 5.2 shows that

$$\frac{d}{dx}(cx^n) = cnx^{n-1}.$$

For example, if $y = 4x^3$, then

$$\frac{dy}{dx} = 12x^2.$$

THEOREM 5.4 *If u and v are differentiable functions of x, then*

$$\frac{d}{dx}(u + v) = \frac{du}{dx} + \frac{dv}{dx}.$$

i.e. the derivative of a sum of two functions is the sum of the derivatives.

Proof. Let $y = u + v$. Change x to $x + \delta x$ and let $y + \delta y$, $u + \delta u$, $v + \delta v$ be the new values of y, u, v. Then

$$y + \delta y = u + \delta u + v + \delta v.$$

By subtraction,

$$\delta y = \delta u + \delta v;$$

consequently,

$$\frac{\delta y}{\delta x} = \frac{\delta u}{\delta x} + \frac{\delta v}{\delta x}.$$

Therefore

$$\frac{dy}{dx} = \lim_{\delta x \to 0} \left(\frac{\delta u}{\delta x} + \frac{\delta v}{\delta x} \right)$$

$$= \lim_{\delta x \to 0} \frac{\delta u}{\delta x} + \lim_{\delta x \to 0} \frac{\delta v}{\delta x}$$

by Theorem 4.1(ii). Since

$$\lim_{\delta x \to 0} \frac{\delta u}{\delta x} = \frac{du}{dx} \quad \text{and} \quad \lim_{\delta x \to 0} \frac{\delta v}{\delta x} = \frac{dv}{dx}$$

the theorem is proved.

The theorem can be immediately extended to a sum of three terms. For, suppose

$$y = u_1 + u_2 + u_3$$

where u_1, u_2 and u_3 are differentiable functions of x. Put $u = u_1 + u_2$, $v = u_3$. Then $y = u + v$ and, by Theorem 5.4,

$$\frac{dy}{dx} = \frac{du}{dx} + \frac{dv}{dx}.$$

Also, applying Theorem 5.4 to $u = u_1 + u_2$ gives

$$\frac{du}{dx} = \frac{du_1}{dx} + \frac{du_2}{dx}$$

so that

$$\frac{dy}{dx} = \frac{du_1}{dx} + \frac{du_2}{dx} + \frac{du_3}{dx}.$$

Further extension is possible. Suppose it has been established for some positive integer m that

$$\frac{d}{dx}(u_1 + u_2 + \ldots + u_m) = \frac{du_1}{dx} + \frac{du_2}{dx} + \ldots + \frac{du_m}{dx}. \tag{6}$$

Let $u = u_1 + u_2 + \ldots + u_m$, $v = u_{m+1}$ and $y = u + v$. Then, by Theorem 5.4,

$$\frac{dy}{dx} = \frac{du}{dx} + \frac{dv}{dx}$$

$$= \frac{du_1}{dx} + \frac{du_2}{dx} + \ldots + \frac{du_m}{dx} + \frac{du_{m+1}}{dx}$$

by (6). Therefore, it follows that, if (6) is true for m terms, it is also true for $m + 1$ terms. Since (6) is known to be valid for 2 terms we conclude that it is valid for 3, and then 4, and then 5, ... terms, i.e. (6) holds for any finite number of terms. Hence

COROLLARY 5.4 *If $y = u_1 + \ldots + u_n$, then*

$$\frac{dy}{dx} = \frac{du_1}{dx} + \ldots + \frac{du_n}{dx}$$

for any finite positive integer n, i.e. the derivative of the sum of a finite number of functions is the sum of the derivatives.

Example 7 Find dy/dx when $y = x^3 - 3x + 4$.

By Corollary 5.4,

$$\frac{dy}{dx} = \frac{d}{dx}(x^3) + \frac{d}{dx}(-3x) + \frac{d}{dx}(4).$$

Introductory Analysis

By Theorem 5.2,

$$\frac{d}{dx}(x^3) = 3x^2.$$

By Theorems 5.3 and 5.2,

$$\frac{d}{dx}(-3x) = (-3)\frac{d}{dx}(x) = -3.1 = -3.$$

By Theorem 5.1,

$$\frac{d}{dx}(4) = 0.$$

Hence

$$\frac{dy}{dx} = 3x^2 - 3.$$

Exercises

28. Write down the derivatives of x^4, x^7, x^{18}, x^{56}.

29. Find the derivatives of (i) $x^2 - 3x + 7$, (ii) $4x^2 - 5x - 3$, (iii) $ax^2 + bx + c$, (iv) $4x^3 - 8x^2 + 9x$, (v) $(x - 3)^2$, (vi) $a_1 x^3 + a_2 x^2 + a_3 x + a_4$, (vii) $x^2(2x^2 - 1)$, (viii) $6x^{10} - 7x^3 + 2$, (ix) $(x^2 - x)(5 - x)$, (x) $(x + 2)^3$, (xi) $(2x - 3)(3x + 4)$, (xii) $\frac{1}{3}x^6 - \frac{1}{2}x^4 + x^2$, (xiii) $(x^2 - 3)^2$.

30. Find dy/dx when $y = f(x) + g(x)$.

31. Find the slope of the curve $y = 3x^2 + 2x - 4$ at the point $(1,1)$.

32. Where is the slope of $y = x^3 - 3x$ parallel to the x-axis?

33. The line $y = 2x$ is a tangent to $y = \frac{1}{2}x^2 + c$. Find c and the point of contact.

34. A ball thrown vertically upwards travels a distance $64t - 16t^2$ ft in t sec. When is its speed zero? With what speed was it projected? What is its speed at a height of 48 ft when it is coming down?

35. The price of certain shares at time x is $2x^3 - 9x^2 + 12x - 8$. In what time intervals is the price (a) increasing, (b) decreasing?

The Derivative

5.5 Newton's method for roots

The derivative of a polynomial that has been found in the preceding section will be useful in many connections. One application is to provide a valuable numerical method of solving equations. Consider the problem of finding the positive value of x which satisfies $x^2 - 6 = 0$. Clearly, an answer is $x_0 = \sqrt{6}$ but we wish to have the answer in decimal form, by a method which will work for more difficult problems.

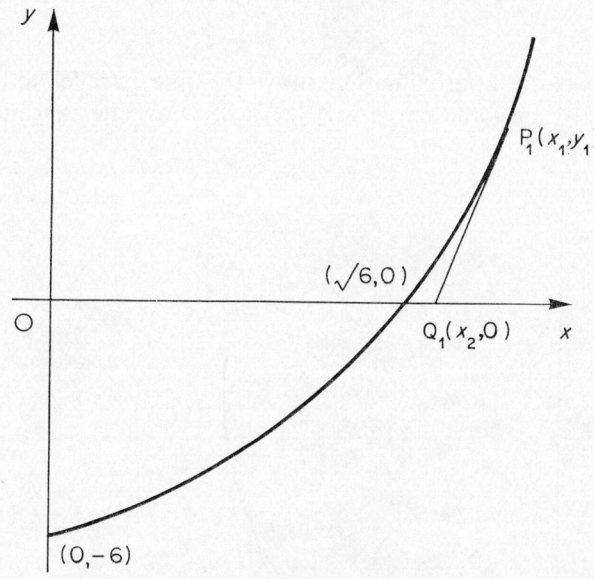

FIGURE 66

Examine the graph of $y = x^2 - 6$ (figure 66). The point $(\sqrt{6}, 0)$ lies on the curve and our object is to devise a method which will locate it conveniently without drawing the graph. Choose some convenient point $P_1(x_1, y_1)$ on the curve to the right of $(\sqrt{6}, 0)$. Draw the tangent to the curve at P_1 and let it cut the x-axis at Q_1. Then the abscissa of Y_1 is closer to $\sqrt{6}$ than that of P_1.

For example, we might choose P_1 as $(3, 3)$. The slope of $y = x^2 - 6$ is $2x$ which is 6 at P_1. Therefore the equation of $Q_1 P_1$ is

$$y - 3 = 6(x - 3).$$

Then $Q_1(x_2, 0)$ is given by $x_2 = 2 \cdot 5$. Accordingly, $2 \cdot 5$ is a better approximation to $\sqrt{6}$ than 3.

In general, the equation of Q_1P_1 is
$$y - y_1 = 2x_1(x - x_1)$$
so that
$$x_2 = x_1 - \frac{y_1}{2x_1}.$$
Since $y_1 = x_1^2 - 6$,
$$x_2 = x_1 - \tfrac{1}{2}x_1 + \frac{3}{x_1}$$
$$= \tfrac{1}{2}\left(x_1 + \frac{6}{x_1}\right). \tag{7}$$

To improve the approximation draw the line parallel to the y-axis through Q_1 to meet the curve in $P_2(x_2,y_2)$. Draw the tangent at P_2 to

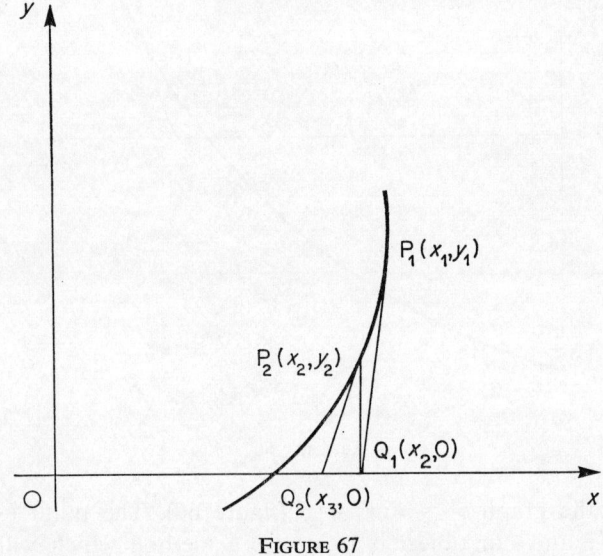

FIGURE 67

meet the x-axis in $Q_2(x_3,0)$. Then x_3 is a still better approximation to $\sqrt{6}$. Since Q_2 is obtained from P_2 in the same way that Q_1 was obtained from P_1 we can use (7) with x_3 in place of x_2, and x_2 in place of x_1. Thus
$$x_3 = \tfrac{1}{2}\left(x_2 + \frac{6}{x_2}\right). \tag{8}$$
If we use the value of 2·5 which we had earlier for x_2 we find
$$x_3 = \tfrac{1}{2}\left(2\cdot 5 + \frac{6}{2\cdot 5}\right) = \tfrac{1}{2}(2\cdot 5 + 2\cdot 4) = 2\cdot 45.$$

The Derivative

The argument can now be repeated. The substitution of x_3 for x_2 in (8) leads to a better approximation x_4 and we can continue the process as often as we like until a sufficiently accurate result is obtained. Thus, with $x_3 = 2{\cdot}45$, x_4 is $2{\cdot}449$. Since both x_3 and x_4 agree as far as $2{\cdot}45$ we conclude that $\sqrt{6} = 2{\cdot}45$ correct to 2 decimal places. To find an approximation correct to more decimal places, say 5, it would be necessary to carry on until an approximation and the one after it agreed to 5 decimal places.

The process described above, which is due to Newton, is often called an *iterative method* and is of extreme importance in modern computing. Since the same formula (7), or (8), connects successive approximations, a digital computer can be programmed to carry out the calculations easily and efficiently; furthermore, instructions can be included so that the calculation ceases when enough decimal figure accuracy has been achieved.

Although the method has been expounded for finding $\sqrt{6}$ in particular, it is clearly applicable more widely because it depends only on the geometrical properties in figures 66 and 67. Therefore, so long as the shape of the curve between P_1 and the point where $y = 0$ is of the same general form as in figure 66 we can expect Newton's method to work. Some account of more detailed conditions under which Newton's method will be successful will be found in section 8.3.

Exercises

36. Use Newton's method to find $\sqrt{7}$ correct to 2 decimal places, starting from $x_1 = 3$.

37. What is the formula corresponding to (6) when finding $\sqrt{a}$ by Newton's method. Draw a flow chart.

38. What is the formula connecting two successive approximations to $a^{\frac{1}{3}}$ in Newton's method. Find the cube root of 3 correct to two decimal places by this method, starting from $x_1 = 2$.

39. The function $x^3 - 2x^2 - 5x + 10$ vanishes for some value between $x = 2{\cdot}1$ and $x = 3$. Use Newton's method to determine this value correct to 2 decimal places, starting from $x_1 = 3$.

40. Check the result of Example 14 of Chapter 4 by Newton's method starting from $x_1 = 0$.

41. Obtain the formula corresponding to (7) for finding the x where $f(x) = 0$. (See also section 8.3.)

42. Draw a flow chart for question 41.

43. Explain what happens geometrically when Newton's method, starting from $x_1 = 2$, is used to solve $x^3 - 6x^2 + 13x - 9 = 0$.

5.6 Derivatives of products and quotients

We now seek rules which will enable us to find the derivatives of more complicated functions than polynomials. The rational functions form one group of interest, but in order to cope with them it proves to be convenient to develop theorems of great power and generality.

Before dealing with them we note an important extension of Theorem 5.2, namely

THEOREM 5.5 *If $v = p/q$, where p and q are integers with $q \geqslant 1$, the derivative of x^v is vx^{v-1}.*

The proof of this theorem is completely analogous to the proof of Theorem 5.2, with the exception that Examples 7(b),(c) of Chapter 4 are used instead of Example 7(a), and so the details will be left to the reader.

The function x^v may not be defined for all values of x; consider for example, $x^{-\frac{1}{2}}$. For all values of v, x^v is well-defined for $x > 0$. If q is even (it is assumed that p and q have no factors in common) x^v is not defined for $x < 0$; but x^v is well-defined for $x < 0$ if q is odd. At $x = 0$ the function is defined unless v is negative. Obviously, the derivative of x^v will not exist at points where x^v is not defined. However, it does exist at all points where x^v is defined with the single exception of $x = 0$ when $v - 1$ is negative.

Thus the derivative of $x^{\frac{1}{3}}$ is $\frac{1}{3}x^{-\frac{2}{3}}$ or $1/3x^{\frac{2}{3}}$ and exists for all x except $x = 0$. Since $x^{\frac{1}{3}}$ is continuous at $x = 0$ this is an example which shows that the continuity of a function at a point is not sufficient to guarantee that it has a derivative at that point.

Indeed it is possible to construct continuous functions which do not possess a derivative at any point. However, we shall be concerned in the following pages only with continuous functions whose derivatives fail to exist at one or two isolated points. The converse statement that a function which has a derivative at a point must be continuous there is proved in Exercise 23 of section 4.5.

Turning now to general theorems we have

THEOREM 5.6 *If $y = uv$, where u and v both have derivatives with respect to x, then*

$$\frac{dy}{dx} = \frac{du}{dx}v + u\frac{dv}{dx}.$$

Observe that the derivative of the product uv is *not* the product of the derivatives. Instead it consists of the sum of two terms. In the first we take

The Derivative

the derivative of u and leave v alone, whereas in the second term we leave u alone and take the derivative of v.

Proof. Change x to $x + \delta x$ and let $y + \delta y$, $u + \delta u$, $v + \delta v$ be the new values of y, u, v, respectively. Then
$$y + \delta y = (u + \delta u)(v + \delta v)$$
$$= uv + u \cdot \delta v + v \cdot \delta u + \delta u \cdot \delta v.$$
Subtract $y = uv$ and then
$$\delta y = v \cdot \delta u + u \cdot \delta v + \delta u \cdot \delta v.$$
Divide by δx to obtain
$$\frac{\delta y}{\delta x} = v \frac{\delta u}{\delta x} + u \frac{\delta v}{\delta x} + \delta u \frac{\delta v}{\delta x}.$$
Hence
$$\frac{\mathrm{d}y}{\mathrm{d}x} = \lim_{\delta x \to 0} \left(v \frac{\delta u}{\delta x} + u \frac{\delta v}{\delta x} + \delta u \frac{\delta v}{\delta x} \right)$$
$$= \lim_{\delta x \to 0} v \frac{\delta u}{\delta x} + \lim_{\delta x \to 0} u \frac{\delta v}{\delta x} + \lim_{\delta x \to 0} \delta u \frac{\delta v}{\delta x}$$
by Theorem 4.1(ii), since
$$\lim_{\delta x \to 0} v \frac{\delta u}{\delta x} = v \lim_{\delta x \to 0} \frac{\delta u}{\delta x} = v \frac{\mathrm{d}u}{\mathrm{d}x},$$
$$\lim_{\delta x \to 0} u \frac{\delta v}{\delta x} = u \lim_{\delta x \to 0} \frac{\delta v}{\delta x} = u \frac{\mathrm{d}v}{\mathrm{d}x}$$
by Theorem 4.1(i) and
$$\lim_{\delta x \to 0} \delta u \frac{\delta v}{\delta x} = \lim_{\delta x \to 0} \delta u \cdot \lim_{\delta x \to 0} \frac{\delta v}{\delta x} = 0 \cdot \frac{\mathrm{d}v}{\mathrm{d}x} = 0$$
by Theorem 4.1(iii). Consequently
$$\frac{\mathrm{d}y}{\mathrm{d}x} = v \frac{\mathrm{d}u}{\mathrm{d}x} + u \frac{\mathrm{d}v}{\mathrm{d}x},$$
the result given in the theorem.

The same technique that was used to derive Corollary 5.4 from Theorem 5.4 may now be employed to establish

COROLLARY 5.6 *If* $y = u_1 u_2 \ldots u_n$ *then*
$$\frac{\mathrm{d}y}{\mathrm{d}x} = \frac{\mathrm{d}u_1}{\mathrm{d}x} u_2 \ldots u_n + u_1 \frac{\mathrm{d}u_2}{\mathrm{d}x} u_3 \ldots u_n + \ldots + u_1 \ldots u_{n-1} \frac{\mathrm{d}u_n}{\mathrm{d}x},$$
i.e. the sum of terms in each of which one factor is differentiated while the other factors are left alone.

Example 8 If $y = x^4 + 1/x^3$ it may be written as
$$y = x^4 + x^{-3}.$$

Then Theorem 5.5 gives
$$\frac{dy}{dx} = 4x^3 - 3x^{-4} = 4x^3 - \frac{3}{x^4}.$$

Example 9 When $y = (x^2 - 1)(x^3 + 2)$ Theorem 5.6 shows that
$$\frac{dy}{dx} = \left[\frac{d}{dx}(x^2 - 1)\right](x^3 + 2) + (x^2 - 1)\frac{d}{dx}(x^3 + 2)$$
$$= 2x(x^3 + 2) + 3x^2(x^2 - 1) = 5x^4 - 3x^2 + 4x.$$

Example 10 If $y = (4x^{\frac{1}{2}} + 1)^2$ we can consider it as
$$y = (4x^{\frac{1}{2}} + 1)(4x^{\frac{1}{2}} + 1)$$
and then, from Theorem 5.6,
$$\frac{dy}{dx} = \left[\frac{d}{dx}(4x^{\frac{1}{2}} + 1)\right](4x^{\frac{1}{2}} + 1) + (4x^{\frac{1}{2}} + 1)\frac{d}{dx}(4x^{\frac{1}{2}} + 1).$$

Now, from Theorems 5.4, 5.3, 5.1 and 5.5,
$$\frac{d}{dx}(4x^{\frac{1}{2}} + 1) = 2x^{-\frac{1}{2}}.$$

Hence
$$\frac{dy}{dx} = 4x^{-\frac{1}{2}}(4x^{\frac{1}{2}} + 1) = 4(4 + x^{-\frac{1}{2}}).$$

Next, the derivative of the quotient of two functions will be obtained. The result to be proved is

THEOREM 5.7 *If $y = u/v$, where u and v both have derivatives with respect to x, then*
$$\frac{dy}{dx} = \frac{v\dfrac{du}{dx} - u\dfrac{dv}{dx}}{v^2}$$
at any point where $v \neq 0$.

The Derivative

Proof. Change x to $x + \delta x$ and let $y + \delta y$, $u + \delta u$, $v + \delta v$ be the new values of y, u, v, respectively. Then

$$y + \delta y = \frac{u + \delta u}{v + \delta v}$$

and so

$$\delta y = \frac{u + \delta u}{v + \delta v} - \frac{u}{v}$$

$$= \frac{v(u + \delta u) - u(v + \delta v)}{v(v + \delta v)}$$

$$= \frac{v \delta u - u \delta v}{v(v + \delta v)}.$$

Hence

$$\frac{\delta y}{\delta x} = \frac{v \dfrac{\delta u}{\delta x} - u \dfrac{\delta v}{\delta x}}{v(v + \delta v)}.$$

Now

$$\lim_{\delta x \to 0} \frac{\delta u}{\delta x} = \frac{du}{dx} \quad \text{and} \quad \lim_{\delta x \to 0} \frac{\delta v}{\delta x} = \frac{dv}{dx}$$

so that, by Theorem 4.1(i) and (ii),

$$\lim_{\delta x \to 0} \left(v \frac{\delta u}{\delta x} - u \frac{\delta v}{\delta x} \right) = v \frac{du}{dx} - u \frac{dv}{dx}.$$

Also

$$\lim_{\delta x \to 0} v(v + \delta v) = v \lim_{\delta x \to 0} (v + \delta v) = v^2.$$

Since $v^2 \neq 0$, Theorem 4.1(iv) gives

$$\lim_{\delta x \to 0} \frac{\delta y}{\delta x} = \frac{\displaystyle\lim_{\delta x \to 0} \left(v \frac{\delta u}{\delta x} - u \frac{\delta v}{\delta x} \right)}{\displaystyle\lim_{\delta x \to 0} v(v + \delta v)} = \frac{v \dfrac{du}{dx} - u \dfrac{dv}{dx}}{v^2}$$

and the theorem is proved.

Example 11 If $y = x/(x^2 + 1)$,

$$\frac{dy}{dx} = \frac{(x^2 + 1) \dfrac{d}{dx} x - x \dfrac{d}{dx}(x^2 + 1)}{(x^2 + 1)^2} = \frac{x^2 + 1 - x \cdot 2x}{(x^2 + 1)^2} = \frac{1 - x^2}{(x^2 + 1)^2}.$$

Example 12 If $y = 1/(x^{\frac{1}{3}} + 1)$ $(x \neq -1)$,

$$\frac{dy}{dx} = \frac{(x^{\frac{1}{3}} + 1)\frac{d}{dx}1 - 1 \cdot \frac{d}{dx}(x^{\frac{1}{3}} + 1)}{(x^{\frac{1}{3}} + 1)^2} = \frac{(x^{\frac{1}{3}} + 1) \cdot 0 - \frac{1}{3}x^{-\frac{2}{3}}}{(x^{\frac{1}{3}} + 1)^2}$$

$$= \frac{-1}{3x^{\frac{2}{3}}(x^{\frac{1}{3}} + 1)^2}.$$

Exercises

44. Write down the derivatives of $x^{1/5}$, $x^{7/4}$, $x^{4/7}$, $x^{1/q}$, $1/x^3$, $1/x^9$, x^{-p}, $1/x^{3/4}$, $1/x^{1/q}$, $x^{-5/8}$, $x^{-1/q}$.

45. If h is small show that an approximation to $(x + h)^\nu$ is $x^\nu + h\nu x^{\nu-1}$. Use this result to determine approximately $\sqrt{101}$, $1/\sqrt{101}$, $1/\sqrt{99}$, $(1{\cdot}001)^8$, $(1020)^{1/5}$, $(217)^{1/3}$.

46. Find the derivative with respect to x of $(x^3 - 1)(x^2 + 2)$, $x^4(1 + 2\sqrt{x})$, $(ax^2 + bx + c)(dx + e)$, $\sqrt{x}[(2/x) + (3/x^3)]$, $x(x^2 - 1)(x^3 + 1)$, $(5x + 1)^3$, $(2x^2 + x + 1)^2$.

47. Knowing the derivative of x, find the derivative of x^n when n is a positive integer by using Corollary 5.6. Use the same method to find the derivative of $[f(x)]^n$.

48. If $y = f(x)g(x)$ find dy/dx.

49. Find the derivative with respect to x of

$$\frac{2x + 3}{3x + 2}, \quad \frac{x^2 + 1}{x^2 + 8}, \quad \frac{3x^2 + 5x + 4}{3x^2 + 5x + 3}, \quad \frac{1 + \sqrt{x}}{1 - \sqrt{x}}, \quad \frac{(x + 2)^2}{x^2 - 2},$$

$$\frac{(x + 1)(x + 3)}{x(x + 2)}.$$

50. Use the method of question 47 to find the derivative of $[f(x)]^{-n}$.

51. If $y = f(x)/g(x)$ find dy/dx.

5.7 The derivative of an implicit function

So far we have been considering relations of the form $y = f(x)$, which gives y explicitly in terms of x. However, it is not uncommon for the relation between x and y to contain terms which involve both x and y. A simple example is
$$xy = 1,$$
the equation of a hyperbola. This example is simple because it is easily rearranged as $y = 1/x$ which is of the form above.

A less simple example is the equation of the circle
$$x^2 + y^2 = 16. \tag{9}$$
If we solve for y we find $y = \pm \sqrt{(16 - x^2)}$. The positive sign corresponds to the points on the semicircle above the x-axis; the negative sign supplies the semicircle below the x-axis.

A much more complicated example is the relation
$$y^6 + 3xy^5 + x^5 = 3. \tag{10}$$
Here, even if we knew how to solve an equation of the sixth degree, the resulting formula for y would be very unwieldy.

Relations of the type considered in this section do determine y, in the sense that if a value of x is substituted, the consequent equation for y determines values of y which correspond to the value of x. Therefore the relations are said to specify y as an *implicit function* of x.

It is now desirable to find the derivative of y with respect to x, when y is given as an implicit function, without the necessity of expressing y explicitly in terms of x. Let us consider (10). Take a derivative with respect to x of both sides; then, by Corollary 5.4,
$$\frac{d}{dx}y^6 + \frac{d}{dx}(3xy^5) + \frac{d}{dx}x^5 = \frac{d}{dx}(3). \tag{11}$$
Now y^6 is the product of six factors, each equal to y, and so, by the rule for the derivative of a product (Corollary 5.6),
$$\frac{d}{dx}y^6 = 6y^5\frac{dy}{dx}.$$
Similarly,
$$\frac{d}{dx}(3xy^5) = 3y^5\frac{d}{dx}(x) + 3x\frac{d}{dx}(y^5) = 3y^5 + 15xy^4\frac{dy}{dx}.$$
Since
$$\frac{d}{dx}x^5 = 5x^4 \quad \text{and} \quad \frac{d}{dx}(3) = 0,$$

(11) becomes
$$6y^5\frac{dy}{dx} + 3y^5 + 15xy^4\frac{dy}{dx} + 5x^4 = 0.$$
Thus, at points where $6y^5 + 15xy^4 \neq 0$,
$$\frac{dy}{dx} = -\frac{3y^5 + 5x^4}{3y^4(y + 5x)}. \tag{12}$$

The expression obtained for dy/dx by this method involves both x and y. If the derivative is required at some known point (x_1, y_1) this causes no difficulty since one merely substitutes x_1 for x and y_1 for y in (12). If we wanted to put dy/dx entirely in terms of x it would be necessary to solve (10) for y as an explicit function of x. This is scarcely possible but, even if it were feasible (as in the case of (9)), one would have to be very careful to choose the correct expression for y from the several which might satisfy the implicit equation.

Example 13 Suppose $y = u^{p/q}$ where p and q are integers with q positive, and u possesses a derivative with respect to x. Then
$$y^q = u^p$$
and, taking a derivative with respect to x as above, we obtain
$$qy^{q-1}\frac{dy}{dx} = pu^{p-1}\frac{du}{dx}.$$

The process is legitimate since p and q are integers. Hence, if $y \neq 0$,
$$\frac{dy}{dx} = \frac{pu^{p-1}}{qy^{q-1}}\frac{du}{dx} = \frac{pu^{p-1}}{q(u^{p/q})^{q-1}}\frac{du}{dx}$$
$$= \frac{pu^{p-1}}{qu^{p-p/q}}\frac{du}{dx} = \frac{p}{q}u^{(p/q)-1}\frac{du}{dx}.$$

It has consequently been shown that
$$\frac{d}{dx}(u^{p/q}) = \frac{p}{q}u^{(p/q)-1}\frac{du}{dx}. \tag{13}$$

The proof was subject to the restriction $y \neq 0$, which is the same as $u \neq 0$. The behaviour at $u = 0$ must be dealt with separately, since y may not exist for both negative and positive u; the formula (12) may or may not hold at $u = 0$.

Example 14 Consider the general equation of a conic discussed in Chapter 2, namely
$$Ax^2 + Bxy + Cy^2 + Dx + Ey + F = 0. \tag{14}$$

The Derivative

This can be regarded as an implicit equation for y and then dy/dx gives the slope of the tangent at a point of the conic. In fact, taking a derivative with respect to x, we have

$$2Ax + By + Bx\frac{dy}{dx} + 2Cy\frac{dy}{dx} + D + E\frac{dy}{dx} = 0$$

so that

$$\frac{dy}{dx} = -\frac{2Ax + By + D}{Bx + 2Cy + E}.$$

Thus the slope of the tangent at the point (x_0, y_0) of the conic is

$$-(2Ax_0 + By_0 + D)/(Bx_0 + 2Cy_0 + E)$$

and the equation of the tangent is

$$\frac{y - y_0}{x - x_0} = -\frac{2Ax_0 + By_0 + D}{Bx_0 + 2Cy_0 + E}$$

or

$$2Axx_0 + Bxy_0 + Dx + Bx_0y + 2Cy_0y + Ey - 2Ax_0^2 - 2Bx_0y_0 - Dx_0 - 2Cy_0^2 - Ey_0 = 0$$

or

$$2[Axx_0 + \tfrac{1}{2}B(xy_0 + x_0y) + Cyy_0 + \tfrac{1}{2}D(x + x_0) + \tfrac{1}{2}E(y + y_0) + F] = 0$$

since (x_0, y_0) lies on the conic and satisfies (14). The reader should check that all the tangents found in Chapter 2 are in agreement with this formula.

Exercises

52. Find dy/dx when x and y are connected by (i) $x^2 + y^2 = a^2$, (ii) $x^3 + y^3 = a^3$, (iii) $y^2 = x^2 - 1$, (iv) $y^2 = (x + 1)/(x - 1)$, (v) $xy = 1$, (vi) $x^{\frac{1}{2}} + y^{\frac{1}{2}} = a^{\frac{1}{2}}$, (vii) $x^2 - 3xy + y^2 = 0$, (viii) $y^2 = x^2 + 1/x^2$, (ix) $x^{\frac{2}{3}} + y^{\frac{2}{3}} = a^{\frac{2}{3}}$, (x) $(x^2 + y^2)^2 = 3x^2y^2$, (xi) $x^5y^2 + x^2y^5 = 1$, (xii) $y^5 = (x + 1)^3$, (xiii) $1/y^2 + 1/x^2 = 1/a^2$, (xiv) $x^2y^2 = (x^2 + 1)^{\frac{2}{3}}$.

53. Find the slopes of the following curves at the points stated:
    (i) $y^4 + 3xy^3 = 4$ at $(1,1)$, (ii) $y + 3 = x^3$ at $(2,5)$,
    (iii) $x^3 + y^3 = 7$ at $(-1,2)$, (iv) $x^2 + y^2 + 2x + 4y + 4 = 0$ at $(0,-2)$.

54. Find the equations of the tangents and normals to the following curves at the points stated: (i) $x^2 + y^2 = 5$ at $(1,2)$, (ii) $x^2 - xy + 3y^2 = 5$ at $(-1,1)$, (iii) $x^2y^2 = 16$ at $(2,-2)$, (iv) $(y + x)^2 = 3y - 2$ at $(0,1)$.

55. The current I and charge Q on a certain capacitor at time t are related by $I = dQ/dt$ and

$$I^2 - I_0^2 = a(Q^2 - Q_0^2)$$

where a, I_0 and Q_0 are constants. Deduce that $dI/dt = aQ$.

56. If two curves pass through the point (x_0,y_0), the angle between their respective tangents at (x_0,y_0) is called the *angle of intersection* of the curves at (x_0,y_0). Find the angle of intersection of (i) $y = x^3$ and $6y = 7 - x^2$ at $(1,1)$, (ii) $xy = 6$ and $x^2y = 12$, (iii) $x^2 + y^2 - 2x = 0$ and $x^2 + y^2 - 8x + 12 = 0$, (iv) the ellipse $4x^2 + y^2 = 4$ and the hyperbola $y^2 - 2x^2 = 2$.

57. The cable of a suspension bridge hangs between supports 250 m apart. The cable forms a parabola and the lowest point is 50 m below the points of suspension. Find the angle between the cable and support at the point of suspension.

58. The tangent at the point P of a hyperbola intersects the asymptotes at T_1 and T_2. Show that $PT_1 = PT_2$.

59. PN is the normal at P, a point of an ellipse with foci S_1, S_2. Prove that $\angle NPS_1 = \angle NPS_2$.

5.8 The chain rule

Suppose we are given that
$$y = 8t^3, \qquad t = x^2 + 3x + 4$$
and we want to find dy/dx. We could, of course, write
$$y = 8(x^2 + 3x + 4)^3 \qquad (15)$$
and evaluate the derivative by rules that have already been given. However, we can calculate dy/dt and dt/dx easily and the question is whether we can determine dy/dx from these two without going through the stage (15).

Let us pose the question more generally. Suppose
$$y = f(t), \qquad t = g(x); \qquad (16)$$
can we find dy/dx without substituting for t? Sometimes the combination (16) is described by saying that y is a *function of a function* or that y is a *composite function* of x.

Make a small change in x so that it becomes $x + \delta x$. Then t will alter to $t + \delta t$ where
$$t + \delta t = g(x + \delta x).$$
As a consequence of the alteration in t, y will become $y + \delta y$ where
$$y + \delta y = f(t + \delta t).$$
Then
$$\delta y = f(t + \delta t) - f(t).$$
But, according to (3),
$$f(t + \delta t) - f(t) = [f'(t) + \varepsilon]\delta t$$

The Derivative

where $\varepsilon \to 0$ as $\delta t \to 0$. Also
$$\delta t = g(x + \delta x) - g(x)$$
so that
$$\delta y = [f'(t) + \varepsilon][g(x + \delta x) - g(x)].$$
Hence
$$\frac{\delta y}{\delta x} = [f'(t) + \varepsilon]\left[\frac{g(x + \delta x) - g(x)}{\delta x}\right].$$
Now
$$\lim_{\delta x \to 0} \frac{g(x + \delta x) - g(x)}{\delta x} = g'(x)$$
and
$$\lim_{\delta x \to 0} [f'(t) + \varepsilon] = f'(t)$$
since $\delta t \to 0$ when $\delta x \to 0$. Therefore, by Theorem 4.1(iii),
$$\frac{dy}{dx} = \lim_{\delta x \to 0} \frac{\delta y}{\delta x} = f'(t)g'(x). \tag{17}$$
The result (17) can also be written as
$$\frac{dy}{dx} = \frac{dy}{dt} \cdot \frac{dt}{dx}. \tag{18}$$

THEOREM 5.8 *If $y = f(t)$ is a differentiable function of t and $t = g(x)$ is a differentiable function of x*
$$\frac{dy}{dx} = f'(t)g'(x) = \frac{dy}{dt} \cdot \frac{dt}{dx}.$$

Example 15 If $y = 8t^3$ and $t = x^2 + 3x + 4$,
$$\frac{dy}{dt} = 24t^2, \qquad \frac{dt}{dx} = 2x + 3.$$

Therefore
$$\frac{dy}{dx} = \frac{dy}{dt} \cdot \frac{dt}{dx} = 24t^2(2x + 3) = 24(2x + 3)(x^2 + 3x + 4)^2.$$

Example 16 If $y = 1/t^{\frac{3}{2}}$ and $t = 1 - x^2$,
$$\frac{dy}{dt} = -\frac{3}{2t^{\frac{5}{2}}}, \qquad \frac{dt}{dx} = -2x.$$

148   *Introductory Analysis*

and
$$\frac{dy}{dx} = \frac{dy}{dt} \cdot \frac{dt}{dx} = \left(-\frac{3}{2t^{\frac{5}{2}}}\right)(-2x) = \frac{3x}{(1-x^2)^{\frac{5}{2}}}$$

or
$$\frac{d}{dx}\frac{1}{(1-x^2)^{\frac{3}{2}}} = \frac{3x}{(1-x^2)^{\frac{5}{2}}}.$$

This result, of course, holds only for $x^2 < 1$.

Example 17 *If, in Theorem 5.8, $dt/dx \neq 0$ we can divide by it and obtain*

$$\frac{dy}{dt} = \frac{dy/dx}{dt/dx}. \tag{19}$$

There are other ways in which this can be expressed. If $y = F(x)$, $x = G(t)$ an interchange of x and t in (19) gives

$$\frac{dy}{dx} = \frac{dy/dt}{dx/dt} \tag{20}$$

and this is useful when $x = G(t)$ is substituted in $F(x)$ so that $y = F[G(t)] = H(t)$ (say). Then

$$\frac{dy}{dx} = \frac{H'(t)}{G'(t)}. \tag{21}$$

Again, if in (20) we take $t = y$, we deduce the important result

$$\frac{dy}{dx} = \frac{1}{dx/dy}. \tag{22}$$

Thus, if $y = 3t^2 + t - 1$ and $x = 3t + 1$,
$$\frac{dy}{dx} = \frac{dy/dt}{dx/dt} = \frac{6t+1}{3} = 2\left(\frac{x-1}{3}\right) + \tfrac{1}{3} = \tfrac{1}{3}(2x-1)$$
and
$$\frac{dx}{dy} = \frac{1}{dy/dx} = \frac{3}{2x-1}.$$

Exercises

60. Calculate the derivatives of (i) $(x^2 + 1)^5$, (ii) $(x^2 - 1)^{\frac{1}{2}}$, (iii) $(3x^2 + 2x + 1)^{-\frac{1}{2}}$, (iv) $1/(4 - x^3)^{\frac{5}{3}}$, (v) $[x/(x^2 + 1)]^{\frac{1}{2}}$, (vi) $x(3x + 1)^{\frac{1}{3}}$, (vii) $(3x^2 + 1)^{\frac{1}{2}}/x$, (viii) $(t^2 - 1)/(t^2 + 1)$ where $t = (x^2 + 2)^{\frac{1}{3}}$.

61. Find dy/dx when (i) $y = t/(1-t)$, $x = t^2$; (ii) $y = t^3 + 4$, $x = t^2 + 2t$; (iii) $y = (1 + t)^{\frac{1}{2}}$, $x = t^{\frac{1}{2}}$.

The Derivative

62. If we attempted to prove (18) directly from

$$\frac{\delta y}{\delta x} = \frac{\delta y}{\delta t} \cdot \frac{\delta t}{\delta x}$$

would the possibility that $\delta t = 0$ cause difficulty?

63. If $y = f(t)$, $t = g(u)$, $u = h(x)$ prove that

$$\frac{dy}{dx} = f'(t)g'(u)h'(x) = \frac{dy}{dt} \cdot \frac{dt}{du} \cdot \frac{du}{dx}.$$

64. If $y = 3x^3 - 2x + 1$ find dx/dy at $x = 0$ and $x = 1$.

65. The position of a particle at time t is $y = t^3 + 3t$, $x = t^4 + 2t^2$. Find the slope of the tangent to the path at time t.

Example 18 *Oil is leaving a conical funnel at a rate of 2 cm³/sec (figure 68). At what rate is the oil level dropping when it is 3 cm deep?*

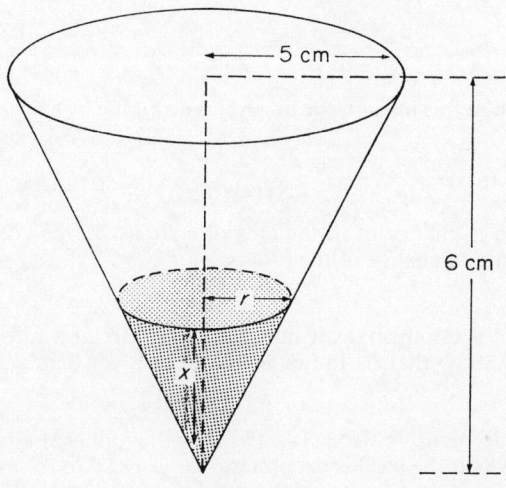

FIGURE 68

Let x, r, V be the height of oil, the radius of the funnel at the height x and the volume of oil at time t. Then the statement that oil is leaving at 2 cm³/sec can be expressed as

$$\frac{dV}{dt} = -2,$$

the minus sign being present because the volume is decreasing as time increases. The relation between the volume and height of a cone is

$$V = \tfrac{1}{3}\pi r^2 x.$$

The variables r and x are connected because, by similar triangles,

$$\frac{r}{x} = \frac{5}{6}.$$

Hence,

$$V = \frac{25}{108}\pi x^3.$$

Therefore

$$\frac{dV}{dt} = \frac{25}{36}\pi x^2 \frac{dx}{dt}$$

and so

$$\frac{dx}{dt} = \frac{36}{25\pi x^2}\frac{dV}{dt}.$$

Hence, when $dV/dt = -2$ and $x = 3$,

$$\frac{dx}{dt} = -\frac{8}{25\pi}$$

which shows that, at this instant, the oil level is dropping by a little over $\tfrac{1}{10}$ cm/sec.

Exercises

66. Find the rate of change of (i) the area of a circle, (ii) the volume of a sphere in terms of the rate of change of the radius.

67. Suppose that a raindrop, as it falls, gathers water at a rate proportional to its surface area. Show that its radius increases at a constant rate.

68. If, in Example 18, oil is also being added to the funnel at a rate of 1 cm³/sec, find the rate at which the level is dropping when it is 2 cm.

69. A man 5 ft tall walks at the rate of 5 ft/sec towards a street light which is 15 ft above the ground. At what rate is the tip of his shadow moving? At what rate is the length of his shadow changing?

70. A particle moves round the circle $x^2 + y^2 = a^2$ in such a way that $dx/dt = -y$. Does it travel clockwise or anti-clockwise when viewed from above?

The Derivative

71. Aircraft A is 20 km west of a beacon and flying east at a constant height with speed 360 k.p.h. Aircraft B is 10 km south of the beacon and flying north at the same height with speed 480 k.p.h. How long elapses before the distance between them is neither increasing nor decreasing?

72. A metal ball of radius 4 cm, surrounded by a protective coating, is placed in a solution which removes the coating at 10 cm^3/hr. How fast is the thickness of the coating decreasing when it is 2 cm thick?

73. A light is at the top of a pole 40 ft high. A ball is dropped from the same height from a point 40 ft from the light. Assuming that the ball falls according to the law $s = 16t^2$, how fast is the shadow of the ball moving along the ground $\frac{1}{2}$ sec later?

5.9 Trigonometric functions

To find dy/dx when $y = \sin x$ and x is measured in radians, let x change to $x + \delta x$ so that y alters to $y + \delta y$. Then

$$y + \delta y = \sin(x + \delta x)$$

and

$$\frac{\delta y}{\delta x} = \frac{\sin(x + \delta x) - \sin x}{\delta x}.$$

Now, from (20) of Chapter 1,

$$\sin(x + \delta x) = \sin x \cos \delta x + \cos x \sin \delta x$$

so that

$$\frac{\delta y}{\delta x} = \frac{\cos \delta x - 1}{\delta x} \sin x + \frac{\sin \delta x}{\delta x} \cos x.$$

In Example 8 of Chapter 4 it has been shown that

$$\lim_{h \to 0} \frac{1 - \cos h}{h} = 0, \quad \lim_{h \to 0} \frac{\sin h}{h} = 1$$

when h is measured in radians. Therefore, putting $h = \delta x$ and using Theorem 4.1, we obtain

$$\frac{dy}{dx} = \lim_{\delta x \to 0} \frac{\delta y}{\delta x} = \cos x.$$

From this result many others can be deduced by means of the chain rule. For example, if

$$y = \sin u, \quad u = f(x),$$

then
$$\frac{dy}{dx} = (\cos u)\frac{du}{dx} = f'(x)\cos[f(x)].$$
Thus,
$$\frac{d}{dx}\sin mx = \frac{d}{dx}(mx)\cdot\cos mx = m\cos mx,$$
$$\frac{d}{dx}\sin x^2 = \frac{d}{dx}(x^2)\cdot\cos x^2 = 2x\cos x^2.$$

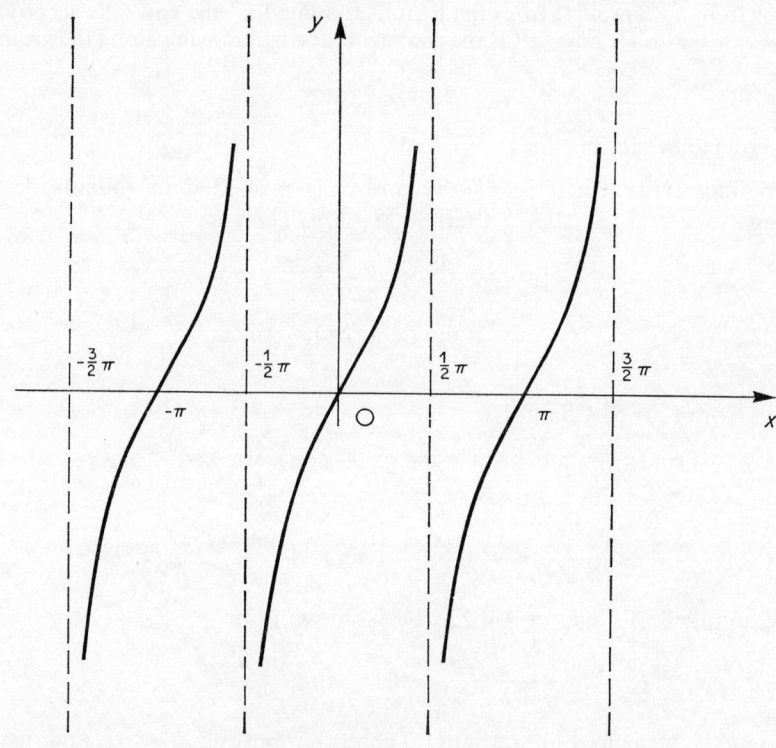

FIGURE 69

Further, if $y = \cos x$, write $y = \sin(\tfrac{1}{2}\pi - x)$. Then, taking $u = \tfrac{1}{2}\pi - x$, we have
$$\frac{d}{dx}\cos x = \frac{d}{dx}(\tfrac{1}{2}\pi - x)\cdot\cos(\tfrac{1}{2}\pi - x) = -\cos(\tfrac{1}{2}\pi - x) = -\sin x$$
since $\cos(\tfrac{1}{2}\pi - x) = \sin x$.

Again, if $y = \tan x$, we write
$$y = \frac{\sin x}{\cos x}$$
and calculate the derivative by means of the rule for quotients. Thus
$$\frac{dy}{dx} = \frac{\cos x \frac{d}{dx}(\sin x) - \sin x \frac{d}{dx}(\cos x)}{\cos^2 x}$$
$$= \frac{\cos x \cdot \cos x - \sin x \cdot (-\sin x)}{\cos^2 x}$$
$$= \frac{\cos^2 x + \sin^2 x}{\cos^2 x} = \frac{1}{\cos^2 x} = \sec^2 x.$$

This derivative will exist at any point where $\cos x \neq 0$. At points where $\cos x = 0$ the derivative does not exist, and this is evident in figure 69 where the graph of $\tan x$ is displayed.

The derivatives of $\cot x$, $\sec x$ and $\text{cosec } x$ may be obtained in a similar way. The results are given in table 2 and should be confirmed by the reader.

TABLE 2

y	dy/dx
$\sin x$	$\cos x$
$\cos x$	$-\sin x$
$\tan x$	$\sec^2 x$
$\cot x$	$-\text{cosec}^2 x$
$\sec x$	$\sec x \tan x$
$\text{cosec } x$	$-\text{cosec } x \cot x$

Example 19 If $y = 5 \sin^3 2x$ find dy/dx.

With $u = \sin 2x$, $y = 5u^3$ so that, by the chain rule,
$$\frac{dy}{dx} = 15u^2 \frac{du}{dx}$$
$$= 15 \sin^2 2x \frac{d}{dx}(\sin 2x)$$
$$= 30 \sin^2 2x \cos 2x.$$

Example 20 If $y = 1/(\sin 4x)^3$ find dy/dx.

One way is to take $u = \sin 4x$, $y = 1/u^3$ and then
$$\frac{dy}{dx} = -\frac{3}{u^4}\frac{du}{dx} = -\frac{3}{(\sin 4x)^4}\frac{d}{dx}(\sin 4x) = -\frac{12 \cos 4x}{(\sin 4x)^4}.$$

Another method is to put $u = \operatorname{cosec} 4x$, $y = u^3$ and then

$$\frac{dy}{dx} = 3u^2 \frac{du}{dx} = 3 \operatorname{cosec}^2 4x \frac{d}{dx}(\operatorname{cosec} 4x)$$

$$= -12 \operatorname{cosec}^2 4x \cdot \operatorname{cosec} 4x \cot 4x = -\frac{12 \cos 4x}{(\sin 4x)^4}.$$

Exercises

74. Find the derivatives of (i) $\sin 4x$, (ii) $\sin(8x - \tfrac{1}{2}\pi)$, (iii) $\tan ax$, (iv) $\sec bx$, (v) $\sin^n x$, (vi) $(\cos x)^{\frac{1}{3}}$, (vii) $\sec^4 x$, (viii) $\cot^3 x$, (ix) $\tan^m ax$, (x) $\cos^2 \tfrac{1}{2}x$, (xi) $x^3 \sin 4x$, (xii) $x^{\frac{1}{2}} \tan x$, (xiii) $(\sin 3x)/x^2$, (xiv) $\sin x \cos x$, (xv) $2\cos^2 3x - 3\sin^2 2x$, (xvi) $\cot^2 \tfrac{1}{2}x$, (xvii) $\tan x^2$, (xviii) $(3 + 4\cos x)^{\frac{1}{2}}$, (xix) $(\operatorname{cosec} 2x)^{\frac{1}{2}}$, (xx) $(1 + \tan x)/(1 - \tan x)$, (xxi) $1/(\sec 2x - 1)^{\frac{3}{2}}$, (xxii) $\sin^3 x \sin 3x$.

75. Find dy/dx when (i) $\sin 2x - \cos 3y = 1$, (ii) $x + \tan(xy) = 0$, (iii) $\sin^2 x + \cos^2 y = a^2$, (iv) $y \tan y = x$.

76. A lighthouse is 900 m from a straight shore and the light makes 4 revolutions per minute. How fast does the beam sweep along the shore at a point 1200 m from the nearest point?

77. Two sides of a triangle are 12 ft and 30 ft long respectively. The angle between them is 60° and increasing at 1° per minute. How fast is the area increasing?

78. Find the tangent to $y = x \sin x$ at (i) $x = \tfrac{1}{2}\pi$, (ii) $x = -\tfrac{1}{2}\pi$.

79. Use Newton's method to find the positive root of $x = \cos x$ correct to 2 decimal places.

80. Find the smallest positive root of

(i) $\tan x = \dfrac{2x}{2 - x^2}$, (ii) $\tan x = \dfrac{x^3 - 9x}{4x^2 - 9}$.

5.10 Small errors and corrections

In many practical investigations data are collected and then inserted into a mathematical formula to give numerical results. These data are often subject to error. For example, the measurement of length cannot be made more accurate than the graduations on the ruler or measuring tape permit. Or again, the estimate of the yield of a field of potatoes from the yield of a square metre will not be correct if the fertility of the field is not uniform or if one potato of the crop from the square metre is missed. Errors in the data produce errors in quantities calculated from them by means of mathematical formulae. The derivative is a valuable device for determining the effect of small errors of measurement upon subsequent calculations. This section gives a few illustrations of how this is done, but the reader should not underrate the importance of this topic. Its application ranges widely and includes the control of rocket trajectories, manufacturing tolerances and the operation of digital computers (of which more will be said in section 8.4), to mention a few examples. The sooner the reader grasps the basic concepts the better.

The necessary theory for our purpose has already been developed in section 5.2 and is essentially contained in equation (5), namely that $f(x) + hf'(x)$ provides a good approximation to $f(x + h)$ when h is small.

Example 21 *A ball is supposed to be a sphere of radius 2 cm. If the radius is made $\frac{1}{10}$ cm too large, what is the error in the volume?*

The volume v of a sphere of radius r is $\frac{4}{3}\pi r^3$. Therefore

$$\frac{dv}{dr} = 4\pi r^2$$

and so, if r is increased by the small quantity h, the volume will increase by $h \cdot 4\pi r^2$ approximately. In this case $r = 2$, $h = \frac{1}{10}$ and the volume will increase by $\frac{1}{10} \times 4\pi \cdot 2^2$ or $1 \cdot 6 \pi$ cm^3 approximately.

Example 22 *The area of a triangle is determined by measuring two sides and the angle between them. Find the error in the area due to a small error in the angle.*

Let b,c be the lengths of the two sides and θ the angle between them. Then the area A is $\frac{1}{2}bc \sin \theta$. Hence

$$\frac{dA}{d\theta} = \tfrac{1}{2}bc \cos \theta$$

and, if θ is increased by the small quantity h, the area will increase by $\frac{1}{2}bch \cos \theta$ approximately.

A measure of how good our approximation is supplied by the *relative error*, which is the ratio of the error to the calculated value. The relative error in the area is

$$\frac{\tfrac{1}{2}bch \cos \theta}{A} = h \cot \theta.$$

So long as $h \cot \theta$ is small the estimate of the error can be expected to be good.

Example 23 *The acceleration due to gravity g is determined by observing the time of oscillation t and length l of a simple pendulum and using the formula* $t = 2\pi\sqrt{(l/g)}$. *Find the percentage error in g if t is in error by* 0·1 *per cent.*

Since $g = 4\pi^2 l/t^2$,

$$\frac{dg}{dt} = -\frac{8\pi^2 l}{t^3}$$

and, a small increase of h in t, will produce an increase of $-8\pi^2 lh/t^3$ in g. Therefore the percentage increase in g is

$$-\frac{8\pi^2 lh/t^3}{4\pi^2 l/t^2} \times 100 = -200h/t.$$

But $h = 0·1t/100$ and $-200h/t = -0·2$. Thus if the error in t is an excess of 0·1 per cent., the calculated value of g will be too small by 0·2 per cent.

Exercises

81. Calculate approximately (i) $\sin 120° 2'$, (ii) $\sec 60° 1'$, (iii) $\tan 45° 1'$, (iv) $\sin^2 29° 57'$.

82. A given volume of plastic is to be made into a circular cylinder of radius 3 cm and height 4 cm. If, in fact, the radius is made 3·1 cm what will be the height?

83. 7·9 m³ of cement are made into a cube. What is the length of an edge of the cube?

84. The side of a square garden is measured as 30 yd. If it is, in fact, $\frac{1}{20}$ ft longer what will be the error in the area?

85. The side c of a triangle is calculated from
$$c^2 = a^2 + b^2 - 2ab \cos C.$$
Find the percentage error due to (i) a 1 per cent. error in a, (ii) a 1 per cent. error in C.

86. The pressure p and volume v of a given mass of gas at constant temperature are connected by $pv = $ constant. If $p = 10$ and $v = 5$ find (i) the volume when the pressure is reduced to 9·9, (ii) the change in pressure which will increase the volume to 5·2.

87. Repeat question 86, assuming that the adiabatic gas law $pv^{1·4} = $ constant is obeyed.

The Derivative

88. The resistance R of a certain wire is related to the temperature T by $R = R_0(1 + aT + bT^2)$ where R_0, a and b are constants. Find the change in resistance when T is increased by the small amount t.

89. If x is the distance of a point on the axis of a lens, from a lens of focal length f the distance y of its image from the lens is given by $1/x + 1/y = 1/f$. If an object has small length h along the axis, show that the length of the image is $y^2 h/x^2$ approximately.

90. The base of a tower is 450 ft away from a point on the same level. The elevation to the top of the tower from this point is measured as $35° \, 30'$. If the height is calculated from this information show that it is in error by 0·99 ft if the elevation is in error by $5'$.

91. The distance of a submarine is calculated by observing that its angle of depression, from an aircraft flying at 120 m, is $15°$. What is the error in the distance if the angle is actually $15·25°$?

92. If $y = k \tan x$, where k is a constant, find the percentage error in y due to an error of $\frac{1}{2}°$ when $x = 45°$.

6
Maxima and minima

6.1 Higher derivatives

The preceding chapter has shown how the derivative can be calculated for some functions of x. The derivative itself is also a function of x and therefore we can consider the possibility of taking a derivative of it.

The derivative of the derivative of y is called the *second derivative* of y with respect to x. It is denoted by d^2y/dx^2. Thus

$$\frac{d^2y}{dx^2} = \frac{d}{dx}\left(\frac{dy}{dx}\right).$$

For example, if $y = x^3$,

$$\frac{dy}{dx} = 3x^2,$$

$$\frac{d^2y}{dx^2} = \frac{d}{dx}(3x^2) = 6x.$$

If a function is represented by $f(x)$ we have denoted its first derivative by $f'(x)$. The second derivative is denoted by $f''(x)$. Thus, if $f(x) = \sin x$,

$$f'(x) = \cos x, \quad f''(x) = -\sin x.$$

Again, the derivative of the second derivative is called the *third derivative* and denoted by d^3y/dx^3 or $f'''(x)$, according to the notation being employed. Thus, when $y = x^3$, $d^3y/dx^3 = 6$ and when $f(x) = \sin x$, $f'''(x) = -\cos x$.

Quite generally, the result of taking n derivatives of y one after the other is called the *nth derivative* of y and denoted by $d^n y/dx^n$.

Similarly, the nth derivative of $f(x)$ is signified by $f^{(n)}(x)$, or by $d^n y/dx^n$. Of course, we are speaking generally here and do not mean to imply that the higher derivatives of a function necessarily exist even if the first derivative does. But we can be sure that $d^n y/dx^n$ does not exist at a point if one of $y, dy/dx, \ldots, d^{n-1}y/dx^{n-1}$ fails to exist there.

Maxima and minima

Example 1 If $y = x^\nu$,

$$\frac{dy}{dx} = \nu x^{\nu-1}, \quad \frac{d^2y}{dx^2} = \nu(\nu-1)x^{\nu-2}, \quad \frac{d^3y}{dx^3} = \nu(\nu-1)(\nu-2)x^{\nu-3},$$

$$\frac{d^ny}{dx^n} = \nu(\nu-1)(\nu-2)\ldots(\nu-n+1)x^{\nu-n}.$$

In particular, if $\nu = n$ a positive integer, we obtain

$$\frac{d^n y}{dx^n} = n(n-1)(n-2)\ldots 1$$

and

$$\frac{d^{n+1}y}{dx^{n+1}} = 0.$$

By using the notation $n(n-1)(n-2) \ldots 1 = n!$ when n is a positive integer we can write

$$\frac{d^n}{dx^n}(x^n) = n!$$

In fact we could write

$$\frac{d}{dx}(x^n) = \frac{n!}{(n-1)!}x^{n-1}, \quad \frac{d^2}{dx^2}(x^n) = \frac{n!}{(n-2)!}x^{n-2}$$

but this is not always convenient.

Example 2 If $f(x) = x^{4/3}$,

$$f'(x) = \frac{4}{3}x^{1/3}, \quad f''(x) = \frac{4}{9}x^{-2/3}, \quad f'''(x) = -\frac{8}{27}x^{-5/3}.$$

This gives an example of a function for which $f(0) = 0, f'(0) = 0$ but yet $f''(0)$ fails to exist.

Example 3 If $y = 1/(1-x)$, Theorems 5.5 and 5.8 give

$$\frac{dy}{dx} = \frac{1}{(1-x)^2}, \quad \frac{d^2y}{dx^2} = \frac{1 \cdot 2}{(1-x)^3}, \quad \frac{d^3y}{dx^3} = \frac{1 \cdot 2 \cdot 3}{(1-x)^4}$$

and, generally,

$$\frac{d^n y}{dx^n} = \frac{n!}{(1-x)^{n+1}}.$$

Exercises

1. Find the first, second, third and nth derivatives of (i) x^9, (ii) $4x^4 + 2x^3 - 3x + 1$, (iii) $1/x^3$, (iv) $\sqrt{x}$, (v) $1/(3x+2)$, (vi) $\cos x$, (vii) $\sin^2 x$.

2. Find the first two derivatives of (i) $(2 - 3x^2)^{\frac{1}{2}}$, (ii) $x \sin x$, (iii) $x^2/(1 + x)$, (iv) $x \tan x$, (v) $x/(x - 1)^{\frac{1}{2}}$.

3. Prove that $f''(x) = \lim\limits_{h \to 0} (1/h^2)[f(x + h) - 2f(x) + f(x - h)]$.

When y is given as a composite function of x or as an implicit function of x, the rules for finding higher derivatives are much more complicated than for the first derivative. Though the rules can be formulated for the first one or two derivatives (see, for example, Exercise 5) it is usually easier to proceed directly with repeated applications of the methods of sections 5.7, 5.8.

Example 4 If $y = 3t^2 + t - 1$ and $x = 3t + 1$, Example 17 of Chapter 5 shows that

$$\frac{dy}{dx} = \frac{dy/dt}{dx/dt} = 2t + \tfrac{1}{3}.$$

Using the rule again, we have

$$\frac{d^2y}{dx^2} = \frac{d}{dx}\left(\frac{dy}{dx}\right) = \frac{d}{dt}\left(\frac{dy}{dx}\right)\bigg/\frac{dx}{dt} = \frac{d}{dt}(2t + \tfrac{1}{3})/3 = \frac{2}{3}.$$

Example 5 Find d^2y/dx^2 at the point $(1,1)$ of the curve $x^3y + 2y^2 - 3 = 0$.

The first derivative of the equation is

$$x^3 \frac{dy}{dx} + 3x^2 y + 4y \frac{dy}{dx} = 0 \qquad (1)$$

and the second derivative is

$$x^3 \frac{d^2y}{dx^2} + 6x^2 \frac{dy}{dx} + 6xy + 4y \frac{d^2y}{dx^2} + 4\left(\frac{dy}{dx}\right)^2 = 0. \qquad (2)$$

Putting $x = 1, y = 1$ in (1) gives $dy/dx = -\tfrac{3}{5}$. Then, substituting $x = 1, y = 1$, $dy/dx = -\tfrac{3}{5}$ in (2) supplies $d^2y/dx^2 = -\tfrac{96}{125}$ which is the required result.

Exercises

4. If $y = t - t^3$ and $x = t - t^2$, find d^2y/dx^2.

Maxima and minima

5. (i) If $x = h(t)$, prove that
$$\frac{d^2x}{dt^2} = \frac{1}{2}\frac{d\dot{x}^2}{dx} \quad \text{where} \quad \dot{x} = \frac{dx}{dt};$$

(ii) If $y = f(t)$ and $t = g(x)$, prove that
$$\frac{d^2y}{dx^2} = \frac{dy}{dt} \cdot \frac{d^2t}{dx^2} + \frac{d^2y}{dt^2}\left(\frac{dy}{dt}\right)^2,$$
$$\frac{d^3y}{dx^3} = \frac{dy}{dt} \cdot \frac{d^3t}{dx^3} + 3\frac{d^2y}{dt^2} \cdot \frac{d^2t}{dx^2} \cdot \frac{dt}{dx} + \frac{d^3y}{dt^3}\left(\frac{dt}{dx}\right)^3.$$

6. In Example 17 of Chapter 5 it is shown that
$$\frac{dx}{dy} = \frac{1}{dy/dx}.$$

Prove that
$$\frac{d^2x}{dy^2} = -\frac{d^2y}{dx^2}\Big/\left(\frac{dy}{dx}\right)^3,$$
$$\frac{d^3x}{dy^3} = \frac{3(d^2y/dx^2)^2 - (dy/dx)(d^3y/dx^3)}{(dy/dx)^5}.$$

7. Find d^2y/dx^2 at $x = 1$ on $x^3y + xy^3 = 2$.

8. Show that, for the circle $x^2 + y^2 = a^2$,
$$\left|\frac{[1 + (dy/dx)^2]^{\frac{3}{2}}}{d^2y/dx^2}\right| = a.$$

6.2 Maxima and minima—theory

Problems involving the greatest and least values which quantities can attain are of great practical interest. For instance, what angle of projection will give the greatest range of a projectile ejected with a given speed? What are the greatest and least distances of the Earth from the Sun? What design of shock absorber will give the passengers in a car the most vibration-free ride? What is the lowest point of a suspension bridge? How should an electrical circuit be designed to give the least generation of heat?

For questions like these the calculus is especially well adapted to provide answers. Consider the function shown in figure 70. The points A and B are clearly of special significance. To be at A would be like standing on a hill, for the curve falls away on either side just as the slopes of a hill. On the other hand, a position at B would be like being at the bottom of a valley with the curve nearby taking the role of the rising walls of land. We say that the function has a *local maximum* at A and a *local minimum* at B. In other words, a function has a local maximum at a point if its value there is larger than its values at nearby points; a function has a local minimum at a point if its value there is less than neighbouring values.

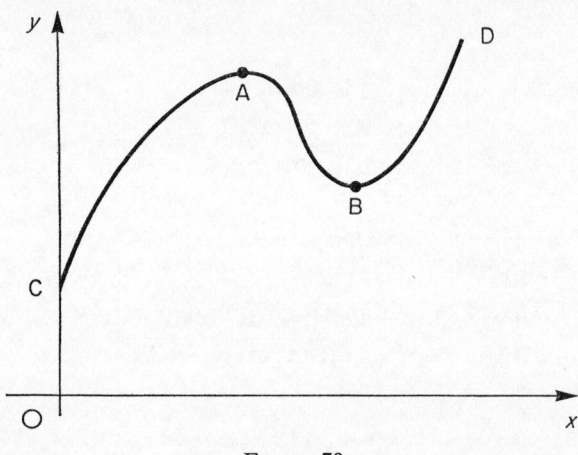

FIGURE 70

The reason for calling A a local maximum rather than a maximum is that A may not be the largest value of the function that occurs in the interval under consideration. In fact, the function has a larger value at D. Similarly it has a smaller value at C than the local minimum at B. So the words 'local maximum' indicate only that the function is larger at a point than it is at points which are immediately on *either* side of it.

According to this definition a function may have any number of local maxima and minima, and it is quite possible for some of the local maxima to be less than some of the minima.

To obtain analytical criteria for locating local maxima and minima we recall that it was shown in section 5.3 that a function is increasing when its derivative is positive and decreasing when its derivative is negative. However, at a local maximum a function is neither increasing nor decreasing. Hence the derivative cannot be either positive or negative at a local maximum. It must, therefore, be zero. Similarly, the derivative vanishes at a local minimum. Thus we have shown

Maxima and minima

THEOREM 6.1 *If $f(x)$ is defined for $a \leqslant x \leqslant b$ and has a local maximum or minimum at $x = c$, where $a < c < b$, then, if f' exists at c,*
$$f'(c) = 0.$$

The reader should note carefully the wording of the theorem. The theorem does not say that $f'(c) = 0$ guarantees that $x = c$ is a local maximum or minimum. In fact, $y = x^3$ has $dy/dx = 3x^2$ which vanishes at $x = 0$ but the graph of $y = x^3$ (see figure 71) does not have a local maximum or minimum at $x = 0$. Any point which satisfies $f'(x) = 0$ is called a *stationary point*. What the theorem states is that a local maximum or minimum must be a stationary point, but the converse is not true.

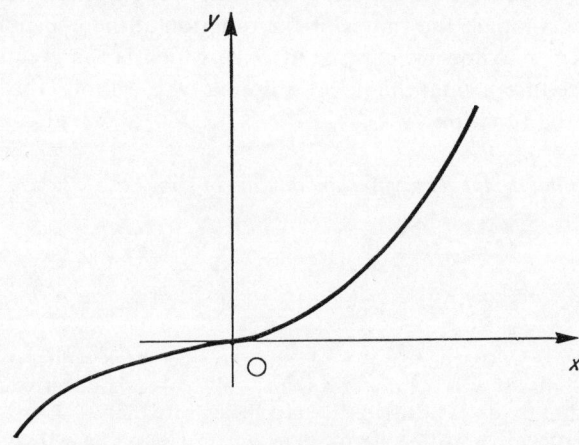

FIGURE 71

How can we ensure that a stationary point is a local maximum or minimum? Looking at figure 70, we see that just before a local maximum f is increasing, and just after a local maximum f is decreasing. So long as we make sure f behaves in this way we obtain a local maximum and we can do it by making f' positive just before and negative just after.

THEOREM 6.2 *If $f'(c) = 0$ and if $f'(x)$ changes from positive to negative values as x increases through c, then f has a local maximum at $x = c$.*

The reader will easily verify the corresponding theorem for a local minimum, namely

THEOREM 6.3 *If $f'(c) = 0$ and if $f'(x)$ changes from negative to positive values as x increases through c, then f has a local minimum at $x = c$.*

The preceding theorems explain what happens when $f'(c)$ exists. If $f'(c)$ does not exist they provide no information, but it would be wrong to

164   Introductory Analysis

conclude that f cannot have a local maximum or minimum. For example, the curve in figure 62 has a local minimum at P but f' does not exist at P.

Further, the theorems require that $a < c < b$ and say nothing about the end-points a and b. The value at D (figure 70) can be greater than values immediately to the left without $f'(b)$ being zero. (Why does the argument leading to Theorem 6.1 not work at an end-point?)

The largest value of a function, whose derivative exists, can be located as follows. The largest value may occur at several points but each one must be either an end-point or an interior point c with $a < c < b$. If the largest value occurs at an interior point c, it must also be a local maximum because the values of the function on either side of c must be smaller. Hence, if we compare the values of the function at the end-points with its values at local maxima we can see at once which is the greatest value. A similar procedure using the local minima will supply the least value attained by the function.

Example 6 Find the local maxima and minima of $y = 2x^3 - 3x^2 - 12x + 2$.

Here

$$\frac{dy}{dx} = 6x^2 - 6x - 12 = 6(x + 1)(x - 2).$$

Therefore $dy/dx = 0$ when $x = -1$ and $x = 2$; these values of x give the stationary points. If x is a little less than -1, $x + 1$ is negative and $x - 2$ is negative so that dy/dx is positive. If x is a little more than -1, $x + 1$ is positive and $x - 2$ is negative so that dy/dx is negative. Hence, as x increases through -1, dy/dx changes from positive to negative values. By Theorems 6.1 and 6.2, y has a local maximum at $x = -1$; its value there is, by direct substitution in the formula for y, 9.

If x is slightly less than 2, $x + 1$ is positive and $x - 2$ is negative so that dy/dx is negative. If x is slightly greater than 2, $x + 1$ is positive and $x - 2$ is positive so that dy/dx is positive. Therefore dy/dx changes from negative to positive as x increases through 2. By Theorems 6.1 and 6.3, y has a local minimum at $x = 2$; its value there is -18.

A rough graph of the function is shown in figure 72.

Example 7 Find the local maxima and minima of $y = x^4 - 6x^2 - 8x + 24$.

Here

$$\frac{dy}{dx} = 4x^3 - 12x - 8 = 4(x+1)^2(x-2).$$

Therefore the stationary points where $dy/dx = 0$ are given by $x = -1$ and $x = 2$. When x is a little less than -1, $(x + 1)^2$ is positive and $x - 2$ is negative

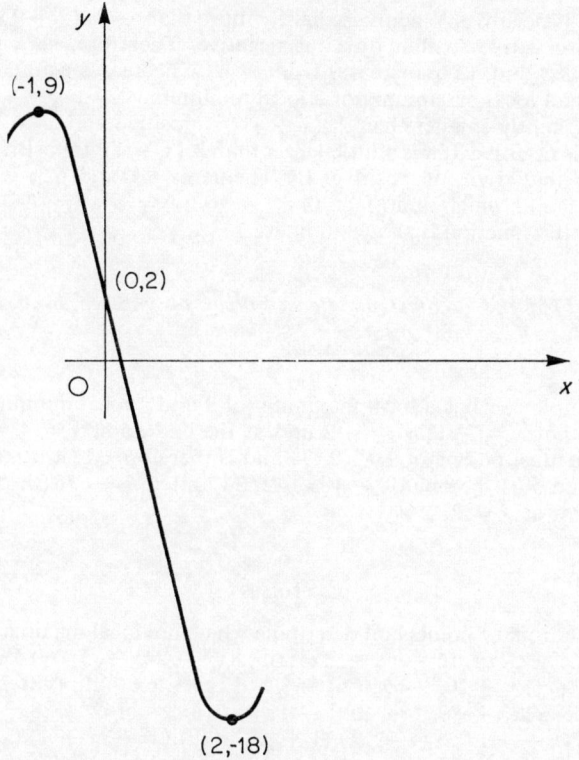

FIGURE 72

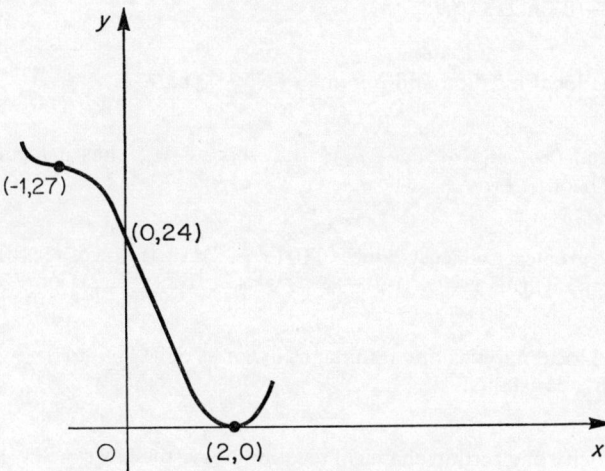

FIGURE 73

so that dy/dx is negative. When x is slightly more than -1, $(x+1)^2$ is positive and $x-2$ is negative so that dy/dx is negative. Therefore, as x increases to -1, y decreases and, as x increases from -1, y decreases again. In this case there is neither a local maximum nor a local minimum.

When x is slightly smaller than 2, $(x+1)^2$ is positive and $x-2$ is negative so that dy/dx is negative. If x is a little larger than 2, $(x+1)^2$ is positive and $x-2$ is positive so that dy/dx is positive. By Theorems 6.1 and 6.3, y has a local minimum at $x=2$ which is, in fact, 0.

A graph of the function is shown in figure 73.

Example 8 Find the greatest and least values of y given in Example 6 for $-2 \leqslant x \leqslant 4$.

From Example 6, y has a local maximum of 9 and a local minimum of -18. At the end-point $x=-2$, $y=-2$ and at the end-point $x=4$, $y=34$. The greatest value must be one of 9, -2, 34 and is therefore 34; it occurs at $x=4$. The least value must be one of -18, -2, 34 and so is -18; it occurs at the local minimum at $x=2$.

Exercises

9. Find the stationary points and determine which are local maxima and minima for (i) $x^2 + 2x - 3$; (ii) $16 - 6x - 3x^2$; (iii) $8x^3 - 24x + 5$; (iv) $4x^3 - 15x^2 + 18x$; (v) $4x^3 + 6x^2 + 20x - 5$; (vi) $8x^3 - 12x^2 + 6x - 1$; (vii) $2x^4 - 8x^3 + 5x^2 + 5$; (viii) $32x^5 - 80x^4 + 40x^3 - 1$; (ix) $16x^4 - 16x^3 + 8x^2 - 12x + 3$; (x) $(2x-1)^2(x-1)$; (xi) $(1-x^2)/(1+x^2)$; (xii) $(x^2 - x + 1)/(x^2 + x + 1)$; (xiii) $x/(x+4)(x+1)$; (xiv) $(x-1)^{\frac{1}{3}}(x-2)^{\frac{2}{3}}$; (xv) $\sin x + \cos x$; (xvi) $\sin 4x - 2x$; (xvii) $8x + \tan 6x$; (xviii) $\sin^3 x \cos x$; (xix) $\sin(x - \frac{1}{3}\pi)\cos(x - \frac{2}{3}\pi)$; (xx) $\tan 2x - 8 \sin 2x$.

10. Find the local maxima and minima of $x^3 - 3ax + b$.

11. Show that $(x-a_1)^2 + (x-a_2)^2 + \ldots + (x-a_n)^2$ has a local minimum when $x = (1/n)(a_1 + a_2 + \ldots + a_n)$.

12. Find the greatest and least values of (i) $(x-2)^2$ on $0 \leqslant x \leqslant 3$; (ii) $(25 - x^2)^{\frac{1}{2}}$ on $-3 \leqslant x \leqslant 3$; (iii) $y = x^2$ on $-2 < x < 2$; (iv) $(x-2)^{\frac{1}{2}}$ on $2 \leqslant x \leqslant 6$.

13. Find the local maxima and minima as ω varies of $[(\Omega^2 - \omega^2)^2 + 4b^2\Omega^2\omega^2]^{\frac{1}{2}}$, Ω and b being constants.

14. The velocity of a certain chemical reaction obeys the law $v = 3(x+1)(2-x)$. Where is the velocity a local maximum?

Maxima and minima

15. The force exerted by a circular current of radius a on a small magnet, distance x above the centre of the circle, is $kx/(x^2 + a^2)^{3/2}$. Show that the force is a maximum at $x = \frac{1}{2}a$.

16. The bending moment of a beam at a distance x from one end is $\frac{1}{2}wlx - \frac{1}{2}wx^2$ where w is the load per unit length and l the length of the beam. Show that the maximum bending moment is at the centre of the beam.

17. Exercise 3 suggests that $(1/h^2)[f(x + h) - 2f(x) + f(x - h)]$ will provide a numerical approximation to $f''(x)$. Use this formula to estimate $f''(1)$ when $f(x) = x^{1/2}$, by taking $h = 0.15, 0.10, 0.08, 0.05, 0.01$. Work with 5 places of decimals. What do your results suggest?

18. A polynomial of the second degree takes the following values:

x	0·60	0·65	0·70	0·75
$f(x)$	0·6221	0·6155	0·6138	0·6170

For what value of x is f a minimum?

19. The derivative of a function has the following values:

x	1·4	1·5	1·6	1·7
$f'(x)$	0·1700	0·0707	− 0·0292	− 0·1288

Estimate, by linear interpolation (section 1.4), where f is stationary and state whether it is a local maximum or minimum.

The rules given in Theorems 6.2 and 6.3 can sometimes be replaced by others which are more convenient to apply. Suppose that the second derivative d^2y/dx^2 is negative at a stationary point where $dy/dx = 0$. Now d^2y/dx^2 is the derivative of dy/dx and so, by section 5.3, dy/dx is decreasing at the stationary point. Because $dy/dx = 0$ at the stationary point and is decreasing there, it must be changing from positive to negative values. Therefore, by Theorem 6.2, y has a local maximum at the stationary point.

Similarly, if d^2y/dx^2 is positive at a stationary point, dy/dx is an increasing function which, because it is zero at the stationary point, must be changing from negative to positive values. Hence, by Theorem 6.3, y has a local minimum at the stationary point.

If d^2y/dx^2 vanishes or does not exist we can draw no conclusions about local maxima and minima, but we have proved

THEOREM 6.4 *If $f'(c) = 0$ and $f''(c) < 0$ there is a local maximum at $x = c$. If $f'(c) = 0$ and $f''(c) > 0$ there is a local minimum at $x = c$.*

Example 9 Find the local maxima and minima of $y = 5 - 12x + 9x^2 - 2x^3$.

Here
$$\frac{dy}{dx} = -12 + 18x - 6x^2 = -6(x-1)(x-2),$$

$$\frac{d^2y}{dx^2} = 18 - 12x.$$

The stationary points at which dy/dx vanishes are $x = 1$ and $x = 2$. At $x = 1$, $d^2y/dx^2 = 6$ which is positive. Therefore, by Theorem 6.4, y has a local minimum at $x = 1$ which is $y = 0$.

Likewise, at $x = 2$, $d^2y/dx^2 = -6$ and y has a local maximum of 1 at $x = 2$.

Exercises

20. Use Theorem 6.4 to find the local maxima and minima of (i) $x^2 + 54/x$, (ii) $(x^2 + x + 1)/(x^2 + 1)$, (iii) $x^4 + 2x^3 - 3x^2 - 4x + 4$.

21. Find the maximum slope of $y = \sin 2x$.

6.3 Maxima and minima—applications

There are numerous problems which can be solved by the theory of maxima and minima. Only a small number of typical examples can be presented here.

Example 10 A piece of string of length l is placed in the form of a rectangle. What arrangement gives the largest area?

Let x and y be the lengths of the two sides of the rectangle. Then
$$2x + 2y = l. \tag{3}$$
The area of the rectangle A is given by
$$A = xy.$$
From (3)
$$y = \tfrac{1}{2}l - x.$$
and so
$$A = x(\tfrac{1}{2}l - x).$$
We want the value of x which makes A a maximum. Now
$$\frac{dA}{dx} = \tfrac{1}{2}l - 2x$$

so that $dA/dx = 0$ when $x = \frac{1}{4}l$. Also, either Theorem 6.2 or Theorem 6.4 shows that A has a local maximum of $\frac{1}{16}l^2$ for this value of x.

In fact, x cannot be less than 0 nor greater than $\frac{1}{2}l$ and for both of these values $A = 0$. Hence the greatest value of A is $\frac{1}{16}l^2$ and occurs when $x = y = \frac{1}{4}l$. The rectangle has the greatest area when it is a square.

Example 11 Two aircraft are flying at the same level on straight lines which intersect at right-angles. One aircraft flying at 200 k.p.h. passes the intersection at the instant when the other aircraft, travelling at 150 k.p.h. towards the intersection, is 100 km away from it. When are the aircraft closest together?

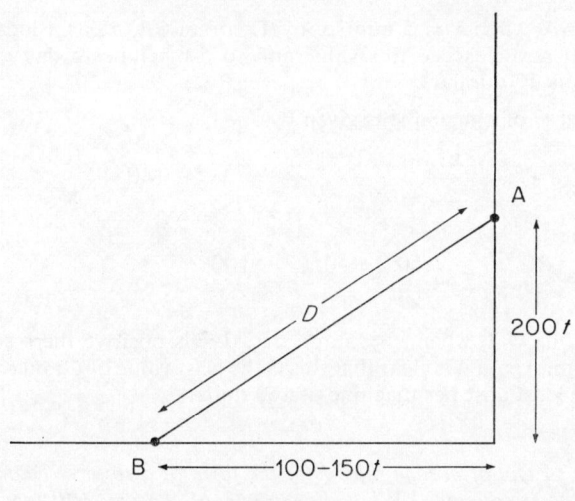

FIGURE 74

After t hours the first aircraft is at A (figure 74) $200t$ km from the intersection and the second, having covered $150t$ km, is at B which is $100-150t$ km from the intersection. Hence the distance D between the aircraft is given by

$$D^2 = (200t)^2 + (100-150t)^2 = 2500(25t^2 - 12t + 4).$$

Therefore

$$\frac{dD^2}{dt} = 2500(50t - 12),$$

$$\frac{d^2D^2}{dt^2} = 2500 \cdot 50.$$

Thus, $dD^2/dt = 0$ when $t = 12/50$ and, because $d^2D^2/dt^2 > 0$, D^2 has a local minimum there. In fact, dD^2/dt is negative if $t < 12/50$ and positive if $t > 12/50$ so that D^2 can never be less than its value at $t = 12/50$. Hence the aircraft are closest at 0·24 hr after the first has passed the intersection and are then 80 km apart.

Introductory Analysis

Example 12 *The total cost of producing x washing machines per day in a certain factory is* $(2x^2 + 140x + 50)$ *dollars and the price at which each machine can be sold is* $(200 - 4x)$ *dollars. What production gives* (i) *the best profit,* (ii) *the least cost per machine?*

(i) The profit p on x machines per day is
$$p = x(200 - 4x) - (2x^2 + 140x + 50) = 60x - 6x^2 - 50.$$
Then
$$dp/dx = 60 - 12x,$$
$$d^2p/dx^2 = -12.$$

Thus $dp/dx = 0$ when $x = 5$ and so, by Theorem 6.4, p has a local maximum. In fact, p can never exceed this value and so 5 machines a day gives the best profit, which is 100 dollars.

(ii) The cost C of a machine is given by
$$C = \frac{2x^2 + 140x + 50}{x} = 2x + 140 + \frac{50}{x}$$
so that
$$\frac{dC}{dx} = 2 - \frac{50}{x^2}, \qquad \frac{d^2C}{dx^2} = \frac{100}{x^3}.$$

Therefore $dC/dx = 0$ when $x = 5$ and d^2C/dx^2 is positive there so that C is a local minimum. Again it is clear that this is the least value of C and so 5 machines/day gives the least cost per machine of 160 dollars.

Example 13 *A can of orange juice is in the form of a right circular cylinder and holds a volume* a^3 *of juice. What dimensions of the can will give the smallest surface area?*

Let r be the radius of the can and h its height. Then
$$\pi r^2 h = a^3. \tag{4}$$
The area of each end is πr^2 and of the sides is $2\pi rh$. Hence the total surface area
$$A = 2\pi r^2 + 2\pi rh.$$
From (4), $h = a^3/\pi r^2$ (why is the division permissible?) and so
$$A = 2\pi r^2 + 2a^3/r.$$
Then
$$\frac{dA}{dr} = 4\pi r - \frac{2a^3}{r^2},$$
$$\frac{d^2A}{dr^2} = 4\pi + \frac{4a^3}{r^3}.$$

Maxima and minima

Thus, $dA/dr = 0$ when $r^3 = a^3/2\pi$ and then $d^2A/dr^2 = 12\pi$. Theorem 6.4 shows that A must have a local minimum at $r = a/(2\pi)^{\frac{1}{3}}$. In fact, because $dA/dr < 0$ when $r < a/(2\pi)^{\frac{1}{3}}$ and $dA/dr > 0$ when $r > a/(2\pi)^{\frac{1}{3}}$, the curve of A against r must be as shown in figure 75. Therefore the local minimum must be the least value of A. Hence the dimensions giving the least surface area are

$$r = a/(2\pi)^{\frac{1}{3}}, \qquad h = 2a/(2\pi)^{\frac{1}{3}} = 2r.$$

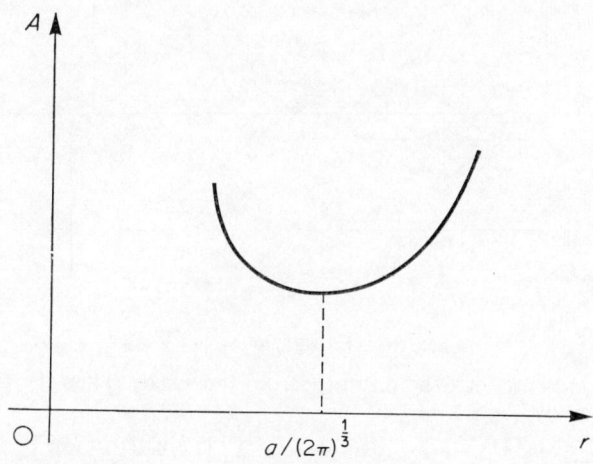

FIGURE 75

Example 14 If light travels in a straight line with speed v_1 find its path if it travels from A to B in the shortest time by reflection at the x-axis. If A and C are on opposite sides of the x-axis and the speed below the x-axis is v_2 find the path which gives the shortest time between A and C (figure 76).

The path in the first case must be APB, and it is required that $AP + PB$ be a minimum. Let A and B be distant a and b, respectively, from the x-axis, and let $MP = x$, $MN = l$. Then

$$y = AP + PB = (a^2 + x^2)^{\frac{1}{2}} + [b^2 + (l - x)^2]^{\frac{1}{2}}.$$

Therefore,

$$\frac{dy}{dx} = \frac{x}{(a^2 + x^2)^{\frac{1}{2}}} - \frac{l - x}{[b^2 + (l - x)^2]^{\frac{1}{2}}}.$$

In terms of the angles θ_1 and θ_2 in figure 76 this is

$$\frac{dy}{dx} = \cos \theta_1 - \cos \theta_2.$$

Thus $dy/dx = 0$ when $\cos \theta_1 = \cos \theta_2$. One solution of this is $\theta_1 = \theta_2$. Now, as P moves from M to N, θ_1 starts at $\frac{1}{2}\pi$ and steadily decreases whereas θ_2

steadily increases to $\tfrac{1}{2}\pi$ at N. Hence dy/dx starts negative with P at M and steadily increases, being positive at N. Consequently, the minimum of y does occur where $\theta_1 = \theta_2$.

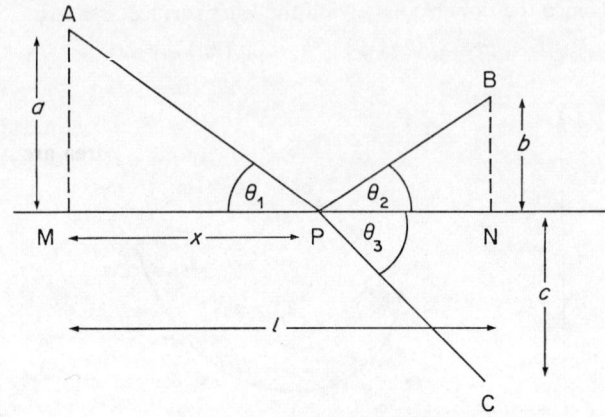

FIGURE 76

In the second case let C be distant c from the x-axis. Then the time T from A to C is given by

$$T = \frac{AP}{v_1} + \frac{PC}{v_2} = \frac{(a^2 + x^2)^{\frac{1}{2}}}{v_1} + \frac{[c^2 + (l - x)^2]^{\frac{1}{2}}}{v_2}.$$

Therefore,

$$\frac{dT}{dx} = \frac{x}{v_1(a^2 + x^2)^{\frac{1}{2}}} - \frac{l - x}{v_2[c^2 + (l - x)^2]^{\frac{1}{2}}}$$

$$= \frac{\cos \theta_1}{v_1} - \frac{\cos \theta_3}{v_2}.$$

Thus $dT/dx = 0$ when

$$\frac{\cos \theta_1}{v_1} = \frac{\cos \theta_3}{v_2}. \tag{5}$$

Again the first term in dT/dx increases as P moves from M to N while the second term decreases. As a result dT/dx is negative when P is at M and positive when P is at N. Consequently there is one minimum given by (5).

These laws of reflection and refraction are known as *Snell's laws* and the idea that light takes the shortest possible time is called *Fermat's principle*.

From these examples it should be plain that the following technique should be adopted in solving problems:

(a) Obtain an equation for the quantity which is to be a maximum or a minimum. If the expression contains more than one variable use

Maxima and minima

relations connecting the variables to eliminate all but one variable (see Examples 10 and 13). The quantity is then expressed as a function of a single variable, say $y = f(x)$.

(b) Determine the stationary points and check whether they are local maxima or minima.

(c) If y is defined for $a \leqslant x \leqslant b$ check the end-points $x = a$ and $x = b$ to see if they are greatest or least values.

(d) Check whether there are points where dy/dx does not exist.

Exercises

22. Verify the statements made about the behaviour of D^2 at the end of Example 11, and about p and C in Example 12 by drawing graphs.

23. The height of a missile above the ground at time t is given by $s = vt - 16t^2$. What is the greatest height?

24. Find the rectangle of given area which has the shortest diagonal.

25. Discuss the sum of a number and its reciprocal.

26. The cost of fuel in running a certain car is proportional to the square of the speed and is 125 p per hour at a speed of 50 m.p.h. Overhead costs are 45 p per hour, whatever the speed. At what speed is the cost per mile a minimum?

27. A triangle has a given base and given area. Show that its perimeter is least when it is isosceles.

28. A right circular cylinder is inscribed in a sphere of radius a. Show that the volume of the cylinder is a maximum when its height is $2a/\sqrt{3}$.

29. A man in a boat is 1 km from the nearest point A of a straight shore. He wishes to reach B, which is $1\frac{1}{3}$ km along the shore from A, in the shortest time. What point C on the shore should he aim for if he walks at 5 k.p.h. and rows at (i) 3 k.p.h., (ii) 4 k.p.h.?

30. A right circular cylinder is being mailed and the postal regulations require that the sum of the length and circumference must not exceed 6 ft. What dimensions give the greatest volume?

31. The strength of a rectangular beam is proportional to its breadth and the square of its depth. Find the width of the strongest beam which can be cut from a circular cylindrical log of radius a. What would be the width if the strength depended on the cube of the depth instead of the square?

32. A military supply depot is to be set up at a point C on a straight railway line to supply two camps A and B. If the shortest distances of A and B from the line are a and b, respectively, and if c is the distance between A and B show that that the sum of the distances of C from A and B cannot be less than $(c^2 + 4ab)^{\frac{1}{2}}$.

33. The centre of mass of a right circular drinking glass is at its centre. The mass of water which fills the glass is three times that of the glass. For what depth of water will the glass be most steady on the table, assuming that this occurs when the centre of mass is lowest? (If the centres of mass of masses m_1, m_2 are at $(x_1, 0), (x_2, 0)$ their combined centre of mass is at $[(m_1 x_1 + m_2 x_2)/(m_1 + m_2), 0]$.)

34. A gas storage vessel consists of a circular cylinder topped by a hemisphere and is of volume v. The cost of construction per square foot of surface area of the hemisphere is twice that of the cylinder. Show that the cost is a minimum when the diameter of the cylinder is $(3v/\pi)^{\frac{1}{3}}$. What will its height be then?

35. A tobacco manufacturer offers a retailer cigarettes at 2400 pence per thousand if the order is 42,000 or less. If, however, the order exceeds 42,000 he reduces the price per thousand for the whole order by $1d.$ for each thousand by which the order exceeds 42,000. Find the order which costs the retailer the most.

36. The deflection y at a distance x from one end of a certain cantilever beam of length l is given by
$$y = Ax^2(x - l)^2$$
where A is a constant. Find where the deflection is greatest.

37. The illumination from a point source of light is proportional to the strength of the source and varies inversely as the square of the distance from the source. At what point on the line joining two sources of light of strengths s_1 and s_2, respectively, will the illumination be a minimum if the distance between the sources is d?

38. What is the shortest distance from the point $(2a, a)$ to the parabola $y^2 = 4ax$?

39. Two missiles are travelling with speeds u and v along straight lines which intersect at an angle θ. If, at a certain instant, they are at distances a and b from the intersection find their minimum distance apart.

40. In which month should a spinster and bachelor marry in order that they pay the least income tax this year? Assume that their salaries do not alter throughout the financial year. Would your conclusion have been the same 10 years ago?

6.4 Curve sketching

Inspection of the graph of a function is often a great help in evaluating its behaviour. However, unless the function is very simple, the drawing of an accurate graph can be a difficult and lengthy procedure. In fact it is probably most expeditiously carried out by using a digital computer with its output connected to a graph plotter. For many purposes, however, it is sufficient to have a broad knowledge of the behaviour of a function without being aware of the fine detail. In other words, knowing the general shape of a curve and one or two points through which it must pass will enable us to draw a rough sketch and obtain a reasonable idea of what the function is doing. It is the aim of this section to indicate how the calculus can help in drawing such sketches.

Clearly, the determination of local maxima and minima will be important. This will involve considering the points where the first derivative vanishes. However, the first derivative also provides other information. For we know that y is increasing where dy/dx is positive and decreasing where dy/dx is negative. So, by finding the ranges of x where dy/dx is positive and where it is negative, we know the regions in which y must rise and those on which y must fall. Rough estimates of the magnitude of dy/dx at various points will also indicate where a rise is faster or slower. An examination of the first derivative will also indicate those points where special consideration is necessary because the derivative does not exist.

The second derivative tells us something about the way a curve is bending. In figure 77 are shown four ways in which a curve can be in relation to its tangent at a point P. In the upper two diagrams the curve is increasing with increasing x; this explains why $dy/dx > 0$ appears underneath them. In the two lower diagrams the curve is decreasing and $dy/dx < 0$. In the two left-hand diagrams the curve lies above its tangent, i.e. if we travel upwards along a line parallel to the y-axis we come first to the tangent and then to the curve. The curve is said to be *concave upward* at P.

In the two right-hand diagrams the curve lies below its tangent and the curve is said to be *concave downward* at P. Concave upward curves bear some resemblance to cups while concave downward curves are more like hats.

To check whether a curve is concave upward or downward we use the second derivative.

Suppose that $d^2y/dx^2 > 0$. Then dy/dx must be an increasing function of x. Therefore the slope of the curve must be less to the left of P than it is to the right. This corresponds to the two possibilities on the left of figure 77. Thus, *if $d^2y/dx^2 > 0$, the curve is concave upward.*

Similarly, if $d^2y/dx^2 < 0$ at P, the slope of the curve is greater to the left of P than it is to the right so that the cases on the right of figure 77 are appropriate. The curve is concave downward.

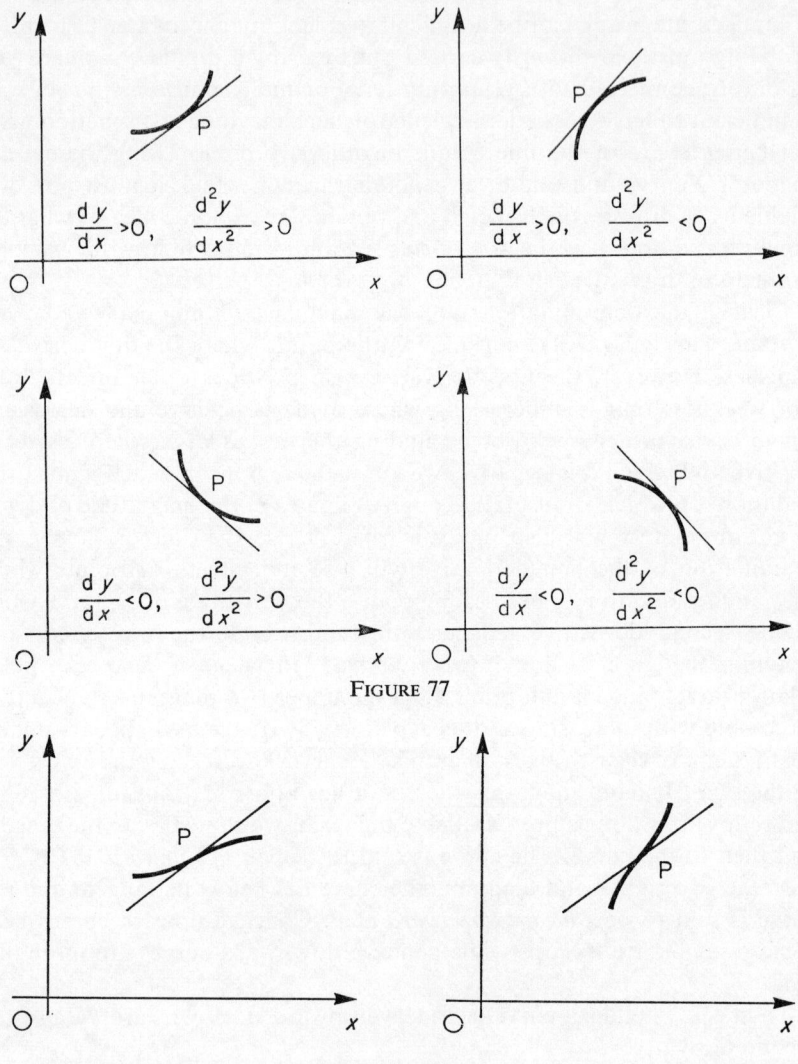

FIGURE 77

FIGURE 78

If $d^2y/dx^2 = 0$ at P, we do not know if the slope is increasing or decreasing. In fact, it is possible that dy/dx has a local maximum or minimum. Now, Theorems 6.2 and 6.3 tell us that this will be true if

Maxima and minima

d^2y/dx^2 changes sign as we pass through P. But this means that the curve changes from concave upward to concave downward or vice versa on passing through P. The two possibilities (for an increasing curve) are shown in figure 78. At such a point the curve crosses the tangent at P because on one side of P it must be above the tangent and on the other side below. A point where a curve changes from concave upward to concave downward or vice versa is called a *point of inflection*.

Figure 71 shows a point of inflection where the tangent is horizontal. In this case the point of inflection is a stationary point. Using Theorems 6.2 and 6.3 we can formulate the following

THEOREM 6.5 *If $f''(c) = 0$ and $f''(x)$ changes from positive to negative values as x goes through c, then the curve has a point of inflection where it changes from concave upward to concave downward.*

Another criterion can be deduced from Theorem 6.4. It is

THEOREM 6.6 *If $f''(c) = 0$ and $f'''(c) < 0$ there is a point of inflection at $x = c$ where the curve changes from concave upward to concave downward as x increases.*

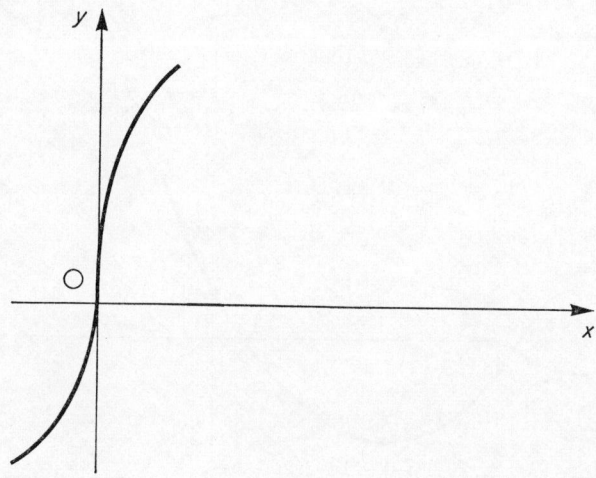

FIGURE 79

We leave the reader to write down the corresponding theorem when $f'''(c) > 0$.

There is one exceptional case of the point of inflection, which is illustrated in figure 79. Here the tangent is vertical at O and both dy/dx and

178   *Introductory Analysis*

d^2y/dx^2 will be infinite there. The preceding theory is then no longer applicable. However, if we think of x as a function of y instead of conversely we can consider the behaviour of dx/dy and d^2x/dy^2, and check that both vanish. Then so long as d^2x/dy^2 changes sign we are sure that we have a point of inflection with a vertical tangent.

Example 15   *Draw a rough sketch of the curve* $y = \frac{1}{3}x^3 + 2x^2 + 3x + \frac{1}{3}$.

Here
$$\frac{dy}{dx} = x^2 + 4x + 3 = (x + 1)(x + 3)$$
and
$$\frac{d^2y}{dx^2} = 2x + 4.$$

In this case dy/dx is positive when $x < -3$ and when $x > -1$. Therefore in these two regions the curve is rising. Between $x = -3$ and $x = -1$, dy/dx is negative, and the curve is falling. By Theorem 6.2 there is a local maximum of $\frac{1}{3}$ at $x = -3$; this is also confirmed by Theorem 6.4 since $d^2y/dx^2 = -2$ there. Similarly, there is a local minimum of -1 at $x = -1$.

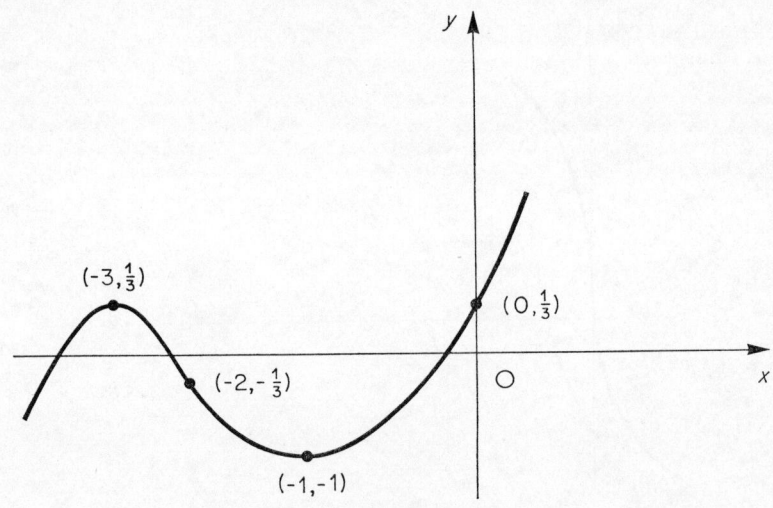

FIGURE 80

At $x = -2$, $d^2y/dx^2 = 0$. To the left of this point d^2y/dx^2 is negative so that the curve is concave downwards. To the right d^2y/dx^2 is positive and the curve is concave upwards. By Theorem 6.5 there is a point of inflection at $x = -2$ where $y = -\frac{1}{3}$ and $dy/dx = -1$. An additional check is provided by noting that $d^3y/dx^3 = 2$ at $x = -2$.

Maxima and minima

This information is sufficient for us to sketch the curve between $x = -3$ and $x = -1$ (see figure 80). Next we note that $y = \frac{1}{3}$ when $x = 0$. Thereafter we know that y continues to increase. In fact, when x is large the term $\frac{1}{3}x^3$ is much larger than the others so that y will be going off to positive infinity in much the same way as $\frac{1}{3}x^3$.

For the portion to the left of $x = -3$, we remark that $y = -1$ when $x = -4$ and that y goes off to negative infinity rather like $\frac{1}{3}x^3$ when x is large and negative.

Putting together all these facts permits one to draw the sketch shown.

Example 16 Draw a sketch of the curve $y = x + 4/x^2$.

Here

$$\frac{dy}{dx} = 1 - \frac{8}{x^3},$$

$$\frac{d^2y}{dx^2} = \frac{24}{x^4}.$$

Firstly, note that $x = 0$ must be specially considered. As x approaches zero $4/x^2$ becomes very large (whether x be positive or negative) while x becomes very small. Hence y goes towards positive infinity as x goes to zero through either positive or negative values.

Next dy/dx is positive when $x < 0$ and when $x > 2$; it is negative for $0 < x < 2$ and vanishes at $x = 2$. Therefore there is a local minimum of 3 at $x = 2$.

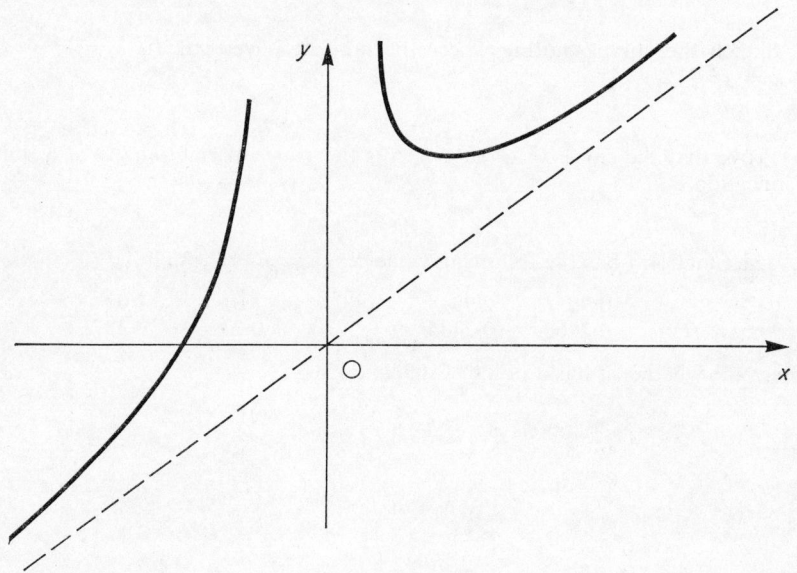

FIGURE 81

180   *Introductory Analysis*

The second derivative d^2y/dx^2 is always positive and so the curve is everywhere concave upward.

At $x = 1$, $y = 5$; at $x = -1$, $y = 3$ and at $x = -2$, $y = -1$. The general shape between $x = -2$ and $x = 2$ can now be sketched in figure 81.

For large values of x, $4x^2$ is small while x is large so that y behaves like x for large x, i.e. $y = x$ is an asymptote. Similarly, $y = x$ is an asymptote for x large and negative.

Exercises

41. For what values of x are the following curves (a) concave upward, (b) concave downward: (i) $y = ax^2 + bx + c$; (ii) $y = 3x^3 + 4x$; (iii) $y = x + 1/x$; (iv) $y = 4\cos x + 3\sin x$; (v) $y = x^4 - 6x^3 + 12x^2 + 5x - 1$; (vi) $y = 1/(1 + x)$.

42. Find the points of inflection (if any) of (i) $y = 2x^3 + 3x^2 + 4x + 5$; (ii) $y = \sin x$; (iii) $y = A(x - a)(x - b)(x - c)$; (iv) $y = \tan x$; (v) $y = x^4$.

43. Sketch the following curves (i) $y = x^3 - 3x$; (ii) $y = x^2(6 - x^2)$; (iii) $y = \cos x - \sin x$; (iv) $y = 4x/(x^2 + 4)$; (v) $y = x/(x - 1)^2$.

44. Sketch a smooth curve $y = f(x)$ for $x > 0$ such that $f(1) = 0$ and $f'(x) = 1/x$ for $x > 0$.

45. Sketch the curves, noting where the tangent is vertical, (i) $x = y^3 - 3y$; (ii) $x = y^2 + 2/y$.

46. Prove that the curve $x^3 - y^3 = 1$ cuts the y-axis at right-angles at a point of inflection.

47. The function f has the following values:

x	0·98	0·99	1·00	1·01	1·02
$f(x)$	0·2468	0·2444	0·2420	0·2396	0·2371

Determine whether it has a point of inflection.

7

The mean-value theorem

7.1 Rolle's theorem

It is easy to believe from a figure such as figure 82 that, if a continuous function vanishes at $x = a$ and $x = b$, there is a point $x = c$ where the function has a horizontal tangent. However, continuity is not enough to

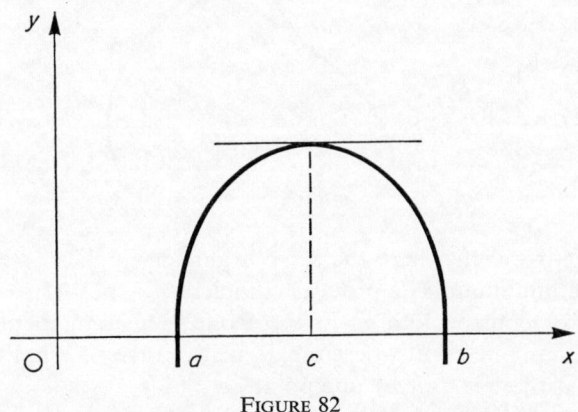

FIGURE 82

ensure this, as can be seen from the continuous function in figure 83. Therefore some extra conditions must be imposed to guarantee a horizontal tangent. These are given by

THEOREM 7.1 (ROLLE'S THEOREM) *If $f(x)$ is continuous on $a \leqslant x \leqslant b$ and has a derivative $f'(x)$ for $a < x < b$, and if $f(a) = f(b) = 0$ then there is at least one number c satisfying $a < c < b$ such that $f'(c) = 0$.*

Proof. If $f(x) = 0$ throughout the interval then $f'(x)$ is also identically zero and the theorem is proved.

If $f(x)$ is not zero at some point it must be positive or negative there. If it is positive we know from section 4.5 that the function will have a maximum positive value somewhere. In other words, there is a point $x = c$ where $f(c)$ is a positive maximum. This point cannot be either a or b since $f(a) = f(b) = 0$ and $f(c) \neq 0$. Therefore $a < c < b$ and Theorem 6.1 shows that $f'(c) = 0$.

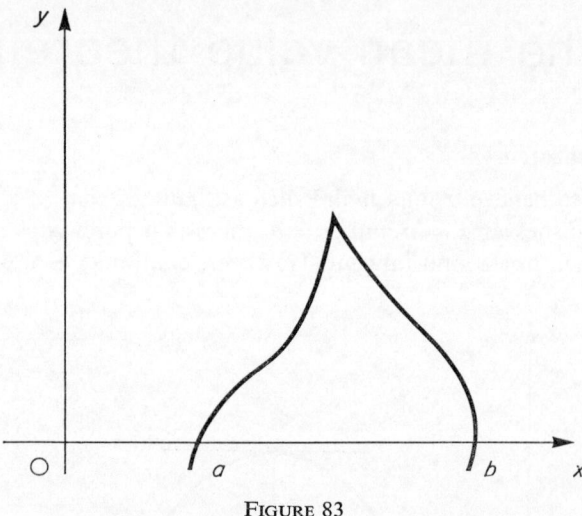

FIGURE 83

If f is negative somewhere the same argument applies except that $f(c)$ is a negative minimum. The proof is complete.

There is, of course, nothing in the Rolle's theorem to prevent there being more than one point where the derivative is zero. What the theorem asserts is that there is *at least* one point.

One interpretation of Rolle's theorem is that between two real roots of $f(x) = 0$ there must be at least one real root of $f'(x) = 0$, assuming that f satisfies the conditions of the theorem. Now, suppose that $f'(c) = 0$ and that $x = d$ is the next number larger than c at which f' vanishes. If there were two values of x between $x = c$ and $x = d$ at which $f(x) = 0$ then $f'(x) = 0$ would have a root between these values, and $x = d$ would not be the next root above $x = c$. Therefore, *between two consecutive points where $f'(x) = 0$ there is at most one point where $f(x) = 0$*. Expressed another way, this states that if $f'(x)$ is always positive (or always negative) between two values of x there is at most one x satisfying $f(x) = 0$. If, at the ends of this interval, f has opposite signs then f must have at least one root in between. Thus, when $f(a)$ and $f(b)$ have opposite signs while $f'(x)$

does not change sign there is precisely one value of x between $x = a$ and $x = b$ such that $f(x) = 0$.

Example 1 When $f(x) = \frac{1}{3}x^3 - x^2 - 3x$, f is a polynomial and so is continuous and differentiable; it therefore complies with the conditions of Rolle's theorem.

Since
$$f'(x) = x^2 - 2x - 3 = (x+1)(x-3)$$
there is at most one root of $f(x) = 0$ between $x = -1$ and $x = 3$. Since $f(-1) = \frac{5}{3}$ and $f(3) = -9$, displaying opposite signs, there is precisely one root. In fact, it is given by $x = 0$.

Since $f(x) \to +\infty$ as $x \to +\infty$ there is another root greater than 3 and, since $f(x) \to -\infty$ as $x \to -\infty$, there is one less than -1.

Example 2 If $f(x) = x^3 - 27x$ on $0 \leqslant x \leqslant 3\sqrt{3}$ find the value of c in Rolle's theorem.

Here $f(0) = 0$, $f(3\sqrt{3}) = 0$ and $f'(x) = 3x^2 - 27$ so that Rolle's theorem does apply. In fact, c must satisfy $3c^2 = 27$ or $c = \pm 3$ of which $c = 3$ is the only one in the given interval.

Exercises

1. Find c when (i) $f(x) = x^2 + 3x + 2$ on $-2 \leqslant x \leqslant -1$, (ii) $f(x) = \sin x$ on $0 \leqslant x \leqslant \pi$.

2. Show that the following equations have one and only one real root on the intervals given (i) $x^3 - 6x^2 + 6 = 0$ on $1 \leqslant x \leqslant 4$, (ii) $x^4 - 8x^2 + 8 = 0$ on $-2 \leqslant x \leqslant -1$.

3. Does Rolle's theorem apply to (i) $x(x-2)/(x+1)$ on $0 \leqslant x \leqslant 2$, (ii) $x(x-2)/(x-1)$ on $0 \leqslant x \leqslant 2$, (iii) $\tan x$ on $0 \leqslant x \leqslant \pi$, (iv) $(4-x^2)^{\frac{2}{3}}$ on $-2 \leqslant x \leqslant 2$, (v) $1/x$ on $-1 \leqslant x \leqslant 2$, (vi) $1 - (2-x)^{\frac{2}{3}}$ on $1 \leqslant x \leqslant 3$?

4. If $f(x)$ is continuous on $a \leqslant x \leqslant b$ and has a derivative $f'(x)$ for $a < x < b$, and if $f(a) = f(b)$, show that there is at least one number c satisfying $a < c < b$ such that $f'(c) = 0$.

5. If $f(x)$ and $f'(x)$ are continuous on $a \leqslant x \leqslant b$, if $f''(x)$ exists for $a < x < b$ and if $f(a) = f(b) = f(c) = 0$ with $a < c < b$, show that there is at least one number d satisfying $a < d < b$ such that $f''(d) = 0$.

6. A polynomial of degree n vanishes at $n + 1$ distinct points. Show that it is the zero polynomial.

7.2 The mean-value theorem

Rolle's theorem makes it possible to derive a result which is useful in determining approximations to the value of a function and forms the basis of some valuable theorems later on. Consider

$$[f(x) - f(a)](b - a) - [f(b) - f(a)](x - a).$$

When $x = a$ this function vanishes. When $x = b$ the function is zero also. Therefore, by Rolle's theorem, there is c, with $a < c < b$, where the derivative vanishes, i.e.

$$f'(c)(b - a) - [f(b) - f(a)] = 0$$

or

$$f(b) - f(a) = (b - a)f'(c).$$

Hence we have proved

THEOREM 7.2 (MEAN-VALUE THEOREM) *If $f(x)$ is continuous on $a \leqslant x \leqslant b$ and has a derivative $f'(x)$ for $a < x < b$, then there is at least one number c, satisfying $a < c < b$, such that*

$$f(b) - f(a) = (b - a)f'(c).$$

The geometrical significance of the mean-value theorem can be seen from figure 84 where the graph of $y = f(x)$ is drawn. A is the point on the

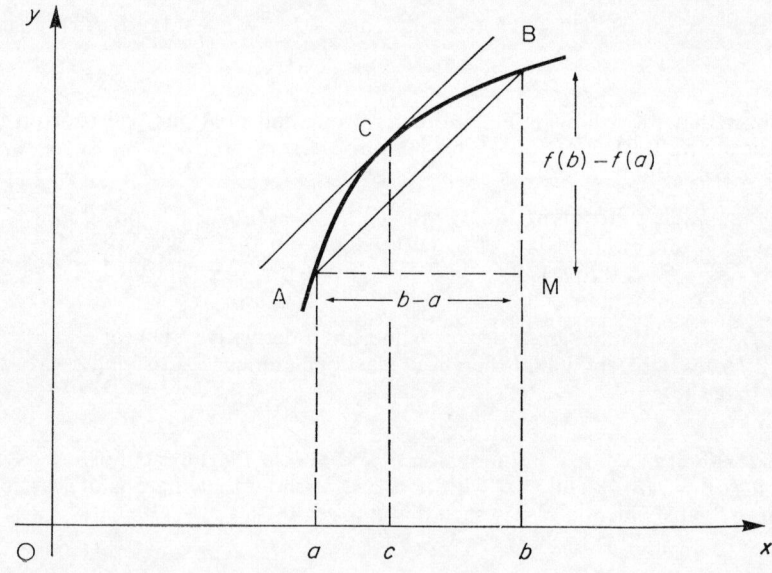

FIGURE 84

graph corresponding to $x = a$; its ordinate is $f(a)$. Similarly, B has ordinate $f(b)$. Then BM $= f(b) - f(a)$ and AM $= b - a$. Hence

$$\frac{f(b) - f(a)}{b - a} = \text{slope of the line AB.}$$

Now, $f'(x)$ is the slope of the tangent to the curve at x. So the mean-value theorem says that there is a point C between A and B at which the slope of the tangent is the same as the slope of the line joining A and B. This is obviously true in the diagram.

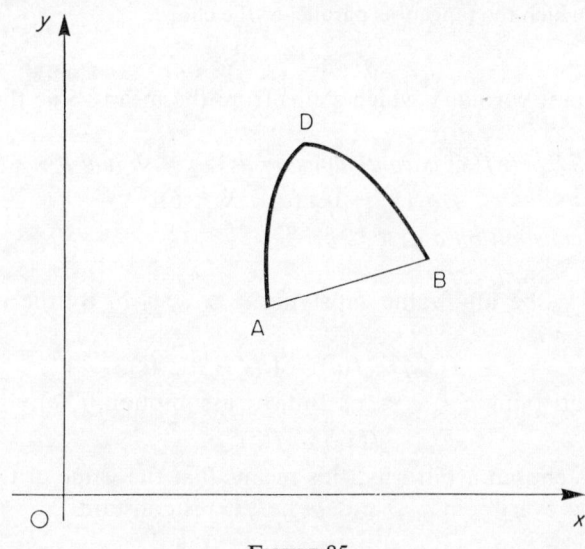

FIGURE 85

If the conditions of the theorem are not met the point C need not exist. See, for example, figure 85 where there is no point C where the tangent is parallel to AB. In fact, the derivative does not exist at D and so the conditions of the mean-value theorem are not met.

The mean-value theorem does not tell us the precise value of c, only that it lies somewhere between a and b. When $f(x)$ is known specifically the theorem may be used to provide an equation to determine c.

Example 3 *If $f(x) = x^3$, find c when (i) $a = 1$, $b = 2$, (ii) $a = -2$, $b = 2$.*

In this case $f'(x) = 3x^2$ and the mean-value theorem states that there is c between a and b such that

$$b^3 - a^3 = 3c^2(b - a).$$

(i) When $a = 1$, $b = 2$ this becomes
$$7 = 3c^2$$
or $c = \pm \sqrt{(7/3)}$, of which $c = \sqrt{(7/3)}$ is the only one in the interval under consideration.

(ii) When $a = -2$, $b = 2$ we have
$$16 = 12c^2$$
or $c = \pm 2/\sqrt{3}$. Both of these points lie in the interval and so either value of c can be used. This does not conflict with the mean-value theorem which merely says that there is *at least one* c. The reader should draw a graph to see where the points are at which the tangent is parallel to the chord.

An important corollary which stems from the mean-value theorem is

COROLLARY 7.2 *If $f(x)$ is continuous on $a \leqslant x \leqslant b$ and if*
$$f'(x) = 0 \quad (a < x < b)$$
then $f(x) =$ constant on $a \leqslant x \leqslant b$.

Proof. Let x_0 be any value satisfying $a \leqslant x_0 \leqslant b$. By the mean-value theorem
$$f(x_0) - f(a) = (x_0 - a)f'(c)$$
for some c satisfying $a < c < x_0$. But, by assumption, $f'(c) = 0$ and so
$$f(x_0) = f(a).$$
Since x_0 was chosen arbitrarily, this means that the value of the function is the same at every x in $a \leqslant x \leqslant b$, i.e. $f(x)$ is constant.

Exercises

7. Find a value of c to satisfy the mean-value theorem when (i) $f(x) = x^2$, $a = 1$, $b = 4$; (ii) $f(x) = px^2 + qx + r$; (iii) $f(x) = 1/x$, $a = 1$, $b = 9$; (iv) $f(x) = (x - 1)^{\frac{2}{3}}$, $a = 1$, $b = 2$.

8(a). If the average speed of a journey without intermediate stops is 40 k.p.h. show, by the mean-value theorem, that at some instant the actual speed must have been 40 k.p.h.

(b). Do you think it is true that, at any instant, there are at least two points on the equator which have the same temperature, assuming that the temperature distribution is differentiable?

9. If $f'(x) > 0$ for $a < x < b$ show, by the mean-value theorem, that $f(b) > f(a)$.

10. Show, by use of the mean-value theorem, that (i) $\sin x < x$ for $x > 0$; (ii) $(1 + x)^{\frac{1}{2}} < 1 + \frac{1}{2}x$ for $x > 0$; (iii) $\tan x > x$ for $0 < x < \frac{1}{2}\pi$; (iv) $|\sin x - \sin a| \leqslant |x - a|$, (v) $\sin x > x - \frac{1}{6}x^3$ for $x > 0$.

11. If $f_1'(x) = f(x)$ and $f_2'(x) = f(x)$ for $a < x < b$, show that $f_1(x) - f_2(x)$ is constant throughout the interval.

12. Does the mean-value theorem apply to (i) $x/(x - 1)$ on $0 \leqslant x \leqslant 2$; (ii) $\tan x$ on $0 \leqslant x \leqslant \pi$; (iii) $x^{\frac{1}{3}}$ on $0 \leqslant x \leqslant 2$; (iv) $|x - 1|^{\frac{1}{2}}$ on $0 \leqslant x \leqslant 2$?

13. If $f(x)$ is continuous on $x \geqslant 0$ and has derivative $f'(x)$ on $x > 0$, prove that
$$f(x) = f(0) + xf'(\theta x)$$
where $0 < \theta < 1$.

14. Let $f(x), f'(x)$ be continuous on $a \leqslant x \leqslant b$ and let $f''(x) > 0$ on $a < x < b$. Let A and B be the points on the curve $y = f(x)$ at which $x = a$ and $x = b$ respectively. Find the point D on the chord whose abscissa is x_0 with $a < x_0 < b$. Apply the mean-value theorem to $f(x)$ on $a \leqslant x \leqslant x_0$ and on $x_0 \leqslant x \leqslant b$. Then use $f''(x) > 0$ and Exercise 9 to show that the point $[x_0, f(x_0)]$ is below D. Deduce that the curve is concave upwards.

7.3 Simple error analysis

In section 5.2 it has been indicated that a first approximation to $f(x + h)$ can be found by using the formula
$$f(x + h) \approx f(x) + hf'(x)$$
and that this approximation will be better the smaller h becomes. The mean-value theorem provides a criterion for estimating the accuracy of formulae of this type.

Suppose that $f'(x)$ is continuous for $a \leqslant x \leqslant b$ and we want to examine the error in the estimate
$$f(b) = f(a) + (b - a)f'(a). \tag{1}$$
From the mean-value theorem
$$f(b) = f(a) + (b - a)f'(c)$$
with $a < c < b$. If M is the greatest value of $f'(x)$ on $a \leqslant x \leqslant b$, then $f'(c) \leqslant M$ and so
$$f(b) - f(a) \leqslant M(b - a).$$
Similarly, if m is the least value of $f'(x)$ on $a \leqslant x \leqslant b$, $f'(c) \geqslant m$ and
$$f(b) - f(a) \geqslant m(b - a).$$

188   *Introductory Analysis*

Therefore
$$[m - f'(a)](b - a) \leqslant f(b) - f(a) - (b - a)f'(a) \leqslant [M - f'(a)](b - a). \tag{2}$$

This inequality provides one estimate of the magnitude of the error in using (1).

Since $f'(a) \geqslant m$, $M - f'(a) \leqslant M - m$ and, since $f'(a) \leqslant M$, $m - f'(a) \geqslant m - M$. An immediate deduction from (2) is that
$$|f(b) - f(a) - (b - a)f'(a)| \leqslant (M - m)(b - a) \tag{3}$$
which gives the maximum error that could possibly be involved.

Example 4   Estimate $(65)^{\frac{1}{3}}$ by (1) and the consequent error.

Here $f(x) = x^{\frac{1}{3}}$ and $a = 64$, $b = 65$. Since $f'(x) = \frac{1}{3}x^{-\frac{2}{3}}$, (1) gives
$$(65)^{\frac{1}{3}} \approx (64)^{\frac{1}{3}} + \tfrac{1}{3}(64)^{-\frac{2}{3}} = 4 + \tfrac{1}{48} = 4\cdot 0208.$$
Now $M = \tfrac{1}{3}(64)^{-\frac{2}{3}}$ and $m = \tfrac{1}{3}(65)^{-\frac{2}{3}}$ so that (2) shows that
$$\tfrac{1}{3}(65)^{-\frac{2}{3}} - \tfrac{1}{3}\cdot 64^{-\frac{2}{3}} \leqslant (65)^{\frac{1}{3}} - 4\cdot 0208 \leqslant 0.$$
The left-hand side can be estimated by applying (1) to $f(x) = \tfrac{1}{3}x^{-\frac{2}{3}}$ and so is $-\tfrac{2}{9}\cdot(64)^{-\frac{5}{3}} = -\tfrac{1}{4608} = -0\cdot 0002$ approximately. Thus it can be concluded that $(65)^{\frac{1}{3}}$ lies between 4·0208 and 4·0206 inclusive.

When the second derivative of f exists it is possible to provide an estimate of the error in (1) which is usually easier to calculate than (2) and just as good. Consider
$$[f(x) - f(b) + (b - x)f'(x)](b - a)^2 - [f(a) - f(b) + (b - a)f'(a)](b - x)^2.$$
This function vanishes at $x = a$ and $x = b$. Therefore, by Rolle's theorem, there is c, with $a < c < b$, where the derivative is zero, i.e.
$$(b - c)f''(c)(b - a)^2 + 2[f(a) - f(b) + (b - a)f'(a)](b - c) = 0.$$
Hence
$$f(b) = f(a) + (b - a)f'(a) + \tfrac{1}{2}(b - a)^2 f''(c) \tag{4}$$
where $a < c < b$. Remember that c will depend on a and b in general.

An alternative way of writing (4) is, by putting $a = x$, $b = x + h$,
$$f(x + h) = f(x) + hf'(x) + \tfrac{1}{2}h^2 f''(x + \theta h)$$
where $0 < \theta < 1$. Changing x or h will usually cause an alteration in θ.

It can be seen from (4) that, if $|f''(x)| \leqslant M_1$ on $a \leqslant x \leqslant b$, the error in using (1) cannot exceed $\tfrac{1}{2}(b - a)^2 M_1$.

The mean-value theorem

Another place where the theorems of this chapter can be used is in *linear interpolation*, which has already been described in section 1.4. In linear interpolation the graph of f over an interval is approximated by a straight line so that, if $f(a)$ and $f(b)$ are known, $f(x)$ is estimated as

$$f(a) + \frac{x-a}{b-a}[f(b) - f(a)].$$

What is the error of this approximation?

Choose any fixed x_0 such that $a < x_0 < b$ and consider

$$(x_0 - a)(x_0 - b)\left\{f(x) - f(a) - \frac{x-a}{b-a}[f(b) - f(a)]\right\} -$$

$$(x - a)(x - b)\left\{f(x_0) - f(a) - \frac{x_0 - a}{b-a}[f(b) - f(a)]\right\}.$$

This function vanishes at $x = a$, $x = x_0$, $x = b$. Therefore, by Rolle's theorem, the derivative vanishes at $x = c_1$ and $x = c_2$ where $a < c_1 < x_0$, $x_0 < c_2 < b$. Hence

$$(x_0 - a)(x_0 - b)\left[f'(x) - \frac{f(b) - f(a)}{b-a}\right] -$$

$$[2x - (a+b)]\left\{f(x_0) - f(a) - \frac{x_0 - a}{b-a}[f(b) - f(a)]\right\}$$

is zero at $x = c_1$ and $x = c_2$. Therefore its derivative vanishes at $x = c$ where $c_1 < c < c_2$, i.e.

$$(x_0 - a)(x_0 - b)f''(c) - 2\left\{f(x_0) - f(a) - \frac{x_0 - a}{b-a}[f(b) - f(a)]\right\} = 0.$$

Since x_0 was selected arbitrarily we can say

$$f(x) - \left\{f(a) + \frac{x-a}{b-a}[f(b) - f(a)]\right\} = \tfrac{1}{2}(x - a)(x - b)f''(c)$$

where $a < c < b$ because $c_1 > a$ and $c_2 < b$.

Consequently, an estimate of the error in linear interpolation is provided by $\tfrac{1}{2}(x - a)(x - b)f''(c)$. To avoid the occurrence of x in the estimate we can note that $(x - a)(x - b)$ must lie between $\pm \tfrac{1}{4}(b - a)^2$ on $a \leqslant x \leqslant b$ (check this statement by the methods of the preceding chapter). Therefore the error cannot exceed

$$\tfrac{1}{8}(b - a)^2 |f''(c)|.$$

If we know that $|f''(x)| \leqslant M_1$ on $a \leqslant x \leqslant b$, we can be sure that the error in linear interpolation does not exceed $\tfrac{1}{8}(b - a)^2 M_1$.

Example 5 In a table of sin x the entries are given for each degree. Find a limit to the error in linear interpolation between two consecutive entries.

Here $f''(x) = -\sin x$, provided that we work in radians, and $b - a$ is one degree or $\pi/180$ radians. Since $|\sin x| \leqslant 1$, $M_1 = 1$ and the error in linear interpolation cannot exceed

$$\frac{1}{8}\left(\frac{\pi}{180}\right)^2 = \frac{\pi^2}{259{,}200} < 0{\cdot}00004.$$

Therefore, linear interpolation would be satisfactory if results were required only to four decimal places.

The estimate can be improved by using the form involving $f''(c)$ in certain parts of the table. Thus, between $4°$ and $5°$, $|f''(x)|$ does not exceed $0{\cdot}1$ and so the error is not greater than

$$0{\cdot}1 \times 0{\cdot}00004 = 0{\cdot}000004$$

in magnitude. Hence, linear interpolation for sin x between $4°$ and $5°$ will be accurate to 5 decimal places if the entries themselves have this accuracy.

Exercises

15. Estimate the following and provide an estimate of the error: (i) $\sqrt{10}$ by taking $f(x) = x^{\frac{1}{2}}$, $a = 9$, $b = 10$; (ii) $(2{\cdot}001)^2$ by taking $f(x) = x^2$, $a = 2$, $b = 2{\cdot}001$; (iii) $1/999$ by taking $f(x) = 1/x$, $a = 1000$, $b = 999$.

16. Show, by means of (4), that cos x can be approximated by 1 with an error less than $0{\cdot}02$ if $|x|$ is less than $0{\cdot}2$ radians.

17. A table of tan x gives entries at intervals of $0{\cdot}002$ radians. Show that linear interpolation is valid (i) to 5 decimal places when $|x| < \frac{1}{4}\pi$; (ii) to 6 decimal places when $|x| < \frac{1}{8}\pi$.

18. At what intervals should entries be given in a table of x^3 on $0 \leqslant x \leqslant 1$ in order that linear interpolation would be accurate to 2 decimal places?

19. The cylinders of a car engine are rebored to have a slightly larger radius. Indicate how you would calculate the reduction in weight of the cylinder block.

20. Estimate the maximum error in using $(1/h)[f(x+h) - f(x)]$ to calculate the numerical value of $f'(x)$ at $x = 1$ when (i) $f(x) = x^{\frac{1}{2}}$; (ii) $f(x) = \sin x$. Assume that f can be calculated exactly.

7.4 Generalized mean-value theorem

There is an extension of the mean-value theorem which will be useful subsequently.

THEOREM 7.3 *If $f(x)$ and $g(x)$ are continuous on $a \leqslant x \leqslant b$, and have derivatives $f'(x)$, $g'(x)$ with $g'(x) \neq 0$ on $a < x < b$, then there is at least one number c, satisfying $a < c < b$, such that*

$$\frac{f(b) - f(a)}{g(b) - g(a)} = \frac{f'(c)}{g'(c)}.$$

Note firstly that this theorem reduces to the mean-value theorem when $g(x) = x$. Secondly, it tells us more than applying the mean-value theorem to f and g separately and dividing the results. For then, we would obtain $f'(c_1)/g'(c_2)$ whereas the theorem has the same c in numerator and denominator.

Proof. Suppose $g(b) = g(a)$; then, by Exercise 4, there is c_0, with $a < c_0 < b$, such that $g'(c_0) = 0$. Since this is contrary to the hypothesis of the theorem, $g(b) \neq g(a)$.

Consider now

$$f(x) - f(b) + [g(b) - g(x)]\frac{f(b) - f(a)}{g(b) - g(a)}.$$

This function is zero at $x = a$ and $x = b$, and satisfies the conditions of Rolle's theorem. Therefore the derivative vanishes for some c such that $a < c < b$, i.e.

$$f'(c) - g'(c) \cdot \frac{f(b) - f(a)}{g(b) - g(a)} = 0$$

and the theorem is proved since $g'(c) \neq 0$.

One illustration of this theorem is

$$\frac{\sin b - \sin a}{\cos b - \cos a} = -\cot c$$

with $a < c < b$, when $0 \leqslant a < b \leqslant \pi$.

Exercise

21. If $f(x)$ and $g(x)$ are continuous on $a \leqslant x \leqslant b$, and have derivatives $f'(x)$, $g'(x)$ on $a < x < b$ prove that there is at least one c, with $a < c < b$, such that

$$[f(b) - f(a)]g'(c) = [g(b) - g(a)]f'(c).$$

7.5 Indeterminate forms

There are many circumstances in which it is desirable to say something about $f(x)/g(x)$ at a point $x = a$ where both $f(x)$ and $g(x)$ vanish. This problem has already been encountered in Chapter 4 in connection with limits; for example, the behaviour of $(\sin x)/x$ near $x = 0$.

The substitution $x = a$ into f and g is not permissible because it leads to 0/0 which is meaningless. Such functions are called *indeterminate forms* although, in fact, they have no value at $x = a$ rather than one which is indeterminate. However, they may possess a limit as x approaches a and it is with the evaluation of such limits that we are concerned.

THEOREM 7.4 (L'HÔPITAL'S RULE) *If $f(x)$ and $g(x)$ are continuous on $a \leqslant x \leqslant b$, and have derivatives $f'(x)$, $g'(x)$ with $g'(x) \neq 0$ on $a < x < b$, and if $f(a) = 0$ and $g(a) = 0$ then*

$$\lim_{x \to a^+} \frac{f(x)}{g(x)} = \lim_{x \to a^+} \frac{f'(x)}{g'(x)}$$

whenever the limit on the right-hand side exists.

Proof. The hypotheses of the theorem satisfy the conditions of Theorem 7.3 so that

$$\frac{f(x) - f(a)}{g(x) - g(a)} = \frac{f'(c)}{g'(c)}$$

for some c such that $a < c < x$. Since $f(a)$ and $g(a)$ are zero

$$\frac{f(x)}{g(x)} = \frac{f'(c)}{g'(c)}.$$

Therefore

$$\lim_{x \to a^+} \frac{f(x)}{g(x)} = \lim_{x \to a^+} \frac{f'(c)}{g'(c)}.$$

But as x approaches a so does c because c must always lie between a and x; consequently

$$\lim_{x \to a^+} \frac{f'(c)}{g'(c)} = \lim_{x \to a^+} \frac{f'(x)}{g'(x)}$$

provided that the limit on the right exists. The proof of the theorem is concluded.

The mean-value theorem

The same method can be used to prove that

$$\lim_{x \to b^-} \frac{f(x)}{g(x)} = \lim_{x \to b^-} \frac{f'(x)}{g'(x)}$$

when $f(b) = 0$ and $g(b) = 0$.

If $\lim_{x \to a} \frac{f'(x)}{g'(x)}$ exists then

$$\lim_{x \to a^+} \frac{f'(x)}{g'(x)} = \lim_{x \to a^-} \frac{f'(x)}{g'(x)}$$

(section 4.2) so that L'Hôpital's rule can be applied to two-sided limits as well as one-sided.

If, in addition to the hypotheses of Theorem 7.4,

$$\lim_{x \to a^+} f'(x) = f'(a),$$

and

$$\lim_{x \to a^+} g'(x) = g'(a) \neq 0 \tag{5}$$

then Theorem 4.1(iv) shows that

$$\lim_{x \to a^+} \frac{f(x)}{g(x)} = \frac{\lim_{x \to a^+} f'(x)}{\lim_{x \to a^+} g'(x)} = \frac{f'(a)}{g'(a)}. \tag{6}$$

However, although (6) is extremely useful, it must be remarked that L'Hôpital's rule is more powerful because (5) need not be met. Suppose, in fact, that $f'(a) = 0$ and $g'(a) = 0$ while $g''(x) \neq 0$ on $a < x < b$. Then, applying L'Hôpital's rule to $f'(x)$, we obtain

$$\lim_{x \to a^+} \frac{f'(x)}{g'(x)} = \lim_{x \to a^+} \frac{f''(x)}{g''(x)}$$

provided that the limit on the right-hand side exists.

Should $g''(a)$ be non-zero then the limit will be known (for the functions in this book, at any rate). If $g''(a)$ and $f''(a)$ are both zero we take a derivative again of the numerator and denominator separately. In fact we continue the process until either the numerator or denominator has a non-zero limit. It may happen that one is zero and the other not. If the numerator is zero, but not the denominator, the limit is zero whereas if the denominator is zero, but not the numerator, then the limit is infinite.

Of course, one is not obliged to evaluate the limit of an indeterminate form by L'Hôpital's rule but it is one of the most powerful and systematic methods available.

194   *Introductory Analysis*

Example 6   Evaluate

$$\lim_{x \to 0} \frac{x + \sin x}{2x - \sin x}.$$

When $x = 0$ both numerator and denominator vanish so that this is an indeterminate form. The conditions of L'Hôpital's rule are met and

$$\lim_{x \to 0} \frac{x + \sin x}{2x - \sin x} = \lim_{x \to 0} \frac{1 + \cos x}{2 - \cos x} = \frac{1 + 1}{2 - 1} = 2.$$

Example 7   Evaluate

$$\lim_{x \to 0} \frac{(1 + 2x)^{\frac{1}{2}} - 1 - x}{x^2}.$$

The form is indeterminate and the conditions of L'Hôpital's rule are met on $x > 0$ and $-\frac{1}{2} < x < 0$. Hence

$$\lim_{x \to 0} \frac{(1 + 2x)^{\frac{1}{2}} - 1 - x}{x^2} = \lim_{x \to 0} \frac{(1 + 2x)^{-\frac{1}{2}} - 1}{2x}.$$

The process has led to another indeterminate form. Apply L'Hôpital's rule again to obtain

$$\lim_{x \to 0} \frac{-(1 + 2x)^{-\frac{3}{2}}}{2} = -\frac{1}{2}.$$

Example 8   Evaluate

$$\lim_{x \to 0} \frac{\sin x}{x^2}.$$

The form is indeterminate and by L'Hôpital's rule the limit is equal to

$$\lim_{x \to 0} \frac{\cos x}{2x}$$

which is clearly infinite. Correspondingly,

$$\lim_{x \to 0} \frac{x^2}{\sin x} = 0.$$

There are several extensions of L'Hôpital's rule that are important. Suppose, firstly, that we are interested in what happens when $x \to +\infty$,

The mean-value theorem

and when both $f(x)$ and $g(x)$ tend to zero. Then we try to apply L'Hôpital's rule by making the change of variable $x = 1/t$ so that

$$\lim_{x \to +\infty} \frac{f(x)}{g(x)} = \lim_{t \to 0^+} \frac{f(1/t)}{g(1/t)}.$$

Then, since

$$\frac{d}{dt} f\left(\frac{1}{t}\right) = f'\left(\frac{1}{t}\right) \frac{d}{dt}\left(\frac{1}{t}\right) = -\frac{1}{t^2} f'\left(\frac{1}{t}\right)$$

by Theorem 5.8,

$$\lim_{x \to +\infty} \frac{f(x)}{g(x)} = \lim_{t \to 0^+} \frac{-(1/t^2) f'(1/t)}{-(1/t^2) g'(1/t)}$$

$$= \lim_{t \to 0^+} \frac{f'(1/t)}{g'(1/t)}$$

$$= \lim_{x \to +\infty} \frac{f'(x)}{g'(x)}.$$

Since we applied L'Hôpital's rule to $f(1/t)$ we can now state

COROLLARY 7.4a *If $f'(x)$ and $g'(x)$ exist, with $g'(x) \neq 0$, on some interval $x > B > 0$ and if $\lim_{x \to +\infty} f(x) = 0$, $\lim_{x \to +\infty} g(x) = 0$ then*

$$\lim_{x \to +\infty} \frac{f(x)}{g(x)} = \lim_{x \to +\infty} \frac{f'(x)}{g'(x)}$$

when the limit on the right-hand side exists.

Indeed, what has been shown is that L'Hôpital's rule can be employed in exactly the same way when $x \to +\infty$ as when $x \to a^+$.

Another class of indeterminate forms occurs when both $f(x)$ and $g(x)$ are infinite at $x = a$. In this case we proceed as follows:

$$\lim_{x \to a^+} \frac{f(x)}{g(x)} = \lim_{x \to a^+} \frac{1/g(x)}{1/f(x)}.$$

Now $1/g(x)$ and $1/f(x)$ must tend to zero as $x \to a^+$. Therefore we can try to apply L'Hôpital's rule with $1/g$ as numerator and $1/f$ as denominator. However, it turns out to give the same result as using L'Hôpital's rule on f/g. A full proof is beyond the scope of this book, but we indicate why the result arises when $\lim_{x \to a^+} f/g$ and $\lim_{x \to a^+} f'/g'$ are neither zero nor infinite.

Then, by Theorem 4.1,

$$\lim_{x\to a^+} \frac{f(x)}{g(x)} = \lim_{x\to a^+} \frac{1/g}{1/f} = \lim_{x\to a^+} \frac{g'/g^2}{f'/f^2}$$

$$= \lim_{x\to a^+} \frac{g'}{f'} \cdot \lim_{x\to a^+} \frac{f^2}{g^2}$$

$$= \lim_{x\to a^+} \frac{g'}{f'} \cdot \left(\lim_{x\to a^+} \frac{f}{g}\right)^2.$$

Dividing through by the left-hand side

$$1 = \lim_{x\to a^+} \frac{g'}{f'} \cdot \lim_{x\to a^+} \frac{f}{g}$$

whence

$$\lim_{x\to a^+} \frac{f}{g} = \frac{1}{\lim_{x\to a^+} (g'/f')} = \lim_{x\to a^+} \frac{f'}{g'}.$$

The result can also be proved to hold when the limit is zero and when it is infinite.

It can also be proved that L'Hôpital's rule can be employed as $x \to +\infty$ when both f and g become infinite.

Example 9 Find

$$\lim_{x\to +\infty} \frac{x^3 + 3}{3x^3 + x}.$$

In this case f and g become infinite as $x \to +\infty$ and L'Hôpital's rule gives

$$\lim_{x\to +\infty} \frac{x^3 + 3}{3x^3 + x} = \lim_{x\to +\infty} \frac{3x^2}{9x^2 + 1}.$$

This is still an indeterminate form so we apply the rule twice more, obtaining

$$\lim_{x\to +\infty} \frac{3x^2}{9x^2 + 1} = \lim_{x\to +\infty} \frac{6x}{18x} = \lim_{x\to +\infty} \frac{6}{18} = \frac{1}{3}.$$

However, the result can be derived more simply by dividing the numerator and denominator of the original fraction by x^3. Then

$$\lim_{x\to +\infty} \frac{x^3 + 3}{3x^3 + x} = \lim_{x\to +\infty} \frac{1 + 3/x^3}{3 + 1/x^2} = \frac{1}{3}$$

since $3/x^3$ and $1/x^2$ tend to zero as $x \to +\infty$.

Example 10 Find
$$\lim_{x \to 0} \frac{\cot x}{\cot 2x}.$$

In this case f and g become infinite as $x \to 0$. Trying L'Hôpital's rule we have
$$\lim_{x \to 0} \frac{\cot x}{\cot 2x} = \lim_{x \to 0} \frac{\operatorname{cosec}^2 x}{2\operatorname{cosec}^2 2x}$$
which is still indeterminate. Therefore we apply the rule again to obtain
$$\lim_{x \to 0} \frac{\cot x}{\cot 2x} = \lim_{x \to 0} \frac{2 \operatorname{cosec}^2 x \cot x}{8 \operatorname{cosec}^2 2x \cot 2x}.$$
It is now clear that L'Hôpital's rule is not helping us because our original fraction is multiplied by $\operatorname{cosec}^2 x / \operatorname{cosec}^2 2x$ which also has both numerator and denominator infinite.

Our next manoeuvre is an attempt to make the fraction have a numerator and denominator which vanish. This can be done by observing that $\cot x = 1/\tan x$. Thus
$$\frac{\cot x}{\cot 2x} = \frac{\tan 2x}{\tan x}$$
and now L'Hôpital's rule can be employed with the result
$$\lim_{x \to 0} \frac{\tan 2x}{\tan x} = \lim_{x \to 0} \frac{2 \sec^2 2x}{\sec^2 x} = 2.$$
Therefore
$$\lim_{x \to 0} \frac{\cot x}{\cot 2x} = 2.$$

This example serves as an illustration that, when both f and g become infinite as $x \to a^+$, it may be preferable to manipulate the fraction so that both numerator and denominator vanish before applying L'Hôpital's rule.

Indeterminate forms can also arise in other ways. For example, one might have $f(x)g(x)$ where f is zero and g infinite when $x = a$; an example is $(x - \frac{1}{2}\pi) \tan x$ when $x = \frac{1}{2}\pi$. Or one might have $f(x) - g(x)$ where both f and g are infinite; an example is $\operatorname{cosec} x - 1/x$ when $x = 0$. In cases such as this we attempt to manipulate the expression so that it becomes one of the forms we have already discussed.

Example 11 Find $\lim_{x \to \frac{1}{2}\pi} (x - \frac{1}{2}\pi) \tan x.$

Now
$$\lim_{x \to \frac{1}{2}\pi} (x - \frac{1}{2}\pi) \tan x = \lim_{x \to \frac{1}{2}\pi} \frac{x - \frac{1}{2}\pi}{\cot x}.$$

In this form both numerator and denominator vanish at $x = \tfrac{1}{2}\pi$. Applying L'Hôpital's rule

$$\lim_{x \to \frac{1}{2}\pi} \frac{x - \tfrac{1}{2}\pi}{\cot x} = \lim_{x \to \frac{1}{2}\pi} \frac{1}{-\cosec^2 x} = -1.$$

Thus

$$\lim_{x \to \frac{1}{2}\pi} (x - \tfrac{1}{2}\pi) \tan x = -1.$$

Example 12 Find

$$\lim_{x \to 0} \left(\cosec x - \frac{1}{x} \right).$$

We have

$$\lim_{x \to 0} \left(\cosec x - \frac{1}{x} \right) = \lim_{x \to 0} \frac{x - \sin x}{x \sin x}.$$

Now, L'Hôpital's rule can be employed because both numerator and denominator vanish. Therefore

$$\lim_{x \to 0} \left(\cosec x - \frac{1}{x} \right) = \lim_{x \to 0} \frac{1 - \cos x}{\sin x + x \cos x}$$

$$= \lim_{x \to 0} \frac{\sin x}{2 \cos x - x \sin x}$$

on using L'Hôpital's rule again. It is now evident that

$$\lim_{x \to 0} \left(\cosec x - \frac{1}{x} \right) = 0.$$

Exercises

22. Evaluate the following limits:

(i) $\lim\limits_{x \to 1} \dfrac{x^2 - 1}{x^2 + x - 2}$; (ii) $\lim\limits_{x \to 0} \dfrac{\sin ax}{\sin bx}$; (iii) $\lim\limits_{x \to a} \dfrac{x - a}{x^2 - a^2}$;

(iv) $\lim\limits_{x \to \pi} \dfrac{x \cos x + \pi}{\sin x}$; (v) $\lim\limits_{x \to 0} \dfrac{\sin x - x}{x^3}$; (vi) $\lim\limits_{x \to 0} \dfrac{x^2}{1 - \cos x}$;

(vii) $\lim\limits_{x \to 0^+} \dfrac{x - \sin x}{(x \sin x)^{3/2}}$; (viii) $\lim\limits_{x \to a} \dfrac{\cos x - \cos a}{x - a}$;

(ix) $\lim\limits_{x \to 0} \dfrac{3 \sin 2x - \sin 6x}{3 \tan 2x - \tan 6x}$; (x) $\lim\limits_{x \to 3} \dfrac{(x - 2)^{1/2} - 1}{x^2 - 9}$.

The mean-value theorem

23. Criticize the following arguments: By L'Hôpital's rule

(a) $\lim\limits_{x \to 0} \dfrac{\frac{1}{2}x^2 - x}{\frac{1}{3}x + 1} = \lim\limits_{x \to 0} \dfrac{x - 1}{\frac{1}{3}} = -3.$

(b) $\lim\limits_{x \to 2} \dfrac{3x^2 + 2x - 16}{x^2 + 4x - 12} = \lim\limits_{x \to 2} \dfrac{6x + 2}{2x + 4} = \lim\limits_{x \to 2} \dfrac{6}{2} = 3.$

24. Criticize the following argument: By L'Hôpital's rule

$$\lim_{x \to 0} \frac{x^2 \sin(1/x)}{x} = \lim_{x \to 0} [2x \sin(1/x) - \cos(1/x)]$$
$$= -\lim_{x \to 0} \cos(1/x)$$

which does not exist. Therefore $\lim\limits_{x \to 0} \dfrac{x^2 \sin(1/x)}{x}$ does not exist.

25. Evaluate (i)

(i) $\lim\limits_{x \to 0} \dfrac{x - \tan x}{ax - \sin x}$; (ii) $\lim\limits_{x \to 0} \dfrac{12 \sin x^2 - 3 \sin^2 2x}{x^4}$;

(iii) $\lim\limits_{x \to 0} \dfrac{\sin(\sin x) - \sin x}{x^3}.$

26. Evaluate

(i) $\lim\limits_{x \to +\infty} x \sin(1/x);$ (ii) $\lim\limits_{x \to +\infty} \dfrac{x^3 + 2x^2}{3x^3 + x};$

(iii) $\lim\limits_{x \to -\infty} \dfrac{x^2 + x}{2x^2 - 1};$ (iv) $\lim\limits_{x \to -\infty} \dfrac{2x}{(x^2 + 1)^{\frac{1}{2}}}.$

27. Evaluate

(i) $\lim\limits_{x \to \frac{1}{2}\pi} \dfrac{\tan 3x}{\tan x};$ (ii) $\lim\limits_{x \to \frac{1}{2}\pi} \dfrac{\sec x + 3}{\tan x - 2}.$

28. Evaluate

(i) $\lim\limits_{x \to +\infty} x^{1/5} \sin\left(\dfrac{1}{x^{\frac{1}{3}}}\right);$ (ii) $\lim\limits_{x \to 0} \left(\cot x - \dfrac{1}{x}\right);$

(iii) $\lim\limits_{x \to 0} \dfrac{1}{x}\left(\operatorname{cosec} x - \dfrac{1}{x}\right);$ (iv) $\lim\limits_{x \to +\infty} [x - \tfrac{1}{2}(4x^2 + x)^{\frac{1}{2}}];$

(v) $\lim_{x \to 0} (\cot 2x - \cot x)$;   (vi) $\lim_{x \to 0} (\cot x - \text{cosec } x)$;

(vii) $\lim_{x \to 2^+} \left[\dfrac{1}{x-2} - \dfrac{1}{(x-2)^{\frac{1}{4}}} \right]$;   (viii) $\lim_{x \to 0} x \text{ cosec } x$;

(ix) $\lim_{x \to \frac{1}{4}\pi} (1 - \tan x) \sec 2x$;   (x) $\lim_{x \to 0} \left(\text{cosec}^2 x - \dfrac{1}{x^2} \right)$.

7.6 Taylor's theorem

One of the most important and fundamental theorems for analysis, numerical work and the practical application of the calculus stems from the mean-value theorem. The theorem is an extension of ideas already introduced earlier in this chapter.

We have seen that, under certain conditions,
$$f(b) = f(a) + (b-a)f'(c)$$
with $a < c < b$ (Theorem 7.2) and
$$f(b) = f(a) + (b-a)f'(a) + \tfrac{1}{2}(b-a)^2 f''(c)$$
(equation (4)). We now try to extend such expansions by including more derivatives, assuming that they exist.

Let
$$R = \frac{1}{(b-a)^n}\left[f(b) - f(a) - \frac{b-a}{1!}f'(a) - \frac{(b-a)^2}{2!}f''(a) - \ldots \right.$$
$$\left. - \frac{(b-a)^{n-1}}{(n-1)!}f^{(n-1)}(a) \right] \quad (7)$$

where, as usual, $n!$ denotes $n(n-1)(n-2) \ldots 2.1$ so that $1! = 1$, $2! = 2$, $3! = 3.2.1 = 6$ for example. (Although we shall not use $0!$, it is convenient to define $0!$ as 1). Consider

$$F(x) = f(x) - f(b) + \frac{b-x}{1!}f'(x) + \frac{(b-x)^2}{2!}f''(x) + \ldots$$
$$+ \frac{(b-x)^{n-1}}{(n-1)!}f^{(n-1)}(x) + (b-x)^n R. \quad (8)$$

Taking $f, f', \ldots, f^{(n-1)}$ continuous on $a \leqslant x \leqslant b$ we see that $F(x)$ is continuous on $a \leqslant x \leqslant b$. Also $F(a) = 0$, because of the choice of R, and $F(b) = 0$. Therefore, so long as $f^{(n)}$ exists for $a < x < b$, Rolle's theorem is applicable and there is a c, with $a < c < b$, such that $F'(c) = 0$, i.e.

$$f'(c) + [(b-c)f''(c) - f'(c)] + \left[\frac{(b-c)^2}{2!}f'''(c) - \frac{b-c}{1!}f''(c)\right] + \ldots$$

$$+ \left[\frac{(b-c)^{n-1}}{(n-1)!}f^{(n)}(c) - \frac{(b-c)^{n-2}}{(n-2)!}f^{(n-1)}(c)\right] - n(b-c)^{n-1}R = 0$$

or

$$\frac{(b-c)^{n-1}}{(n-1)!}f^{(n)}(c) - n(b-c)^{n-1}R = 0$$

or

$$R = \frac{1}{n!}f^{(n)}(c)$$

since $b \neq c$. On substituting for R in (7) we have proved

THEOREM 7.5 (TAYLOR'S THEOREM) *If $f(x), f'(x), \ldots, f^{(n-1)}(x)$ are continuous on $a \leq x \leq b$ and if $f^{(n)}(x)$ exists for $a < x < b$, then there is at least one number c, satisfying $a < c < b$, such that*

$$f(b) = f(a) + \frac{b-a}{1!}f'(a) + \frac{(b-a)^2}{2!}f''(a) + \ldots$$

$$+ \frac{(b-a)^{n-1}}{(n-1)!}f^{(n-1)}(a) + \frac{(b-a)^n}{n!}f^{(n)}(c).$$

Often the right-hand side in Theorem 7.5 is called a *Taylor expansion* of f. To indicate that the nth derivative is the last one which occurs it is sometimes said that the formula in Theorem 7.5 is the *Taylor expansion to n terms*. Again, to indicate that all, but the last, derivatives in the expansion are calculated at a the expansion is described as being *centred on a*. The last term is different from all the others in that c enters it; in view of this difference it is known as the *remainder term*. To be more precise, the term $(b-a)^n f^{(n)}(c)/n!$ is known as *Lagrange's form of the remainder*. It can be put into various other forms. For instance, suppose that instead of defining $F(x)$ by (8) we had taken

$$F(x) = f(x) - f(b) + \frac{b-x}{1!}f'(x) + \ldots$$

$$+ \frac{(b-x)^{n-1}}{(n-1)!}f^{(n-1)}(x) + (b-x)(b-a)^{n-1}R.$$

Then Rolle's theorem could still be employed and
$$\frac{(b-c)^{n-1}}{(n-1)!}f^{(n)}(c) - (b-a)^{n-1}R = 0.$$
Substitution for R from (7) leads to

COROLLARY 7.5 *Under the conditions of Theorem 7.5, there is c, satisfying $a < c < b$, such that*
$$f(b) = f(a) + \frac{b-a}{1!}f'(a) + \ldots$$
$$+ \frac{(b-a)^{n-1}}{(n-1)!}f^{(n-1)}(a) + \frac{(b-a)(b-c)^{n-1}}{(n-1)!}f^{(n)}(c).$$

The term $(b-a)(b-c)^{n-1}f^{(n)}(c)/(n-1)!$ is called *Cauchy's form of the remainder*. In general, the c of Corollary 7.5 will be different from the c of Theorem 7.5. In fact, it is important to remember that c may be changed if, a, b or n is altered in a Taylor expansion. Even if a and b are kept fixed but the number of terms in the expansion is varied by increasing or decreasing n, c may alter.

Lagrange's and Cauchy's forms are not the only possible ways of expressing the remainder (see Exercise 36). In general, one chooses the form which is most convenient to the problem on hand. Sometimes Cauchy's form is most suitable and sometimes Lagrange's (see sections 7.7, 9.5) and sometimes neither is convenient.

Example 13 Let $f(x) = 1/(1 + x)$.

Then
$$f'(x) = -\frac{1}{(1+x)^2}, \quad f''(x) = \frac{2}{(1+x)^3},$$
$$f'''(x) = -\frac{3!}{(1+x)^4}, \ldots, f^{(n)}(x) = \frac{n!(-1)^n}{(1+x)^{n+1}}.$$

Thus the conditions of Taylor's theorem are met provided that the point $x = -1$ is not in the interval under consideration. Let x play the part of b in the expansion of Theorem 7.5 and take $a = 1$. With $a = 1$,
$$f(a) = \tfrac{1}{2}, \quad f'(a) = -\tfrac{1}{4}, \quad f''(a) = \tfrac{1}{4}, \ldots, f^{(n)}(a) = n!(-1)^n/2^{n+1}.$$
Hence, writing x for b, we have
$$\frac{1}{1+x} = \frac{1}{2} - \frac{1}{4}\cdot(x-1) + \frac{1}{4}\cdot\frac{(x-1)^2}{2!} - \frac{(x-1)^3}{(1+c)^4}$$

The mean-value theorem

where $1 < c < x$. If $x < 1$ the same formula holds but now $x < c <$ must impose the further restriction that $x > -1$ in order that the co Taylor's theorem may be satisfied. Remember that c depends on x.

If more terms are taken in the expansion, then

$$\frac{1}{1+x} = \frac{1}{2} - \frac{1}{4}(x-1) + \frac{1}{8}(x-1)^2 - \ldots$$
$$+ \frac{(-1)^{n-1}}{2^{n+1}}(x-1)^{n-1} + \frac{(-1)^n n(x-1)^n}{(1+c)^{n+1}}$$

where c lies between 1 and x, and $x > -1$.

If we had used Corollary 7.5 instead of Theorem 7.5 we would have obtained

$$\frac{1}{1+x} = \frac{1}{2} - \frac{1}{4}(x-1) + \ldots$$
$$+ \frac{(-1)^{n-1}}{2^{n+1}}(x-1)^{n-1} + \frac{(-1)^n n(x-1)(x-c)^{n-1}}{(1+c)^{n+1}}.$$

The expansion of Taylor's theorem can be put into a number of different forms. For example, the substitution $b = a + h$ and $c = a + \theta h$ gives

$$f(a+h) = f(a) + hf'(a) + \frac{h^2}{2!}f''(a) + \ldots$$
$$+ \frac{h^{n-1}}{(n-1)!}f^{(n-1)}(a) + \frac{h^n}{n!}f^{(n)}(a+\theta h) \qquad (9)$$

where $0 < \theta < 1$. If Cauchy's form of the remainder is used the last term is replaced by

$$\frac{h^n(1-\theta)^{n-1}}{(n-1)!}f^{(n)}(a+\theta h).$$

Put $a = 0$ and $h = x$ in (9); the result is

$$f(x) = f(0) + xf'(0) + \frac{x^2}{2!}f''(0) + \ldots$$
$$+ \frac{x^{n-1}}{(n-1)!}f^{(n-1)}(0) + \frac{x^n}{n!}f^{(n)}(\theta x) \qquad (10)$$

where $0 < \theta < 1$. This particular version of Taylor's theorem is usually referred to as *Maclaurin's theorem*. Again, it is important to observe that, since θ was derived from c, the actual values of θ in (9) and (10) need not be, and rarely are, the same. The value of θ in (9) will vary if a, h or n are altered. Similarly, changing x or n in (10) will, in general, cause an alteration of θ.

Example 14 *Apply Maclaurin's theorem to* $\sin x$.

In this case $f'(x) = \cos x$, $f''(x) = -\sin x$, $f'''(x) = -\cos x$, $f^{(iv)}(x) = \sin x$ so that $f(0) = 0$, $f'(0) = 1$, $f''(0) = 0$, $f'''(0) = -1$, $f^{(iv)}(0) = 0$, etc. All the even derivatives are alternately 1 and -1. Therefore, (10) becomes

$$\sin x = x - \frac{x^3}{3!} + \frac{x^5}{5!} - \frac{x^7}{7!} + \dots$$

$$+ \frac{(-1)^{n-1} x^{2n-1}}{(2n-1)!} + \frac{(-1)^n x^{2n}}{(2n)!} \sin(\theta x).$$

By choosing $n = 1$ we see that

$$\frac{\sin x}{x} = 1 - \frac{1}{2} x \sin(\theta x)$$

from which it is evident that $(\sin x)/x \to 1$ as $x \to 0$ in accordance with previous results.

Exercises

29. If m is a positive integer show that

$$(a+h)^m = a^m + m a^{m-1} h + \frac{m(m-1)}{2!} a^{m-2} h^2 + \dots + m a h^{m-1} + h^m.$$

30. (i) Expand $(1+x)^\mu$ by Maclaurin's theorem when $x > -1$;

(ii) for what values of c does

$$\text{(a)} \quad \frac{1}{x+1} = \frac{1}{2} - \frac{1}{4}(x-1) + \frac{(x-1)^2}{(1+c)^3}$$

and

$$\text{(b)} \quad \frac{1}{x+1} = \frac{1}{2} - \frac{x-1}{(1+c)^2}$$

when $x = -\frac{3}{4}, 15$?;

(iii) for what values of θ does

$$\frac{1}{2+x} = \frac{1}{2} - \frac{1}{4}x + \frac{x^2}{(2+\theta x)^3}$$

when $x = -\frac{7}{4}, 14$?

31. Show that

$$\cos x = 1 - \frac{x^2}{2!} + \frac{x^4}{4!} - \frac{x^6}{6!} + \dots + \frac{(-1)^n x^{2n}}{(2n)!} + \frac{(-1)^{n+1} x^{2n+1}}{(2n+1)!} \sin(\theta x)$$

where $0 < \theta < 1$.

32. Give the first few terms in Taylor's theorem for (i) $\sin x$, (ii) $\cos x$, (iii) $\tan x$.

33. Give the first three terms of Maclaurin's expansion for (i) tan x; (ii) sec x; (iii) x cosec x; (iv) tan $(x + \tfrac{1}{4}\pi)$.

34. Show that
$$\tan x = 1 + 2(x - \tfrac{1}{4}\pi) + 2(x - \tfrac{1}{4}\pi)^2 + \tfrac{1}{3}(2\tan^2 c + \sec^2 c)(x - \tfrac{1}{4}\pi)^3 \sec^2 c$$
where c lies between $\tfrac{1}{4}\pi$ and x. Would you put any restrictions on x?

35. If $f^{(n)}(x)$ is continuous on $a \leqslant x \leqslant b$, show that Taylor's theorem can be written
$$f(b) = f(a) + (b - a)f'(a) + \ldots + \frac{(b - a)^n}{n!}f^{(n)}(a) + \frac{(b - a)^n}{n!}g(b)$$
where $g(b) \to 0$ as $b \to a$.

36. If, in (8), $(b - x)^n R$ is replaced by $(b - x)^p(b - a)^{n-p}R$ show that
$$\frac{(b - a)^p(b - c)^{n-p}}{(n - 1)!\,p}f^{(n)}(c)$$
is the form of the remainder in Taylor's theorem. ($p = n$ is Lagrange's form, and $p = 1$ Cauchy's).

37. Prove that
$$(a + h)^{\tfrac{5}{2}} = a^{\tfrac{5}{2}} + \frac{5}{2}a^{\tfrac{3}{2}}h + \frac{15}{8}a^{\tfrac{1}{2}}h^2 + \frac{5}{16}\frac{h^3}{(a + \theta h)^{\tfrac{1}{2}}}$$
where $0 < \theta < 1$. For what value of θ is this true when $a = 0$? What happens when further terms are taken in the expansion?

7.7 Numerical work

The tremendous asset of Taylor's theorem is that it represents a function $f(x)$ as a polynomial
$$f(a) + (x - a)f'(a) + \ldots + \frac{(x - a)^{n-1}}{(n - 1)!}f^{(n-1)}(a)$$
together with a remainder term $(x - a)^n f^{(n)}(c)/n!$. Now polynomials are much the simplest functions and so are much more easily handled than other functions. Therefore Taylor's theorem supplies us with a method of approximating complicated functions by manageable ones provided that the remainder term is of no importance.

Apart from the analytical value of such approximations they are also extremely useful in numerical work and well adapted for calculations on digital computers. For instance,
$$\frac{1}{1 - x} = 1 + x + x^2 + \ldots + x^{n+1} + \frac{x^{n+2}}{(1 - \theta x)^{n+3}}$$

according to (10) and a simple flow-chart (see figure 86) can be constructed for calculating the polynomial portion. The computation is stopped at the value of n that makes $|x^{n+1}| < \varepsilon$ and the idea is that if ε is chosen sufficiently small the result will be a good approximation to the value of $1/(1-x)$. Now this will be valid provided that the remainder term $x^{n+2}/(1-\theta x)^{n+3}$ is small, and so an investigation is needed as to when this is true.

If $|x| > 1$, x^{n+1} gets larger and larger as n increases and so the criterion for stopping could not be met. Nor can it be met when $x = \pm 1$ since x^{n+1} never becomes small. Therefore the method has no chance of succeeding unless $|x| < 1$.

Now suppose $0 < x < \frac{1}{2}$; then $1 - \theta x > 1 - x$ since $0 < \theta < 1$ and so

$$\frac{x^{n+2}}{(1-\theta x)^{n+3}} < \frac{1}{1-x} \cdot \left(\frac{x}{1-x}\right)^{n+2}. \tag{11}$$

Also $x/(1-x) < 1$ if $x < \frac{1}{2}$. Therefore as the remainder becomes smaller and smaller as n increases we can be sure that our polynomial will be a good approximation if n is taken sufficiently large. The criterion of the flow-chart gives one method of ensuring that n is large; an additional check would be provided by calculating the right-hand side of (11).

When $-1 < x \leq 0$, $1 - \theta x \geq 1$ and

$$\left|\frac{x^{n+2}}{(1-\theta x)^{n+3}}\right| < |x^{n+2}| \leq \varepsilon$$

so that in this case it is certain that if we stop when $|x|^{n+1} < \varepsilon$, the error in the polynomial approximation will be less than ε, i.e. the process works satisfactorily.

It is of interest to see what conclusions can be drawn from the Cauchy form of the remainder. Here

$$\frac{1}{1-x} = 1 + x + x^2 + \ldots + x^{n+1} + \frac{(n+2)(1-\theta)^{n+1}x^{n+2}}{(1-\theta x)^{n+3}}$$

where $0 < \theta < 1$. Now $1 - \theta < 1 - \theta x$ for $x < 1$ and therefore the remainder does not exceed

$$\frac{(n+2)x^{n+2}}{(1-|x|)^2}\left(\frac{1-\theta}{1-\theta x}\right)^{n+1}$$

which will certainly be less than ε if n is sufficiently large when $|x| < 1$. This shows that the polynomial approximation can definitely be used for $|x| < 1$; it will work better for $x < 0$ than for $x > 0$ in the sense that fewer terms of the polynomial will need to be computed when x is negative. Observe that the remainder is less than $(n+2)x\varepsilon/(1-|x|)^2$ when $|x^{n+1}| < \varepsilon$ which is not such a good estimate as above.

The last point is an important general one. The estimate of the error involves the nth derivative and for complicated functions a general expression for this derivative cannot be obtained as a rule. Therefore it is desirable that any approximate representation of a function should require

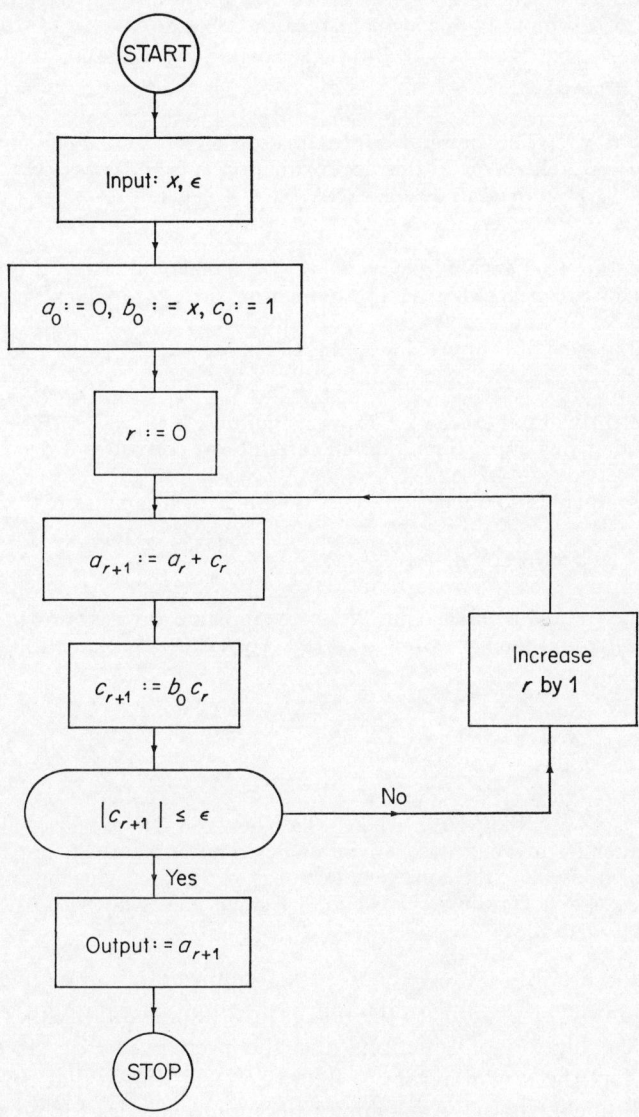

FIGURE 86

the determination of as few derivatives as possible. The added advantage that the minimum time will be demanded of the computer also accrues.

Suppose, in fact, that it was desired to find the minimum number of terms in the polynomial which would give accurate values of sin x. It pays to take the expansion of Example 14 one stage further and obtain

$$\sin x = x - \frac{x^3}{3!} + \ldots + \frac{(-1)^{n-1}x^{2n-1}}{(2n-1)!} + \frac{(-1)^n x^{2n+1}}{(2n+1)!} \cos(\theta x)$$

where $0 < \theta < 1$. The simplest approximation to $\sin x$, namely $\sin x = x$, is given by $n = 1$. The error in this approximation is $(x^3/3!)\cos(\theta x)$ which does not exceed $\frac{1}{6}|x|^3$ in magnitude since $|\cos \theta x| \leq 1$. Therefore

$$\sin x = x$$

will be accurate to 3 decimal places if $\frac{1}{6}|x|^3 < 0\cdot0004$, i.e. if $|x| < 0\cdot134$. Since $0\cdot134$ radians are nearly $8°$ the approximation $\sin x = x$ will be correct to 3 decimal places for angles betweeen $-7\frac{1}{2}°$ and $7\frac{1}{2}°$.

The next most simple approximation is to choose

$$\sin x = x - \tfrac{1}{6}x^3,$$

when the error will not exceed $|x|^5/5!$ in magnitude. If $|x| = 0\cdot5$ this is less than $0\cdot0003$, so that this approximation will certainly be correct to 3 decimal places for angles between $-29°$ and $29°$.

Of course, there is no obligation to use the Maclaurin expansion. For example,

$$\cos x = \cos a - (x-a)\sin a - \frac{(x-a)^2}{2!} \cos c$$

is useful when $x - a$ is small. Thus, if x is the radian equivalent of $44°$, we might choose $a = \frac{1}{4}\pi$ and then $x - a = -\pi/180$. Then our approximation to $\cos 44°$ would be

$$\cos 45° + \frac{\pi}{180}\sin 45° = 0\cdot70711 + 0\cdot01234$$

$$= 0\cdot71945.$$

The error is $-\frac{1}{2}(\pi/180)^2 \cos c$ where c lies between $44°$ and $45°$. Since $\cos c$ does not alter by much between these angles our estimate will not be out by much if we choose $45°$; the consequent error is $-0\cdot00011$. This tells us that our value for $\cos 44°$ is certainly correct to 3 decimal places but is possibly a unit high in the fourth place.

It should now be obvious to the reader how important Taylor's theorem is in the calculation of functions and the preparation of tables. As a matter of fact the flow diagram of figure 86 is so simple that it is usually easier and quicker to ask a machine to compute a series for a value of x than to ask it to store a large table and look up the appropriate answer.

The mean-value theorem

The error committed by using a series is that due to neglecting the remainder in Taylor's theorem; it is often called the *truncation error*. In essence, ε (figure 86) is a measure of the truncation error which the calculator is prepared to accept.

In actual calculations there may be an additional error because a machine can handle only a finite number of decimal places and therefore does not compute each term of the series precisely. The source of inaccuracy, called *round-off error*, will be discussed more fully in section 8.4.

Before leaving our discussion of the remainder term in Taylor's theorem we should remark that it can be useful in evaluating limits and may achieve results more rapidly than L'Hôpital's rule. Thus,

$$\lim_{x \to 0} \frac{x - \sin x}{x - x \cos x} = \lim_{x \to 0} \frac{x - \left[x - \frac{1}{6}x^3 + \frac{x^5}{120}\cos(\theta_1 x)\right]}{x - x\left[1 - \frac{1}{2}x^2 + \frac{x^4}{24}\cos(\theta_2 x)\right]}$$

$$= \lim_{x \to 0} \frac{x^3\left[\frac{1}{6} - \frac{x^2}{120}\cos(\theta_1 x)\right]}{x^3\left[\frac{1}{2} - \frac{x^2}{24}\cos(\theta_2 x)\right]}$$

$$= \lim_{x \to 0} \frac{\frac{1}{6} - \frac{x^2}{120}\cos(\theta_1 x)}{\frac{1}{2} - \frac{x^2}{24}\cos(\theta_2 x)} = \frac{1}{3}$$

since $x^2 \cos(\theta x) \to 0$ as $x \to 0$.

Exercises

38. If in Taylor's theorem $|f^{(n)}(x)| \leqslant C$ for $a < x < b$ show that the error in neglecting Lagrange's remainder does not exceed $C|b - a|^n/n!$ in magnitude.

39. For what angles can $\sin x$ be replaced by x with an error which does not exceed 0·0005 in magnitude?

40. For what angles can $\sin x$ be replaced by $x - \frac{1}{6}x^3$ with an error not greater than 0·001?

41. If $\cos x$ was calculated as $1 - \frac{1}{2}x^2$ when $x = 0·1$, how many decimal places would you expect to be correct?

42. The approximation $(1 + x)^{\frac{1}{2}} = 1 + \frac{1}{2}x$ is used to calculate $(1·01)^{\frac{1}{2}}$. Will the result be correct to 4 decimal places?

43. Expand $4x^2 - \frac{1}{16}x^4$ in powers of $(x - 4)$ by Taylor's theorem and use it to calculate the value to two decimal places at $x = 3.4$.

44. Determine correct to 3 decimal places (i) $\cos 61°$; (ii) $\sin 137°$; (iii) $\tan 46°$; (iv) $(72)^{\frac{1}{3}}$; (v) $\tan 55°$; (vi) $(\frac{8}{9})^{1/5}$.

45. Solve $\cos x = 2x^2$ with an error not greater than 0.01. Compare your answer with that given by Newton's method. (Hint: $|x|$ must be less than $1/\sqrt{2}$).

46. Estimate the truncation error in calculating $f'(x)$ by (i) $[f(x + h) - f(x)]/h$; (ii) $[f(x) - f(x - h)]/h$; (iii) $[f(x + h) - f(x - h)]/2h$; (iv) $[f(x - 2h) - 8f(x - h) + 8f(x + h) - f(x + 2h)]/12h$. Compare the errors of (iii) and (iv) in calculating $f'(1.1)$ with $h = 0.05$ and $f(x) = x^{\frac{1}{2}}$.

47. What is the truncation error in using $(1/h^2)[f(x + h) - 2f(x) + f(x - h)]$ for $f''(x)$?

48. If $F(h) = [f(x + h) - f(x - h)]/h$ calculate $f'(x) - F(h)$ as far as h^4, and show that the truncation error starts at h^2. Deduce that using $\frac{1}{3}[4F(\frac{1}{2}h) - F(h)]$ for $f'(x)$ has a truncation error starting at h^4 so that greater accuracy can be obtained by combining less accurate results.

49. Find an expansion in powers of x for

$$\frac{2a}{3}(8 \sin \frac{1}{2}x - \sin x) - 2ax.$$

50. Use a Taylor expansion to evaluate

$$\lim_{x \to 0} \frac{\sin x^2 - \sin^2 x}{x^4}.$$

7.8 Interpolation

Taylor's theorem provides an excellent method of finding a polynomial which approximates a function in the interval surrounding a particular point. In fact, let us construct a polynomial $P_0(x)$ by neglecting the remainder term in Taylor's theorem for $f(x)$, i.e.

$$P_0(x) = f(a) + (x - a)f'(a) + \ldots + \frac{(x - a)^n}{n!} f^{(n)}(a).$$

It is immediately evident that

$$P_0(a) = f(a), \quad P_0'(a) = f'(a), \ldots, \quad P_0^{(n)}(a) = f^{(n)}(a).$$

Thus the polynomial takes the same value as the function at $x = a$ and each of its first n derivatives agrees with the corresponding derivative of the function at $x = a$. Consequently, we expect P_0 to be a good approximation to f near $x = a$ and to be better the larger n. On the other hand, we must recognize that,

The mean-value theorem

since P_0 has been designed to be accurate near $x = a$, there is a distinct possibility that it will be a poor approximation to f when x is some distance from a. Therefore, the problem arises of trying to find a polynomial which agrees with f over a wider range; clearly one must be prepared to sacrifice the property that the first n derivatives of the polynomial equal those of f at some point.

Instead we try, if the values of f are known at the points $x_1, x_2, ..., x_n$, to find a polynomial which has the same values as f at these points. The answer to this problem has already been given in Exercise 30 of Chapter 3 and the requisite polynomial $P(x)$ is

$$P(x) = f(x_1)P_1(x) + f(x_2)P_2(x) + ... + f(x_n)P_n(x) \tag{12}$$

where

$$P_k(x) = \frac{(x - x_1)(x - x_2) ... (x - x_{k-1})(x - x_{k+1}) ... (x - x_n)}{(x_k - x_1)(x_k - x_2) ... (x_k - x_{k-1})(x_k - x_{k+1}) ... (x_k - x_n)}. \tag{13}$$

Notice that the factor $x - x_k$ is omitted from the numerator and the corresponding factor $x_k - x_k$ (which would be zero) from the denominator.

Now, if we put $x = x_1$, $P_k(x_1) = 0$ unless $k = 1$ and $P_1(x_1) = 1$ so that $P(x_1) = f(x_1)$. Similarly, $P(x_2) = f(x_2), ..., P(x_n) = f(x_n)$. Therefore $P(x)$ is a polynomial of degree $n - 1$ which takes the same values as f at $x_1, x_2, ..., x_n$.

There is no other polynomial of degree $n - 1$ with this property. For, suppose $\hat{P}(x)$ were another polynomial of degree $n - 1$ which took the same values as f at $x_1, x_2, ..., x_n$. Then

$$P(x_1) - \hat{P}(x_1) = 0, \quad P(x_2) - \hat{P}(x_2) = 0, ..., \quad P(x_n) - \hat{P}(x_n) = 0.$$

Therefore $P(x) - \hat{P}(x)$ is a polynomial of degree $n - 1$ which vanishes at n distinct points and so, by Exercise 6, is zero, i.e. $P(x) = \hat{P}(x)$.

The polynomial (12) is known as *Lagrange's interpolation polynomial* and the $P_k(x)$ are called *Lagrange interpolation coefficients*. When the points $x_1, ..., x_n$ are equally spaced the Lagrange coefficients can be considerably simplified and extensive tables of them have been calculated.

The Lagrange interpolation polynomial agrees with f at $x_1, ..., x_n$, but if the values of f are altered at points other than these the polynomial would not be changed. Therefore f and P could differ substantially at points which were not one of $x_1, ..., x_n$. An estimate of the error involved in this approximation is therefore highly desirable. It is supplied by

THEOREM 7.6 *Let $a \leqslant x_1 < x_2 < ... < x_n \leqslant b$. If $f(x), f'(x), ..., f^{(n-1)}(x)$ are continuous on $a \leqslant x \leqslant b$ and if $f^{(n)}(x)$ exists for $a < x < b$, then there is at least one number c, satisfying $a < c < b$, such that*

$$f(x) - P(x) = (x - x_1)(x - x_2) ... (x - x_n)f^{(n)}(c)/n!$$

when $a \leqslant x \leqslant b$.

Proof. If $x = x_1$ the equation is satisfied for any c such that $a < c < b$ since both sides are zero. The same is true if x takes one of the values $x_2, ..., x_n$. Therefore the theorem is proved if x is an interpolation point.

Introductory Analysis

Now let $x_0 (a \leqslant x_0 \leqslant b)$ be a point which is not an interpolation point and define $F(x)$ by

$$F(x) = (x_0 - x_1)(x_0 - x_2) \ldots (x_0 - x_n)[f(x) - P(x)] -$$
$$(x - x_1)(x - x_2) \ldots (x - x_n)[f(x_0) - P(x_0)].$$

Obviously $F(x_0) = 0$, $F(x_1) = 0, \ldots, F(x_n) = 0$. Therefore $F(x)$ is a continuous function for $a \leqslant x \leqslant b$ which vanishes at $n + 1$ distinct points. It possesses a derivative $F'(x)$ and so, by applying Rolle's theorem to each interval with two consecutive zeros as end-points, there must be n distinct points between a and b at which F' vanishes. Applying the same argument to F' we see that F'' is zero at $n - 1$ distinct points between a and b. Carrying on in this way we find that there is at least one point between a and b at which $F^{(n)}$ vanishes, i.e.

$$F^{(n)}(c) = 0$$

with $a < c < b$.

Since $P^{(n)}(x) = 0$, because P is a polynomial of degree $n - 1$,

$$(x_0 - x_1) \ldots (x_0 - x_n) f^{(n)}(c) = n! [f(x_0) - P(x_0)].$$

This is the same as the statement made in the theorem because x_0 was chosen arbitrarily and so the proof is complete.

The reader should note that c may be changed if x is altered or if n is altered.

It must not be supposed that Lagrange's formula is the only one used for interpolation. There are many others in active use but a discussion of them would take us too far afield. It should be remarked that Theorem 7.6 gives an estimate of the error only on $a \leqslant x \leqslant b$; it states nothing about what happens outside the interval. In fact, the polynomial when used outside the interval, i.e. in *extrapolation*, can be utterly unreliable. Inaccuracy can also result when a derivative of Lagrange's polynomial is used to approximate a derivative of f.

Exercises

51. If $x_k = x_1 + (k - 1)h$ $(k = 2,3,\ldots,n)$ show that

$$P_k(x_1 + th) = \frac{(-1)^{n-1-k}}{(n-1-k)!k!} t(t-1) \ldots (t-k+1)$$
$$\times (t - k - 1) \ldots (t - n + 1)$$

where $ht = x - x_1$.

52. Write down the values of $\sin x$ at $x = 0.77$, 0.78, 0.80, 0.81 (radians, remember) from a table and then determine $P(x)$ to agree with $\sin x$ at these four points. Calculate $P(0.79)$ and use Theorem 7.6 to estimate the error of the approximation to $\sin 0.79$. Compare with the table, and with the first 2 terms of Maclaurin's expansion.

If a graph plotter is available, compare the graphs of $\sin x$, $P(x)$ and Maclaurin's expansion over $0 \leqslant x \leqslant \pi$. Repeat the exercise given $\sin x$ at 4 points on either side of 0.79.

53. Estimate $(2.57)^{\frac{1}{2}}$ from a table giving $x^{\frac{1}{2}}$ at intervals of 0.1 by using Lagrange's interpolation formula with $n = 3$.

8
Sequences

The reader now has at his disposal a number of important theorems concerning functions and their derivatives. So far, however, the specific functions to which these theorems have been applied are polynomials, rational functions, trigonometric functions and (in some cases) algebraic functions. The more functions that are available the more extensive will be the area of usefulness of the calculus. Therefore we have reached a point where it is desirable to provide a systematic method of constructing functions, especially transcendental ones (see section 3.3). If, in addition, the method is well adapted for calculations on modern computers (as it is) we shall achieve two purposes at one time.

The preceding chapter has shown (in Taylor's theorem, for example) how useful polynomials are for approximating functions and suggests that the approximation improves as the degree of the polynomial increases. It is therefore tempting to conjecture that functions could be constructed by steadily increasing the degree of polynomials. So we are led to the question: what happens to polynomials as their degree is increased without limit?

The answer to this question is not simple and only a partial solution can be provided in this book; even that will require the erection of a substantial framework of ideas. But the framework proves to be fundamental for modern work in analysis and numerical methods. As with the building of all frameworks the purpose of the initial steps is not always clear except to those who have built them before. For this reason, the reader may not understand why he is learning the earlier sections of this chapter until later sections when he sees numerical applications; in the next two chapters the methods will be used to provide building blocks for functions.

8.1 Sequences

Consider the numbers
$$\frac{1}{2}, \frac{3}{4}, \frac{7}{8}, \frac{15}{16}.$$
They can be rewritten as
$$1 - \frac{1}{2},\ 1 - \left(\frac{1}{2}\right)^2,\ 1 - \left(\frac{1}{2}\right)^3,\ 1 - \left(\frac{1}{2}\right)^4$$

and so have an obvious rule of formation. Indeed, by following the same rule, we could supply further numbers such as 31/32, 63/64 which would be recognized as belonging to the same family. This could be indicated by writing

$$\frac{1}{2}, \frac{3}{4}, \frac{7}{8}, \frac{15}{16}, \frac{31}{32}, \ldots, 1 - \left(\frac{1}{2}\right)^n, \ldots$$

to show that the numbers are generated by a rule which gives any one of them on substitution of the appropriate value of n.

Again, writing

$$1, \frac{1}{2}, \frac{1}{3}, \ldots, \frac{1}{n}, \ldots$$

demonstrates another rule whereby numbers can be generated.

A general notation for such sets of numbers is

$$u_1, u_2, u_3, \ldots, u_n, \ldots$$

The first case above is covered by taking $u_1 = \frac{1}{2}$, $u_2 = \frac{3}{4}$, $u_3 = \frac{7}{8}$, $u_n = 1 - (\frac{1}{2})^n$; the second case is supplied by taking $u_1 = 1$, $u_2 = \frac{1}{2}$, $u_3 = \frac{1}{3}$, $u_n = 1/n$. Other sets of numbers can be obtained by prescribing different rules.

An ordered set of numbers

$$u_1, u_2, u_3, \ldots, u_n, \ldots$$

for which a rule is provided to determine each member is called an *infinite sequence* or briefly *sequence*. A sequence has a *first term* u_1, a *second term* u_2 and, in general, u_n is called the *nth term*. A sequence does not end because, whatever the value of the positive integer n, the rule tells us what u_n is.

A very simple sequence is provided by $u_n = 1$, namely

$$1, 1, 1, \ldots, 1, \ldots$$

But there is no reason why the rule should not be much more complicated. For example,

$$u_{2n-1} = -1, \quad u_{2n} = n^2 + 1$$

gives a sequence whose first few terms are

$$-1, 2, -1, 5, -1, 10, -1, 17.$$

Another method of specifying the rule, of frequent occurrence in computing, is one which tells one how to calculate u_n from terms which have gone before. Thus

$$u_1 = 1, \quad u_2 = 1, \quad u_n = u_{n-1} + u_{n-2} \quad (n \geqslant 3)$$

supplies the sequence

$$1, 1, 2, 3, 5, 8, 13, 21, \ldots.$$

A relation such as $u_n = u_{n-1} + u_{n-2}$ is called a *recursion formula*.

Sequences

The first question of interest concerning sequences is to enquire whether, as we go further along the sequence, the terms approach any particular number, i.e. we wish to assess the behaviour of u_n as n becomes large.

Now, if the sequence is

$$1, \frac{1}{2}, \frac{1}{3}, \ldots$$

so that $u_n = 1/n$ we see that u_n becomes steadily smaller as n increases. In fact, u_n can be made as close to zero as we like by choosing n large enough; for example, if $n = 10^3$, $u_n = 0\cdot001$ and if $n = 10^6$, $u_n = 0\cdot000001$. In view of this, it seems a good idea to write

$$\lim_{n \to \infty} u_n = 0.$$

Similarly, for the sequence

$$\frac{1}{2}, \frac{3}{4}, \frac{7}{8}, \ldots$$

where $u_n = 1 - (\frac{1}{2})^n$ we have

$$\lim_{n \to \infty} u_n = 1.$$

However, not every sequence has the property that u_n gets closer and closer to some finite number as n increases. For example, in the sequence

$$0, 3, 8, \ldots, n^2 - 1, \ldots$$

u_n becomes larger and larger as n increases. Whatever finite number we pick, a larger u_n can be found by choosing n large enough. In this case u_n is becoming positively infinite as $n \to \infty$.

There are other ways in which a sequence may fail to have a limit. Thus the sequence

$$0, 2, 0, 2, \ldots, 1 + (-1)^n, \ldots$$

has all its terms finite. Its odd terms are zero and its even terms are 2. Not *all* the terms come closer and closer to the same number and so the sequence does not have a limit.

Again, the sequence

$$-1, 2, -1, 5, \ldots$$

considered above does not have a limit. Nor do all terms become positively infinite.

Exercises

1. Write down the first four terms of the sequence whose nth terms is (i) n, (ii) $(n^2 + 3n)/5n$, (iii) $\cos \frac{1}{2}n\pi$, (iv) $[1 + (-1)^n]/n$, (v) $1/\sqrt{n}$, (vi) $(n + \sin \frac{1}{2}n\pi)/(2n + 1)$.

2. Write down the nth term of the sequence

(i) $\dfrac{1}{2}, \dfrac{1}{4}, \dfrac{1}{8}, \dfrac{1}{16}, \ldots,$ (ii) $0, \dfrac{1}{2}, \dfrac{2}{3}, \dfrac{3}{4}, \dfrac{4}{5}, \ldots,$ (iii) $\dfrac{1}{1.2}, \dfrac{1}{2.3}, \dfrac{1}{3.4}, \ldots,$

(iv) $\dfrac{3}{2}, \dfrac{5}{4}, \dfrac{9}{8}, \dfrac{17}{16}, \ldots$ (v) $-1, \dfrac{1}{2}, -\dfrac{1}{6}, \dfrac{1}{24}, -\dfrac{1}{120}\ldots$

3. Say which of the sequences in question 1 has a finite limit.

We now examine more carefully what is meant by saying that a sequence has a limit. Consider the sequence

$$0, \dfrac{3}{2}, \dfrac{2}{3}, \dfrac{5}{4}, \ldots, 1 + \dfrac{(-1)^n}{n}, \ldots$$

in which u_n gets closer and closer to 1 as n increases. The terms in the sequence are plotted in figure 87 with the vertical coordinate showing the value of u_n and the horizontal coordinate the value of n. From this diagram

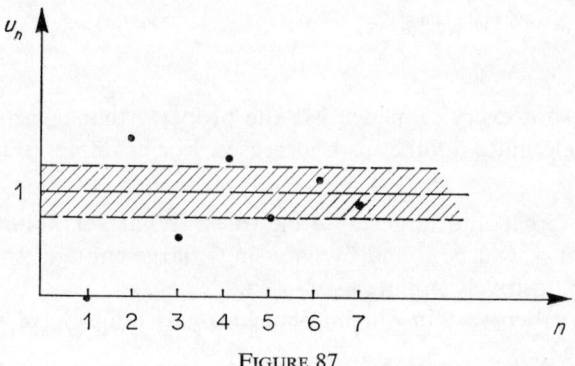

FIGURE 87

it can be seen that, if a horizontal band of width $\tfrac{2}{5}$ is drawn symmetrically about the value 1, all the terms from u_6 onwards would lie inside this band. Similarly, if we made the width of the band $2/1000$, all the terms from u_{1001} onwards would lie inside it. In fact, we can see that, no matter how narrow the band, we should eventually reach a term in the sequence at which all subsequent terms were inside it.

On the other hand the sequence (figure 88)

$$1, 4, 9, \ldots, n^2, \ldots$$

is such that no band of finite width will contain all its terms from some point onwards. Also the sequence (figure 89)

$$0, 2, 0, 2, \ldots, 1 + (-1)^n, \ldots$$

is such that no band of width less than 2 can contain all terms following any given term.

From this we deduce that sequences which have a limit are characterized by the possibility of finding a band which, no matter how narrow a band is chosen, contains all the terms from some term onwards. To express this property mathematically, suppose that the width of the band is 2ε and that the limit is L. Then all numbers in the band lie between $L - \varepsilon$ and

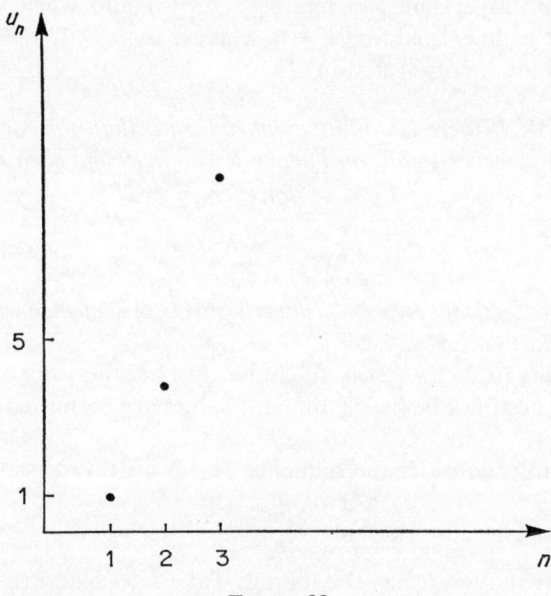

FIGURE 88

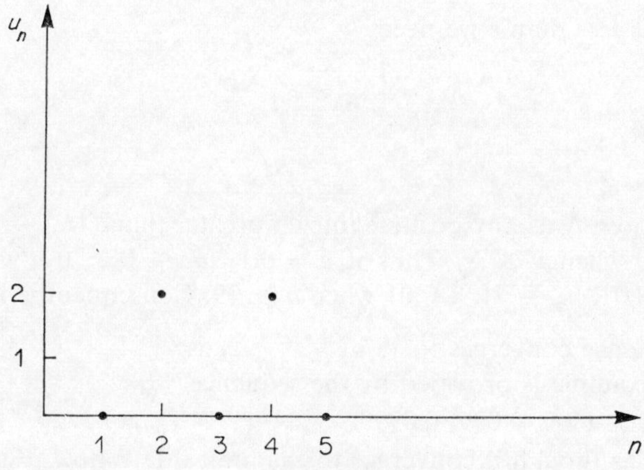

FIGURE 89

$L + \varepsilon$ and we want all the terms $u_N, u_{N+1}, u_{N+2}, \ldots$, to be in this band, i.e. we want
$$L - \varepsilon < u_n < L + \varepsilon \tag{1}$$
for $n = N, N + 1, \ldots$. Moreover, since we wish the band to be as narrow as we like, we need this to be true for some N no matter how small ε is so long as it is positive. Thus, in figure 87, $L = 1$ and when $\varepsilon = \frac{1}{5}$ it was found that the band existed for $N = 6$, whereas for $\varepsilon = 1/1000$, $N = 1001$.

These ideas are incorporated in

DEFINITION 8.1 *If there is a finite number L such that, for each given positive number ε, however small, an integer N can be found such that*
$$|u_n - L| < \varepsilon \quad \text{for every } n \geqslant N$$
then we write
$$\lim_{n \to \infty} u_n = L$$
and say that the sequence $u_1, u_2, \ldots$ converges to L. A sequence which does not converge is said to diverge.

The inequality $|u_n - L| < \varepsilon$ is, of course, just another way of writing (1). The integer N need not be fixed; different values are permitted for different choices of ε.

As an example, consider the sequence
$$\frac{2}{3}, \frac{3}{4}, \frac{4}{5}, \ldots, \frac{n+1}{n+2}, \ldots$$
which looks as though it has the limit 1. Take $L = 1$ and then
$$|u_n - L| = |u_n - 1| = \frac{1}{n+2}.$$
To make this less than ε we need
$$\frac{1}{n+2} < \varepsilon$$
or
$$n + 2 > 1/\varepsilon.$$
So, if we choose N as any positive integer greater than $(1/\varepsilon) - 2$, we have $|u_n - L| < \varepsilon$ when $n \geqslant N$. Thus if $\varepsilon = 0.1$, $|u_n - 1| < 0.1$ when $n \geqslant 9$ and, if $\varepsilon = 0.01$, $|u_n - 1| < 0.01$ when $n \geqslant 99$. Consequently $\lim_{n \to \infty} u_n = 1$ and the sequence converges to 1.

Another example is provided by the sequence
$$0.09, 0.099, 0.0999, \ldots, \tfrac{1}{10}(1 - \tfrac{1}{10^n}), \ldots$$
which looks as though it converges to the limit 0.1. With $L = 0.1$,
$$|u_n - L| = 1/10^{n+1}$$

Sequences

and this is less than ε if $1/10^{n+1} < \varepsilon$ or
$$10^{n+1} > 1/\varepsilon.$$
This can always be satisfied by taking n large enough. Thus, if $\varepsilon = 0\cdot 1$, $N = 1$ and $|u_n - 0\cdot 1| < 0\cdot 1$ for $n \geqslant 1$ while if $\varepsilon = 10^{-6}$, $N = 6$ and $|u_n - 0\cdot 1| < 10^{-6}$ for $n \geqslant 6$. Hence the sequence is convergent and
$$\lim_{n \to \infty} u_n = 0\cdot 1.$$
Note that we are not obliged to take the least possible value for N. All that is required is that N be finite. Often a calculation can be simplified by working with N much greater than the least possible value.

One question to be settled is: if a sequence converges can it have two or more limits? The answer is furnished by

THEOREM 8.1 *If a sequence converges then it has only one limit.*

Proof. We assume that there are two possible limits L_1 and L_2, which are different, and show that this leads to a contradiction. Given positive ε there is a positive integer N_1 such that
$$|u_n - L_1| < \varepsilon \quad \text{for} \quad n \geqslant N_1 \tag{2}$$
and a positive integer N_2 such that
$$|u_n - L_2| < \varepsilon \quad \text{for} \quad n \geqslant N_2. \tag{3}$$
Now
$$|L_1 - L_2| = |L_1 - u_n + u_n - L_2| \leqslant |u_n - L_1| + |u_n - L_2|$$
by (4) of Chapter 3. Therefore, if N is the larger of N_1 and N_2, (2) and (3) give
$$|L_1 - L_2| < 2\varepsilon$$
when $n \geqslant N$. If, however, $L_1 \neq L_2$, the number $L_1 - L_2$ is positive and ε could be chosen as $\tfrac{1}{4}|L_1 - L_2|$, leading to
$$|L_1 - L_2| < \tfrac{1}{2}|L_1 - L_2|.$$
This is false and so the assumption that the sequence has two different limits cannot be valid. Therefore the limit of a convergent sequence is unique.

Exercises

4. Show that the following sequences converge according to Definition 8.1 and indicate possible choices for N when $\varepsilon = 0\cdot 01$, $0\cdot 0001$: (i) $u_n = 1/n$, (ii) $u_n = 2n/(n+1)$, (iii) $u_n = 3 + (-1)^n/n$, (iv) $u_n = (1/n) \cos \tfrac{1}{2}n\pi$. Can you indicate the procedure you used on a flow-chart?

5. Let r be any number and choose a positive integer $p > |r|$. Show that
$$\left|\frac{r^n}{n!}\right| < \left|\frac{r^p}{p!}\right|\left(\frac{|r|}{p}\right)^{n-p}$$

when $n > p$. Deduce that
$$\lim_{n \to \infty} \frac{r^n}{n!} = 0.$$

6. If $\lim_{n \to \infty} u_n = L$ and $v_n = u_{n+1}$ prove that $\lim_{n \to \infty} v_n = L$.

8.2 Properties of sequences

The next few theorems are to help in calculations with limits of sequences.

DEFINITION 8.2 *A sequence is said to be bounded if there is a finite number M such that*
$$|u_n| \leq M$$
for all n.

For example, the sequence with $u_n = 1/n$ is bounded since $|u_n| \leq 1$ for all n so that we could take $M = 1$. The sequence $u_n = \sin \tfrac{1}{4}n\pi$ is bounded with $M = 1$; this last example demonstrates that a bounded sequence may not converge. However, the converse is true, namely

THEOREM 8.2 *A sequence which converges is bounded.*

Proof. Since the sequence converges we know, from Definition 8.1, that if we put $\varepsilon = 1$ in the definition there is an integer N such that
$$|u_n - L| < 1$$
for $n \geq N$. Therefore
$$|u_n| = |u_n - L + L| \leq |u_n - L| + |L| < |1 + L|$$
when $n \geq N$. This shows that $u_N, u_{N+1}, \ldots$ are bounded. As regards $u_1, u_2, \ldots, u_{N-1}$ there is only a finite number of them and so, if we take M_1 to be the greatest of them,
$$|u_n| \leq M_1 \quad (1 \leq n \leq N - 1).$$
Choosing M as the larger of M_1 and $1 + |L|$ we have $|u_n| \leq M$ for all n and the theorem is proved.

If an unbounded sequence converged, Theorem 8.2 would be contradicted. Therefore we have

COROLLARY 8.2 *Every unbounded sequence diverges.*

Example 1 Let
$$u_n = \frac{2n - 3}{n + 3}.$$

Then
$$u_n = 2 - \frac{9}{n + 3}$$

and the sequence converges to the limit 2. Now
$$|u_n - 2| = \frac{9}{n + 3} < 1$$

if $n > 6$; therefore $N = 7$. To find M_1, examine $u_1, u_2, u_3, u_4, u_5, u_6$; they are
$$u_1 = -\tfrac{1}{4}, \quad u_2 = \tfrac{1}{5}, \quad u_3 = \tfrac{1}{2}, \quad u_4 = \tfrac{5}{7}, \quad u_5 = \tfrac{7}{8}, \quad u_6 = 1.$$
The absolute values are $\tfrac{1}{4}, \tfrac{1}{5}, \tfrac{1}{2}, \tfrac{5}{7}, \tfrac{7}{8}, 1$ and the maximum of these is 1. Thus $M_1 = 1$ and, since $|L| + 1 = 3$, we take $M = 3$. Hence
$$\left|\frac{2n - 3}{n + 3}\right| \leqslant 3$$

for all positive integers n.

THEOREM 8.3 *If* $\lim_{n \to \infty} u_n = L_1$ *and* $\lim_{n \to \infty} v_n = L_2$, *then*
$$\lim_{n \to \infty} (u_n + v_n) = L_1 + L_2.$$

Proof. Choose an arbitrary positive number ε. Given the positive number $\tfrac{1}{2}\varepsilon$ there are integers N_1 and N_2 such that
$$|u_n - L_1| < \tfrac{1}{2}\varepsilon$$
when $n \geqslant N_1$, and
$$|v_n - L_2| < \tfrac{1}{2}\varepsilon$$
when $n \geqslant N_2$. Now
$$|(u_n + v_n) - (L_1 + L_2)| = |(u_n - L_1) + (v_n - L_2)| \leqslant |u_n - L_1| + |v_n - L_2|.$$
Therefore, if N is the larger of N_1 and N_2,
$$|(u_n + v_n) - (L_1 + L_2)| < \tfrac{1}{2}\varepsilon + \tfrac{1}{2}\varepsilon = \varepsilon$$
when $n \geqslant N$. But this is the condition that the sequence with term $u_n + v_n$ has the limit $L_1 + L_2$, and the theorem is proved.

If $\lim_{n \to \infty} v_n = L_2$, then $\lim_{n \to \infty} (-v_n) = -L_2$ and so we have

COROLLARY 8.3 *If* $\lim_{n \to \infty} u_n = L_1$ *and* $\lim_{n \to \infty} v_n = L_2$, *then*
$$\lim_{n \to \infty} (u_n - v_n) = L_1 - L_2.$$

It should perhaps be emphasized that Theorem 8.3 works only one way round. The converse is not available, i.e. $\lim_{n\to\infty} (u_n + v_n) = L_1 + L_2$ does not imply that $\lim_{n\to\infty} u_n = L_1$ and $\lim_{n\to\infty} v_n = L_2$.

Example 2 Let
$$u_n = \frac{3n}{n+1}, \qquad v_n = \frac{n+1}{n}.$$

Then
$$u_n = 3 - \frac{3}{n+1} \quad \text{and} \quad v_n = 1 + \frac{1}{n}$$
so that $\lim_{n\to\infty} u_n = 3$ and $\lim_{n\to\infty} v_n = 1$. Since
$$u_n + v_n = \frac{4n^2 + 2n + 1}{n(n+1)}$$
Theorem 8.3 shows that
$$\lim_{n\to\infty} \frac{4n^2 + 2n + 1}{n(n+1)} = 3 + 1 = 4.$$

To see how the details of the proof of Theorem 8.3 might be verified in this case take $\varepsilon = 0\cdot 1$. Then $|u_n - 3| < \tfrac{1}{2}\varepsilon$ if $n + 1 > 6/\varepsilon$ and so $N_1 = 60$; also, $|v_n - 1| < \tfrac{1}{2}\varepsilon$ if $n > 2/\varepsilon$ and so $N_2 = 21$. N is the larger of N_1 and N_2 and so is 60. Therefore we can be sure
$$\left| \frac{4n^2 + 2n + 1}{n(n+1)} - 4 \right| < 0\cdot 1$$
when $n \geqslant 60$.

THEOREM 8.4 *If* $\lim_{n\to\infty} u_n = L_1$ *and* $\lim_{n\to\infty} v_n = L_2$ *then*
$$\lim_{n\to\infty} u_n v_n = L_1 L_2.$$

Proof. Let ε be any positive number. We need to show that there is a positive integer N such that
$$|u_n v_n - L_1 L_2| < \varepsilon$$
when $n \geqslant N$. Now
$$\begin{aligned}
|u_n v_n - L_1 L_2| &= |(u_n - L_1)v_n + L_1(v_n - L_2)| \\
&\leqslant |(u_n - L_1)v_n| + |L_1(v_n - L_2)| \\
&\leqslant |u_n - L_1||v_n| + |L_1||v_n - L_2|
\end{aligned} \qquad (4)$$

Sequences

by equations (4) and (2) of Chapter 3. If we can make each of the terms on the right-hand side less than $\frac{1}{2}\varepsilon$ we shall have achieved our aim. This can be done for the second term if
$$|v_n - L_2| \leq \varepsilon/2|L_1|.$$
To avoid the difficulty that L_1 might be zero, we decide to make the denominator $1 + |L_1|$ instead of $|L_1|$. Since v_n converges to L_2 there is a positive integer N_2 such that
$$|v_n - L_2| < \frac{\varepsilon}{2(|L_1| + 1)}$$
when $n \geq N_2$. Therefore, when $n \geq N_2$,
$$|L_1| |v_n - L_2| < \tfrac{1}{2}\varepsilon \tag{5}$$
since $|L_1|/(1 + |L_1|) < 1$.

Also, by Theorem 8.2, there is M such that
$$|v_n| \leq M$$
for every positive integer n. There is a positive integer N_1 such that
$$|u_n - L_1| < \frac{\varepsilon}{2(M + 1)}$$
when $n \geq N_1$. Therefore, when $n \geq N_1$,
$$|u_n - L_1| |v_n| < \tfrac{1}{2}\varepsilon \tag{6}$$
since $M/(1 + M) < 1$.

Select N as the larger of N_1 and N_2. Then both (5) and (6) hold when $n \geq N$ and (4) gives
$$|u_n v_n - L_1 L_2| < \varepsilon.$$
The proof is finished.

If k is any number the sequence in which $v_n = k$ converges with limit k. In fact $v_n - k$ is zero for every n and so, if $\varepsilon > 0$,
$$|v_n - k| < \varepsilon$$
for $n \geq 1$. If this sequence is used in Theorem 8.4 we obtain

COROLLARY 8.4 *If* $\lim_{n \to \infty} u_n = L_1$ *and* k *is any number, then*
$$\lim_{n \to \infty} ku_n = kL_1.$$

Example 3 Let
$$u_n = \frac{3n}{n + 1}, \qquad v_n = \frac{n + 2}{n}.$$

Then $u_n = 3 - \dfrac{3}{n+1}, \quad v_n = 1 + \dfrac{2}{n}.$

so that $\lim_{n\to\infty} u_n = 3$ and $\lim_{n\to\infty} v_n = 1$. Since

$$u_n v_n = 3(n+2)/(n+1)$$

Theorem 8.4 shows that

$$\lim_{n\to\infty} \frac{3(n+2)}{n+1} = 3.1 = 3.$$

Let us follow the steps in the proof of Theorem 8.4 when $\varepsilon = 0{\cdot}1$. Since $L_1 = 3, L_2 = 1$ we want N_2 so that

$$|v_n - 1| < \varepsilon/2.4,$$

i.e. $2/n < \varepsilon/8$ or $n > 16/\varepsilon$. Therefore N_2 can be taken as 161.

Next $|v_n| \leqslant 3$ for all n and so $M = 3$. Hence, to find N_1 we need

$$|u_n - 3| < \varepsilon/2.4$$

or $3/(n+1) < \varepsilon/8$, i.e. $n+1 > 24/\varepsilon$. Therefore N_1 can be taken as 240.

N, being the larger of N_1 and N_2, is 240. Consequently

$$|u_n v_n - 3| < 0{\cdot}1$$

for $n \geqslant 240$. Notice that 240 is not necessarily the smallest value of n for which the inequality is true but that does not matter so long as there is some finite N which meets Definition 8.1.

Theorem 8.5 *If a sequence converges to a non-zero limit L then*

$$\lim_{n\to\infty} \frac{1}{u_n} = \frac{1}{L}.$$

Proof. Since $\lim_{n\to\infty} u_n = L$ there is N_1 such that

$$|u_n - L| < \tfrac{1}{2}|L|$$

for $n \geqslant N_1$ (remember $L \neq 0$). Therefore

$$|L| = |L - u_n + u_n| \leqslant |L - u_n| + |u_n| < \tfrac{1}{2}|L| + |u_n|$$

or

$$|u_n| > \tfrac{1}{2}|L| \qquad (7)$$

when $n \geqslant N_1$.

There is also N_2 such that, if $\varepsilon > 0$,

$$|u_n - L| < \tfrac{1}{2}\varepsilon|L|^2 \qquad (8)$$

when $n \geqslant N_2$. Choose N the larger of N_1 and N_2. Then when $n \geqslant N$ both (7) and (8) are valid so that

$$\left|\frac{1}{u_n} - \frac{1}{L}\right| = \left|\frac{u_n - L}{u_n L}\right| = \frac{|u_n - L|}{|u_n|\,|L|} < \frac{\tfrac{1}{2}\varepsilon|L|^2}{\tfrac{1}{2}|L|\,|L|} = \varepsilon$$

when $n \geqslant N$. The proof is complete.

Example 4 Let $u_n = 3n/(n + 1)$.

By Example 3, $\lim_{n\to\infty} u_n = 3$. Therefore, by Theorem 8.5,

$$\lim_{n\to\infty} \frac{n+1}{3n} = \frac{1}{3}.$$

The reader should go through the steps of Theorem 8.5 finding N_1, N_2 and N when $\varepsilon = 0 \cdot 1$.

COROLLARY 8.5 *If* $\lim_{n\to\infty} u_n = L_1$ *and* $\lim_{n\to\infty} v_n = L_2 \neq 0$, *then*

$$\lim_{n\to\infty} \frac{u_n}{v_n} = \frac{L_1}{L_2}.$$

Proof. By Theorem 8.5, $\lim_{n\to\infty} 1/v_n = 1/L_2$ and then Theorem 8.4 applied to $u_n(1/v_n)$ gives

$$\lim_{n\to\infty} \frac{u_n}{v_n} = \lim_{n\to\infty} u_n \lim_{n\to\infty} \frac{1}{v_n} = L_1 \cdot \frac{1}{L_2}.$$

The reader should compare Theorems 8.3, 8.4, 8.5 and their attendant corollaries with Theorem 4.1.

Example 5 Find

$$\lim_{n\to\infty} \frac{n^2 + 3n - 8}{2n^2 + 1}.$$

We cannot apply Corollary 8.5 immediately because neither the sequence $n^2 + 3n - 8$ nor $2n^2 + 1$ converges. However, if we divide numerator and denominator by n^2 we obtain

$$\frac{n^2 + 3n - 8}{2n^2 + 1} = \frac{1 + \dfrac{3}{n} - \dfrac{8}{n^2}}{1 + \dfrac{1}{n^2}}.$$

Now, with $u_n = 1 + 3/n - 8/n^2$, $v_n = 2 + 1/n^2$

$$\lim_{n\to\infty} u_n = 1, \quad \lim_{n\to\infty} v_n = 2.$$

Therefore, by Corollary 8.5,

$$\lim_{n\to\infty} \frac{n^2 + 3n - 8}{2n^2 + 1} = \frac{1}{2}.$$

If a sequence is such that, whatever positive number M is chosen, there is a positive integer N such that $u_n > M$ for $n \geq N$ we often write
$$\lim_{n \to \infty} u_n = +\infty.$$

Exercises

7. Improve the estimate in Example 1 to $|u_n| \leq 2$ for all positive integers n.

8. If $\lim_{n \to \infty} u_n = L$ prove that $\lim_{n \to \infty} |u_n| = |L|$.

9. If $\lim_{n \to \infty} u_n = L$ show that $\lim_{n \to \infty} u_n = L^2$.

10. (i) If $u_n \neq 0$ and $\lim_{n \to \infty} u_n = +\infty$, prove that $\lim_{n \to \infty} 1/u_n = 0$; (ii) if $\lim_{n \to \infty} u_n = L \neq 0$, prove that there are positive δ, N such that $|u_n| \geq \delta > 0$ for every $n \geq N$.

11. If $k > 1$ prove that $\lim_{n \to \infty} k^n = +\infty$.

12. Which of the following sequences are bounded and which convergent?

(i) $u_n = n + \dfrac{1}{n}$, (ii) $u_n = \dfrac{2 + 3(-1)^n}{n}$, (iii) $u_n = \cos \tfrac{1}{4}n\pi$,

(iv) $u_n = \dfrac{n!}{10^n}$, (v) $u_n = n^{2/3}$.

13. Evaluate

(i) $\lim_{n \to \infty} \dfrac{4 + \sin n}{n + 1}$; (ii) $\lim_{n \to \infty} \dfrac{4n^2 + \sin n}{2n^2 - 1}$; (iii) $\lim_{n \to \infty} \left(\dfrac{4n^2 - 1}{n^2 + 1} \right)^{1/2}$;

(iv) $\lim_{n \to \infty} (1 + n)^{1/2} - n^{1/2}$.

14. If a sequence with nth term u_n is bounded (but not necessarily convergent) and $\lim_{n \to \infty} v_n = 0$, prove that $\lim_{n \to \infty} u_n v_n = 0$.

15. If, for $n \geq N_0$, $u_n \leq v_n \leq w_n$ and $\lim_{n \to \infty} u_n = \lim_{n \to \infty} w_n = L$, prove that $\lim_{n \to \infty} v_n = L$.

16. If $u_n \geq 0$ and $\lim_{n \to \infty} u_n = L$, show that $\lim_{n \to \infty} u_n^{\frac{1}{2}} = L^{\frac{1}{2}}$.

   (Hint: $u_n^{\frac{1}{2}} - L^{\frac{1}{2}} = (u_n - L)/(u_n^{\frac{1}{2}} + L^{\frac{1}{2}})$.)

One useful criterion for deciding whether a sequence is convergent is:

CRITERION *If $u_n \leq u_{n+1} \leq M$ for all n or if $u_n \geq u_{n+1} \geq m$ for all n, then the sequence converges to a limit L such that $L \leq M$ in the first case or $L \geq m$ in the second case.*

The proof of this result is beyond the scope of this book.

8.3 Iteration

One of the most important practical applications of the theory of sequences is in the discussion of the iterative processes used in computing nowadays. Two examples of iteration have already been encountered. The first—the bisection method for solving $f(x) = 0$—is described in section 4.6. In this method a sequence is generated from two starting values a_0, b_0 by putting $c_r = \frac{1}{2}(a_{r-1} + b_{r-1})$ and then defining $a_r = c_r$, $b_r = b_{r-1}$ if $f(c_r)f(a_{r-1}) > 0$ or $a_r = a_{r-1}$, $b_r = c_r$ if $f(c_r)f(a_{r-1}) < 0$. It is shown in Theorem 4.4(iii) that

$$\left|x_0 - \tfrac{1}{2}(a_r + b_r)\right| < \frac{1}{2^{r+1}}(b - a)$$

where x_0 is a root of $f(x) = 0$. An immediate conclusion is that

$$\lim_{r \to \infty} \tfrac{1}{2}(a_r + b_r) = x_0$$

i.e. the bisection method furnishes a sequence with terms $\frac{1}{2}(a_r + b_r)$ which converges to a root of $f(x) = 0$.

Thus, in this case, one can be sure the iteration converges to the desired end. Generally, the analysis of iterative processes is concerned with deciding whether a certain sequence converges and, if so, to what limit it tends. It is also important to know how fast the convergence is and whether some iterative procedures will provide answers quicker than others. In this respect Theorem 4.4(iii) gives an indication of the number of iterations (the size of r) necessary to obtain a specified accuracy. Corresponding criteria for other numerical methods of finding the root of an equation enable the comparison of various methods and a decision as to their relative efficiency.

Newton's iterative method of finding the root of an equation has been met in section 5.5, where it was examined from a geometrical point of view. To see how the method arises analytically suppose $f(x_0) = 0$ and let u_r be an approximation to x_0. If $x_0 = u_r + h$, Taylor's theorem says that

$$f(x_0) = f(u_r) + hf'(u_r) + \tfrac{1}{2}h^2 f''(u_r + \theta h) \tag{9}$$

where $0 < \theta < 1$. If u_r is a good approximation, h can be expected to be small and then, if f'' is not too large, the last term can be neglected, i.e.

$$f(x_0) \approx f(u_r) + hf'(u_r)$$

the sign $\approx$ being used to indicate approximate equality. This will make $f(x_0)$ zero if h is chosen so that

$$h = -f(u_r)/f'(u_r).$$

In other words, if u_r is an approximation to x_0, $u_r - f(u_r)/f'(u_r)$ should be a better one. If this new approximation is called u_{r+1},

$$u_{r+1} = u_r - \frac{f(u_r)}{f'(u_r)} \tag{10}$$

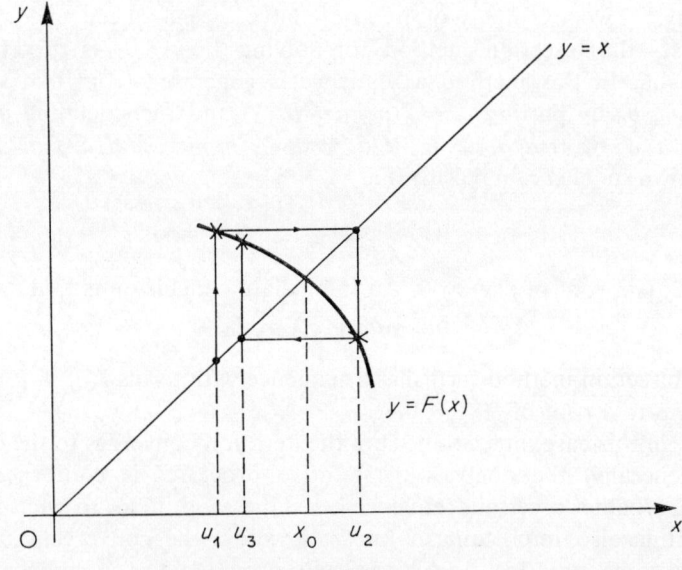

FIGURE 90

and an iterative process is obtained giving a sequence of steadily improving (it is hoped) approximations to x_0. Equation (10) is the iteration described in section 5.5 with a different notation.

If, in (10), the sequence u_r tends to a limit L it looks as though (10) approaches in the limit $L = L - f(L)/f'(L)$, i.e. $f(L) = 0$. Thus we hope that the iteration will converge to a root of our equation.

Instead of discussing the convergence of (10) we shall consider the more general iteration

$$u_{r+1} = F(u_r), \qquad (11)$$

which we hope will enable us to find numerical solutions to

$$x = F(x).$$

If $\lim_{r \to \infty} u_r = x_0$, then $\lim_{r \to \infty} u_{r+1} = x_0$ and we expect $\lim_{r \to \infty} F(u_r) = F(x_0)$ so that the iteration should lead to $x_0 = F(x_0)$ in the limit, i.e. (11) would be expected to provide a solution of the equation $x = F(x)$. Unfortunately, this is not always true. In figure 90 the intersection of the straight line

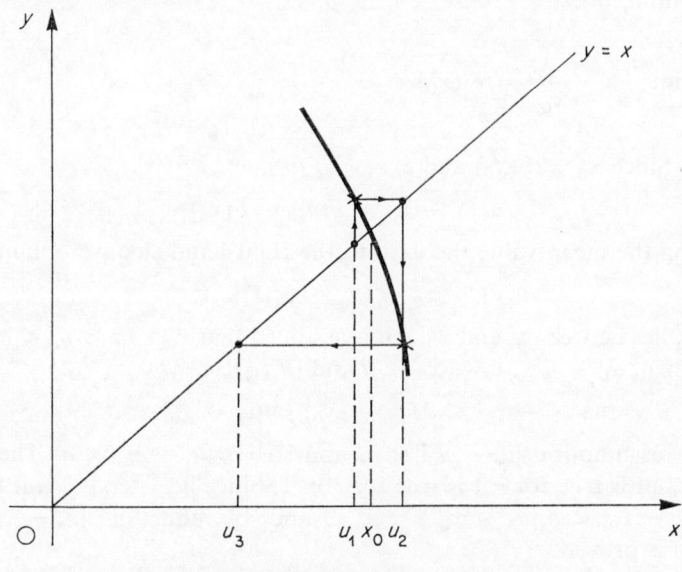

FIGURE 91

$y = x$ and the curve $y = F(x)$ provides the root x_0. Then, if u_1 is an approximation to x_0, $F(u_1)$ is the cross on $y = F(x)$ above the dot corresponding to u_1 on $y = x$. The dot corresponding to u_2 is found by drawing the horizontal through $F(u_1)$ to intersect $y = x$. The vertical through u_2 intersects $y = F(x)$ in $F(u_2)$ and the procedure continues. It will be seen that the process spirals in to x_0. On the other hand, we see that in figure 91 the same procedure not only does not spiral into x_0 but in fact gets further away from it.

Indeed, the figures indicate that spiralling in can be anticipated only when u_{r+1} is closer to x_0 than u_r and this spiralling cannot happen geometrically when $F'(x) < -1$ for values of x in the interval under consideration. The reader should satisfy himself that the same phenomenon may occur when $F'(x) > 1$. It will now be proved that the process converges if $|F'(x)| < 1$.

THEOREM 8.6 *Let $x_0 = F(x_0)$ where $F(x)$ is continuous on $x_0 - a \leqslant x \leqslant x_0 + a$ ($a > 0$). Assume that $F'(x)$ exists and that $|F'(x)| \leqslant M < 1$ on $x_0 - a < x < x_0 + a$. If $x_0 - a < u_1 < x_0 + a$ and $u_{r+1} = F(u_r)$, then*

(i) $x_0 - a < u_r < x_0 + a$,

(ii) $|u_{r+1} - x_0| \leqslant M^r |u_1 - x_0|$,

(iii) $\lim\limits_{r \to \infty} u_r = x_0$,

(iv) $\lim\limits_{r \to \infty} \dfrac{u_{r+1} - x_0}{u_r - x_0} = F'(x_0)$.

Proof. Since $x_0 = F(x_0)$ and $u_{r+1} = F(u_r)$,
$$u_{r+1} - x_0 = F(u_r) - F(x_0).$$
Applying the mean-value theorem to the right-hand side we obtain
$$u_{r+1} - x_0 = (u_r - x_0) F'(c_r) \tag{12}$$
where c_r lies between u_r and x_0, if it is assumed that $x_0 - a < u_r < x_0 + a$. Since, then, $x_0 - a < c_r < x_0 + a$ and $|F'(c_r)| \leqslant M$
$$|u_{r+1} - x_0| \leqslant M|u_r - x_0| < |u_r - x_0|. \tag{13}$$
But, by assumption $|u_r - x_0| < a$, and so $|u_{r+1} - x_0| < a$. Therefore, if the result is true for r it is true for $r + 1$. Since $|u_1 - x_0| < a$ it follows that $|u_2 - x_0| < a, |u_3 - x_0| < a, \ldots$ and, by induction, $|u_r - x_0| < a$. Thus (i) is proved.

From (13)
$$|u_r - x_0| \leqslant M|u_{r-1} - x_0|$$
and so
$$|u_{r+1} - x_0| \leqslant M^2 |u_{r-1} - x_0|.$$
Repeating the process we arrive eventually at
$$|u_{r+1} - x_0| \leqslant M^r |u_1 - x_0|$$
and (ii) is proved.

Since $M < 1$, an immediate deduction is that
$$\lim_{r \to \infty} |u_{r+1} - x_0| = 0$$
which proves (iii).

Sequences

Finally, since $u_r = x_0 + h_r$ where $\lim_{r \to \infty} h_r = 0$,

$$\lim_{r \to \infty} \frac{u_{r+1} - x_0}{u_r - x_0} = \lim_{r \to \infty} \frac{F(u_r) - F(x_0)}{u_r - x_0}$$
$$= \lim_{r \to \infty} \frac{F(x_0 + h_r) - F(x_0)}{h_r}$$
$$= F'(x_0) \tag{14}$$

by the definition of a derivative. The proof of the theorem is complete.

Although Theorem 8.6 tells us that the iteration converges to the required limit it is not very useful in determining the rate of convergence because (ii) involves the unknown quantity x_0. To overcome this difficulty write (12), with $r = 1$, as

$$u_2 - x_0 = (u_1 - u_2 + u_2 - x_0)F'(c_1)$$

or

$$(u_2 - x_0)[1 - F'(c_1)] = (u_1 - u_2)F'(c_1)$$

whence

$$u_2 - x_0 = \frac{(u_1 - u_2)F'(c_1)}{1 - F'(c_1)}.$$

Now $|F'(c_1)| \leqslant M$ and $1 - F'(c_1) \geqslant 1 - M$; hence

$$|u_2 - x_0| \leqslant \frac{M}{1 - M}|u_2 - u_1|.$$

Therefore, since $|u_{r+1} - x_0| \leqslant M^{r-1}|u_2 - x_0|$,

$$|u_{r+1} - x_0| \leqslant \frac{M^r}{1 - M}|u_2 - u_1|.$$

If it happens that F' is negative then $1 - F'(c_1) \geqslant 1$ and the $1 - M$ in the denominator can be replaced by unity. Accordingly, we have

COROLLARY 8.6 *Under the conditions of Theorem 8.6*

$$|u_{r+1} - x_0| \leqslant \frac{M^r}{1 - M}|u_2 - u_1|$$

and, if $F'(x) \leqslant 0$ *on* $|x - x_0| < a$,

$$|u_{r+1} - x_0| \leqslant M^r|u_2 - u_1|.$$

Example 6 *Find the root of* $27x^3 + 18x - 25 = 0$ *between 0 and 1 by using the iteration*

$$u_{r+1} = \tfrac{1}{3}(25 - 18u_r)^{\frac{1}{3}}.$$

Writing the given equation as $x^3 = \tfrac{1}{27}(25 - 18x)$ or $x = \tfrac{1}{3}(25 - 18x)^{\frac{1}{3}}$ suggests the iteration proposed. Here

$$F(x) = \tfrac{1}{3}(25 - 18x)^{\frac{1}{3}} \quad \text{and} \quad F'(x) = -2(25 - 18x)^{-\frac{2}{3}}.$$

Therefore, for $0 \leqslant x \leqslant 1$,
$$|F'(x)| \leqslant 2/7^{\frac{2}{3}} \leqslant 2/3 \cdot 66$$
which shows that, so long as we do not stray outside the interval (0,1), Theorem 8.6 can be applied with $M = 0.56$.

If we take $u_1 = 0$, we find $u_2 = 0.974$, $u_3 = 0.652$, $u_4 = 0.789$, $u_5 = 0.737$, $u_6 = 0.757$, $u_7 = 0.749$, $u_8 = 0.753$, $u_9 = 0.751$, $u_{10} = 0.752$ from which it is clear that the root is 0.75_1.

Another iteration procedure that might have been tried for the equation is
$$u_{r+1} = \tfrac{1}{18}(25 - 27u_r^3).$$
In this case $F'(x) = -9x^2/2$ so that $|F'(x)| > 1$ when $x > \tfrac{1}{2}$. Thus Theorem 8.6 is not applicable in the interval of interest. However, that does not mean that the sequence fails to converge. To see what happens try $u_1 = 0$; then $u_2 = 1.39$, $u_3 = -2.72$, $u_4 \approx 28$ and it is fairly evident that the iteration diverges.

Example 7 *Find the smaller root of $x^2 - 3x + 1 = 0$ by using the iteration*
$$u_{r+1} = \tfrac{1}{3}(u_r^2 + 1).$$

Since $x^2 - 3x + 1$ is positive at $x = 0$, negative at $x = 1$ and positive at $x = 3$ the smaller root lies between 0 and 1, while the larger lies between 1 and 3.

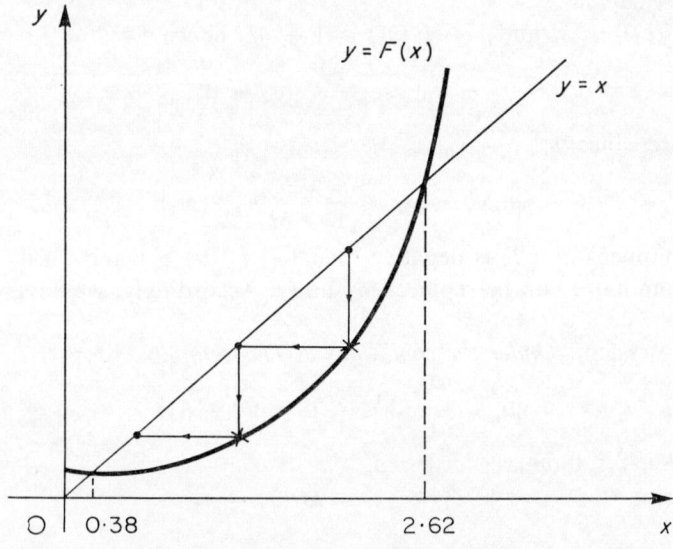

FIGURE 92

Now $F(x) = \tfrac{1}{3}(x^2 + 1)$ and $F'(x) = 2x/3$. Therefore Theorem 8.6 can certainly be used for $|x| < \tfrac{3}{2}$. Try $u_1 = 1$ and then $u_2 = 0.667$, $u_3 = 0.481$, $u_4 = 0.411$, $u_5 = 0.390$, $u_6 = 0.384$, $u_7 = 0.382$.

If we had started with $u_1 = 1.8$ we should have obtained $u_2 = 1.41$, $u_3 = 0.996$ and thereafter the terms in the sequence differ little from those in starting from 1.

Thus it is possible to start at a point where $|F'(x)| > 1$ and still have a convergent sequence. The reason can be seen from figure 92, which also shows that any positive u_1 less than 2·6 will lead to the smaller root. On the other hand any u_1 greater than 2·62 will lead to a divergent iteration. An iteration scheme to provide the larger root is

$$u_{r+1} = 3 - \frac{1}{u_r}.$$

It is important to realize that Theorem 8.6 states that the iteration converges if $|F'| < 1$. It does not state that $|F'| > 1$ implies that the iteration diverges. This point has already been illustrated in Example 7 and figure 92. One set of conditions that covers this sort of case is the assumption that $x - F(x) > 0$, $F'(x) > 0$ for $a + x_0 \geqslant x > x_0$. Then, if $a + x_0 \geqslant u_r > x_0$,

$$u_{r+1} = F(u_r) < u_r$$

while (12) shows that $u_{r+1} > x_0$. Therefore, if $a + x_0 \geqslant u_1 > x_0$, induction shows that $x_0 < u_{r+1} < u_r$ for all r and so, by the criterion at the end of the preceding section, $\lim_{r \to \infty} u_r = L$ where $L \geqslant x_0$. But $L = F(L)$ and so $L = x_0$. Thus the sequence converges to x_0.

Similarly it can be shown that $F(x) - x > 0$, $F'(x) > 0$ for $x_0 - a \leqslant x < x_0$ gives a sequence converging to x_0 if $x_0 - a \leqslant u_1 < x_0$.

Exercises

17. Solve $8x^3 - 4x - 5 = 0$ by using the iteration scheme

$$u_{r+1} = \tfrac{1}{2}(4u_r + 5)^{\frac{1}{3}}.$$

18. Solve $x^3 + x - 1 = 0$ by using

$$u_{r+1} = \frac{1}{1 + u_r^2}.$$

19. What algorithm does Theorem 8.6 suggest? Illustrate with a flow-chart.

20(a). Let x_0 and x_1 be the roots of $x^2 - bx + c = 0$ where $b > 0$, $\tfrac{1}{4}b^2 > c > 0$ and $x_0 < x_1$. Discuss the iteration schemes

(i) $u_{r+1} = (u_r^2 + c)/b$,

(ii) $u_{r+1} = b - (c/u_r)$.

(b) Examine the three iteration schemes for solving $2x^3 + 3x - 3 = 0$ by writing it as (i) $x = 1 - \tfrac{2}{3}x^3$, (ii) $x = \tfrac{3}{2}(\tfrac{3}{2} + x^2)^{-1}$, (iii) $x = (\tfrac{3}{2} - \tfrac{3}{2}x)^{\frac{1}{3}}$.

21. Show that Theorem 8.6 applies to Newton's method if $|ff''/f'^2| < 1$. (Remember that, in Newton's method, $F(u) = u - f(u)/f'(u)$.)

22. What happens when Newton's method is applied to $x^2 - 2x + 2 = 0$?

The inequality in Exercise 21 contains f in the numerator so that as the iteration proceeds this quantity should become smaller and smaller (provided that f'' does not become too large nor f' too small). Since this plays the role of M in Theorem 8.6, one expects the Newton method to converge more rapidly than a general iteration.

Another way of seeing this is to note that (9) can be written as

$$f(u_r) + (x_0 - u_r)f'(u_r) + \tfrac{1}{2}(x_0 - u_r)^2 f''(u_r + \theta h) = 0.$$

Substitution for $f(u_r)$ from (10) gives

$$(u_r - u_{r+1})f'(u_r) + (x_0 - u_r)f'(u_r) + \tfrac{1}{2}(x_0 - u_r)^2 f''(u_r + \theta h) = 0$$

or

$$u_{r+1} - x_0 = (x_0 - u_r)^2 \frac{f''(u_r + \theta h)}{2f'(u_r)}.$$

If u_r differs from x_0 by a small quantity, u_{r+1} will differ from x_0 by the square of that small quantity multiplied by $f''/2f'$. So long as $f''/2f'$ is not large u_{r+1} will be appreciably closer to x_0 than u_r. The contrast with (12) where the differences are merely proportional to one another is very noticeable.

It is convenient to distinguish between these possibilities by saying that, in a convergent iteration, if

$$\lim_{r \to \infty} \frac{u_{r+1} - x_0}{(u_r - x_0)^p} = b$$

where b is non-zero and finite, *the iterative method is of order p.*

It seems from what has gone before that Newton's method is likely to be of order 2, if f' does not vanish. In order to verify this we prove

THEOREM 8.7 *Let* $\lim_{r \to \infty} u_r = x_0$ *where* $u_{r+1} = F(u_r)$ *and $F(x)$ is continuous on* $x_0 - a \leqslant x \leqslant x_0 + a$ $(a > 0)$.

  (i) *If $F'(x)$ exists on $x_0 - a < x < x_0 + a$ and $F'(x_0) \neq 0$, then the iterative method is of order 1.*

  (ii) *If $F'(x_0) = 0$ and $F''(x)$ is continuous on $x_0 - a < x < x_0 + a$, and if $F''(x_0) \neq 0$, then the iterative method is of order 2.*

Proof. Since $\lim_{r \to \infty} u_r = x_0$,

$$\lim_{r \to \infty} \frac{u_{r+1} - x_0}{u_r - x_0} = F'(x_0)$$

exactly as in the derivation of (14). Therefore, when $F'(x_0) \neq 0$, the iterative method is of order 1 and (i) is proved.

In case (ii), note that $\lim_{r \to \infty} u_r = x_0$ implies that all the u_r from some r onwards will certainly lie between $x_0 - a$ and $x_0 + a$. For such r Taylor's theorem gives

$$F(u_r) = F(x_0) + (u_r - x_0)F'(x_0) + \tfrac{1}{2}(u_r - x_0)^2 F''(c_r)$$

where c_r lies between u_r and x_0. Since $F'(x_0) = 0$,

$$F(u_r) - F(x_0) = \tfrac{1}{2}(u_r - x_0)^2 F''(c_r)$$

and

$$\lim_{r \to \infty} \frac{u_{r+1} - x_0}{(u_r - x_0)^2} = \lim_{r \to \infty} \frac{F(u_r) - F(x_0)}{(u_r - x_0)^2}$$
$$= \lim_{r \to \infty} \tfrac{1}{2} F''(c_r)$$
$$= \tfrac{1}{2} F''(x_0)$$

because F'' is continuous and $c_r \to x_0$, since $u_r \to x_0$. Since $F''(x_0) \neq 0$ the proof of the theorem is complete.

Choose, as for Newton's method,

$$F(x) = x - \frac{f(x)}{f'(x)}. \tag{15}$$

Then

$$F'(x) = \frac{f(x)f''(x)}{[f'(x)]^2}$$

and

$$F''(x) = \frac{f''(x)}{f'(x)} + \frac{f(x)f'''(x)}{[f'(x)]^2} - \frac{2f(x)[f''(x)]^2}{[f'(x)]^3}.$$

Since $f(x_0) = 0$, $F'(x_0) = 0$ as long as $f'(x_0) \neq 0$; also

$$F''(x_0) = f''(x_0)/f'(x_0).$$

Consequently (ii) applies provided that $f''(x_0) \neq 0$ and the derivatives of f are suitably continuous. Therefore we have

COROLLARY 8.7a *If $f'''(x)$ is continuous on an interval including x_0, if $f'(x_0) \neq 0$ and if $f''(x_0) \neq 0$, then Newton's method of solving $f(x) = 0$ is of order 2.*

If $f'(x_0) = 0$, then
$$f(x) = (x - x_0)^q g(x) \tag{16}$$
where $q > 1$ and $g(x_0) \neq 0$. Therefore
$$f'(x) = q(x - x_0)^{q-1} g(x) + (x - x_0)^q g'(x)$$
and now (14) becomes
$$F(x) = x - \frac{(x - x_0)g(x)}{qg(x) + (x - x_0)g'(x)}$$
so that
$$F'(x) = 1 - \frac{g(x)}{qg(x) + (x - x_0)g'(x)} +$$
$$\frac{(x - x_0)g(x)[(g + 1)g' + (x - x_0)g'']}{[qg(x) + (x - x_0)g'(x)]^2}.$$
Therefore
$$F'(x_0) = 1 - 1/q \neq 0$$
and Theorem 8.7(i) applies. Hence we have

COROLLARY 8.7b *If $f''(x)$ is continuous on an interval including x_0 and if $f'(x_0) = 0$, then Newton's method of solving $f(x) = 0$ is of order 1.*

Exercises

23. If $f(x)$ is given by (16) and
$$F(x) = x - q\frac{f(x)}{f'(x)}$$
show that the iteration process is of order 2.

24. If $F(x) = x + h(x)f(x)$ choose $h(x)$ so that the iteration method is of order 2. (This choice of F gives an iteration to solve $f(x) = 0$.)

25. To calculate $\sqrt{x_0}$ when $x_0 > 0$ the following iteration method is suggested:
$$u_{r+1} = \frac{u_r^3 + 3x_0 u_r}{3u_r^2 + x_0}.$$
Show that it is of order 3.

26. If $F'(x_0) = 0$, $F''(x_0) = 0$ and $F'''(x_0) \neq 0$ show that the iterative method is of order 3. Deduce that
$$F(x) = x - \frac{f(x)}{f'(x)} - \frac{f''(x)[f(x)]^2}{[f'(x)]^3}$$
is of order 3 for solving $f(x) = 0$ when $f'(x_0) \neq 0$.

27. In a certain process of order 1
$$u_{r+1} - x_0 = b(u_r - x_0)$$
and, in a process of order 2,
$$u_{r+1} - x_0 = c(u_r - x_0)^2.$$
If $b = c = \frac{1}{2}$ and $u_s - x_0 = 1$, compare $u_{s+2} - x_0$ in the two processes and also $u_{s+7} - x_0$. Repeat the calculations when $b = c = 0.9$. (This shows that as soon as one is reasonably near x_0 it pays to use a process of order 2 rather than one of order 1.)

28. In an iterative process of order 1
$$u_{r+1} - x_0 = b_r(u_r - x_0), \qquad u_{r+2} - x_0 = b_{r+1}(u_{r+1} - x_0).$$
If $b_r = (1 + \eta)b_{r+1}$ show that
$$(u_{r+2} + u_r - 2u_{r+1})(x_0 - u_{r+2}) = -(u_{r+2} - u_{r+1})^2 + \eta(u_r - x_0)(u_{r+2} - x_0).$$
Deduce that, if η and $u_r - x_0$ are both small,
$$u_{r+2} - \frac{(u_{r+2} - u_{r+1})^2}{u_{r+2} + u_r - 2u_{r+1}}$$
is a better approximation to x_0 than u_r, u_{r+1} or u_{r+2}.

29. The following iteration scheme is devised for calculating $u_1, u_2, \ldots$ from u_0 (compare question 28): $v_{r+1} = F(u_r)$, $v_{r+2} = F(v_{r+1})$,
$$u_{r+1} = v_{r+2} - \frac{(v_{r+2} - v_{r+1})^2}{v_{r+2} + u_r - 2v_{r+1}}.$$
If x_0 is the desired root of $x = F(x)$ show that
$$u_{r+1} - x_0 = \frac{\varepsilon(x_0 - v_{r+2}) - (v_{r+1} - x_0)^2}{v_{r+2} + u_r - 2v_{r+1}}$$
where $\varepsilon = x_0 - u_r$. If F''' is continuous prove that
$$v_{r+1} = x_0 + \delta, \qquad v_{r+2} = x_0 + a\delta + \tfrac{1}{2}b\delta^2 + \tfrac{1}{6}d\delta^3$$
where $\delta = -a\varepsilon + \tfrac{1}{2}b\varepsilon^2 - \tfrac{1}{6}c\varepsilon^3$, $a = F'(x_0)$, $b = F''(x_0)$.

By expressing v_{r+2} in powers of ε as far as ε^3 deduce that
$$\lim_{r \to \infty} \frac{u_{r+1} - x_0}{(u_r - x_0)^2} = \frac{-ab}{2(1 - a)}$$
if $a \neq 1$. This shows that the iteration scheme is of order 2 if neither a nor b is zero. (The scheme is known as *Aitken's δ^2-method* for speeding up the convergence of a sequence.)

If $a = 1$ the scheme is of order 1.

Corollaries 8.7(a), 8.7(b) and Exercise 23 indicate that different formulae must be used to obtain a method of order 2 when $f'(x_0) \neq 0$ and when

$f'(x_0) = 0$. This is a clear-cut distinction, mathematically speaking, but in actual numerical calculations the situation is more complicated because the errors introduced in any computation will often convert a small number to zero or vice versa. For this reason, it is probably best to arrange that, whenever f' appears to be getting small, the computer draws your attention to this fact so that you can decide what is the best action to undertake.

8.4 Round-off error

There is one source of error that is present in any numerical calculation. A computer, whether human or machine, can handle only a finite quantity of digits and any number which requires more digits for an exact representation has to be approximated. For example, if only three digits to the right of the decimal place can be retained, the number 0·93242 cannot be held in its entirety. The nearest numbers with three digits are 0·932 and 0·933 and, of these, 0·932 is the nearer. If, therefore, it was arranged that the computer represented 0·93242 by 0·932 an error of 0·00042 would be introduced. The conversion of 0·93242 to 0·932 is called *rounding off*, and the error so introduced is called *round-off error*.

The precise rule employed for rounding off determines the error that arises and so we use one which involves the least error. Suppose that rounding off occurs at the nth place. If the digits to be discarded represent a number less than half a unit in the nth place, make no change in the nth digit. On the other hand, if the discarded digits represent a number greater than half a unit in the nth place, increase the nth digit by one. Thus, 0·93242 is 0·932 rounded off to the 3rd place because 0·00042 is less than half of 0·001, whereas 0·93268 becomes 0·933 when rounded off because 0·00068 is greater than 0·0005.

It is possible that the discarded digits give a number which is exactly half a unit in the nth place. In that case the nth digit is left unaltered if it is even, but increased by one if it is odd. The object of this rule is to guarantee that, when an error of half a unit in the nth place is committed, numbers are rounded up and down in equal quantities on the average. Thus, 0·9325 is rounded to 0·932 but 0·9335 becomes 0·934.

There are two main methods for specifying the place where rounding off is to occur. In the first, the working is carried out to *n decimal places*. This means that the nth digit to the right of the decimal point is rounded. For instance, 0·9326 is 0·933 to 3 decimal places.

In the second method the working is carried out to *n significant figures*. By this is meant that, moving from the left to the right of a number, we count n digits starting from the first one which is non-zero. Thus, to 3 significant figures

$$932 \cdot 6 \text{ is } 933$$
$$0 \cdot 9326 \text{ is } 0 \cdot 933$$
$$0 \cdot 009326 \text{ is } 0 \cdot 00933.$$

Note, however, that $0 \cdot 009326$ is $0 \cdot 009$ to 3 decimal places.

When considering a number such as $845 \cdot 01$ the reader should distinguish between 845, which is correct to 3 significant figures, and $845 \cdot 0$ which is correct to 4 significant figures.

Computations on electronic machines are frequently carried out by means of *floating-point*. In floating-point every number is written in the form $10^a \cdot b$ where a, the *exponent*, is an integer and b, the fractional part, may be positive or negative but must satisfy

$$0 \cdot 1 \leqslant |b| < 1.$$

For example

$$547 = 10^3 \cdot (0 \cdot 547), \quad a = 3, \quad b = 0 \cdot 547$$
$$54 \cdot 7 = 10^2 \cdot (0 \cdot 547), \quad a = 2, \quad b = 0 \cdot 547$$
$$0 \cdot 0547 = 10^{-1} \cdot (0 \cdot 547), \quad a = -1, \quad b = 0 \cdot 547$$
$$100 = 10^3 \cdot (0 \cdot 1), \quad a = 3, \quad b = 0 \cdot 1.$$
$$-1 = 10 \cdot (-0 \cdot 1), \quad a = 1, \quad b = -0 \cdot 1.$$

In most machines the number of digits to the right of the decimal point in b is fixed and is the same throughout any calculation. If this number of digits is n the machine is effectively working to n significant figures.

There is no round-off problem with a, which is always an integer, unless the numbers being handled are so large or so small that a is too large in magnitude to be conveniently stored. Such an occurrence is so rare that often it can be safely disregarded.

Nevertheless the size of numbers to be expected in a calculation must not be ignored in planning a piece of numerical work. Suppose, for example, we proposed to find the square root of a by the iteration

$$u_{r+1} = \frac{1}{2}\left(\frac{a}{u_r} + u_r\right)$$

and that we agree that a good approximation will have been reached when u_{r+1} and u_r differ by less than 10^{-6}, so that the iteration can be stopped then.

Now, if $a = 10^{-20}$, the iteration will stop before the correct value 10^{-10} is reached; in fact, the process will cease as soon as u_r is a little less than 10^{-6}. On the other hand, if $a = 3.10^{20}$, u_r will have an error of less

240   Introductory Analysis

than 10^{-6} only when u_r is correct to 16 significant figures; if the computer does not carry this number of figures the iteration will, in general, continue indefinitely.

It is therefore always worth while giving special consideration to any very large or very small numbers which are likely to occur in a computation.

Exercises

30. Write the following numbers (a) correct to 4 significant figures, (b) correct to 2 decimal places: 842·23, 7·565, 5402·036, 402·036, 108·375, 69732·117.

31. To 0·48 add 0·0041. Round-off to 2 significant figures. To the result add 0·0037 and round-off to 2 significant figures.

Compare your result with adding together 0·48, 0·0041, 0·0037 and then rounding off.

32. Express 332, 25·2, 0·106, − 0·0026, 1, − 0·1, 5000 in floating-point form.

It is when operations with numbers are involved that we see the big difference between numerical and mathematical descriptions. For a mathematician $x + y + z$ has a precise meaning but, numerically speaking, different answers may be obtained by different methods of round-off (see Exercise 31).

Again, the sequence $u_{n+1} = 0.8u_n$ with $u_1 = 1$ is such that $\lim_{n \to \infty} u_n = 0$ according to the mathematician. But if the sequence is calculated by a machine which rounds off to 1 significant figure the following terms are obtained

$$u_1 = 1, u_2 = 0.8, u_3 = 0.6, u_4 = 0.5, u_5 = 0.4, u_6 = 0.3, u_7 = 0.2,$$
$$u_8 = 0.2.$$

Thus the numerical process leads to $\lim_{n \to \infty} u_n = 0.2$. This grave discrepancy between what is predicted mathematically and what happens numerically when round-off error is present makes it very important to be able to estimate the size of the error. Where no estimate can be provided the computer has little alternative to working to as many digits as are available and hoping that errors will not creep up to the first few.

Suppose that x is an exact number and that numerical work obliges one to make the approximation X to it. Then $x = X + \varepsilon$ and ε is the error introduced. The *relative error* is ε/x; in many cases ε is very much smaller than X and then the relative error can be taken as ε/X without serious loss of accuracy.

Now let x_1, x_2 be two exact numbers, with approximations X_1, X_2 and errors ε_1, ε_2. Then

$$x_1 x_2 = (X_1 + \varepsilon_1)(X_2 + \varepsilon_2) = X_1 X_2 + \varepsilon_1 X_2 + \varepsilon_2 X_1 + \varepsilon_1 \varepsilon_2.$$

Thus the error in taking $X_1 X_2$ as an approximation to $x_1 x_2$ is $\varepsilon_1 X_2 + \varepsilon_2 X_1 + \varepsilon_1 \varepsilon_2$ and the relative error is

$$\frac{\varepsilon_1}{X_1} + \frac{\varepsilon_2}{X_2} + \frac{\varepsilon_1 \varepsilon_2}{X_1 X_2}.$$

Assuming that the relative errors are so small that the last term can be neglected, we can take the relative error as

$$\frac{\varepsilon_1}{X_1} + \frac{\varepsilon_2}{X_2},$$

i.e. the relative error of a product of numbers is the *sum* of the separate relative errors of the terms in the product. Consequently, we can expect that, when numbers with small relative errors are multiplied together, the resulting product also has a small relative error.

However, when the numbers are added,

$$x_1 + x_2 = X_1 + X_2 + \varepsilon_1 + \varepsilon_2$$

and the relative error is now

$$\frac{\varepsilon_1 + \varepsilon_2}{X_1 + X_2}.$$

If $X_1 + X_2$ is small, as when X_1 and X_2 are nearly equal in magnitude but opposite in sign, the relative error can be large and can certainly greatly exceed the separate relative errors of X_1 and X_2. One of the reasons for the rule for rounding off decimals to a given number of places is to encourage the cancellation of errors when numbers are added.

For instance, suppose $x_1 = 0.1036$, $x_2 = -0.1024$ and $X_1 = 0.104$, $X_2 = -0.102$. Then $\varepsilon_1 = -0.0004$, $\varepsilon_2 = -0.0004$ and the separate relative errors are -0.004 to a first approximation. However, $X_1 + X_2 = 0.002$ and the relative error is now $-0.0008/0.002 = -0.4$ which is a hundred times the separate relative error. In fact, $x_1 + x_2 = 0.0012$ which is 0.001 when rounded off. In this case, although X_1 and X_2 are correct to 3 significant figures, their sum has only one significant digit and that is in error by a unit.

Exercises

33. One hundred numbers, each correct to two decimal places, are added together. What is the maximum possible error?

34. Using the following approximations for $\sqrt{2}$ and $\sqrt{3}$ calculate $\sqrt{6}$: (i) 1·4, 1·7; (ii) 1·41, 1·73; (iii) 1·414, 1·732. It will be found that the product has about the same accuracy as its factors.

35. If $X_1 = 4\cdot42$ and $X_2 = 3\cdot29$ correct to two decimal places estimate the relative error in $X_1 + X_2$ and $X_1 X_2$.

Obviously in a complicated piece of numerical work, with many multiplications and additions, it would be impossible to follow through the round-off error produced at each stage. What is often found in practice is that, if $f(x)$ is the exact value of a function, and $f_c(x)$ is the value obtained by using a computer working to a given number of significant figures (i.e. the computer retains a given number of significant figures at each step in its evaluation of f for a given value of x)

$$|f_c(x) - f(x)| < \varepsilon, \qquad (17)$$

where ε is small, for a reasonable range of values of x. This relation tells us that, although we cannot say what is the precise difference between f_c and f, we can be sure that f_c will be somewhere between $f - \varepsilon$ and $f + \varepsilon$. Actually, since the numerical process gives us f_c and we want to know whether this can be used as a reasonable approximation for f we usually say that f must be somewhere between $f_c - \varepsilon$ and $f_c + \varepsilon$.

The determination of ε may be achieved by analysis or by experiment; the method will not concern us here.

Let us now consider what happens to an iteration scheme $u_{r+1} = F(u_r)$ in a numerical calculation. At the given stage u_r will be replaced by the computed value U_r. The next iterate would be $F(U_r)$ if the computation could be carried out exactly. As this is not achieved, what is actually calculated is U_{r+1} where

$$U_{r+1} = F(U_r) + \varepsilon_r \qquad (18)$$

and ε_r is the error made in the calculation.

In general there is no reason why the U_r should tend to the same limit as the u_r; there is a very simple example above (p. 240) where the limits are different. Indeed, there is no guarantee that $\lim_{r \to \infty} U_r$ exists and often it does not. In spite of this, it can be shown that U_r can provide an estimate of the value x_0, which satisfies $x_0 = F(x_0)$, as accurately as the computation will permit.

It will be assumed that (17) is valid for (18) in the form
$$|\varepsilon_r| < \varepsilon \tag{19}$$
where ε does not depend on r. Then we have

THEOREM 8.8 *Assume that F satisfies the conditions of Theorem 8.6 and that $(1 - M)a > \varepsilon$. Then, if*

$$x_0 - a + \frac{\varepsilon}{1 - M} < u_1 < x_0 + a - \frac{\varepsilon}{1 - M},$$

(i) $x_0 - a < U_r < x_0 + a$ $(r = 2, 3, \ldots)$,

(ii) $|x_0 - U_{r+1}| \leq M^r |x_0 - u_1| + \dfrac{\varepsilon}{1 - M}$.

Proof. We use induction, assuming that $U_2, U_3, \ldots, U_r$ all satisfy (i). Then, from (18) and $x_0 = F(x_0)$,
$$U_{r+1} - x_0 = F(U_r) + \varepsilon_r - F(x_0).$$
Applying the mean-value theorem we obtain
$$U_{r+1} - x_0 = (U_r - x_0)F'(c_r) + \varepsilon_r$$
where c_r lies between x_0 and U_r. Hence
$$|x_0 - U_{r+1}| \leq M|x_0 - U_r| + |\varepsilon_r| \leq M|x_0 - U_r| + \varepsilon$$
from (19). Therefore
$$|x_0 - U_{r+1}| \leq M[M|x_0 - U_{r-1}| + \varepsilon] + \varepsilon$$
and, by repetition,
$$|x_0 - U_{r+1}| \leq M^r |x_0 - u_1| + (M^{r-1} + M^{r-2} + \ldots + 1)\varepsilon$$
$$\leq M^r |x_0 - u_1| + \frac{1 - M^r}{1 - M}\varepsilon. \tag{20}$$
Since $M < 1$, (20) implies that
$$|x_0 - U_{r+1}| < |x_0 - u_1| + \frac{\varepsilon}{1 - M} < a$$
by the assumption on u_1. Thus (i) follows by induction and, since (20) is then true for all r, (ii) is an immediate deduction. The proof of the theorem is finished.

This theorem should be compared with Theorem 8.6. There is no analogue to Theorem 8.6(iii) because (20) differs from Theorem 8.6(ii) by the presence of the term involving ε. Therefore, it is not possible to deduce that U_r tends to a limit as $r \to \infty$.

Secondly, the right-hand side of Theorem 8.8(ii) can never be less than $\varepsilon/(1 - M)$. There is consequently no point in prolonging the iteration unduly; certainly, there would be little attraction in taking r beyond the value at which $M^r|x_0 - u_1|$ was less than $\varepsilon/10(1 - M)$.

As an illustration of the theorem consider the iteration $u_{n+1} = 0 \cdot 8 u_n$ with $u_1 = 1$ discussed earlier. With rounding off to 1 significant figure ε can be taken as $0 \cdot 05$. Since $F(x) = 0 \cdot 8x$, $F'(x) = 0 \cdot 8$ and therefore $M = 0 \cdot 8$. All computed iterates after the seventh are $0 \cdot 2$ and so $\lim_{r \to \infty} U_r = 0 \cdot 2$. It follows from Theorem 8.8(ii) that

$$|x_0 - 0 \cdot 2| \leqslant \frac{0 \cdot 05}{1 - 0 \cdot 8} = 0 \cdot 25,$$

i.e.

$$-0 \cdot 05 \leqslant x_0 \leqslant 0 \cdot 45.$$

This is correct, because $x_0 = 0$, but indicates the lack of precision which can arise on account of round-off.

Exercises

36. If, in Theorem 8.8, $F' < 0$ show that

$$0 < 1 + F'(c_r) + F'(c_r)F'(c_{r-1}) + \ldots + F'(c_r)F'(c_{r-1}) \ldots F'(c_1) < 1.$$

Deduce that, if $\varepsilon_r = \varepsilon$ $(r = 1, 2, \ldots)$,

$$|x_0 - U_{r+1}| \leqslant M^r|x_0 - u_1| + \varepsilon.$$

37. Prove that, under the conditions of Theorem 8.8,

$$|x_0 - U_{r+1}| \leqslant [M^r|U_2 - u_1| + \varepsilon]/(1 - M).$$

Can the denominator $1 - M$ be replaced by 1 when $F' < 0$?

38. An example demonstrated by Professor B. Noble is

$$u_{r+1} = 1000[(u_r + 0 \cdot 2)^{\frac{1}{2}} - (u_r + 0 \cdot 1)^{\frac{1}{2}}]$$

with $u_1 = 10$ and the working carried out in floating point with b correct to 4 significant figures. Show that u_r eventually takes the values $13 \cdot 00$ and $14 \cdot 00$ alternately and, if $\varepsilon = 0 \cdot 85$, deduce that $12 \cdot 30 \leqslant x_0 \leqslant 14 \cdot 70$. (In fact, $x_0 = 13 \cdot 52$.)

36. If, in the bisection method of section 4.6, $|A_r - B_r| \leqslant 2\lambda$ show that

$$|x_0 - \tfrac{1}{2}(A_r + B_r)| \leqslant \lambda + \varepsilon/m$$

where m is the smallest value of $|f'(x)|$ on $a \leqslant x \leqslant b$.

9
Infinite series

9.1 Infinite series of constant terms

Let us consider what happens, when we add together the numbers
$$\frac{1}{1.2}, \frac{1}{2.3}, \frac{1}{3.4}, \ldots,$$
which are the terms of a sequence in which the general term is $1/n(n+1)$.
$$\frac{1}{1.2} = \frac{1}{2} = 1 - \frac{1}{2},$$
$$\frac{1}{1.2} + \frac{1}{2.3} = \frac{2}{3} = 1 - \frac{1}{3},$$
$$\frac{1}{1.2} + \frac{1}{2.3} + \frac{1}{3.4} = \frac{3}{4} = 1 - \frac{1}{4}.$$

These results suggest that, when n terms are added together, the total is $1 - 1/(n + 1)$. To prove that this is in fact so let $a_n = 1/n(n + 1)$; then a_1, a_2 and a_3 are the first three terms given above. Also
$$\frac{1}{n(n + 1)} = \frac{1}{n} - \frac{1}{n + 1}$$
so that
$$a_1 = \frac{1}{1} - \frac{1}{2},$$
$$a_2 = \frac{1}{2} - \frac{1}{3},$$
$$a_3 = \frac{1}{3} - \frac{1}{4},$$
$$\vdots$$
$$a_{n-1} = \frac{1}{n - 1} - \frac{1}{n},$$
$$a_n = \frac{1}{n} - \frac{1}{n + 1}.$$

When these are added together the term with a minus sign cancels a term in the row below so that the only terms which are left are the term with the plus sign in the first row and the term with the minus sign in the last row. Hence

$$a_1 + a_2 + \ldots + a_n = 1 - \frac{1}{n+1}.$$

As n becomes larger and larger $1/(n + 1)$ becomes smaller and smaller so that it seems reasonable to say that an infinite number of terms on the left-hand side would give the answer 1, i.e.

$$\frac{1}{1.2} + \frac{1}{2.3} + \ldots + \frac{1}{n(n+1)} + \ldots = 1.$$

If we do say this then we have asserted that the sum of an infinite number of terms is $\lim_{n \to \infty} (a_1 + a_2 + \ldots + a_n)$. This suggests how we might define the sum of an infinite number of terms in general.

DEFINITION 9.1 *Let s_n be a term of the sequence defined by*

$$s_1 = a_1, \qquad s_2 = a_1 + a_2, \ldots, \qquad s_n = a_1 + a_2 + \ldots + a_n.$$

If the sequence converges to S, i.e. $\lim_{n \to \infty} s_n = S$, *then we write*

$$a_1 + a_2 + \ldots + a_n + \ldots = S. \tag{1}$$

The left-hand side of (1) is called an infinite series and the limit S is called the sum of the infinite series.

When (1) holds it is often said that the *series is convergent*. If $\lim_{n \to \infty} s_n$ does not exist the series is said to be divergent. The sequence $s_1, s_2, \ldots, s_n, \ldots$ is sometimes called the *sequence of partial sums*.

There is, of course, no guarantee that an infinite series converges. Thus, if $a_n = 1$ for every n, $s_n = n$ and $\lim_{n \to \infty} s_n$ does not exist as a finite number. Consequently, Definition 9.1 does not assign a sum to the infinite series

$$1 + 1 + \ldots + 1 + \ldots.$$

As another example, the series

$$1 - 1 + 1 - 1 + 1 - 1 + \ldots$$

has $s_1 = 1$, $s_2 = 0$, $s_3 = 1, \ldots$ and generally $s_{2n+1} = 1$, $s_{2n} = 0$. Again $\lim_{n \to \infty} s_n$ does not exist and the series is divergent.

Infinite series

Example 1 Consider the arithmetical progression in which $a_n = a + nd$ where a and d are constants.

Then
$$s_n = a + (a + d) + (a + 2d) + \ldots + [a + (n - 1)d]. \tag{2}$$
Also, by writing the terms in the reverse order, we have
$$s_n = [a + (n - 1)d] + [a + (n - 2)d] + [a + (n - 3)d] + \ldots + a. \tag{3}$$
Adding together (2) and (3) we obtain
$$2s_n = [2a + (n - 1)d] + [2a + (n - 1)d] + [2a + (n - 1)d] + \ldots \\ + [2a + (n - 1)d]$$
$$= n[2a + (n - 1)d].$$
Hence
$$s_n = \tfrac{1}{2}n[2a + (n - d1)]$$
or
$$a + (a + d) + \ldots + [a + (n - 1)d] = \tfrac{1}{2}n[2a + (n - 1)d]. \tag{4}$$

As $n \to \infty$, s_n must tend to either $+\infty$ or $-\infty$ (except in the trivial case when both a and d are zero). Hence an infinite arithmetical progression is always divergent.

Example 2 In the geometric series
$$1 + a + a^2 + \ldots + a^{n-1} + \ldots,$$
$$s_n = 1 + a + \ldots + a^{n-1}.$$

If $a = 1$, then $s_n = n$ and the series diverges.
If $a \neq 1$, note that
$$as_n = a + a^2 + \ldots + a^{n-1} + a^n$$
so that, by subtraction, $(1 - a)s_n = 1 - a^n$.
Now $1 - a \neq 0$, since $a \neq 1$, so that division by $1 - a$ is permissible and
$$s_n = \frac{1 - a^n}{1 - a} \quad (a \neq 1). \tag{5}$$

If $|a| < 1$, $\lim_{n \to \infty} a^n = 0$ so that
$$\lim_{n \to \infty} s_n = \frac{1}{1 - a}$$
and the series converges to the sum $1/(1 - a)$.
If $a = -1$, $s_n = 1$ or 0 according as n is odd or even and the series is divergent.
If $a > 1$,
$$s_n = \frac{1}{1 - a} + \frac{a^n}{a - 1} \to +\infty$$
as $n \to \infty$ and the series is divergent.

248   *Introductory Analysis*

If $a < -1$, put $a = -b$ where $b > 1$. Then
$$s_n = \frac{1}{1+b} - \frac{(-1)^n b^n}{1+b}$$
so that $s_{2n} \to -\infty$ and $s_{2n+1} \to +\infty$. Consequently, the series diverges.

Hence, *the geometric series converges if and only if* $|a| < 1$, and
$$1 + a + a^2 + \ldots + a^{n-1} + \ldots = \frac{1}{1-a} \qquad (6)$$
when $|a| < 1$.

The geometric series enables one to attach a meaning to *recurring decimals*. Thus
$$0 \cdot 666\ldots = \frac{6}{10} + \frac{6}{10^2} + \frac{6}{10^3} + \ldots = \frac{6/10}{1 - 1/10}$$
by (6). Therefore
$$0 \cdot 666\ldots = \frac{6}{9} = \frac{2}{3}.$$

Similarly
$$0 \cdot 232323\ldots = \frac{23}{100} + \frac{23}{100^2} + \frac{23}{100^3} + \ldots = \frac{23/100}{1 - 1/100} = \frac{23}{99}.$$

For the general decimal
$$0 \cdot c_1 c_2 c_3 \ldots = \frac{c_1}{10} + \frac{c_2}{10^2} + \frac{c_3}{10^3} + \ldots$$
it will be seen later (Theorem 9.2 (i)) that the series always converges though it is not always expressible as the ratio of two integers.

Exercises

1. Find the sums of the following series
$$\frac{1}{1.3} + \frac{1}{2.4} + \frac{1}{3.5} + \ldots + \frac{1}{n(n+2)} + \ldots,$$
$$\frac{1}{1.3} + \frac{1}{3.5} + \frac{1}{5.7} + \ldots + \frac{1}{(2n-1)(2n+1)} + \ldots.$$

2. Which of the following series converges?
   (i) $1 + \frac{1}{3} + \frac{1}{9} + \frac{1}{27} + \ldots,$
   (ii) $1 + 2 + 4 + 8 + \ldots,$
   (iii) $2 - \frac{1}{2} + \frac{1}{8} - \frac{1}{32} + \ldots,$
   (iv) $5 + 0 \cdot 5 + 0 \cdot 05 + 0 \cdot 005 + \ldots,$
   (v) $0 \cdot 1 + 0 \cdot 2 + 0 \cdot 3 + 0 \cdot 4 + \ldots.$

What is the sum of the series which converge?

3. Show that if an infinite series converges its sum can have only one value. (Hint: Theorem 8.1.)

4. Show that
$$1 + 2a + 3a^2 + \ldots + na^{n-1} + \ldots = \frac{1}{(1-a)^2}$$
if $|a| < 1$ but diverges for other values of a. (Hint: Form $s_n - as_n$).

5. If $a_1 + a_2 + \ldots = S_1$ and $b_1 + b_2 + \ldots = S_2$ show that
$$ka_1 + ka_2 + \ldots + ka_n + \ldots = kS_1,$$
$$(a_1 + b_1) + (a_2 + b_2) + \ldots + (a_n + b_n) + \ldots = S_1 + S_2.$$
(Hint: Theorem 8.3 and Corollary 8.4.)

6. If $a_1 + a_2 + \ldots + a_n + \ldots$ diverges, prove that $ka_1 + ka_2 + \ldots + ka_n + \ldots$ diverges if $k \neq 0$.

7. When a ball is dropped from a height h above a horizontal tennis racquet it rebounds to a height $\tfrac{3}{5}h$. If a ball is dropped from a height of 4 ft find the total distance it will cover.

8. Express the recurring decimals
$$0{\cdot}451451451\ldots, \qquad 1{\cdot}8636363\ldots$$
as the ratio of two integers. Can every recurring decimal be expressed as the ratio of two integers?

9. If $1 - 1 + 1 - 1 + \ldots$ were allotted the sum S we might write
$$S = 1 - (1 - 1 + 1 - 1 + \ldots) = 1 - S$$
whence $S = \tfrac{1}{2}$. Devise also methods which purport to show that $S = 0$ and that $S = 1$. What is wrong?

It is desirable to have some criteria which will test as simply as possible the convergence or divergence of an infinite series. One of the simplest tests, though not always the most useful, is a test of divergence and is

THEOREM 9.1 *If $a_1 + a_2 + \ldots + a_n + \ldots$ converges, then*
$$\lim_{n \to \infty} a_n = 0.$$

Proof. Let the series converge to the sum S. Then
$$\lim_{n \to \infty} s_n = S.$$
According to Definition 8.1, given $\varepsilon > 0$, N can be found such that
$$|s_n - S| < \varepsilon$$
for $n \geqslant N$. Hence, if $n \geqslant N + 1$,
$$|a_n| = |s_n - s_{n-1}| = |s_n - S + S - s_{n-1}| \leqslant |s_n - S| + |S - s_{n-1}| \leqslant 2\varepsilon.$$
By Definition 8.1, this implies that $\lim_{n \to \infty} a_n = 0$ and the theorem is proved.

The theorem says that, if the series converges, a_n must tend to zero or, alternatively, that if a_n does not approach zero the series must diverge. The theorem does not tell us that if a_n tends to zero the series must converge. Thus Theorem 9.1 is really most useful in indicating divergence.

Example 3 The series
$$\frac{2}{3} + 1 + \frac{8}{7} + \ldots + \frac{3n - 1}{2n + 1} + \ldots$$
is such that $\lim_{n \to \infty} a_n = \frac{3}{2}$. Since the limit is not zero Theorem 9.1 shows that the series must diverge.

Example 4 The series
$$1 + \frac{1}{2} + \frac{1}{3} + \ldots + \frac{1}{n} + \ldots$$
is called the *harmonic series*. In this series
$$\lim_{n \to \infty} a_n = 0$$
but we cannot conclude from Theorem 9.1 that the series converges. In fact, since $\frac{1}{3} > \frac{1}{4}$,
$$s_4 > 1 + \frac{1}{2} + \frac{1}{4} + \frac{1}{4} = 1 + \frac{1}{2} + \frac{1}{2}.$$
Also $\frac{1}{5} + \frac{1}{6} + \frac{1}{7} + \frac{1}{8} > \frac{4}{8}$ so that
$$s_8 > 1 + \frac{1}{2} + \frac{1}{2} + \frac{1}{2}.$$
Similarly
$$s_{16} > 1 + \frac{1}{2} + \frac{1}{2} + \frac{1}{2} + \frac{1}{2}.$$
and, in general, if $p = 2^n$
$$s_p > 1 + \frac{n}{2}.$$
Since $\frac{1}{2}n \to \infty$ as $n \to \infty$ the series is divergent.

Infinite series

Exercises

10. Are the following series convergent?

(i) $\dfrac{1}{3}+\dfrac{2}{5}+\dfrac{3}{7}+\dfrac{4}{9}+\ldots,$ (ii) $\dfrac{1}{2}+\dfrac{1}{4}+\dfrac{1}{6}+\ldots,$ (iii) $1+\dfrac{1}{2}+\dfrac{1}{4}+\dfrac{1}{8}+\ldots,$

(iv) $\dfrac{3}{2}+\dfrac{5}{4}+\dfrac{9}{8}+\dfrac{17}{16}+\ldots.$

11. If $a_1 + a_2 + \ldots + a_n + \ldots$ is convergent and p is a fixed integer prove that
$$\lim_{n \to \infty} (a_n + a_{n+1} + \ldots + a_{n+p}) = 0.$$

12. In the series $a_1 + a_2 + \ldots + a_n + \ldots$ the terms $a_1, a_2, \ldots, a_m$ where m is a fixed finite number are changed to $b_1, b_2, \ldots, b_m$. Show that the new series is convergent or divergent according as the original series is convergent or divergent.

9.2 Series of positive terms

One of the fundamental methods for deciding whether a series is convergent or not is to compare it with series which are already known to be convergent or divergent. The method is described by

THEOREM 9.2 (COMPARISON TEST) Let $a_1 + a_2 + \ldots + a_n + \ldots$, $b_1 + b_2 + \ldots + b_n + \ldots$ both be series, each term of which is positive.

(i) If $a_n \leqslant b_n$ for each value of n and $b_1 + b_2 + \ldots + b_n + \ldots$ is convergent, then $a_1 + a_2 + \ldots + a_n + \ldots$ is convergent.

(ii) If $a_n \geqslant b_n$ for each value of n and $b_1 + b_2 + \ldots + b_n + \ldots$ is divergent, then $a_1 + a_2 + \ldots + a_n + \ldots$ is divergent.

Proof. Let
$$s_n = a_1 + \ldots + a_n, \quad \sigma_n = b_1 + \ldots + b_n.$$

(i) When $a_n \leqslant b_n$ for each value of n, $s_n \leqslant \sigma_n$. Since $b_1 + \ldots + b_n + \ldots$ is convergent, σ_n is a term of a bounded sequence (Therem 8.2) and because $b_1, b_2, \ldots$ are all positive $\sigma_n \leqslant \Sigma$ where Σ is the sum of $b_1 + \ldots + b_n + \ldots$. Therefore $s_n \leqslant \Sigma$.

Also $s_{n+1} \geqslant s_n$ because all the terms in the series are positive. Therefore, by the criterion at the end of section 8.2, $\lim_{n \to \infty} s_n$ exists and $\lim_{n \to \infty} s_n \leqslant \Sigma$. Thus $a_1 + \ldots + a_n + \ldots$ is convergent and its sum does not exceed the sum of $b_1 + \ldots + b_n + \ldots$.

(ii) To prove this part suppose that $a_1 + \ldots + a_n + \ldots$ is convergent. Then, since $a_n \geqslant b_n$, (i) tells us that $b_1 + \ldots + b_n + \ldots$ must be convergent. But we are given that $b_1 + \ldots + b_n + \ldots$ is divergent. Therefore $a_1 + \ldots + a_n + \ldots$ cannot be convergent. The proof of the theorem is complete.

Example 5 The general (*or* nth) *term of the series*

$$\frac{1}{4} + \frac{1}{10} + \frac{1}{28} + \ldots + \frac{1}{3^n + 1} + \ldots$$

satisfies $\frac{1}{3^n + 1} < \frac{1}{3^n}$. The series $\frac{1}{3} + \frac{1}{9} + \ldots + \frac{1}{3^n} + \ldots$ is a convergent geometric series (Example 2) and so, by Theorem 9.2(i), the given series is convergent.

Example 6 Consider the *generalized harmonic series*

$$1 + \frac{1}{2^p} + \frac{1}{3^p} + \ldots + \frac{1}{n^p} + \ldots .$$

We know from Example 5 that this series is divergent when $p = 1$. Suppose now that $p < 1$; then

$$\frac{1}{n^p} > \frac{1}{n}$$

since $n^{1-p} > 1$. Hence, by Theorem 9.2(ii) the series is divergent.

When $p > 1$,

$$\frac{1}{2^p} + \frac{1}{3^p} < \frac{1}{2^p} + \frac{1}{2^p} = \frac{2}{2^p} = \frac{1}{2^{p-1}},$$

$$\frac{1}{4^p} + \frac{1}{5^p} + \frac{1}{6^p} + \frac{1}{7^p} < \frac{1}{4^p} + \frac{1}{4^p} + \frac{1}{4^p} + \frac{1}{4^p} = \frac{1}{4^{p-1}} = \left(\frac{1}{2^{p-1}}\right)^2,$$

$$\frac{1}{8^p} + \frac{1}{9^p} + \ldots + \frac{1}{15^p} < \frac{1}{8^{p-1}} = \left(\frac{1}{2^{p-1}}\right)^3,$$

Proceeding in this way, we see that the generalized harmonic series is less than

$$1 + \frac{1}{2^{p-1}} + \left(\frac{1}{2^{p-1}}\right)^2 + \left(\frac{1}{2^{p-1}}\right)^3 + \ldots$$

which is a geometric series (Example 2) with $a = 1/2^{p-1}$ and $a < 1$ since $p > 1$. Thus the geometric series converges and so, by Theorem 9.2(i), does the generalized harmonic series. Consequently, we have shown that *the generalized harmonic series converges for $p > 1$ and diverges for $p \leq 1$.*

Exercises

13. Determine whether the following series are convergent or divergent:

(i) $\frac{1}{6} + \frac{5}{10} + \frac{13}{18} + \ldots + \frac{2^{n+1} - 3}{2^{n+1} + 2} + \ldots$, (ii) $\frac{1}{2^p} + \frac{1}{4^p} + \frac{1}{6^p} + \ldots$,

(iii) $1 + \dfrac{1}{\sqrt{8}} + \dfrac{1}{\sqrt{27}} + \ldots + \dfrac{1}{\sqrt{n^3}} + \ldots,$

(iv) $\dfrac{1}{1.2} + \dfrac{1}{2.3} + \dfrac{1}{3.4} + \ldots + \dfrac{1}{n(n+1)} + \ldots,$

(v) $\sqrt{\tfrac{2}{3}} + \sqrt{\tfrac{3}{4}} + \sqrt{\tfrac{4}{5}} + \ldots + \sqrt{\left(\dfrac{n+1}{n+2}\right)} + \ldots,$

(vi) $\dfrac{2}{1^p} + \dfrac{3}{2^p} + \dfrac{4}{3^p} + \dfrac{5}{4^p} + \ldots,$ (vii) $1 + \dfrac{2^p}{2!} + \dfrac{3^p}{3!} + \ldots + \dfrac{n^p}{n!} + \ldots,$

(viii) $1 + \dfrac{1}{2^2} + \dfrac{2^2}{3^3} + \dfrac{3^3}{4^4} + \ldots,$

(ix) $(\sqrt{2} - 1) + (\sqrt{5} - 2) + \ldots + [\sqrt{(n^2+1)} - n] + \ldots,$

(x) $\dfrac{a}{1.2} + \dfrac{a^2}{3.4} + \dfrac{a^3}{5.6} + \ldots \;(a \geq 0),$

(xi) $\dfrac{a}{3} + \dfrac{a^2}{6} + \dfrac{a^3}{11} + \ldots + \dfrac{a^n}{n^2+2} + \ldots \;(a \geq 0),$

(xii) $\dfrac{a}{6} + \dfrac{5a^2}{10} + \ldots + \dfrac{2^{n+1} - 3}{2^{n+1} + 2}a^n + \ldots \;(a \geq 0).$

14. If, for $n \geq N$, $\dfrac{a_{n+1}}{a_n} < \dfrac{b_{n+1}}{b_n}$ and all of a_n and b_n are positive show that

$$a_{n+1} < \dfrac{b_{n+1}}{b_N} a_N.$$

Deduce that, if $b_1 + b_2 + \ldots$ converges, so does $a_1 + a_2 + \ldots$. If $a_1 + a_2 + \ldots$ diverges prove that $b_1 + b_2 + \ldots$ diverges.

If, in Exercise 14, $b_n = a^n$ where $a > 0$ the series $b_1 + b_2 + \ldots$ is the geometric series and converges for $a < 1$. Consequently, the series $a_1 + a_2 + \ldots$ converges if, for $n \geq N$, $a_{n+1}/a_n \leq a < 1$.

On the other hand, suppose that $a_{n+1}/a_n \geq c > 1$ for $n \geq N$. Then, for $n \geq N$,

$$\dfrac{a_n}{a_N} = \dfrac{a_n}{a_{n-1}} \cdot \dfrac{a_{n-1}}{a_{n-2}} \cdot \ldots \cdot \dfrac{a_{N+1}}{a_N} \geq c^{n-N}.$$

Since $c^{n-N} \to \infty$ as $n \to \infty$ it follows that $a_n \to \infty$ and the series must be divergent. Thus we have proved

THEOREM 9.3 (RATIO TEST) *If, for $n \geqslant N$,*
$$a_{n+1}/a_n \leqslant c_1 < 1$$
the series of positive terms $a_1 + a_2 + ...$ is convergent. If, for $n \geqslant N$,
$$a_{n+1}/a_n \geqslant c_2 > 1$$
the series of positive terms $a_1 + a_2 + ...$ is divergent.

There is a useful corollary to this theorem, namely

COROLLARY 9.3(a) *If*
$$\lim_{n \to \infty} \frac{a_{n+1}}{a_n} = c$$
the series of positive terms $a_1 + a_2 + ...$ is convergent if $c < 1$, divergent if $c > 1$. If $c = 1$, no conclusion can be drawn.

Proof. If $c < 1$ choose λ so that $c < \lambda < 1$. Since a_{n+1}/a approaches c as $n \to \infty$, all the values of a_{n+1}/a_n will be very close to c when n is sufficiently large. Therefore there is an N such that, for $n \geqslant N$,
$$a_{n+1}/a_n \leqslant \lambda.$$
Since $\lambda < 1$ it follows from the Ratio Test that the series is convergent.

When $c > 1$, choose μ so that $c > \mu > 1$. Again, a large enough N can be found for
$$a_{n+1}/a_n \geqslant \mu$$
and the Ratio Test implies that the series is divergent. The proof of the Corollary is complete.

Example 7 In the series
$$\frac{1}{1!} + \frac{1}{2!} + ... + \frac{1}{n!} + ... ,$$
$$\frac{a_{n+1}}{a_n} = \frac{n!}{(n+1)!} = \frac{1}{n+1} \to 0$$
as $n \to \infty$. Therefore, by Corollary 9.3(a), the series is convergent.

Example 8 In the series
$$\frac{3}{1^2} + \frac{3^2}{2^2} + ... + \frac{3^n}{n^2} + ... ,$$
$$\frac{a_{n+1}}{a_n} = \frac{3^{n+1}}{(n+1)^2} \cdot \frac{n^2}{3^n} = \frac{3}{(1+1/n)^2} \to 3$$
as $n \to \infty$. Hence, Corollary 9.3(a) shows that the series is divergent.

Infinite series

The last sentence of Corollary 9.3(a) can be justified by means of examples. For instance, in

$$1 + \frac{1}{2} + \ldots + \frac{1}{n} + \ldots,$$

$$\lim_{n\to\infty} \frac{a_{n+1}}{a_n} = \lim_{n\to\infty} \frac{n}{n+1} = 1.$$

Since the series is known to diverge, this is a case in which a divergent series leads to the limit 1. On the other hand, the convergent series (Example 6)

$$1 + \frac{1}{2^2} + \ldots + \frac{1}{n^2} + \ldots$$

gives

$$\lim_{n\to\infty} \frac{a_{n+1}}{a_n} = \lim_{n\to\infty} \frac{n^2}{(n+1)^2} = \lim_{n\to\infty} \frac{1}{(1+1/n)^2} = 1.$$

Thus

$$\lim_{n\to\infty} \frac{a_{n+1}}{a_n} = 1$$

does not distinguish between convergent and divergent series.

Sometimes the problem can be resolved by

COROLLARY 9.3(b) *If*

$$\lim_{n\to\infty} n\left(\frac{a_n}{a_{n+1}} - 1\right) = b$$

the series $a_1 + a_2 + \ldots$ is divergent if $b < 1$, convergent if $b > 1$. If $b = 1$, no conclusion can be drawn.

Proof. If $b < 1$, choose λ so that $b < \lambda < 1$. Then there is N such that, for $n \geq N$,

$$n\left(\frac{a_n}{a_{n+1}} - 1\right) \leq \lambda.$$

Now, by the Taylor's theorem (Theorem 7.5), if λ_1 is any number

$$\left(1 + \frac{1}{n}\right)^{\lambda_1} = 1 + \frac{\lambda_1}{n} + \frac{\lambda_1(\lambda_1 - 1)}{2!n^2}(1+c)^{\lambda_1-2}$$

where $0 < c < 1/n$. We deduce that

$$\lim_{n\to\infty} n\left[\left(1 + \frac{1}{n}\right)^{\lambda_1} - 1\right] = \lambda_1.$$

Taking $1 > \lambda_1 > \lambda$, we can be sure that there is N_1 such that

$$n\left[\left(1 + \frac{1}{n}\right)^{\lambda_1} - 1\right] > \lambda$$

for $n \geqslant N_1$. Hence, if N_0 is the greater of N and N_1,

$$n\left(\frac{a_n}{a_{n+1}} - 1\right) < n\left[\left(1 + \frac{1}{n}\right)^{\lambda_1} - 1\right]$$

for $n \geqslant N_0$. Hence

$$\frac{a_{n+1}}{a_n} > \frac{1}{[1 + (1/n)]^{\lambda_1}} = \frac{1/(n+1)^{\lambda_1}}{1/n^{\lambda_1}}$$

for $n \geqslant N_0$. But the series with general term $1/n^{\lambda_1}$ is divergent when $\lambda_1 < 1$ and so, by Exercise 14, the series $a_1 + a_2 + \ldots$ is divergent.

A similar proof applies when $b > 1$, with $b > \mu > \mu_1 > 1$ and the various inequalities reversed.

Example 9 In the series

$$1 + \frac{1}{2} \cdot \frac{1}{3} + \frac{1 \cdot 3}{2 \cdot 4} \cdot \frac{1}{5} + \frac{1 \cdot 3 \cdot 5}{2 \cdot 4 \cdot 6} \cdot \frac{1}{7} + \ldots,$$

$$a_n = \frac{1 \cdot 3 \cdot 5 \ldots (2n-3)}{2 \cdot 4 \cdot 6 \ldots (2n-2)} \cdot \frac{1}{2n-1}.$$

Hence

$$\frac{a_n}{a_{n+1}} = \frac{2n(2n+1)}{(2n-1)(2n-1)} \to 1$$

as $n \to \infty$. Therefore no conclusion can be drawn from Corollary 9.3(a). However

$$\lim_{n \to \infty} n\left(\frac{a_n}{a_{n+1}} - 1\right) = \lim_{n \to \infty} \frac{n(6n-1)}{(2n-1)^2} = \frac{3}{2}.$$

Therefore, by Corollary 9.3(b), the series is convergent.

Exercises

15. Determine whether the following series are convergent or divergent:

(i) $\dfrac{1}{2} + \dfrac{2}{2^2} + \ldots + \dfrac{n}{2^n} + \ldots,$ (ii) $\dfrac{1}{4} + 2\left(\dfrac{1}{4}\right)^2 + \ldots + n\left(\dfrac{1}{4}\right)^n + \ldots,$

(iii) $\dfrac{1}{3} + \dfrac{2!}{3^2} + \ldots + \dfrac{n!}{3^n} + \ldots,$

(iv) $1 + \frac{1}{2} \cdot \frac{a^2}{4} + \frac{1 \cdot 3 \cdot 5}{2 \cdot 4 \cdot 6} \cdot \frac{a^4}{8} + \frac{1 \cdot 3 \cdot 5 \cdot 7 \cdot 9}{2 \cdot 4 \cdot 6 \cdot 8 \cdot 10} \cdot \frac{a^6}{12} + \ldots,$

(v) $\frac{1}{2} + \frac{1 \cdot 2}{2 \cdot 4} + \ldots + \frac{n!}{2 \cdot 4 \ldots 2n} + \ldots,$

(vi) $\frac{1}{1 \cdot 2} + \frac{1}{2 \cdot 3} + \ldots + \frac{1}{n(n+1)} + \ldots,$

(vii) $1 + \frac{\alpha \cdot \beta}{1 \cdot \gamma} a + \frac{\alpha(\alpha+1)\beta(\beta+1)}{1 \cdot 2 \cdot \gamma(\gamma+1)} +$

$$\frac{\alpha(\alpha+1)(\alpha+2)\beta(\beta+1)(\beta+2)}{1 \cdot 2 \cdot 3 \cdot \gamma(\gamma+1)(\gamma+2)} a^3 + \ldots$$

$(a > 0, \gamma \neq \alpha + \beta, \gamma \neq 0, -1, -2,\ldots).$

16. If $\lim_{n \to \infty} (a_n)^{1/n} = c$ prove that $a_1 + a_2 + \ldots$ is convergent if $c < 1$, divergent if $c > 1$. (This is known as *Cauchy's test*.)

17. Determine whether the following series are convergent or divergent:

(i) $\frac{\sqrt{2}}{3} + \frac{\sqrt{3}}{7} + \ldots + \frac{\sqrt{(n+1)}}{n^2+n+1} + \ldots,$

(ii) $\frac{1}{\sqrt{3}} + \frac{1}{\sqrt{7}} + \ldots + \frac{1}{\sqrt{(n^2+n+1)}} + \ldots,$

(iii) $\frac{1 \cdot 2}{3} + \frac{2 \cdot 3}{3^2} + \ldots + \frac{n(n+1)}{3^n} + \ldots,$

(iv) $\frac{1^3}{1!} + \frac{2^3}{2!} + \ldots + \frac{n^3}{n!} + \ldots,$

(v) $\frac{1}{\sqrt{2}} + \frac{1}{\sqrt{5}} + \ldots + \frac{1}{\sqrt{(n^2+1)}} + \ldots,$

(vi) $\frac{2}{2} + \frac{3}{5} + \ldots + \frac{n+1}{n^2+1} + \ldots,$

(vii) $\frac{\sqrt{1}}{2} + \frac{\sqrt{2}}{5} + \ldots + \frac{\sqrt{n}}{n^2+1} + \ldots,$

(viii) $\frac{1}{7} + \frac{2}{10} + \ldots + \frac{n}{n^2+6} + \ldots.$

(ix) $\dfrac{1}{2^2 - 1^2} + \dfrac{1}{3^2 - 2^2} + \dfrac{1}{4^2 - 3^2} + \ldots,$

(x) $\dfrac{5}{1 \cdot 3} + \dfrac{8}{2 \cdot 4} + \dfrac{11}{3 \cdot 5} + \ldots,$

(xi) $\dfrac{2}{3} + \dfrac{5}{21} + \ldots + \dfrac{n^2 + 1}{n^4 + n^2 + 1} + \ldots,$

(xii) $\dfrac{2}{1^{10}} + \dfrac{4}{2^{10}} + \ldots + \dfrac{2^n}{n^{10}} + \ldots,$

(xiii) $\dfrac{1}{1 + a} + \dfrac{1}{2 + a} + \dfrac{1}{2^2 + a} + \dfrac{1}{2^3 + a} + \ldots \ (a > 0).$

9.3 Absolute convergence

The tests in the preceding section are for series in which every term is positive. They can be used for a series in which every term is negative by treating it as the negative of a series with positive terms. But, when some terms are positive and some are negative, the problem must be looked at afresh.

The simplest series for which a test is available is one in which the terms are alternatively positive and negative, e.g.

$$1 - \frac{1}{2} + \frac{1}{4} - \frac{1}{8} + \ldots.$$

Such series are called *alternating series*.

THEOREM 9.4 *If a_n is positive and $a_n > a_{n+1}$ for each value of n and if $\lim\limits_{n \to \infty} a_n = 0$, then the alternating series $a_1 - a_2 + a_3 - a_4 + \ldots$ is convergent.*

Proof Let
$$s_{2m} = a_1 - a_2 + a_3 - a_4 + \ldots + a_{2m-1} - a_{2m}.$$
By writing
$$s_{2m} = (a_1 - a_2) + (a_3 - a_4) + \ldots + (a_{2m-1} - a_{2m})$$
we see that $s_{2m} > 0$ because each of the expressions in brackets is positive from the conditions imposed on a_n in the theorem. It also follows that $s_{2m+2} > s_{2m}$. Moreover, since
$$s_{2m} = a_1 - (a_2 - a_3) - (a_4 - a_5) - \ldots - (a_{2m-2} - a_{2m-1}) - a_{2m},$$
it is clear that $s_{2m} < a_1$. Thus $s_{2m} < s_{2m+2} < a_1$ and so, by the criterion at the end of section 8.2, the sequence $s_2, s_4, \ldots$ converges to a limit $L < a_1$, i.e. $\lim\limits_{m \to \infty} s_{2m} = L.$

Infinite series

As regards $s_1, s_3, \ldots$ note that
$$s_{2m+1} = s_{2m} + a_{2m+1}.$$
Since $\lim_{m \to \infty} s_{2m} = L$ and $\lim_{m \to \infty} a_{2m+1} = 0$, Theorem 4.1(ii) gives
$$\lim_{m \to \infty} s_{2m+1} = L.$$
By combining these two results we obtain
$$\lim_{m \to \infty} s_m = L$$
whether m be odd or even, i.e. the alternating series converges and the theorem is proved.

Next, observe that
$$L - s_{2m} = a_{2m+1} - a_{2m+2} + \ldots.$$
The right hand-side is an alternating series whose first term is a_{2m+1}. Therefore, the above proof tells us that
$$0 < L - s_{2m} < a_{2m+1}. \tag{7}$$
Similarly we can prove that
$$0 < s_{2m+1} - L < a_{2m+2}. \tag{8}$$
These two inequalities provide a valuable estimate of the error made in computing the sum of a convergent alternating series by retaining only the first n terms. If n is even the approximation is too small but is not in error by more than the magnitude of the first neglected term, according to (7). On the other hand, (8) shows that, if n is odd, the approximation is too large but is not in error by more than the magnitude of the first neglected term. Hence we have

COROLLARY 9.4 *In a convergent alternating series the error in the sum committed by neglecting all terms after the nth has the same sign as the $(n + 1)$th term and is less in magnitude than it.*

Example 10 The series
$$1 - \frac{1}{2} + \frac{1}{3} - \frac{1}{4} + \ldots$$
is an alternating series in which $a_n = 1/n$. *Therefore* $a_n > a_{n+1}$ *and* $\lim_{n \to \infty} a_n = 0$.
So the series is convergent by Thereom 9.4. If L is the sum of the series the approximation $1 - \frac{1}{2} + \frac{1}{3} - \ldots - \frac{1}{10}$ will be less than L but will not differ from it by more than $\frac{1}{11}$, from Corollary 9.4. Similarly $1 - \frac{1}{2} - \ldots - \frac{1}{10} + \frac{1}{11}$ exceeds L but does not differ from it by more than $\frac{1}{12}$.

The alternating series in Example 10 is of interest because we know that if all the signs in the series were positive, it would be divergent, being the harmonic series (Example 4). On the other hand, the geometric series
$$1 - \frac{1}{2} + \frac{1}{4} - \frac{1}{8} + \ldots$$

converges and so does the geometric series.
$$1 + \frac{1}{2} + \frac{1}{4} + \frac{1}{8} + \ldots .$$
Thus there are convergent series of positive and negative terms which become divergent when the signs of the negative terms are altered, and there are convergent series of positive and negative terms which remain convergent when the signs of the negative terms are changed. These two possibilities motivate

DEFINITION 9.2 *The series $a_1 + a_2 + \ldots$ is said to be absolutely convergent if*
$$|a_1| + |a_2| + \ldots + |a_n| + \ldots$$
is convergent. If the series $a_1 + a_2 + \ldots$ is convergent but not absolutely convergent it is said to be conditionally convergent.

Thus $1 - \frac{1}{2} + \frac{1}{3} - \frac{1}{4} + \ldots$ is conditionally convergent and $1 - \frac{1}{2} + \frac{1}{4} - \frac{1}{8} + \ldots$ is absolutely convergent.

The idea of absolute convergence is important because it provides a mechanism for saying something about the convergence of a series with positive and negative terms by considering the convergence of a series of positive terms. The relevant theorem is

THEOREM 9.5 *If a series is absolutely convergent then it is also convergent.*

Proof. Let
$$\sigma_n = |a_1| + |a_2| + \ldots + |a_n|.$$
Then $\sigma_{n+1} \geq \sigma_n$. By assumption, $\lim_{n \to \infty} \sigma_n$ exists and is σ, say. Hence
$$\sigma_n \leq \sigma_{n+1} \leq \sigma \qquad (9)$$
for all n.

Now let
$$\rho_n = (a_1 + |a_1|) + (a_2 + |a_2|) + \ldots + (a_n + |a_n|).$$
Since $a_m + |a_m| \geq 0$ for all m, $\rho_{n+1} \geq \rho_n$. Also
$$\rho_n \leq (|a_1| + |a_1|) + (|a_2| + |a_2|) + \ldots + (|a_n| + |a_n|),$$
i.e.
$$\rho_n \leq 2\sigma_n \leq 2\sigma$$
from (9). Hence, by the criterion at the end of section 8.2, $\lim_{n \to \infty} \rho_n$ exists; let the limit be ρ. Then, if
$$s_n = a_1 + a_2 + \ldots + a_n,$$
$$\lim_{n \to \infty} s_n = \lim_{n \to \infty} (\rho_n - \sigma_n) = \lim_{n \to \infty} \rho_n - \lim_{n \to \infty} \sigma_n = \rho - \sigma$$
on using Theorem 4.1(ii). The proof of the theorem is complete.

Infinite series

One consequence of the theorem is that the convergence of a series can be tested by testing the corresponding series of absolute values—for this purpose the tests for series with positive terms developed in the preceding section can be used. If the series of absolute values converges then so does the original series. If the series of absolute values does not converge, no conclusion can be drawn about the original series if it contains varying signs—it may still be conditionally convergent.

Exercises

18. Determine whether the following series are convergent or divergent:

(i) $1 - \dfrac{1}{2^2} + \dfrac{1}{3^2} - \dfrac{1}{4^2} + \ldots ,$ (ii) $1 - \dfrac{1}{2!} + \dfrac{1}{3!} - \dfrac{1}{4!} + \ldots ,$

(iii) $\dfrac{1}{3} - \dfrac{1}{6} + \dfrac{1}{11} - \dfrac{1}{18} + \ldots ,$ (iv) $2 - 1\dfrac{1}{2} + 1\dfrac{1}{3} - 1\dfrac{1}{4} + \ldots ,$

(v) $1 - \dfrac{1}{\sqrt{2}} + \dfrac{1}{\sqrt{3}} - \dfrac{1}{\sqrt{4}} + \ldots ,$

(vi) $1 - \dfrac{4}{3} + \dfrac{3}{2} - \dfrac{8}{5} + \ldots + \dfrac{(-1)^{n+1}2n}{n+1} + \ldots ,$

(vii) $1 - \dfrac{2}{4} + \dfrac{3}{4^2} - \dfrac{4}{4^3} + \ldots ,$ (viii) $\dfrac{1}{2} - \dfrac{1}{2^{\frac{1}{2}}} + \dfrac{1}{2^{\frac{1}{3}}} - \dfrac{1}{2^{\frac{1}{4}}} + \ldots ,$

(ix) $\dfrac{1}{\sqrt{1 \cdot 2}} - \dfrac{1}{\sqrt{2 \cdot 3}} + \dfrac{1}{\sqrt{3 \cdot 4}} - \dfrac{1}{\sqrt{4 \cdot 5}} + \ldots .$

19. In each of the convergent series of question 18 the sum is approximated by neglecting all terms after (a) the 4th, (b) the 7th. Give a bound for the error in each case.

20. Draw a flow-chart for computing the sum of a convergent alternating series.

21. If
$$\lim_{n \to \infty} \left| \dfrac{a_{n+1}}{a_n} \right| = c$$
show that $a_1 + a_2 + \ldots$ is absolutely convergent if $c < 1$, but divergent if $c > 1$.

22. If
$$\lim_{n \to \infty} n \left(\left| \dfrac{a_n}{a_{n+1}} \right| - 1 \right) = b$$
show that $a_1 + a_2 + \ldots$ is absolutely convergent if $b > 1$. Must $a_1 + a_2 + \ldots$ be divergent if $b < 1$? Consider the series with $a_n = (-1)^{n+1}/n^{\frac{1}{2}}$.

23. Give an analogue to Theorem 9.2(i) for absolute convergence.

24. Which of the series in question 18 are absolutely convergent?

25. Examine the following series for absolute or conditional convergence:

(i) $1 - \dfrac{1}{27} + \dfrac{1}{125} - \ldots + \dfrac{(-1)^{n+1}}{(2n-1)^3} + \ldots,$

(ii) $\cos a + \dfrac{\cos 2a}{2^2} + \dfrac{\cos 3a}{3^2} + \ldots + \dfrac{\cos na}{n^2} + \ldots,$

(iii) $\dfrac{1}{\sqrt{3}} - \dfrac{1}{\sqrt{10}} + \ldots + \dfrac{(-1)^{n+1}}{[n(2n+1)]^{\frac{1}{2}}} + \ldots,$

(iv) $\dfrac{1}{4} - \dfrac{1}{7} + \ldots + \dfrac{(-1)^{n+1}}{n^2+3} + \ldots,$

(v) $\sin a + \dfrac{\sin 2a}{2^{\frac{3}{2}}} + \ldots + \dfrac{\sin na}{n^{\frac{3}{2}}} + \ldots,$

(vi) $\dfrac{1!3}{2!} - \dfrac{2!3^2}{4!} + \ldots + (-1)^{n+1}\dfrac{n!3^n}{(2n)!} + \ldots,$

(vii) $\dfrac{a}{1+a^2} + \dfrac{a^2}{1+a^4} + \dfrac{a^3}{1+a^6} + \ldots \; (|a| < 1),$

(viii) $1 - \dfrac{5}{17} + \ldots + (-1)^{n+1}\dfrac{n^2+1}{n^4+1} + \ldots,$

(ix) $\dfrac{1}{1!a} + \dfrac{1}{2!a^2} + \ldots + \dfrac{1}{n!a^n} + \ldots \; (a \neq 0),$

(x) $\dfrac{(1!)^2 3}{2!} - \dfrac{(2!)^2 3^2}{4!} + \ldots + (-1)^{n+1}\dfrac{(n!)^2 3^n}{(2n)!} + \ldots,$

(xi) $\dfrac{(1!)^2 5}{2!} - \dfrac{(2!)^2 5^2}{4!} + \ldots + (-1)^{n+1}\dfrac{(n!)^2 5^n}{(2n)!} + \ldots,$

(xii) $\dfrac{1}{a^2-1} + \dfrac{a^2}{a^4-1} + \ldots + \dfrac{(a^2)^{n-1}}{(a^2)^n - 1} + \ldots \; (a^2 < 1).$

So far we have been considering a single series. It is sometimes desirable to combine series in certain ways and in order to do this we need theorems covering addition, subtraction and multiplication.

Infinite series

THEOREM 9.6 *If $a_1 + a_2 + \ldots$, $b_1 + b_2 + \ldots$ are convergent with sums s_a, s_b respectively then $(a_1 + b_1) + (a_2 + b_2) + \ldots$ and $(a_1 - b_1) + (a_2 - b_2) + \ldots$ are convergent with sums $s_a + s_b$ and $s_a - s_b$ respectively. If $a_1 + a_2 + \ldots \pm b_1 + b_2 + \ldots$ are absolutely convergent so are $(a_1 \pm b_1) + (a_2 \pm b_2) + \ldots$.*

Proof. Let $s_n = a_1 + a_2 + \ldots + a_n$, $\sigma_n = b_1 + b_2 + \ldots + b_n$. Then
$$\lim_{n \to \infty} [(a_1 + b_1) + (a_2 + b_2) + \ldots + (a_n + b_n)] = \lim_{n \to \infty} (s_n + \sigma_n)$$
$$= \lim_{n \to \infty} s_n + \lim_{n \to \infty} \sigma_n$$
$$= s_a + s_b$$

by Theorem 4.1(ii). A similar proof applies to the difference between two series.

Also, since
$$|a_n \pm b_n| \leqslant |a_n| + |b_n|,$$
it is clear that if the two series are absolutely convergent then so are the series obtained by addition or subtraction. The proof is complete.

The reader must be careful not to read into this theorem more than is there. It does not give a licence to reorder the terms of a convergent series in any fashion one likes before carrying out a calculation. For example, the convergent series $1 - \frac{1}{2} + \frac{1}{3} - \frac{1}{4} + \ldots$ might be grouped as $(1 - \frac{1}{2}) + (\frac{1}{3} - \frac{1}{4}) + \ldots = \frac{1}{2} + \frac{1}{12} + \ldots$. But if the terms were grouped in a different order as
$$\left(1 - \frac{1}{2} - \frac{1}{4}\right) + \left(\frac{1}{3} - \frac{1}{6} - \frac{1}{8}\right) + \ldots + \left(\frac{1}{2n-1} - \frac{1}{4n-2} - \frac{1}{4n}\right) + \ldots$$
$$= \frac{1}{4} + \frac{1}{24} + \ldots = \frac{1}{2}\left(\frac{1}{2} + \frac{1}{12} + \ldots\right)$$
we obtain a different result. Thus, in general, the order of terms in a series must be preserved if contradictions are not to arise. However, there is one important exception to this rule, given by

THEOREM 9.7 *The sum of an absolutely convergent series is not changed if the order of the terms is altered.*

Proof. The proof is carried out in two parts, firstly for series of positive terms and secondly for general series.

(a) Suppose that $a_1 + a_2 + \ldots$ consists entirely of positive terms and that, by altering the order, the series $b_1 + b_2 + \ldots$ is obtained. Let
$$s_n = a_1 + a_2 + \ldots + a_n,$$
$$\sigma_n = b_1 + b_2 + \ldots + b_n.$$

Given σ_m, we can choose n so large that every term in σ_m occurs in s_n because every b is some a. Also $s_n \leqslant s$ where $s = \lim_{n \to \infty} s_n$. Hence
$$\sigma_m \leqslant s_n \leqslant s.$$
Since $\sigma \geqslant_{m+1} \sigma_m$, it follows from the criterion at the end of section 8.2 that $\lim_{m \to \infty} \sigma_m$ exists; denote it by σ. Then
$$\sigma \leqslant s.$$

On the other hand, the series $a_1 + a_2 + \ldots$ is an alteration of order of the series $b_1 + b_2 + \ldots$ which has been proved convergent. Therefore, by repeating the argument with a and b interchanged, we find
$$s \leqslant \sigma.$$
The two inequalities are consistent only if $s = \sigma$, i.e. the sum of the series is not altered by changing the order of the terms.

(b) Suppose now that $a_1 + a_2 + \ldots$, an absolutely convergent series contains terms of both signs. Let there be μ positive and ν negative terms in s_n and let their sums be P_μ and N_ν respectively. Then
$$s_n = P_\mu + N_\nu.$$
Let
$$\rho_n = |a_1| + |a_2| + \ldots + |a_n|;$$
then
$$\rho_n = P_\mu - N_\nu.$$
Hence
$$P_\mu = \tfrac{1}{2}(\rho_n + s_n), \qquad N_\nu = \tfrac{1}{2}(s_n - \rho_n).$$
As $n \to \infty$, $s_n \to s$ and $\rho_n \to \rho$ because the series is absolutely convergent. Hence
$$P_\mu \to \tfrac{1}{2}(\rho + s), \qquad N_\nu \to \tfrac{1}{2}(s - \rho).$$
Thus the series of positive a's is convergent and so is the series of negative a's. By part (a) both parts can be altered in order without affecting the sum. Hence the sum of the original series is unaltered by a change in order; the proof is finished.

The illustration before Theorem 9.7 shows that the sum of a conditionally convergent series can be adjusted by a change of order. In fact, there are rearrangements which diverge.

Turning now to multiplication we have

THEOREM 9.8 *Let* $a_1 + a_2 + \ldots$, $b_1 + b_2 + \ldots$ *be absolutely convergent with sums* s_a, s_b, *respectively. Let*
$c_1 = a_1 b_1, c_2 = a_1 b_2 + a_2 b_1, \ldots,$
$$c_n = a_1 b_n + a_2 b_{n-1} + \ldots + a_{n-1} b_2 + a_n b_1, \ldots.$$
Then the series $c_1 + c_2 + \ldots$ *is absolutely convergent and*
$$c_1 + c_2 + \ldots = s_a s_b.$$

Proof. Let
$$s_n = a_1 + a_2 + \ldots + a_n, \quad \rho_n = b_1 + b_2 + \ldots + b_n, \quad \rho_n = c_1 + c_2 + \ldots + c_n$$

(a) Let all the a's and b's be positive. The product $s_n \sigma_n$ contains all terms of the type $a_p b_q$ which occur in ρ_n and possibly more. Hence
$$\rho_n \leqslant s_n \sigma_n.$$
Since neither s_n nor σ_n decreases as n increases and $\lim_{n \to \infty} s_n \sigma_n = s_a s_b$ (Theorem 4.1(iii)) we have $\rho_n \leqslant s_a s_b$. Since $\rho_{n+1} \geqslant \rho_n$ the criterion at the end of section 8 demonstrates that $\lim_{n \to \infty} \rho_n$ exists; denote the limit by ρ. Then
$$\rho \leqslant s_a s_b. \tag{10}$$
Further, ρ_{2n-1} contains all terms $a_p b_q$ for which $p + q \leqslant 2n$. Therefore
$$\rho_{2n-1} \geqslant s_n \sigma_n.$$
Letting $n \to \infty$ we have $\rho \geqslant s_a s_b$. Combining this with (10) we have
$$\rho = s_a s_b$$
so that the theorem is true for series of positive terms.

(b) When the terms are not all positive let
$$S_n = |a_1| + |a_2| + \ldots + |a_n|, \quad \Sigma_n = |b_1| + |b_2| + \ldots + |b_n|,$$
$$R_n = |c_1| + |c_2| + \ldots + |c_n|.$$

Now
$$|s_n \sigma_n - \rho_n| \leqslant \text{sum of moduli of all terms } a_p b_q \text{ which are in } s_n \sigma_n \text{ but not in } \rho_n$$
$$\leqslant S_n \Sigma_n - R_n.$$
But, by (a), $S_n \Sigma_n - R_n \to 0$ as $n \to \infty$. Therefore
$$s_n \sigma_n - \rho_n \to 0.$$
Since $s_n \sigma_n \to s_a s_b$, the theorem is proved.

Example 11 For the series
$$1 + \frac{1}{1!} + \frac{1}{2!} + \ldots, \quad 1 + \frac{2}{1!} + \frac{2^2}{2!} + \ldots$$
we have
$$a_n = \frac{1}{(n-1)!}, \quad b_n = \frac{2^{n-1}}{(n-1)!}$$
with the understanding that $0! = 1$. Then
$$c_n = 1 \cdot \frac{2^{n-1}}{(n-1)!} + \frac{1}{1!} \cdot \frac{2^{n-2}}{(n-2)!} + \frac{1}{2!} \cdot \frac{2^{n-3}}{(n-3)!} + \ldots + \frac{1}{(n-1)!} \cdot 1$$
$$= \frac{1}{(n-1)!} \left[2^{n-1} + (n-1)2^{n-2} + (n-1)(n-2)\frac{2^{n-3}}{2!} + \ldots + 1 \right]$$
$$= \frac{1}{(n-1)!}(2+1)^{n-1} = \frac{3^{n-1}}{(n-1)!}.$$

Hence
$$\left(1 + \frac{1}{1!} + \frac{1}{2!} + \ldots\right)\left(1 + \frac{2}{1!} + \frac{2^2}{2!} + \ldots\right) = 1 + \frac{3}{1!} + \frac{3^2}{2!} + \ldots.$$

Exercises

26. If $a_1 + a_2 + \ldots$, $b_1 + b_2 + \ldots$ are convergent and k_1, k_2 are constants show that $(k_1 a_1 + k_2 b_1) + (k_1 a_2 + k_2 b_2) + \ldots$ is convergent and has sum
$$k_1(a_1 + a_2 + \ldots) + k_2(b_1 + b_2 + \ldots).$$

27. If $a_1 + a_2 + \ldots$ is convergent and $b_1 + b_2 + \ldots$ is divergent prove that $(a_1 + b_1) + (a_2 + b_2) + \ldots$ is divergent. Deduce that
$$\frac{2+1}{1 \cdot 1} + \frac{2^2 + 2}{2^2 \cdot 2} + \frac{2^3 + 3}{2^3 \cdot 3} + \ldots + \frac{2^n + n}{2^n \cdot n} + \ldots$$
is divergent.

28. Give an example of two divergent series whose sum is divergent and whose difference is convergent.

29. If $a_1 + a_2 + \ldots$ is absolutely convergent prove that $a_1^2 + a_2^2 + \ldots$ is convergent.

30. If $a_1 + a_2 + \ldots$ is conditionally convergent prove that
$$(a_1 + |a_1|) + (a_2 + |a_2|) + \ldots$$
is divergent.

31. Show that
$$\left(1 + \frac{3}{1!} + \frac{3^2}{2!} + \ldots\right)^2 = 1 + \frac{6}{1!} + \frac{6^2}{2!} + \ldots.$$

32. Prove that, if $|a| < 1$,
$$\left(a - \frac{a^2}{2} + \frac{a^3}{3} - \frac{a^4}{4} + \ldots\right)^2 =$$
$$2\left[\frac{1}{2}a^2 - \frac{1}{3}\left(1 + \frac{1}{2}\right)a^3 + \frac{1}{4}\left(1 + \frac{1}{2} + \frac{1}{3}\right)a^4 - \ldots\right].$$

33. Prove that
$$\left(1 + \frac{2^2}{2!} + \frac{2^4}{4!} + \ldots\right)\left(2 + \frac{2^3}{3!} + \frac{2^4}{5!} + \ldots\right) = \frac{1}{2}\left(4 + \frac{4^3}{3!} + \frac{4^5}{5!} + \ldots\right).$$

34. The series $\frac{1}{2} + \frac{1}{5} + \ldots + \frac{1}{n^2 + 1} + \ldots$ is convergent but nearly 5,800 terms are needed to calculate the sum correct to 3 decimal places. By subtracting the series $1 + \frac{1}{4} + \ldots + \frac{1}{n^2} + \ldots$, which is known to have sum 1·64493, obtain a series in which the first 10 terms lead to the sum 1·077 for the first series.

35. Use the method of question 34 to calculate
$$\frac{1}{1 \cdot 2} + \frac{1}{2^2 \cdot 5} + \ldots + \frac{1}{n^2(n^2 + 1)} + \ldots$$
given that $1 + \frac{1}{2^4} + \ldots + \frac{1}{n^4} + \ldots = 1 \cdot 08234$.

36. Calculate $1 + \frac{1}{2^3} + \ldots + \frac{1}{n^3}$ by writing
$$\frac{1}{n^3} = \frac{1}{2}\left[\frac{1}{n(n-1)} - \frac{1}{n(n+1)}\right] - \frac{1}{n^3(n^2-1)}$$
when $n \neq 1$.

9.4 Power series

So far we have been considering infinite series of constant terms. Such series are of importance in mathematics and computation. However, it is when the terms are taken to be functions of x that the real strength of the theory of infinite series is seen. The simplest illustration is provided by the series
$$c_0 + c_1 x + c_2 x^2 + \ldots + c_n x^n + \ldots \tag{11}$$
where $c_0, c_1, \ldots$ are constants. Such a series is called a *power series in x*. Note that the series has been started at $n = 0$ so that the same n can be attached to both c and x.

If a particular value of x, say $x = \frac{1}{3}$, is inserted in the power series a series of constant terms is obtained. Should this series converge, the value of (11) at $x = \frac{1}{3}$ is known. For another value of x, say $x = \frac{2}{3}$, the series of constant terms may diverge. Since a meaning has been attached to an infinite series only when it converges, (11) will have a meaning only for those values of x for which it converges. Thus (11) represents a function of x whose domain consists of those values of x for which (11) converges. Consequently, it is necessary to investigate the convergence of (11) for different values of x. Fortunately, it will turn out that we need not test every value of x to determine the convergence properties of (11). This is because of

Introductory Analysis

THEOREM 9.9 *If the power series*

$$c_0 + c_1 x + c_2 x^2 + \ldots$$

converges for $x = a$ ($a \neq 0$) then it is absolutely convergent for any x satisfying $|x| < |a|$. If the power series diverges for $x = b$ then it diverges for any x such that $|x| > |b|$.

Proof. When the series

$$c_0 + c_1 a + c_2 a^2 + \ldots$$

converges, Theorem 9.1 shows that $\lim_{n \to \infty} c_n a^n = 0$. Therefore there is N such that

$$|c_n a^n| < 1$$

for all $n \geq N$, i.e.

$$|c_n| < \frac{1}{|a|^n} \quad (n \geq N). \tag{12}$$

Choose any x such that $|x| < |a|$. Then, for $n \geq N$,

$$|c_n x^n| < \frac{|x|^n}{|a|^n}$$

by (12). But the series

$$\left|\frac{x}{a}\right|^N + \left|\frac{x}{a}\right|^{N+1} + \ldots$$

is a geometric series which converges because $|x/a| < 1$ (Example 2). Hence, by Theorem 9.2, the series

$$|c_0| + |c_1 x| + |c_2 x^2| + \ldots$$

converges, i.e. the original series converges absolutely. The first half of the theorem is proved.

As far as the second half is concerned suppose that the power series diverged at $x = b$ but converged at $x = c$ where $|c| < |b|$. Taking $a = c$ in the first half of the theorem we would conclude that the power series converged absolutely at $x = b$. Since the series cannot converge absolutely and diverge at the same time (Theorem 9.5) we deduce that it cannot converge at $x = c$; the theorem is proved.

One consequence of Theorem 9.9 is that there are only three ways in which a power series can behave:

(i) it always converges for $x = 0$ but may not converge for any other value of x,

(ii) it may converge absolutely for all finite x,

(iii) there may be a positive number R such that the power series converges absolutely for $|x| < R$ and diverges for $|x| > R$.

Setting aside case (i), where the series converges for only one value of x, we see that there is always a domain of values of x for which the power series converges. This domain is called the *interval of convergence*. If the end-points are ignored, it is symmetrically placed about the origin and is of length $2R$ (if R is infinite case (ii) arises). Consequently, a rule which determines R tells us most of what we need to know about the convergence of a given power series.

One such rule is given by

THEOREM 9.10 *If* $\lim_{n \to \infty} |c_{n+1}/c_n|$ *exists and is denoted by* $1/R$, *the power series*
(11) *converges absolutely for* $|x| < R$ *and diverges for* $|x| > R$.

Proof. Choose a definite value of x, say x_0, such that $|x_0| < R$. In the series of constants
$$c_0 + c_1 x_0 + c_2 x_0^2 + \dots$$
put $a_n = c_{n-1} x_0^{n-1}$. Then
$$\lim_{n \to \infty} \left| \frac{a_{n+1}}{a_n} \right| = \lim_{n \to \infty} \left| \frac{c_n x_0}{c_{n-1}} \right| = |x_0| \lim_{n \to \infty} \left| \frac{c_n}{c_{n-1}} \right|$$
since x_0 does not involve n. By the assumption in the theorem the right-hand side is $|x_0|/R$. Since this is less than 1, Corollary 9.3(a) shows that the series
$$|a_1| + |a_2| + \dots$$
is convergent. The only restriction that was imposed on x_0 was that $|x_0| < R$. Consequently (11) is absolutely convergent for $|x| < R$ and the first half of the theorem is proved.

For the second half of the theorem take $|x_0| > R$ and repeat the analysis. Then Corollary 9.3(a) demonstrates that (11) is not absolutely convergent for $|x| > R$. If (11) were convergent (even though not absolutely convergent) at $x = R + h\,(h > 0)$ then Theorem 9.9 implies absolute convergence at $x = R + \tfrac{1}{2}h$. Since absolute convergence is excluded (11) must be divergent for $|x| > R$. The proof is terminated.

The theorem does not say what happens at $x = R$ or at $x = -R$. In fact, there are series which converge at both points, others which diverge at both, and yet others which converge at one end-point and diverge at the other.

Example 12 In the power series
$$x - \frac{1}{2}x^2 + \frac{1}{3}x^3 - \dots + (-1)^{n-1}\frac{x^n}{n} + \dots,$$

$c_n = (-1)^{n-1}/n$. Therefore

$$\lim_{n \to \infty} \left|\frac{c_{n+1}}{c_n}\right| = \lim_{n \to \infty} \frac{n}{n+1} = 1.$$

Hence, by Theorem 9.10, the power series is absolutely convergent for $|x| < 1$ and divergent for $|x| > 1$.

When $x = 1$, the series becomes

$$1 - \frac{1}{2} + \frac{1}{3} - \cdots$$

which is an alternating series, and is convergent by Theorem 9.4.

When $x = -1$, the series is

$$-\left(1 + \frac{1}{2} + \frac{1}{3} + \cdots\right)$$

and is divergent (Example 4).

The interval of convergence for the power series is $-1 < x \leqslant 1$.

Example 13 In the power series

$$1 + x + \frac{x^2}{2!} + \cdots + \frac{x^n}{n!} + \cdots,$$

$c_n = 1/n!$. Hence

$$\lim_{n \to \infty} \left|\frac{c_{n+1}}{c_n}\right| = \lim_{n \to \infty} \frac{1}{n+1} = 0.$$

In this case R can be taken as large as we please. In other words, the power series converges for any finite value of x.

Example 14 In the power series

$$1 - \frac{2x}{1^2} + \frac{(2x)^2}{2^2} - \cdots + \frac{(-1)^n (2x)^n}{n^2} + \cdots,$$

$c_n = (-2)^n/n^2$. Therefore

$$\lim_{n \to \infty} \left|\frac{c_{n+1}}{c_n}\right| = \lim_{n \to \infty} \frac{2n^2}{(n+1)^2} = \lim_{n \to \infty} \frac{2}{(1 + 1/n)^2} = 2.$$

In this case $R = \frac{1}{2}$ and the power series converges absolutely for $|x| < \frac{1}{2}$ but diverges for $|x| > \frac{1}{2}$.

When $x = \pm \frac{1}{2}$ comparison with the series

$$1 + \frac{1}{1^2} + \frac{1}{2^2} + \cdots + \frac{1}{n^2} + \cdots$$

(Example 6) shows that the series is convergent.

The interval of convergence is $-\frac{1}{2} \leqslant x \leqslant \frac{1}{2}$.

Infinite series

The series
$$c_0 + c_1(x - a) + c_2(x - a)^2 + \ldots + c_n(x - a)^n + \ldots \qquad (13)$$
is called a *power series in $x - a$*. It is really the same as (11) slightly disguised. For, putting $X = x - a$, gives
$$c_0 + c_1 X + c_2 X^2 + \ldots + c_n X_n + \ldots \qquad (14)$$
which is of the same form as (11). Suppose that the interval of convergence of (14) is $-R < X \leqslant R$. Then (13) converges when
$$-R < x - a \leqslant R.$$
Hence the interval of convergence of (13) is
$$-R + a < x \leqslant R + a.$$
In this way the interval of convergence of (13) can always be determined from that for (11).

Exercises

37. Determine the interval of convergence of the following power series, stating where the series converges absolutely:

(i) $1 + x + x^2 + \ldots + x^n + \ldots$, (ii) $1 + x + \dfrac{x^2}{2^2} + \ldots + \dfrac{x^n}{n^2} + \ldots$,

(iii) $1 - \dfrac{3x}{2} + \dfrac{5x^2}{3} - \ldots + \dfrac{(-1)^n(2n-1)}{n}x^{n-1} + \ldots$,

(iv) $1 + 2x + 3x^2 + \ldots + nx^{n-1} + \ldots$,

(v) $\dfrac{1!x}{2!} - \dfrac{2!x^2}{4!} + \ldots + \dfrac{(-1)^{n-1}n!x^n}{(2n)!} + \ldots$,

(vi) $1 \cdot 2 + 2 \cdot 3x + \ldots + n(n+1)x^{n-1} + \ldots$,

(vii) $\dfrac{2}{3} + \dfrac{3}{5}x + \ldots + \dfrac{2^n+1}{2^{n+1}+1}x^n + \ldots$,

(viii) $\dfrac{2x}{2} + \dfrac{5}{17}x^2 + \ldots + \dfrac{(n^2+1)}{n^4+1}x^n + \ldots$,

(ix) $1 - \dfrac{2x}{1!} + \dfrac{3x^2}{2!} - \ldots + (-1)^n \dfrac{(n+1)}{n!}x^n \ldots$,

(x) $\dfrac{px}{1^p} + \dfrac{(px)^2}{2^p} + \ldots + \dfrac{(px)^n}{n^p} + \ldots$ (p the ratio of two positive integers),

(xi) $\dfrac{(1!)^2}{2!}x + \dfrac{(2!)^2}{4!}x^2 + \ldots + \dfrac{(n!)^2}{(2n)!}x^n + \ldots$,

(xii) $\dfrac{2!x \sin \frac{1}{4}\pi}{(1!)^2} + \dfrac{4!}{(2!)^2}x^2 \sin \frac{1}{8}\pi + \ldots + \dfrac{(2n)!}{(n!)^2}x^n \sin \dfrac{\pi}{4n} + \ldots,$

(xiii) $1 + 1!x + 2!x^2 + \ldots + n!x^n + \ldots,$

(xiv) $\dfrac{x}{4} - \dfrac{x^2}{2 \cdot 4^2 \cdot 7} + \dfrac{x^3}{3 \cdot 4^3 \cdot 7^2} - \ldots + \dfrac{(-1)^{n-1}x^n}{n \cdot 4^n \cdot 7^{n-1}} + \ldots,$

(xv) $1 + x + x^4 + x^9 + \ldots + x^{n^2} \ldots.$

(Ignore the behaviour at the end-points in (xi) and (xii).)

38. Determine the interval of convergence of

(i) $\dfrac{x-3}{1} + \dfrac{(x-3)^2}{2} + \ldots + \dfrac{(x-3)^n}{n} + \ldots,$

(ii) $1 - \dfrac{3x-2}{2} + \dfrac{(3x-2)^2}{3} - \ldots + \dfrac{(-1)^n(3x-2)^n}{n+1} + \ldots,$

(iii) $1 + \dfrac{x+1}{2} + \dfrac{(x+1)^2}{2!2^2} + \ldots + \dfrac{(x+1)^n}{n!2^n} + \ldots,$

(iv) $1 + 1!(x+2) + 2!(x+2)^2 + \ldots + n!(x+2)^n + \ldots,$

(v) $\dfrac{1}{\sqrt{9}}\left(\dfrac{x-2}{3}\right)^2 + \dfrac{\sqrt{2}}{\sqrt{28}}\left(\dfrac{x-2}{3}\right)^3 + \ldots + \dfrac{\sqrt{(n-1)}}{\sqrt{(n^3+1)}}\left(\dfrac{x-2}{3}\right)^n + \ldots,$

(vi) $\dfrac{2x+1}{\sqrt{1}} - \dfrac{(2x+1)^2}{\sqrt{2}} + \ldots + \dfrac{(-1)^{n-1}(2x+1)^n}{\sqrt{n}} + \ldots,$

(vii) $1 + \dfrac{x-2}{1^2} + \dfrac{(x-2)^2}{2^2} + \ldots + \dfrac{(x-2)^n}{n^2} + \ldots,$

(viii) $x + 2 - \dfrac{(x+2)^3}{3} + \ldots + (-1)^n\dfrac{(x+2)^{2n+1}}{2n+1} + \ldots.$

39. (i) Deduce from Example 13 that $\lim_{n \to \infty} x^n/n! = 0$ for every real x. (ii) Deduce from Exercise 16 that if $\lim |c_n|^{1/n} = c$, the power series (11) converges for $|x| < 1/c$ and diverges for $|x| > 1/c$.

There are various operations which can be carried out with power series. For example,

THEOREM 9.11 *If*

$$f(x) = c_0 + c_1x + c_2x^2 + \ldots + c_nx^n + \ldots,$$
$$g(x) = d_0 + d_1x + d_2x^3 + \ldots + d_nx^n + \ldots,$$

Infinite series

both series converging absolutely for $|x| < R$, *then*
$$f(x) \pm g(x) = c_0 \pm d_0 + (c_1 \pm d_1)x + \ldots + (c_n \pm d_n)x^n + \ldots,$$
$$f(x)g(x) = c_0 d_0 + (c_0 d_1 + c_1 d_0)x + \ldots + (c_0 d_n + c_1 d_{n-1} + \ldots$$
$$+ c_n d_0)x^n + \ldots$$
for $|x| < R$.

This theorem is an immediate consequence of Theorem 9.6 and 9.8, and shows that two power series may be added, subtracted or multiplied in any common domain of absolute convergence. The rule for multiplication is the same as multiplying the series term-by-term and collecting together like powers of x. The division of a power series by another power series may also be represented as a power series provided that the denominator does not vanish in the interval under consideration. However, the problem of division will not be discussed further here.

One useful theorem about power series is

THEOREM 9.12 *If a power series converges absolutely for* $|x| < R$, *then it represents a continuous function of x on* $|x| < R$.

Proof. Choose a so that $-R < a < R$ and then choose a positive b such that $|a| < b < R$. Since the power series is absolutely convergent for $x = b$ (Theorem 9.9)
$$|c_0| + |c_1|b + |c_2|b^2 + \ldots + |c_n|b^n + \ldots$$
is convergent. Therefore, by the definition of convergence, given $\varepsilon > 0$ there is N such that
$$|c_{N+1}|b^{N+1} + |c_{N+2}|b^{N+2} + \ldots < \varepsilon, \tag{15}$$
since the sequence of partial sums is tending to the sum and so the remainder is tending to zero. Let the 'remainder' $r_n(x)$ be defined by
$$r_n(x) = c_{n+1}x^{n+1} + c_{n+2}x^{n+2} + \ldots.$$
Then, if $|x| \leq b$,
$$|r_N(x)| \leq |c_{N+1}x^{N+1}| + |c_{N+2}x^{N+2}| + \ldots$$
$$\leq |c_{N+1}|b^{N+1} + |c_{N+2}|b^{N+2} + \ldots$$
$$\leq \varepsilon, \tag{16}$$
from (15).

Choose any h such that $-b < a + h < b$. Then
$$|c_0 + c_1(a+h) + c_2(a+h)^2 + \ldots - c_0 - c_1 a - c_2 a^2 - \ldots|$$
$$= |c_0 + c_1(a+h) + \ldots + c_N(a+h)^N - c_0 - c_1 a - \ldots$$
$$- c_N a^N + r_N(a+h) - r_N(a)|$$
$$\leq |c_0 + c_1(a+h) + \ldots + c_N(a+h)^N - c_0 - c_1 a - \ldots - c_N a^N| +$$
$$|r_N(a+h)| + |r_N(a)|.$$

Because of the choice of h, (16) shows that each of the last two terms does not exceed ε. Since $c_0 + c_1 x + \ldots + c_N x^N$ is a polynomial it is continuous and therefore h can certainly be made small enough for the first term to be less than ε. Hence the right-hand side can be made as small as desired by taking h sufficiently small, i.e. the power series is continuous at $x = a$. Since a was arbitrary (provided only that $-R < a < R$) the proof is complete.

One of the most valuable properties of a power series is that one can obtain its derivatives by adding together the derivatives of the separate terms in the series. In other words, Corollary 5.4 remains true even if an infinite number of terms is involved, so long as it is a power series which is under consideration.* This process of obtaining the derivative of a power series is often called 'taking derivatives term-by-term'. The relevant theorem, which shows that a power series is not only continuous but differentiable, is

THEOREM 9.13 *If $c_0 + c_1 x + \ldots + c_n x^n + \ldots$ converges absolutely for $|x| < R$, then its derivative is*
$$c_1 + 2c_2 x + \ldots + nc_n x^{n-1} + \ldots$$
for $|x| < R$.

Proof. First we show that the power series, which is stated to be the derivative, is absolutely convergent for $|x| < R$. Choose any positive a such that $0 < a < R$ and let h be a small positive number such that $0 < a < a + h < R$. Let
$$f(x) = c_0 + c_1 x + \ldots + c_n x^n + \ldots.$$
The series for $f(a)$ and $f(a + h)$ are absolutely convergent and so (Theorem 9.6)
$$\frac{f(a+h) - f(a)}{h} = c_1 \frac{a+h-a}{h} + \ldots + c_n \frac{(a+h)^n - a^n}{h} + \ldots$$
and the series on the right is absolutely convergent. By the mean-value theorem (Theorem 7.2)
$$(a+h)^n - a^n = hn(a + \theta_n h)^{n-1}$$
where $0 < \theta_n < 1$. Therefore the series
$$c_1 + c_2 2(a + \theta_2 h) + \ldots + c_n n(a + \theta_n h)^{n-1} + \ldots$$
is absolutely convergent. Since $a + \theta_n h > a$ it follows that
$$c_1 + 2c_2 a + \ldots + nc_n a^{n-1} + \ldots$$
is absolutely convergent, which is what we wished to demonstrate.

Let
$$F(x) = c_1 + 2c_2 x + \ldots + nc_n x^{n-1} + \ldots,$$
$$\rho_n(x) = (n+1)c_{n+1} x^n + (n+2)c_{n+2} x^{n+1} + \ldots.$$

* This result is not necessarily true for infinite series which are not power series.

Choose a positive b such that $0 < b < R$. Since $F(x)$ is absolutely convergent for $x = b$, there is N_1 (as in deriving (16)) such that
$$|\rho_{N_1}(x)| \leqslant \varepsilon \tag{17}$$
for $|x| \leqslant b$. Similarly, if
$$\sigma_n = c_{n+1}(n+1)(x + \theta_{n+1}h) + c_{n+2}(n+2)(x + \theta_{n+2}h)^{n+1} + \ldots$$
there is N_2 such that
$$|\sigma_{N_2}| \leqslant \varepsilon \tag{18}$$
for $|x + h| < b$. Let N be the greater of N_1 and N_2.

Choose $|x| < b$ and any non-zero h such that $0 < |x + h| < b$. Then
$$\left| F(x) - \frac{f(x+h) - f(x)}{h} \right|$$
$$= \left| c_1 + \ldots + Nc_N x^{N-1} - c_1 \frac{x + h - x}{h} - \ldots \right.$$
$$\left. - c_N \frac{(x+h)^N - x^N}{h} + \rho_N(x) - \sigma_N \right|$$
$$\leqslant \left| c_1 + \ldots + Nc_N x^{N-1} - c_1 \frac{x + h - x}{h} - \ldots - c_N \frac{(x+h)^N - x^N}{h} \right| + 2\varepsilon$$

from (17) and (18). By taking h small enough the first term on the right can be made less than ε, since the derivative of a polynomial of degree N is being calculated. An immediate deduction is that
$$\lim_{h \to 0} \left| F(x) - \frac{f(x+h) - f(x)}{h} \right| = 0.$$
But
$$\lim_{h \to 0} \frac{f(x+h) - f(x)}{h} = f'(x)$$
and so
$$f'(x) = F(x).$$
The proof of the theorem is complete.

Example 15 The power series
$$1 + x + x^2 + \ldots + x^n + \ldots$$
is absolutely convergent for $|x| < 1$ (*Example 2*). Therefore, by Theorem 9.13, its derivative is
$$1 + 2x + \ldots + nx^{n-1} + \ldots$$
for $|x| < 1$. Since
$$1 + x + \ldots + x^n + \ldots = \frac{1}{1-x}$$

for $|x| < 1$, it follows that
$$1 + 2x + \ldots + nx^{n-1} + \ldots = \frac{1}{(1-x)^2}$$
for $|x| < 1$.

Since the derivative of a power series is a power series itself, we can apply Theorem 9.13 to the power series representing the derivative. Hence
$$\frac{d^2}{dx^2}(c_0 + c_1 x + \ldots + c_n x^n + \ldots)$$
$$= \frac{d}{dx}(c_1 + 2c_2 x + \ldots + nc_n x^{n-1} + \ldots)$$
$$= 2c_2 + 6c_3 x + \ldots + n(n-1)c_n x^{n-2} + \ldots .$$
In this way a derivative of any order can be determined for a power series.

COROLLARY 9.13 *If*
$$c_0 + c_1 x + \ldots + c_n x^n + \ldots = 0$$
for $|x| < R$ ($R > 0$) then $c_0 = 0, c_1 = 0, \ldots, c_n = 0, \ldots$. (19)

Proof. Putting $x = 0$ (permissible by Theorem 9.12) in (19) gives $c_0 = 0$. By Theorem 9.13 the derivative of (19) for $|x| < R$ is
$$c_1 + 2c_2 x + \ldots + nc_n x^{n-1} + \ldots = 0.$$
Putting $x = 0$ gives $c_1 = 0$. Another derivative, followed by $x = 0$, supplies $c_2 = 0$ and this procedure may be followed indefinitely.

Exercises

40. Find the derivatives of the series in the odd sections of question 37 and state where your results are sure to be valid.

41. Show that if $c_1 + 2c_2 x + \ldots + nc_n x^{n+1} + \ldots$ converges absolutely for $|x| < R$ then so does $c_0 + c_1 x + \ldots + c_n x^n + \ldots$.

42. If, for $|x| < R, f(x) = a_0 + a_1 x + \ldots + a_n x^n + \ldots$ and
$$f(x) = b_0 + b_1 x + \ldots + b_n x^n + \ldots$$
show that
$$a_0 = b_0, \quad a_1 = b_1, \ldots, \quad a_n = b_n, \ldots .$$

9.5 Taylor series

In the preceding section the properties of power series regarded as functions have been examined and the results derived will be extremely pertinent to the next chapter. There is, however, a converse question which needs a little study, namely: 'Can a given function be expanded in a power series?'

Suppose we assume that a function f can be expanded in a power series for $|x| < R$ so that
$$f(x) = c_0 + c_1 x + \dots + c_n x^n + \dots .$$
Then putting $x = 0$ gives $c_0 = f(0)$. Taking a derivative and using Theorem 9.13 provides
$$f'(x) = c_1 + \dots + n c_n x^{n-1} + \dots$$
whence $x = 0$ gives $f'(0) = c_1$. Carrying on in this way we obtain
$$f''(0) = 2! c_2, \dots, \qquad f^{(n)}(0) = n! c_n, \dots .$$
Therefore it must be concluded that, if f can be expanded in a power series, the expansion must be of the form
$$f(x) = f(0) + x f'(0) + \frac{x^2}{2!} f''(0) + \dots + \frac{x^n}{n!} f^{(n)}(0) + \dots .$$

It may be demonstrated in the same way that, if f can be expanded in a power series of $x - a$ for $|x - a| < R$, the series must be
$$f(x) = f(a) + (x - a) f'(a) + \frac{(x-a)^2}{2!} f''(a) + \dots$$
$$+ \frac{(x-a)^n}{n!} f^{(n)}(a) + \dots . \qquad (20)$$

An expansion which bears a distinct resemblance to this has been derived earlier in Taylor's theorem (Theorem 7.5), namely
$$f(x) = f(a) + (x - a) f'(a) + \dots$$
$$+ \frac{(x-a)^{n-1}}{(n-1)!} f^{(n-1)}(a) + \frac{(x-a)^n}{n!} f^{(n)}(c) \qquad (21)$$

where c lies between x and a and may depend on n. Comparison of (20) and (21) reveals that the power series (20), i.e. the infinite Taylor series, converges to $f(x)$ if and only if
$$\lim_{n \to \infty} \frac{(x-a)^n}{n!} f^{(n)}(c) = 0. \qquad (22)$$

(This may be stated alternatively by using Cauchy's form of remainder.) It is often difficult in practice to confirm whether or not this result is true.

Some possibilities are covered by the following theorem which gives some conditions under which it is certain that a function can be expressed as a Taylor series.

THEOREM 9.14 *If f has derivatives of every order for $|x - a| < R$ and if there is a constant C such that*

$$|f^{(n)}(x)| \leq C \quad (all\ n)$$

for $|x - a| < R$, then the infinite Taylor expansion is valid for $|x - a| < R$.

Proof.

$$\left|\frac{(x-a)^n}{n!} f^{(n)}(c)\right| \leq \frac{C|x-a|^n}{n!}.$$

But, by Exercise 39, the right-hand side tends to zero as $n \to \infty$. Hence (22) is satisfied and the theorem proved.

Exercises

43. Prove that the following Taylor expansions hold:

(i) $\cos x = 1 - \dfrac{x^2}{2!} + \dfrac{x^4}{4!} - \ldots + (-1)^n \dfrac{x^{2n}}{(2n)!} + \ldots$ (all finite x),

(ii) $\sin x = x - \dfrac{x^3}{3!} + \ldots + (-1)^{n-1} \dfrac{x^{2n-1}}{(2n-1)!} + \ldots$ (all finite x),

(iii) $\dfrac{1}{3-x} = \dfrac{1}{3} + \dfrac{x}{9} + \ldots + \dfrac{x^n}{3^{n+1}} + \ldots$ ($|x| < 3$),

(iv) $\sin^2 x = x^2 - \dfrac{2^3}{4!} x^4 + \ldots + (-1)^{n+1} \dfrac{2^{2n-1}}{(2n)!} x^{2n} + \ldots$ (all finite x),

(v) $\cos x = \dfrac{1}{\sqrt{2}} \left[1 - (x - \tfrac{1}{4}\pi) - \dfrac{(x - \tfrac{1}{4}\pi)^2}{2!} + \dfrac{(x - \tfrac{1}{4}\pi)^3}{3!} + \ldots \right]$

(all finite x).

Consider the function $f_\mu(x)$ defined by

$$f_\mu(x) = 1 + \mu x + \frac{\mu(\mu-1)}{2!} x^2 + \ldots + \frac{\mu(\mu-1)\ldots(\mu-n+1)}{n!} x^n + \ldots. \tag{23}$$

Infinite series

When μ is the positive integer m, all the terms with $n = m + 1, m + 2, \ldots$, disappear because a bracket in the numerator vanishes. Therefore

$$f_m(x) = 1 + mx + \frac{m(m-1)}{2!}x^2 + \ldots + mx^{m-1} + x^m.$$

It follows from Exercise 29 of Chapter 7 that

$$f_m(x) = (1 + x)^m \quad \text{(all } x\text{)}.$$

For other values of μ observe that

$$\lim_{n\to\infty} \left| \frac{\mu(\mu-1)\ldots(\mu-n)}{(n+1)!} x^{n+1} \frac{n!}{\mu(\mu-1)\ldots(\mu-n+1)x^n} \right|$$

$$= \lim_{n\to\infty} \left| \frac{\mu - n}{n + 1} x \right| = |x|.$$

Hence, by Theorem 9.10, the power series converges absolutely for $|x| < 1$ and $f_\mu(x)$ is well-defined for these values of x. It has now to be ascertained whether or not f_μ is $(1 + x)^\mu$.

First, the convergence of (23) when $|x| < 1$ implies that

$$\lim_{n\to\infty} \frac{\mu(\mu-1)\ldots(\mu-n+1)}{n!} x^n = 0 \quad (|x| < 1). \tag{24}$$

Next, Taylor's theorem (Theorem 7.5) gives

$$(1 + x)^\mu = 1 + \mu x + \ldots + \frac{\mu(\mu-1)\ldots(\mu-n+2)}{(n-1)!} x^{n-1} +$$

$$\frac{\mu(\mu-1)\ldots(\mu-n+1)}{n!} x^n (1 + \theta_n x)^{\mu-n} \tag{25}$$

where $0 < \theta_n < 1$. When $x \geqslant 0$,

$$0 \leqslant \frac{x}{1 + \theta_n x} \leqslant x$$

and so, when $0 \leqslant x < 1$, (24) gives

$$\lim_{n\to\infty} \frac{\mu(\mu-1)\ldots(\mu-n+1)}{n!} x^n (1 + \theta_n x)^{\mu-n} = 0.$$

Hence, when $0 \leqslant x < 1$,

$$f_\mu(x) = (1 + x)^\mu. \tag{26}$$

This result can be improved by using Cauchy's form of the remainder (Corollary 7.5). The last term of (25) is then replaced by

$$\frac{\mu(\mu-1)\ldots(\mu-n+1)(1-\theta_n)^{n-1}}{(n-1)!} x^n (1 + \theta_n x)^{\mu-n}.$$

280   *Introductory Analysis*

Since $0 \leqslant (1 - \theta_n)/(1 + \theta_n x) \leqslant 1$ for $|x| < 1$, (24) demonstrates immediately that the remainder tends to zero as $n \to \infty$, which leads to the conclusion that (26) is true when $|x| < 1$. Hence

$$(1 + x)^\mu = 1 + \mu x + \ldots + \frac{\mu(\mu - 1) \ldots (\mu - n + 1)}{n!} x^n + \ldots \quad (|x| < 1), \tag{27}$$

$$(1 + x)^m = 1 + mx + \ldots + mx^{m-1} + x^m \quad (\text{all } x; \quad m = 0,1,2,\ldots). \tag{28}$$

These results are often called the *binomial expansion for any index*.

Exercises

44. If $\mu \geqslant 0$ show that (27) holds when $x = \pm 1$. (For $-1 < \mu < 0$, (27) is valid for $x = 1$ but not for $x = -1$. When $\mu \leqslant -1$, the power series diverges at $x = \pm 1$.)

45. Find the first three terms in the power series expansions of the following and specify R so that the power series is absolutely convergent for $|x| < R$:

(i) $(1 + x)^{\frac{1}{2}}$, (ii) $(1 - x)^{\frac{2}{3}}$, (iii) $(1 + x^2)^{-2}$, (iv) $(1 + x^2)^2$, (v) $(1 + 2x)^{-\frac{1}{2}}$, (vi) $(1 + \frac{1}{3}x)^{-3}$, (vii) $(8 + 12x)^{\frac{2}{3}}$, (viii) $(4a - 8x)^{-\frac{1}{2}}$ $(a > 0)$.

46. Find correct to three places of decimals, (i) $(998)^{\frac{1}{3}}$, (ii) $(2400)^{\frac{1}{4}}$, (iii) $1/(128)^{\frac{1}{3}}$, (iv) $(630)^{-\frac{3}{4}}$.

47. If x is so small that its square and higher powers may be neglected in binomial expansions, simplify

(i) $(1 - 7x)^{\frac{1}{3}}(1 + 2x)^{-\frac{3}{4}}$, (ii) $\dfrac{(8 + 3x)^{\frac{2}{3}}}{(2 + 3x)(4 - 5x)^{\frac{1}{2}}}$,

(iii) $\dfrac{(1 - 3x)^{\frac{2}{3}} + [1 + (25x/6)]^{-6}}{(1 + 10x)^{\frac{1}{3}} + [1 - (5x/2)]^{1/5}}$.

48. If $s_n = 1 + 2 + \ldots + n$ prove that
$$(1 - x)^{-3} = s_1 + s_2 x + \ldots + s_n x^{n-1} + \ldots$$
when $|x| < 1$, and that
$$s_1 s_{2n} + s_2 s_{2n-1} + \ldots + s_n s_{n+1} = \frac{1}{2} \cdot \frac{(2n + 4)!}{5!(2n - 1)!}.$$

Infinite series

49. By considering $(1 - x^2)^{-\frac{1}{2}}$ prove that

$$a_{2n} - a_1 a_{2n-1} + a_2 a_{2n-2} - \ldots + (-1)^{n-1} a_{n-1} a_{n+1} = \tfrac{1}{2} a_n [1 + (-1)^{n-1} a_n]$$

where

$$a_n = \frac{1 \cdot 3 \cdot 5 \cdot \ldots \cdot (2n-1)}{2 \cdot 4 \cdot 6 \cdot \ldots \cdot 2n}.$$

50. By using $(1 + x)^\mu (1 + x)^\nu = (1 + x)^{\mu+\nu}$ prove that

$$\frac{\nu(\nu - 1) \ldots (\nu - n + 1)}{n!} + \frac{\mu \nu(\nu - 1) \ldots (\nu - n + 2)}{(n-1)! \, 1!} +$$

$$\frac{\mu(\mu - 1)\nu(\nu - 1) \ldots (\nu - n + 3)}{(n-2)! \, 2!} + \ldots + \frac{\mu(\mu - 1) \ldots (\mu - n + 1)}{n!} =$$

$$\frac{(\mu + \nu)(\mu + \nu - 1) \ldots (\mu + \nu - n + 1)}{n!},$$

n being a positive integer.

51. If $\lim\limits_{n \to \infty} \left| \dfrac{c_{n+1}}{c_n} \right|$ exists and is denoted by c show that

$$c_0 + \frac{c_1}{x} + \frac{c_2}{x^2} + \ldots + \frac{c_n}{x^n} + \ldots$$

converges absolutely for $|x| > c$ and diverges for $|x| < c$.

52. If x is so large that $1/x^2, 1/x^3,\ldots$ can be neglected, simplify

(i) $\dfrac{(x^2 + 1)^{5/24}(x - 7)^{\frac{1}{3}}}{(x + 2)^{3/4}}$, (ii) $\dfrac{(x^5 + 2)^{1/6}(8x + 3)}{(2x + 3)(4x - 5)^{\frac{1}{2}}}$.

For the use of power series in computation the reader should refer to section 7.7. However, it is necessary to emphasize once again the difference between the mathematical and computational points of view. As far as analysis is concerned the infinite series

$$1 + \frac{1}{2} + \frac{1}{3} + \ldots + \frac{1}{n} + \ldots$$

is divergent and does not have a finite sum. On the other hand the computer, calculating the sum as $1, 1 + \tfrac{1}{2}, 1 + \tfrac{1}{2} + \tfrac{1}{3},\ldots$, eventually reaches an n which is so large that $1/n$ is too small to be read by the computer as other than zero. Thus the computer will ascribe a finite sum to this divergent series. Those preparing programs for summing infinite series by computers must take cognisance of this sort of behaviour.

9.6 Summation notation

There is a compact notation for series which considerably simplifies the writing down when used properly. It employs the symbol Σ which is the capital letter sigma from the Greek alphabet. For example,
$$s_n = a_1 + a_2 + \ldots + a_n$$
is written in this notation as
$$s_n = \sum_{m=1}^{n} a_m.$$
Here a_m signifies a typical term in the series and the index m of this typical term is required first to take the value $m = 1$ (as shown underneath the Σ) and then to take successively the values 2,3,4,... until finally it takes the value n (shown at the top of the Σ); all these terms are to be added together. Thus
$$\sum_{m=1}^{n} a_m = a_1 + a_2 + \ldots + a_n.$$
Put briefly the rule is: add together all the terms starting from $m = 1$ and working up, one at a time, to $m = n$.

The notation $\sum_{m=1}^{n} a_m$ is read as 'sigma from $m = 1$ to n of a_m'.

Some examples of the use of the notation are
$$1 + 2 + 3 + \ldots + 50 = \sum_{m=1}^{50} m,$$
$$1^3 + 2^3 + 3^3 + \ldots + 50^3 = \sum_{m=1}^{50} m^3,$$
$$a + (a + d) + (a + 2d) + \ldots + [a + (n-1)d] = \sum_{m=0}^{n-1} (a + md),$$
$$1 + a + a^2 + \ldots + a^n = \sum_{m=0}^{n} a^m,$$
$$\frac{a}{3} + \frac{a^2}{6} + \ldots + \frac{a^n}{n^2 + 2} = \sum_{m=1}^{n} \frac{a^m}{m^2 + 2},$$
$$1 + \frac{1}{2} + \frac{1}{2} + \ldots + \frac{1}{13} = \sum_{m=1}^{13} \frac{1}{m},$$
$$c_0 + c_1 x + c_2 x^2 + \ldots + c_n x^n = \sum_{m=0}^{n} c_m x^m.$$

Infinite series

The sigma notation may be used in any mathematical expression when it is convenient. Thus, if $a_1, a_2, \ldots, b_1, b_2, \ldots$ are positive and $a_m \leqslant b_m$ for every m

$$a_1 + a_2 + \ldots + a_n \leqslant b_1 + b_2 + \ldots + b_n$$

which may be written as

$$\sum_{m=1}^{n} a_m \leqslant \sum_{m=1}^{n} b_m.$$

When the sum of an infinite series is being discussed the symbol ∞ is placed at the top of Σ, e.g.

$$\sum_{m=1}^{\infty} a_m = a_1 + a_2 + \ldots + a_n + \ldots.$$

Consequently, a power series can be written

$$\sum_{m=0}^{\infty} c_m x^m = c_0 + c_1 x + c_2 x^2 + \ldots + c_n x^n + \ldots.$$

Expressed in this notation, the results of Theorem 9.11 could be formulated as

$$\sum_{m=0}^{\infty} c_m x^m \pm \sum_{m=0}^{\infty} d_m x^m = \sum_{m=0}^{\infty} (c_m \pm d_m) x^m,$$

$$\left(\sum_{l=0}^{\infty} c_l x^l\right)\left(\sum_{m=0}^{\infty} d_m x^m\right) = \sum_{m=0}^{\infty} (c_0 d_m + c_1 d_{m-1} + \ldots + c_m d_0) x^m$$

$$= \sum_{m=0}^{\infty} \left(\sum_{r=0}^{m} c_r d_{m-r}\right) x^m.$$

Notice how two different letters were used on the Σ's on the left-hand side of the last equation to make sure that the two sums were not confused.

Again, Theorem 9.13 could be expressed as

$$\frac{d}{dx} \sum_{m=0}^{\infty} c_m x^m = \sum_{m=1}^{\infty} m c_m x^{m-1}.$$

Exercises

53. Verify the following:

(i) $\sum_{m=1}^{4} (5 a_m) = 5\left(\sum_{m=1}^{4} a_m\right)$,   (ii) $\sum_{m=0}^{n} (a_{m+1} - a_m) = a_{n+1} - a_0$,

(iii) $\sum_{m=1}^{n} (2m + 1) = \sum_{m=1}^{n} \{(m+1)^2 - m^2\} = n^2 + 2n$,

(iv) $\sum_{m=1}^{n} \frac{1}{m(m+1)} = 1 - \frac{1}{n+1}$,   (v) $\sum_{m=0}^{11} (3 + 2m) = 168$,

(vi) $\sum_{m=0}^{n-1} a^m = \dfrac{1-a^n}{1-a} \;(a \neq 1),$ (vii) $\sum_{m=0}^{\infty} \dfrac{1}{2^m} = 2,$

(viii) $\sum_{m=0}^{\infty} \dfrac{(-1)^m}{3^m} = \dfrac{3}{4},$ (ix) $2\sin\theta \sum_{m=1}^{n} \sin 2m\theta = \cos\theta - \cos(2n+1)\theta.$

54. Prove that

(i) $\cos x = \sum_{m=0}^{\infty} \dfrac{(-1)^m x^{2m}}{(2m)!}$ (all finite x),

(ii) $\sum_{m=1}^{\infty} \dfrac{x^{m-1}}{4^m} = \dfrac{1}{4-x}$ ($|x| < 4$),

(iii) $\sum_{m=1}^{\infty} m^2 x^m = \dfrac{x^2 + x}{(1-x)^3}$ ($|x| < 1$),

(iv) $\sum_{m=0}^{\infty} \dfrac{(x - \frac{1}{4}\pi)^m}{m!} \cos(\tfrac{1}{2}m + \tfrac{1}{4})\pi = \cos x$ (all finite x).

10
Functions defined by series

10.1 The exponential function

Let $f(x)$ be defined by

$$f(x) = 1 + x + \frac{x^2}{2!} + \ldots + \frac{x^n}{n!} + \ldots. \tag{1}$$

According to Example 13 of Chapter 9 the power series converges for every finite value of x. Therefore $f(x)$ is defined for every finite value of x.

Now

$f(x) \cdot f(y)$

$$= \left(1 + x + \frac{x^2}{2!} + \ldots + \frac{x^n}{n!} + \ldots\right)\left(1 + y + \frac{y^2}{2!} + \ldots + \frac{y^n}{n!} + \ldots\right)$$

$$= 1 + (x + y) + \frac{(x+y)^2}{2!} + \ldots$$

$$+ \left[\frac{x^n}{n!} + \frac{x^{n-1}y}{(n-1)!1!} + \frac{x^{n-2}y^2}{(n-2)!2!} + \ldots + \frac{y^n}{n!}\right] + \ldots,$$

by Theorem 9.6, for any finite x and any finite y. Also

$$\frac{x^n}{n!} + \frac{x^{n-1}y}{(n-1)!1!} + \frac{x^{n-2}y^2}{(n-2)!2!} + \ldots + \frac{y^n}{n!}$$

$$= \frac{1}{n!}\bigg[x^n + nx^{n-1}y + \frac{n(n-1)}{2!}x^{n-2}y^2 + \ldots$$

$$+ \frac{n(n-1)\ldots(n-r+1)}{r!}x^{n-r}y^r + \ldots + y^n\bigg]$$

$$= \frac{1}{n!}(x+y)^n$$

by the binomial expansion (section 9.5). Hence
$$f(x) \cdot f(y) = 1 + (x + y) + \frac{(x + y)^2}{2!} + \ldots + \frac{(x + y)^n}{n!} + \ldots$$
$$= f(x + y). \tag{2}$$

This result can be cast into a very convenient form by making a small change in notation. Let us agree to write $f(x)$ as f^x. Then (2) becomes
$$f^x \cdot f^y = f^{x+y}. \tag{3}$$
Written in this way it looks like the rule for multiplying together the powers of a quantity f. This idea will be made more explicit.

When $x = 1$, f^x has a definite value given by the series in (1). Denote this value by e; then
$$e = f^1 = 1 + 1 + \frac{1}{2!} + \ldots + \frac{1}{n!} + \ldots.$$
The value of e can be calculated as accurately as one wishes by using the methods which have already been indicated for series. It is found that, correct to 11 decimal places,
$$e = 2 \cdot 71828\ 18284\ 6.$$

By putting $y = x$ in (3) we find $(f^x)^2 = f^{2x}$; then $y = 2x$ in (3) gives $f^{3x} = f^x \cdot f^{2x} = f^x \cdot (f^x)^2 = (f^x)^3$. Carrying on this procedure supplies
$$f^{nx} = (f^x)^n \tag{4}$$
for every positive integer n. In particular, when $x = 1$,
$$f^n = (f^1)^n = e^n \tag{5}$$
whereas $x = 1/n$ provides
$$f^1 = (f^{1/n})^n$$
or, taking the $1/n$th root,
$$f^{1/n} = e^{1/n}. \tag{6}$$
Finally, if $x = 1/m$ where m is a positive integer, (4) becomes
$$f^{n/m} = (f^{1/m})^n = (e^{1/m})^n = e^{n/m}$$
from (6). Thus, whenever x is the ratio of two positive integers, f^x can be regarded as the (n/m)th power of the number e.

In fact, this is also true when x is negative. Put $x = -y$, with y positive, in (3) so that
$$f^{-y} \cdot f^y = f^0 = 1$$
from (1). But f^y cannot be zero for $y \geqslant 0$ because no term in (1) will be negative and so
$$f^{-y} = \frac{1}{f^y} = \frac{1}{e^y} \tag{7}$$
from the property already proved.

Functions defined by series

In view of these properties it is eminently reasonable to write

$$e^x = 1 + x + \frac{x^2}{2!} + \ldots + \frac{x^n}{n!} + \ldots \qquad (8)$$

for all real finite values of x, which is really a definition of e^x. Expressed in terms of e the results obtained above are

$$e^x \cdot e^y = e^{x+y}, \qquad (9)$$

$$e^0 = 1, \qquad e^1 = e = 2 \cdot 718\ldots, \qquad (10)$$

$$e^{-x} = 1/e^x, \qquad (11)$$

$$(e^x)^r = e^{rx} = (e^r)^x. \qquad (12)$$

r being a rational number

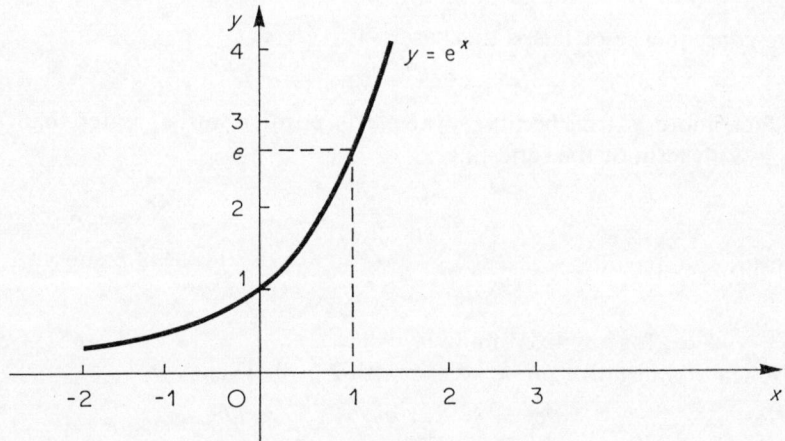

FIGURE 93

As x increases from 0 each term in (8) is positive and increasing; therefore e^x must be positive and increasing. Since $e^0 = 1$, it follows that $e^x > 1$ when $x > 0$. When x is negative, put $x = -z$, and then (11) gives

$$e^x = e^{-z} = 1/e^z.$$

As we go away from the origin, z increases and so $1/e^z$ decreases, i.e. e^x decreases steadily as x goes from 0 to $-\infty$. Hence the graph of e^x is as shown in figure 93.

The function e^x, defined by (8), is called the *exponential function* of x and read as 'e to the x'; sometimes x is described as the *exponent* of the exponential. Occasionally e^x is written as $\exp(x)$, especially when the exponent x is complicated.

The exponential function is obviously a single-valued function of x. Moreover, the exponential function increases steadily as x goes from $-\infty$

to $+\infty$ so that $y = e^x$ would define x as a single-valued function of y, when y is positive.

On account of Theorem 9.12, e^x is a continuous function of x for all finite x. Alternatively, it may be said that, if $h \to 0$,

$$e^{x+h} \to e^x.$$

Yet another way of expressing this property is

$$\lim_{y \to x} e^y = e^x. \tag{13}$$

Each term in (8) is positive when x is positive and so e^x is certainly greater than the sum of the first two terms, i.e.

$$e^x > 1 + x \quad (x > 0). \tag{14}$$

One consequence of this is that, as $x \to +\infty$,

$$e^x \to +\infty.$$

In fact, more is true because, when x is positive, e^x is greater than the $(m+2)$th term of the series, i.e.

$$e^x > \frac{x^{m+1}}{(m+1)!}.$$

It follows that

$$e^x/x^m \to +\infty \tag{15}$$

as $x \to +\infty$, for any fixed finite m.

When x is negative put $x = -z$ with $z > 0$. Then

$$x^m e^x = (-z)^m e^{-z} = (-z)^m/e^z$$

from (11). As $x \to -\infty$, $z \to +\infty$ and so (15) shows that the right-hand side tends to zero. Hence

$$\lim_{x \to -\infty} x^m e^x = 0 \tag{16}$$

for any fixed finite m. From (16) and (15) can be deduced, by changing the sign of x,

$$\lim_{x \to +\infty} x^m e^{-x} = 0 \tag{17}$$

and

$$e^{-x}/|x|^m \to +\infty \quad (x \to -\infty). \tag{18}$$

These results may be stated as: (i) e^x tends to infinity, as $x \to +\infty$, faster than any power of x and (ii) e^x tends to zero, as $x \to -\infty$, faster than any inverse power of x. Some idea of the rapidity of growth may be gained from the values

$$e^5 = 148, \quad e^{10} = 22{,}026, \quad e^{-5} = 0{\cdot}007, \quad e^{-10} = 0{\cdot}00005.$$

Functions defined by series

Since e^x is defined by means of a power series its derivative can be calculated by taking derivatives term-by-term (Theorem 9.13). Hence

$$\frac{d}{dx}e^x = \frac{d}{dx}\left(1 + x + \frac{x^2}{2!} + \frac{x^3}{3!} + \ldots + \frac{x^n}{n!} + \ldots\right)$$

$$= \frac{d}{dx}1 + \frac{d}{dx}x + \frac{d}{dx}\frac{x^2}{2!} + \frac{d}{dx}\frac{x^3}{3!} + \ldots + \frac{d}{dx}\frac{x^n}{n!} + \ldots$$

$$= 1 + x + \frac{x^2}{2!} + \ldots + \frac{x^{n-1}}{(n-1)!} + \ldots$$

$$= e^x. \tag{19}$$

Thus the exponential function has the remarkable characteristic that its derivative is itself. Since

$$\frac{d^2}{dx^2}e^x = \frac{d}{dx}\left(\frac{d}{dx}e^x\right) = \frac{d}{dx}(e^x) = e^x \tag{20}$$

we can see that any derivative of e^x is e^x.

The exponential function can be used more widely than has so far been indicated. Since it is defined for any finite value of the exponent it will be defined if the exponent is $g(x)$ so long as the function g takes only finite values for the values of x under consideration. For example,

$$e^{g(x)} = 1 + g(x) + \frac{[g(x)]^2}{2!} + \ldots + \frac{[g(x)]^n}{n!} + \ldots$$

for all x at which g has a definite finite value. By the chain rule (Theorem 5.8) and (19),

$$\frac{d}{dx}e^g = \frac{de^g}{dg}\frac{dg}{dx} = e^g\frac{dg}{dx}. \tag{21}$$

For instance

$$\frac{d}{dx}e^{x^3} = 3x^2 e^{x^3}.$$

Exercises

1. Find $\lim_{x \to 0} (e^x - 1)/x$.

2. Prove that
$$\frac{1}{e} = \frac{2}{3!} + \frac{4}{5!} + \frac{6}{7!} + \ldots .$$
(Hint: Theorem 9.7.)

3. Expand $(e^{4x} + e^{2x})/e^{3x}$ as a power series in x.

4. Find dy/dx when y is given by: (i) $y = e^{3x}$, (ii) $y = x^2 e^x$, (iii) $y = e^{-x}$, (iv) $y = e^{-x} \cos x$, (v) $y = e^{\cos 2x}$, (vi) $y = (e^x - e^{-x})/(e^x + e^{-x})$, (vii) $y = e^{1/x}$, (viii) $y = \tan e^{3x}$, (ix) $e^{ax} = \cos(x + 2y)$, (x) $y = e^{ax}/x^2$, (xi) $y = e^{e^x}$.

5. Show that the curves $y = e^{4x}$ and $y = e^{4x} \cos 4x$ touch at $x = \tfrac{1}{2}\pi$.

6. Find where $y = x e^x$ is stationary and determine its nature there. Show that there is a point of inflection on $x = -2$ and that the curve is concave downwards to the left of this point.

7. Find the points of inflection of the probability distribution $y = A e^{-x^2/b^2}$ and sketch the curve.

8. Show that the successive values of y where $y = e^{-\frac{1}{2}x} \sin 2x$ touches the curves $y = \pm e^{-\frac{1}{2}x}$ form a geometric series.

9. If $dy/dx = ay$, put $y = z e^{ax}$ and use Corollary 7.2 to prove that $y = A e^{ax}$, where A is a constant.

10. The rate of growth of the number of bacteria in a culture is proportional to the number of bacteria. Show that the bacteria increase exponentially.

11. If $y = A e^{ax} + B e^{-ax}$ where A, B, a are constants, prove that

$$\frac{d^2 y}{dx^2} = a^2 y.$$

12. In a chemical reaction at temperature T, $I = I_0 e^{-\frac{1}{2}a(T - T_0)/T_0 T}$ where I_0, a, T_0 are constants. If a small error h is made in T what is the relative error in I?

13. Find the local maximum of $|x| e^{-|x-1|}$ in $x < 1$.

14. Which of the following sequences are convergent and what is the limit of each convergent sequence? (i) $e^{-1}, e^{-2}, \dots, e^{-n}, \dots$, (ii) $e^1, e^2, \dots, e^n, \dots$, (iii) $e^{-1}, e^{-9}, \dots, e^{-n^2}, \dots$, (iv) $\dfrac{e^1 - e^{-1}}{e^1 + e^{-1}}, \dfrac{e^2 - e^{-2}}{e^2 + e^{-2}}, \dots, \dfrac{e^n - e^{-n}}{e^n + e^{-n}}, \dots$, (v) $\dfrac{1 + e^{-1}}{2}, \dfrac{2 + 4e^{-2}}{3}, \dots, \dfrac{n + n^2 e^{-n}}{n + 1}, \dots$.

15. Test for convergence (i) $e^{-1} + e^{-9} + \ldots + e^{-n^2} + \ldots$,
(ii) $(1 - e^{-1}) + (\frac{1}{2} - e^{-2}) + \ldots + [(1/n) - e^{-n}] + \ldots$,
(iii) $(1 - e^{-1}) - (\frac{1}{2} - e^{-2}) + \ldots + (-1)^{n-1}[(1/n) - e^{-n}] + \ldots$,
(iv) $1 + xe + x^2 e^2/2! + \ldots + x^n e^n/n! + \ldots$.

16. Evaluate

(i) $\lim\limits_{x \to 0} \left(\dfrac{1}{x} - \dfrac{1}{e^x - 1} \right)$, (ii) $\lim\limits_{x \to 0} \dfrac{x e^x}{1 - e^x}$, (iii) $\lim\limits_{x \to 0} \dfrac{e^x - 1}{\tan 3x}$,

(iv) $\lim\limits_{x \to 3} \dfrac{e^x - e^3}{x - 3}$, (v) $\lim\limits_{x \to 0} \dfrac{e^{3x} - e^{-3x}}{\sin x}$, (vi) $\lim\limits_{x \to +\infty} \dfrac{e^x + x^4}{3e^x + 4x^3}$.

(vii) $\lim\limits_{x \to \frac{1}{2}\pi^-} e^{-\tan x} \sec^2 x$, (viii) $\lim\limits_{x \to 0} \dfrac{e^x + e^{-x} - x^2 - 2}{x^2 - \sin^2 x}$,

(ix) $\lim\limits_{x \to 0} \dfrac{x(e^x + 1) - 2(e^x - 1)}{x^3}$, (x) $\lim\limits_{x \to 0} \dfrac{e^{2x} - e^x - x}{x^2}$,

(xi) $\lim\limits_{x \to 0} \dfrac{e^{-1/x^2}}{x^{15}}$, (xii) $\lim\limits_{x \to +\infty} \dfrac{x^{17}}{e^x}$.

17. Show the first terms of Maclaurin's expansion for $e^{\cos x - 1}$ are
$$1 - \frac{1}{2}x^2 + \frac{1}{6}x^4 - \frac{31}{720}x^6,$$
and for $e^x \sin x$ are $x + x^2 + \dfrac{x^3}{3} - \dfrac{x^5}{30}$.

18. Prove that
$$(x^2 + x)e^x = x + \frac{4}{2!}x^2 + \ldots + \frac{n^2}{n!}x^n + \ldots.$$

19. In certain biological situations the growth at time t is given by
$$G = a \exp(-b e^{-ct})$$
where a, b and c are constant. Show that G has no maxima or minima but has a point of inflection at time t_0 where $e^{ct_0} = b$.

20. It is found that the concentration C of a drug in the body at time t satisfies
$$\frac{dC}{dt} = -C.$$
Initially, there is no drug in the body and then a dose C_0 is given. Thereafter a dose C_0 is given at regular intervals of time t_0. Find the concentration in the body at the end of the nth interval and show that it tends to $C_0/(e^{t_0} - 1)$ as $n \to \infty$. (Hint: question 10.)

21. Use Newton's method to find the smallest root of $e^x \sin x = 1$ correct to 2 decimal places.

22. What is the greatest error in using the approximation $1 + x$ for e^x when $0 \leqslant x \leqslant 0{\cdot}01$? For what values of x can e^x be replaced by $1 + x + \frac{1}{2}x^2$ with an error of magnitude $0{\cdot}0005$ at most?

23. Calculate e^{-1} and e^{-2} from the series for e^x correct to 4 decimal places. (Hint: Corollary 9.4.)

24. Express e^x as a Taylor series with remainder. Estimate how many terms of the series will be needed if the remainder is to be less than (i) $0{\cdot}0001$ if $x = 1$, (ii) $0{\cdot}0005$ if $x = \frac{1}{2}$. (Assume $1 < e < 3$.)

If, in (i), the terms are calculated to 5 decimal places show that
$$2{\cdot}71822 < e < 2{\cdot}71828$$
when round-off error is ignored and $2{\cdot}71826 < e < 2{\cdot}71833$ when some allowance is made for round-off error. What can be deduced about the value of e?

25. Find a root of $x = e^{-x}$ by iteration starting with $x_0 = \frac{1}{2}$.

26. Write a computer program for calculating e^x by means of iteration from the defining series. Arrange for the number of iterations needed for a particular value to be recorded. What is this number on your computer when
$$x = \pm\, 0{\cdot}1,\ \pm 1,\ \pm 2,\ \pm 5,\ \pm 20,\ \pm 50?$$
Compare your values with those given by the computer's own program.

27. Draw a flow chart for determining the integer n so that, for a given positive number a, $e^{n-1} \leqslant a < e^n$.

28. Given a positive number a, solve the equation $e^x = a$ by iteration.

29. Write a computer program to calculate $\sum_{n=1}^{\infty} \frac{1}{n} e^{-n^2 x^2} \sin n\pi x$ stopping the series when the exponential is less than 10^{-6}.

30. If
$$\frac{x}{e^x - 1} = B_0 + \frac{B_1}{1!}x + \frac{B_2}{2!}x^2 + \ldots + \frac{B_n}{n!}x^n + \ldots,$$
prove that
$$B_0 = 1,\ \ B_1 = -\tfrac{1}{2},\ \ B_2 = \tfrac{1}{6},\ \ B_3 = 0,\ \ B_4 = -\tfrac{1}{30},\ \ B_5 = 0,\ \ B_6 = \tfrac{1}{42},$$
$$B_7 = 0,\ \ B_8 = -\tfrac{1}{30}.$$
(These are called *Bernoulli numbers*.)

10.2 The natural logarithm

As y increases from $-\infty$ to $+\infty$, e^y increases steadily from 0 to $+\infty$. Therefore the equation $x = e^y$ defines y as a single-valued function of x when $x > 0$; the relation expressing y as a function of x is written

$$y = \ln x \quad \text{or} \quad y = \log_e x.$$

Then y is called the *natural logarithm* of x or the logarithm of x to the base e. Natural logarithms are also called *Napierian logarithms* and, occasionally, e is called the natural base of logarithms. Exactly the same information is conveyed by $y = \ln x$ and by $x = e^y$.

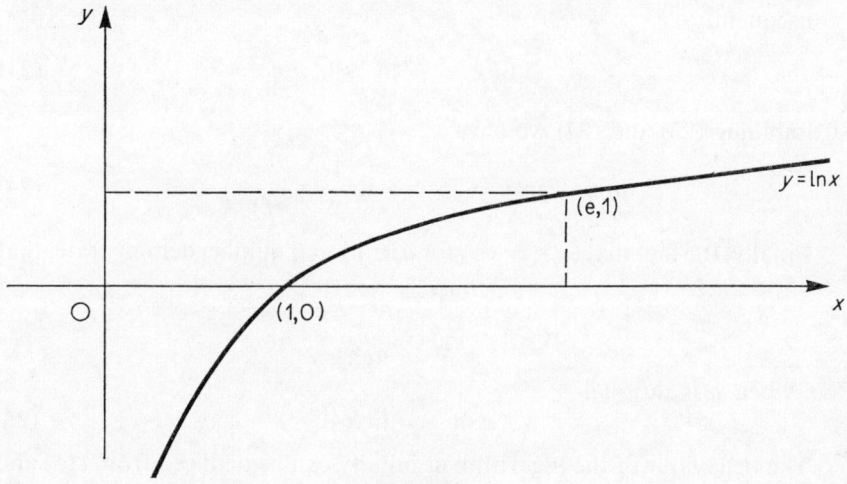

FIGURE 94

Since $e^0 = 1$ and $e^1 = e$ we must have

$$\ln 1 = 0, \qquad \ln e = 1$$

by taking first $x = 1$, $y = 0$ and then $x = e$, $y = 1$. Also when y is negative, $e^y < 1$ and so $\ln x$ is negative when $0 < x < 1$. As $y \to -\infty$, $e^y \to 0$ and so $\ln x \to -\infty$ as $x \to 0^+$. When $y > 0$, e^y is greater than unity and so $\ln x$ is positive when $x > 1$. Since $e^y \to +\infty$ as $y \to +\infty$ it follows that $\ln x \to +\infty$ as $x \to +\infty$. The graph of $y = \ln x$ is therefore as shown in figure 94.

Let x and y be positive numbers and let $z = \ln x$, $w = \ln y$. Then

$$x = e^z, \qquad y = e^w.$$

Therefore

$$xy = e^z \cdot e^w = e^{z+w}$$

from (9). The definition of a logarithm implies that
$$z + w = \ln xy.$$
Hence
$$\ln x + \ln y = \ln xy. \qquad (22)$$

Furthermore, (11) yields
$$e^{-z} = \frac{1}{e^z} = \frac{1}{x}$$
whence
$$-z = \ln \frac{1}{x};$$
consequently
$$\ln x = -\ln \frac{1}{x}. \qquad (23)$$

Combining (23) and (22) we have
$$\ln x - \ln y = \ln x + \ln \frac{1}{y} = \ln \frac{x}{y}. \qquad (24)$$

Finally, the fact that $(e^z)^a = e^{az}$, for a a rational number demonstrates that
$$e^{az} = x^a$$
or
$$az = \ln x^a,$$
i.e. when a is rational
$$a \ln x = \ln x^a. \qquad (25)$$

The behaviour of the logarithm at infinity can be deduced from (15) and (16). Suppose that $e^x = y$, so that $x = \ln y$. Since e^x tends to infinity faster than x^a as $x \to +\infty$ it follows that y tends to infinity faster than $(\ln y)^a$. This may be interpreted as

$$\frac{y^b}{\ln y} \to +\infty \quad \text{or} \quad \frac{1}{y^b} \ln y \to 0 \qquad (26)$$

as $y \to +\infty$ when $b > 0$. Putting $y = 1/z$ and using (23) we obtain
$$z^b \ln z \to 0 \quad (b > 0) \qquad (27)$$
as $z \to 0^+$.

To find the derivative of $\ln x$ take the derivative with respect to x of
$$e^y = x.$$
From (21),
$$e^y \frac{dy}{dx} = 1.$$

Functions defined by series

Since $e^y = x$ and $x > 0$,
$$\frac{dy}{dx} = \frac{1}{x}.$$

But $y = \ln x$, and so
$$\frac{d}{dx} \ln x = \frac{1}{x}. \tag{28}$$

An immediate consequence is that, if g is a positive function of x,
$$\frac{d}{dx} \ln g = \frac{1}{g} \cdot \frac{dg}{dx}. \tag{29}$$

Example 1 Expand $\ln(1 + x)$ as a Maclaurin series with remainder.

Denote $\ln(1 + x)$ by $f(x)$. Then, from (29), with $g = 1 + x$,
$$f'(x) = \frac{1}{1 + x}, \quad f''(x) = -\frac{1}{(1 + x)^2},$$
$$f^{(n)}(x) = \frac{(n-1)!(-1)^{n-1}}{(1 + x)^n} \quad (n = 2, 3, \ldots).$$

Thus the conditions of Theorem 7.5 are met for $x > -1$. Since $\frac{1}{n!} f^{(n)}(0) = (-1)^{n-1}/n$ and $\ln 1 = 0$,
$$\ln(1 + x) = x - \frac{x^2}{2} + \frac{x^3}{3} - \ldots + \frac{(-1)^{n-2} x^{n-1}}{n-1} + \frac{(-1)^{n-1} x^n}{n(1 + \theta_n x)^n} \tag{30}$$

where $0 < \theta_n < 1$. This is the Maclaurin series with Lagrange's form of the remainder. When $x \geq 0$, $1 + \theta_n x \geq 1$ and, when $0 \leq x \leq 1$, $x^n \leq 1$. Hence, if $0 \leq x \leq 1$,
$$\frac{x^n}{n(1 + \theta_n x)^n} \leq \frac{1}{n} \to 0$$

as $n \to \infty$. Therefore the remainder in (30) tends to zero as $n \to \infty$ and $\ln(1 + x)$ can be represented by a power series. (Note that the argument is not legitimate when $x > 1$ because θ_n may approach zero as n increases.)

When x is negative the remainder in (30) is not convenient and it is preferable to use Cauchy's form as given in Corollary 7.5; then
$$\ln(1 + x) = x - \frac{x^2}{2} + \ldots + \frac{(-1)^{n-2} x^{n-1}}{n-1} + \frac{(-1)^{n-1} x^n (1 - \theta_n)^{n-1}}{(1 + \theta_n x)^n}. \tag{31}$$

Now $1 - \theta_n \leqslant 1 + \theta_n x$ for $-1 < x \leqslant 0$ and so the remainder does not exceed

$$\frac{|x|^n}{1+x}$$

in magnitude. This tends to zero as $n \to \infty$ when $|x| < 1$. Combining this information with that for positive x we have

$$\ln(1+x) = x - \frac{x^2}{2} + \frac{x^3}{3} - \ldots + \frac{(-1)^{n-1}x^n}{n} + \ldots \quad (-1 < x \leqslant 1).$$
(32)

This series provides an alternative method of computing the logarithms of numbers between 0 and 2, and shows, in particular, that

$$\ln 2 = 1 - \frac{1}{2} + \frac{1}{3} - \frac{1}{4} + \ldots. \qquad (33)$$

Exercises

31. Show that $(d/dx) \ln|x| = 1/x \quad (x \neq 0)$.

32. Find the derivative of (i) $\ln(3x-1)$, (ii) $\ln(3-x)$, (iii) $\ln(a+bx)$, (iv) $\ln(1-x^2)$, (v) $\ln(a-bx)$, (vi) $\ln(x^2-3x+2)$, (vii) $\ln(x^3+x)$, (viii) $\ln(2+5\sin x)$, (ix) $\ln \tan x$, (x) $\ln(a-b\cos^2 x)$, (xi) $x^n \ln(x^2-1)$, (xii) $(\ln x)/x$, (xiii) $[\ln(ax+b)]/x^{\frac{1}{2}}$, (xiv) $\ln(\sin x)^{\frac{1}{2}}$, (xv) $\ln\dfrac{(x+1)(2-x)}{(3-x)^2}$, (xvi) $\ln\dfrac{a+b\sin x}{p+q\cos x}$, (xvii) $\ln[x+(x^2+1)^{\frac{1}{2}}]$, (xviii) $\ln\dfrac{e^x}{1+e^x}$, (xix) $\ln(1+e^{2x})^{\frac{1}{2}}$, (xx) $x\sin(\ln x)$, (xxi) $\ln(\ln x)$, (xxii) $\ln(\ln \tan x)$, (xxiii) $x(\ln x)^2$.

33. Find the second derivative of (i) $\ln(2-x)$, (ii) $e^{-x}\ln x$.

34. Certain bacterial colonies grow according to the law

$$G = G_0 \frac{1+b}{1+b\,e^{-at}}$$

where b_0, a and b are positive constants. Prove that

$$\frac{d}{dt}\ln G = \frac{ab}{e^{at}+b}$$

and deduce that G has no local maxima or minima.

Functions defined by series

35. Prove that the curve $y = x^2 \ln x$ has a local minimum at $x = 1/e^{\frac{1}{2}}$ and a point of inflection at $x = 1/e^{\frac{3}{2}}$, the curve being concave downwards to the left of this point.

36. In selling a blend of tea composed of grades A and B, it is found that the profits are proportional to $x \ln(1/x)$ where x is the proportion of grade A in the blend. What proportion gives the most profit?

37. Find
$$\lim_{x \to 0} \frac{\ln(1 - \frac{1}{2}x)}{x}.$$

38. Show that the series
$$\left[\ln\left(1 + \frac{x}{2}\right) - \frac{x}{2}\right] + \left[\ln\left(1 + \frac{x}{3}\right) - \frac{x}{3}\right] + \ldots + \left[\ln\left(1 + \frac{x}{n}\right) - \frac{x}{n}\right] + \ldots$$
is absolutely convergent for $x > -2$.

39. Find the first three terms of the power series in x for $[\ln(1 + x)]^2$.

40. Prove that, if $|x| < 1$,
$$\ln \frac{1+x}{1-x} = 2\left(x + \frac{x^3}{3} + \ldots + \frac{x^{2n+1}}{2n+1} + \ldots\right).$$

41. Prove that, if $|x| < \frac{1}{3}$,
$$\ln \frac{1+3x}{1-2x} = 5x - \frac{5x^2}{2} + \frac{35x^3}{3} - \ldots$$
and find the general term of the series.

42. By using Maclaurin series with remainder show that, for $x > 0$,
$$x - \tfrac{1}{2}x^2 < \ln(1 + x) < x.$$

43. Evaluate

(i) $\lim\limits_{x \to 0} \dfrac{\ln(1 + x) - x}{1 - \cos x}$, (ii) $\lim\limits_{x \to 0} \dfrac{\ln(\cos ax)}{\ln(\cos bx)}$, (iii) $\lim\limits_{x \to +\infty} \dfrac{\ln(\ln x)}{\ln x}$,

(iv) $\lim\limits_{x \to 0+} \dfrac{\ln \sin x}{\ln \tan x}$, (v) $\lim\limits_{x \to +\infty} \dfrac{3x + 2 \ln x}{x + 3 \ln x}$, (vi) $\lim\limits_{x \to 1} \left(\dfrac{1}{\ln x} - \dfrac{x}{x - 1}\right).$

44. Is the series $\dfrac{1}{\ln 2} + \dfrac{1}{\ln 3} + \ldots + \dfrac{1}{\ln n} + \ldots$ convergent or divergent?

45. What is the limit of the sequence $\tfrac{1}{2}\ln 2$, $\tfrac{1}{3}\ln 3,\ldots, (1/n)\ln n,\ldots$?

46. Obtain the first 3 terms of a Taylor expansion about $x = \tfrac{1}{3}\pi$ of $\ln \cos x$.

47. Show that
$$\ln x = \ln \tfrac{1}{2} + \tfrac{1}{2}(x-2) - \tfrac{1}{4}\dfrac{(x-2)^2}{2!} + \ldots$$
for $0 < x \leqslant 4$.

48. Use (32) to calculate $\ln 1\cdot 25$ correct to 3 decimal places.

49. How accurate is the approximation x for $\ln(1+x)$ when $|x| < 0\cdot 02$?

50. Show that
$$\ln 2 = 2\ln \tfrac{24}{25} - 7\ln \tfrac{9}{10} + 3\ln \tfrac{81}{80}.$$
Use
$$\ln \tfrac{9}{10} = \ln(1 - \tfrac{1}{10}), \quad \ln \tfrac{24}{25} = \ln(1 - \tfrac{4}{100}), \quad \ln \tfrac{81}{80} = \ln(1 + \tfrac{1}{80})$$
to calculate $\ln 2$ correct to 3 places of decimals.

51. By putting $x = 1/(2m+1)$ in question 40 deduce that
$$\ln \dfrac{m+1}{m} = 2\left[\dfrac{1}{2m+1} + \dfrac{1}{3(2m+1)^3} + \dfrac{1}{5(2m+1)^5} + \ldots\right].$$
The formula $\ln(m+1) = \ln m + \ln \dfrac{m+1}{m}$ enables the logarithms of the integers to be determined recursively. Draw a flow-chart for a suitable computational program.

52. Write a computer program for calculating $\ln(1+y)$ from its series expansion when $-1 < y \leqslant 1$. Indicate how then $\ln x$ could be calculated for any positive x. Compare the values you obtain by this method with those you obtain by question 28.

Functions defined by series

53. Deduce from question 51 that, if x is positive and $a \geqslant 0$,
$$\ln(x + a) = \ln x + 2\left[\frac{a}{2x + a} + \frac{a^3}{3(2x + a)^3} + \frac{a^5}{5(2x + a)^5} + \cdots\right].$$
Taking $a = 0\cdot01$ and $\ln 2 = 0\cdot693147$, use this formula to calculate the natural logarithms of 2, 2·01, 2·02,..., 4.

54. Find the point of inflection of
$$y = e^{-x} \ln x.$$

55. Obtain the root of
$$2x^2 + \tfrac{1}{2}\ln 2x = 1$$
correct to 3 decimal places.

56. In the iterative procedure
$$u_{r+1} = u_r - \frac{f(u_r)\ln|f(u_r)|}{f'(u_r)\ln|f(u_r)/f'(u_r)|},$$
u_r converges to a root x_0 of $f(x_0) = 0$ such that $f'(x_0) \neq 0$. Prove that
$$\lim_{r \to \infty} \frac{u_{r+1} - x_0}{u_r - x_0} \ln|u_r - x_0| = -\ln|f'(x_0)|.$$
Does the definition before Theorem 8.7 assign an order to this iterative method?

10.3 The function a^x

If a is positive and
$$b = \ln a \tag{34}$$
then
$$e^b = a. \tag{35}$$
If the value of b in (34) be substituted in (35) the result is
$$a = e^{\ln a}.$$
If both sides of this equation were raised to the power n we would obtain
$$a^n = e^{n \ln a}.$$
In order that this result be true even when n is not a positive integer we introduce

DEFINITION 10.1 *If a is positive, a^x is defined by*
$$a^x = e^{x \ln a}.$$

It is an immediate consequence of the definition that a^x is positive and that
$$\ln a^x = x \ln a \tag{37}$$

which is consistent with (25) and is true whether x is rational or not. Since the exponential is never zero neither is a^x.

On account of (10) and (36)
$$a^0 = 1, \quad a^1 = a. \tag{38}$$
Also, from (9),
$$a^x \cdot a^y = e^{x \ln a} \cdot e^{y \ln a} = e^{(x+y) \ln a} = a^{x+y}. \tag{39}$$
It follows from (39) and (38) that
$$a^x \cdot a^{-x} = a^{x-x} = a^0 = 1$$
whence
$$a^{-x} = \frac{1}{a^x}. \tag{40}$$
Furthermore
$$(a^x)^y = e^{y \ln a^x} = e^{yx \ln a}$$
from (37). Therefore
$$(a^x)^y = a^{xy}. \tag{41}$$
If b is positive
$$(ab)^x = e^{x \ln ab} = e^{x(\ln a + \ln b)} = e^{x \ln a} \cdot e^{x \ln b}$$
from (22) and (9). Hence
$$(ab)^x = a^x \cdot b^x. \tag{42}$$

The determination of the derivative of a^x is a straightforward application of (21) with $g = x \ln a$. Since $dg/dx = \ln a$, we have
$$\frac{d}{dx} a^x = \frac{d}{dx} e^{x \ln a} = e^{x \ln a} \ln a$$
or
$$\frac{d}{dx} a^x = a^x \ln a. \tag{43}$$

The meaning to be attached to $[f(x)]^{g(x)}$, where f is positive, now presents no difficulties; it is
$$[f(x)]^{g(x)} = e^{g(x) \ln f(x)}. \tag{44}$$
Properties (37)–(41) continue to hold with f, g replacing a and x, respectively. As regards the derivative
$$\frac{d}{dx}[f(x)]^{g(x)} = e^{g(x) \ln f(x)} \frac{d}{dx}[g(x) \ln f(x)]$$
$$= [f(x)]^{g(x)} \left[g'(x) \ln f(x) + \frac{g(x) f'(x)}{f(x)} \right]. \tag{45}$$

Functions defined by series

The logarithm can sometimes be used to advantage in calculating derivatives. The method, which employs the *logarithmic derivative*, for the equation

$$y = f(x)$$

is to take the natural logarithm of both sides so that

$$\ln y = \ln f(x).$$

Since

$$\frac{d}{dx} \ln y = \frac{1}{y} \frac{dy}{dx},$$

$$\frac{dy}{dx} = y \frac{d}{dx} \ln f(x).$$

So long as $\ln f$ can be simplified, by means of the properties (22)–(25), it may be possible to evaluate the derivative on the right more easily than the original.

Example 2 Find dy/dx when $y = x^x$ for $x > 0$.

By (37)

$$\ln y = \ln x^x = x \ln x.$$

Therefore

$$\frac{1}{y}\frac{dy}{dx} = \ln x + x \cdot \frac{1}{x}$$

or

$$\frac{dy}{dx} = y(1 + \ln x) = x^x(1 + \ln x).$$

Example 3 Find dy/dx when

$$y = \frac{x(2 - x^2)^2}{(1 + x^2)^{\frac{1}{2}}}.$$

In this case, (24) gives

$$\ln y = \ln [x(2 - x^2)^2] - \ln (1 + x^2)^{\frac{1}{2}}$$
$$= \ln x + 2 \ln (2 - x^2) - \tfrac{1}{2} \ln (1 + x^2) \tag{46}$$

from (22) and (25). Hence
$$\frac{1}{y}\frac{dy}{dx} = \frac{1}{x} - \frac{4x}{2-x^2} - \frac{x}{1+x^2}$$
so that
$$\frac{dy}{dx} = \frac{x(2-x^2)^2}{(1+x^2)^{\frac{1}{2}}}\left[\frac{1}{x} - \frac{4x}{2-x^2} - \frac{x}{1+x^2}\right]. \tag{47}$$

Note. It is necessary to impose the restriction $x > 0$ in order that $\ln x$ is properly defined in (46). When $x < 0$, multiply the equation for y by (-1) before taking logarithms; then, y and x being negative, $-y$ and $-x$ are positive and $\ln(-y)$, $\ln(-x)$ are well defined. Again, if $x^2 > 2$, the second logarithm in (46) is not well defined but this difficulty can be overcome by the observation that $(2-x^2)^2 = (x^2-2)^2$ and $\ln(x^2-2)$ is well defined. The net result of these manoeuvres is that (47) holds without restriction on x.

Exercise

57. Use the logarithmic derivative to find the derivative of (i) $[x(x+2)]^{\frac{1}{2}}$ $(x > 0)$, (ii) $x^2 e^{-3x} \sin 2x$, (iii) $\left[\dfrac{(x+2)(x^2-1)}{(2x+5)(x^2+2x+2)}\right]^{\frac{1}{3}}$ $(x > 1)$, (iv) $x^{\ln x}$ $(x > 0)$, (v) $(\ln x)^x$ $(x > 1)$, (vi) a^{a^x} $(a > 0)$, (vii) x^{x^x} $(x > 0)$, (viii) $x^{e^{-x}}$ $(x > 0)$, (ix) $2^{\tan x}$ $(|x| < \tfrac{1}{2}\pi)$, (x) $x^{\sin x}$ $(x > 0)$, (xi) $(\sec x)^{\sin x}$ $(|x| < \tfrac{1}{2}\pi)$.

Various indeterminate forms have already been discussed in section 7.5. The power law defined in this section introduces further possibilities. For example, in
$$\lim_{x \to a^+} [f(x)]^{g(x)}$$
it can happen that $f(a^+) = 0$, $g(a^+) = 0$. Then the indeterminate form 0^0 arises. Again, if $f(x) \to +\infty$ as $x \to a^+$ and $g(a^+) = 0$, we are led to the indeterminate form ∞^0. Another form is 1^∞, when $f(a^+) = 1$ and $g \to \infty$ as $x \to a^+$. However, by taking the natural logarithm, we obtain
$$g(x) \ln f(x)$$
and, in each of the three cases, this is an indeterminate form of the type $\infty . 0$. The methods described in section 7.5 then becomes available for determining the limit.

It only remains to check that, by finding the result for the logarithm, we can derive our original limit. Now, by (44),
$$\lim_{x \to a^+} [f(x)]^{g(x)} = \lim_{x \to a^+} e^{g(x) \ln f(x)}.$$
If
$$\lim_{x \to a^+} g(x) \ln f(x) = b \tag{48}$$

(13) shows that
$$\lim_{x \to a^+} e^{g(x) \ln f(x)} = e^b.$$
Hence
$$\lim_{x \to a^+} [f(x)]^{g(x)} = e^b. \tag{49}$$

Thus so long as the limiting process (48) can be carried out, the necessary limit can be found from (49).

Example 4 Find $\lim_{x \to 0^+} x^x$.

Since
$$\lim_{x \to 0^+} \ln x^x = \lim_{x \to 0^+} x \ln x = 0,$$
b in (48) is zero. Hence, from (49),
$$\lim_{x \to 0^+} x^x = e^0 = 1.$$

Example 5 Find $\lim_{x \to 0} (1 + ax)^{1/x}$.

In this case
$$\ln (1 + ax)^{1/x} = \frac{\ln (1 + ax)}{x}.$$
When $x = 0$ both numerator and denominator vanish, so that an application of l'Hôpital's rule (Theorem 7.4) can be considered. Then
$$\lim_{x \to 0} \frac{\ln (1 + ax)}{x} = \lim_{x \to 0} \frac{a/(1 + ax)}{1} = a.$$
Alternatively, the same result could be achieved by using (31) with $n = 2$.
It has now been shown that
$$\lim_{x \to 0} \ln (1 + ax)^{1/x} = a.$$
Hence
$$\lim_{x \to 0} (1 + ax)^{1/x} = e^a. \tag{50}$$
By putting $x = 1/m$, we derive
$$\lim_{m \to +\infty} \left(1 + \frac{a}{m}\right)^m = e^a. \tag{51}$$

Exercise

58. Evaluate (i) $\lim_{x \to 0} (1 - 3x)^{2/x}$, (ii) $\lim_{x \to 0} (1 + 2x)^{-3/x}$, (iii) $\lim_{x \to 1} x^{1/(1-x)}$ (iv) $\lim_{x \to 0^+} x^{\sin x}$, (v) $\lim_{x \to \frac{1}{2}\pi^-} (\tan x)^{\cos x}$, (vi) $\lim_{x \to 0} (e^x - 3x)^{1/x}$, (vii) $\lim_{x \to +\infty} (e^{2x} + 2)^{3/x}$, (viii) $\lim_{x \to 0^+} (x + \sin x)^{\sin x}$, (ix) $\lim_{x \to 0} (2 \sin x + \cos x)^{\operatorname{cosec} x}$, (x) $\lim_{x \to \frac{1}{2}\pi} (2x/\pi)^{\tan x}$, (xi) $\lim_{x \to 1^-} x^{-1/(1-x^2)}$, (xii) $\lim_{x \to 0^+} x^{a/\ln x}$, (xiii) $\lim_{x \to +\infty} x^{1/x}$.

10.4 The general logarithm

When e is raised to the power $\ln x$ the number x is obtained. For this reason e is called the *base* of natural logarithms. This idea may be generalized so as to give logarithms with other bases.

DEFINITION 10.2 *If a is positive number but $a \neq 1$, and if*
$$y = a^x, \tag{52}$$
then x is called the logarithm of y to the base a and we write
$$x = \log_a y. \tag{53}$$

In this notation, $\ln x$ may also be written $\log_e x$ because $y = \ln x$ means that $x = e^y$.

Since $3^2 = 9$, Definition 10.2 with $a = 3$, $x = 2$ and $y = 9$ gives
$$\log_3 9 = 2.$$
Similarly, since $5^{-2} = 1/25$,
$$\log_5 25 = -2.$$
To evaluate a quantity such as $\log_{27} 9$ we need to express 9 as a power of 27. In fact, if $\log_{27} 9 = x$,
$$9 = 27^x$$
or
$$3^2 = 3^{3x}.$$
Therefore $3x = 2$ so that $27^{\frac{2}{3}} = 9$ and
$$\log_{27} 9 = \tfrac{2}{3}.$$

Taking the natural logarithm of (52) we have
$$\ln y = \ln a^x = x \ln a$$
from (37). With $a \neq 1$, $\ln a$ is not zero and so
$$x = \frac{\ln y}{\ln a}.$$
But x is also given by (53). Hence
$$\log_a y = \frac{\ln y}{\ln a} \tag{54}$$
when a is positive and not equal to unity.

As a consequence of (54)

$$\log_a x + \log_a y = \frac{\ln x + \ln y}{\ln a} = \frac{\ln xy}{\ln a} = \log_a xy \tag{55}$$

where (22) has been used, and then (54) again. In a similar way one can establish, from (24) and (37),

$$\log_a \frac{x}{y} = \log_a x - \log_a y, \tag{56}$$

$$\log_a x^y = y \log_a x. \tag{57}$$

Again, we see from (29) and (54) that

$$\frac{\mathrm{d}}{\mathrm{d}x} \log_a g = \frac{1}{g \ln a} \frac{\mathrm{d}g}{\mathrm{d}x}. \tag{58}$$

Formula (58) demonstrates why natural logarithms are used more frequently in calculus than logarithms to any other base; the formula takes its simplest form when $a = \mathrm{e}$ because $\ln \mathrm{e} = 1$.

The most commonly occurring bases are $a = \mathrm{e}$, $a = 10$ and $a = 2$. The reason for the popularity of e has already been explained. The base $a = 10$ is used for tables of common logarithms, but (58) now contains the awkward factor $\ln 10$ or $2 \cdot 3026 \ldots$. The logarithms with base 2 do not avoid the awkwardness with (58) ($\ln 2 = 0 \cdot 6931 \ldots$) but arise naturally in the theory of information, where $-\log_2 p$ is the information content (in bits) of a choice of probability p.

The construction of slide rules is based on rule (55). A scale is made by marking the numbers 1, 2, 3, etc., so that the distance from 1 to 2 is $\log_a 2$, and the distance from 1 to 3 is $\log_a 3$. In general, the distance from the point marked x to the point marked 1 is made $\log_a x$. Suppose there are two such scales, made in exactly the same way, denoted by A and B. Let 1 on B be placed opposite x on A. Look at the number on A which is opposite y on B. Its distance from 1 on the A scale will be $\log_a x + \log_a y$ or $\log_a xy$. It will therefore be marked xy. Thus multiplication is simply and effectively carried out by means of a slide rule.

Some slide rules also provide scales which enable powers of the exponential to be written down; they are called log-log slide rules. If the line on the visor is placed on a number of the D scale, say 3, then the LL1 scale gives $\mathrm{e}^{0 \cdot 03}$, the LL2 gives $\mathrm{e}^{0 \cdot 3}$ and the LL3 gives e^3.

A slide rule is a form, crude perhaps, of computer in that it stores information in a manner which is convenient for rapid calculation. As the storage capacity of a computer increases the necessity for tables, especially common logarithms, become less.

306   Introductory Analysis

Actually, most tables of common logarithms list only the logarithms of numbers between 1 and 10. Thus the common logarithm of 4 is given as 0·6021. This tells us that $4 = 10^{0·6021}$. The portion to the left of the decimal point is called the *characteristic* and the remainder of the logarithm, the decimal part, is called the *mantissa*. Because the logarithm of 4 is 0·6021, the characteristic is 0 and the mantissa is ·6021. Now

$$400 = 4.100 = 10^{0·6021} \cdot 10^2 = 10^{2·6021}.$$

Therefore the common logarithm of 400 is 2·6021; its characteristic is 2 and its mantissa is ·6021. Thus multiplying 4 by a positive power of 10 alters the characteristic of the logarithm but not the mantissa. Clearly it is sufficient in a table to provide the mantissa of every number from 1 to 10; the logarithm of any number larger than 10 can be obtained by adjusting the characteristic.

This is also true for numbers between 0 and 1. Suppose we want the logarithm of 0·04. Since

$$0·04 = 4 \cdot 10^{-2}$$

we have

$$0·04 = 10^{0·6021} \cdot 10^{-2}.$$

The procedure to be followed now depends upon the purpose of the calculation. As far as arithmetic is concerned it is most convenient to combine 0·6021 and -2 as $\bar{2}·6021$, the bar being placed over the 2 to demonstrate that only the 2 is negative. For analysis, we add 0·6021 and -2 in the usual way to obtain $-1·3979$. Whatever notation is used, the common logarithm of a number between 0 and 1 is negative.

Exercises

59. Prove that $\log_a 1 = 0$, $\log_a a = 1$.

60. Evaluate (i) $\log_2 8$, (ii) $\log_9 27$, (iii) $\log_8 128$, (iv) $\log_6 \frac{1}{216}$, (v) $\log_2 0·0625$, (vi) $\log_{27} \frac{1}{81}$, (vii) $\log_{343} 49$, (viii) $\log_{0·01} 1000$.

61. Solve the equations (i) $5^x = 4 \cdot 3^x$, (ii) $3^{2x} \cdot 2^{3x} = 5^5$, (iii) $2^{\log_2 8} \cdot 5^{\log_5 3} = 3^{\log_3 x}$.

62. Prove that (i) $\log_b x = \log_b a \log_a x$, (ii) $\log_b x = \log_a x / \log_a b$, (iii) $\log_b a \log_a b = 1$ where a, b, x are positive numbers with $a \neq 1, b \neq 1$.

63. Show that

$$\frac{1}{\log_2 x} + \frac{1}{\log_3 x} + \ldots + \frac{1}{\log_n x} = \frac{1}{\log_{n!} x}.$$

64. Prove that

$$a^x = 1 + x \ln a + \frac{x^2 (\ln a)^2}{2!} + \ldots + \frac{x^n (\ln a)^n}{n!} + \ldots.$$

Functions defined by series 307

65. Given that $\log_{10} 5\cdot 9 = 0\cdot 7709$ write down the common logarithms of 59, 5900, 0·0059, 0·000059.

66. Given that $\log_{10} 2 = 0\cdot 3010$, $\log_{10} 7 = 0\cdot 8451$ find (i) $\log_{10} 64$, (ii) $\log_{25} 200$, (iii) $\log_7 \sqrt{2}$.

67. Draw a flow-chart for the calculation of a^x. Do you think a bisection method might give $\log_a y$? If so, make necessary modifications to your flow-chart to permit this computation.

68. Show that the number of iterations to reduce the error of an iterative method of order 1 by a factor of 0·1 is approximately $-1/\log_{10}|F'(x_0)|$ where x_0 is the required root. (Hint: Theorem 8.6(iv).)

10.5 Trigonometric functions

Consider the functions

$$c(x) = 1 - \frac{x^2}{2!} + \frac{x^4}{4!} + \ldots + (-1)^n \frac{x^{2n}}{(2n)!} + \ldots, \qquad (59)$$

$$s(x) = x - \frac{x^3}{3!} + \ldots + (-1)^{n-1} \frac{x^{2n-1}}{(2n-1)!} + \ldots . \qquad (60)$$

Since

$$\lim_{n \to \infty} \frac{1}{(2n+2)!} \cdot (2n)! = \lim_{n \to \infty} \frac{1}{(2n+2)(2n+1)} = 0$$

and

$$\lim_{n \to \infty} \frac{1}{(2n+1)!} \cdot (2n-1)! = \lim_{n \to \infty} \frac{1}{(2n+1)2n} = 0$$

Theorem 9.10 tells us that both series converge absolutely for all finite x. By Theorem 9.12 both $c(x)$ and $s(x)$ are continuous functions for all finite x and

$$c(0) = 1, \quad s(0) = 0, \quad c(-x) = c(x), \quad s(-x) = -s(x). \qquad (61)$$

On account of Theorem 9.13

$$\frac{d}{dx} c(x) = -\frac{x}{1!} + \frac{x^3}{3!} + \ldots + (-1)^n \frac{x^{2n-1}}{(2n-1)!} + \ldots$$

$$= -s(x) \qquad (62)$$

and, similarly,

$$s'(x) = c(x). \qquad (63)$$

Furthermore

$$\frac{d}{dx}(c^2 + s^2) = 2cc' + 2ss' = -2cs + 2sc = 0$$

from (62) and (63). Therefore Corollary 7.2 shows that

$$[c(x)]^2 + [s(x)]^2 = \text{constant}.$$

By putting $x = 0$ and using (61) we find that the constant is unity and so

$$[c(x)]^2 + [s(x)]^2 = 1. \tag{64}$$

The same result could have been obtained directly from (59) and (60) by Theorem 9.11. A similar argument, the derivative of

$$[s(x + a) - s(x)c(a) - s(a)c(x)]^2 + [c(x + a) - c(x)c(a) + s(x)s(a)]^2$$

being taken, may be employed to show that

$$s(x + a) = s(x)c(a) + s(a)c(x), \tag{65}$$

$$c(x + a) = c(x)c(a) - s(x)s(a). \tag{66}$$

The properties (61)–(66) are exactly the same as those of the trigonometric functions $\cos x$ and $\sin x$. Thus (59) and (60) can be used as the *starting-point* of an analytical treatment of trigonometric functions which makes no initial reference to geometry. There are two difficulties about this approach. The first is to show that $c(x)$ and $s(x)$ are periodic; this can be achieved by showing that $c(x) = 0$ has a smallest positive root and calling it $\frac{1}{2}\pi$ (this is actually a definition of the number π) and then periodicity of c and s follows from (64)–(66). The second difficulty is to demonstrate that the functions defined analytically have geometrical significance. It is possible to overcome both difficulties but details will not be given here.

Exercises

69. Show that, when $0 \leqslant x \leqslant \sqrt{2}$,

$$1 - \frac{x^2}{2!} \geqslant 0, \quad \frac{x^4}{4!} - \frac{x^6}{6!} \geqslant 0, \quad \frac{x^8}{8!} - \frac{x^{10}}{10!} \geqslant 0, \ldots.$$

Deduce that $c(x) > 0$ on $0 \leqslant x \leqslant \sqrt{2}$.

Prove that $c(2) < -\frac{1}{3}$ and deduce that $c(x) = 0$ has at least one root in $\sqrt{2} < x < 2$. (This demonstrates that the number $\frac{1}{2}\pi$ lies between $\sqrt{2}$ and 2.)

70. If $f''(x) + f(x) = 0$ and $g(x) = f(x) - Ac(x) - Bs(x)$ show that A and B can be chosen so that

$$\frac{d}{dx}\{[g'(x)]^2 + [g(x)]^2\} = 0.$$

Deduce that $f(x) = Ac(x) + Bs(x)$.

10.6 Hyperbolic functions

There are certain combinations of the exponentials e^x and e^{-x} which occur so frequently in the applications of mathematics that they have been given special names. They are the *hyperbolic cosine*, defined by

$$\cosh x = \tfrac{1}{2}(e^x + e^{-x}), \tag{67}$$

and the *hyperbolic sine*, defined by

$$\sinh x = \tfrac{1}{2}(e^x - e^{-x}). \tag{68}$$

The pronunciation of cosh follows its spelling but sinh is pronounced by some as shine and by others as sinsh.

By substituting the exponential series (8) in (67) and (68) (and remembering Theorem 9.11) we obtain

$$\cosh x = 1 + \frac{x^2}{2!} + \frac{x^4}{4!} + \ldots + \frac{x^{2n}}{(2n)!} + \ldots, \tag{69}$$

$$\sinh x = x + \frac{x^3}{3!} + \ldots + \frac{x^{2n-1}}{(2n-1)!} + \ldots. \tag{70}$$

These series bear a distinct resemblance to the series in (59) and (60), so that we may anticipate that the hyperbolic functions have some properties which are similar to those of the trigonometric functions. However, it is usually easier to derive these properties from the definitions (67) and (68).

For instance

$$\cosh^2 x - \sinh^2 x = \tfrac{1}{4}(e^{2x} + 2 + e^{-2x}) - \tfrac{1}{4}(e^{2x} - 2 + e^{-2x}) \tag{71}$$
$$= 1$$

which is the hyperbolic analogue of (64).

If $x = \cos \theta$, $y = \sin \theta$ the point (x,y) lies on the circle $x^2 + y^2 = 1$. On the other hand, if $x = \cosh \theta$, $y = \sinh \theta$ the point (x,y) lies on the hyperbola $x^2 - y^2 = 1$. This geometric attribute, coupled with formulae such as (71) which are analogous to corresponding trigonometric formulae, accounts for the nomenclature of the hyperbolic functions.

The hyperbolic functions have so many important applications that tables of their numerical values have been prepared.

The graph of $y = \cosh x$ is shown in figure 95. At $x = 0$, $\cosh x = 1$ from (10). The hyperbolic cosine is an even function of x and is symmetric about the y-axis, i.e.

$$\cosh(-x) = \cosh x$$

as is immediately evident from (67). The curve $y = \cosh x$ is a particular case of a curve which occurs in mechanics and is called the *catenary*, because a uniform chain hanging in a vertical plane under its own weight

adopts this shape. The graph of $y = \sinh x$ is given in figure 96. Note that $\sinh 0 = 0$ and
$$\sinh(-x) = -\sinh x$$
from (68).

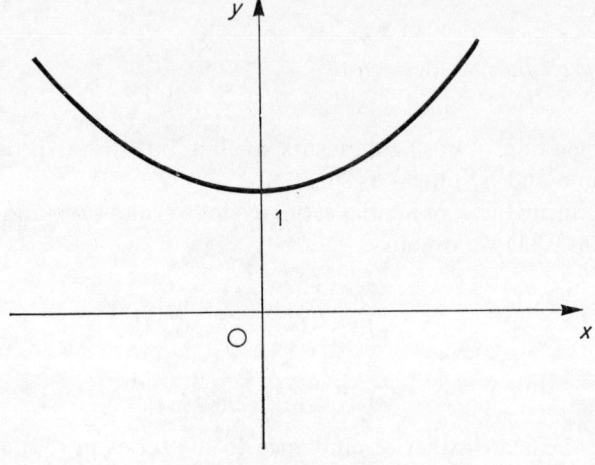

FIGURE 95

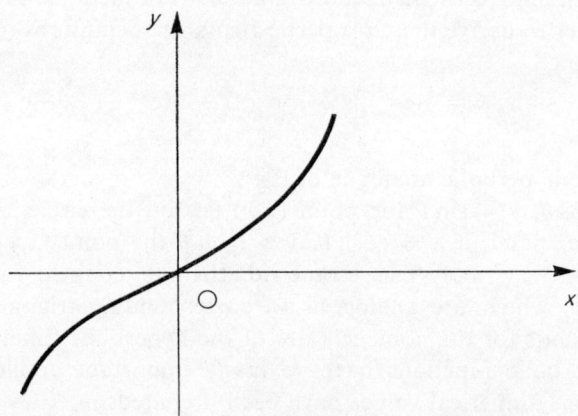

FIGURE 96

Although there are similarities between the hyperbolic functions and the trigonometric functions there are significant differences. Every time that x is increased by 2π, $\sin x$ and $\cos x$ are unaltered, but this is not true of $\sinh x$ and $\cosh x$ as can be seen from the graphs. In fact, both $\sin x$ and $\cos x$ oscillate between the values -1 and $+1$ whereas $\sinh x$

Functions defined by series

increases from $-\infty$ to $+\infty$ and $\cosh x$ varies from $+\infty$ to 1 to $+\infty$ as x increases from $-\infty$ to $+\infty$.

Again, from (15) and (17), we see that
$$\cosh x \sim \tfrac{1}{2}e^x, \quad \sinh x \sim \tfrac{1}{2}e^x \tag{72}$$
as $x \to +\infty$, where $\sim$ means 'behaves like'. Also
$$\cosh x \sim \tfrac{1}{2}e^{-x}, \quad \sinh x \sim -\tfrac{1}{2}e^{-x} \tag{73}$$
as $x \to -\infty$. These properties have no parallel with the sine and cosine.

Next, since
$$\frac{d}{dx}e^x = e^x \quad \text{and} \quad \frac{d}{dx}e^{-x} = -e^{-x},$$

$$\frac{d}{dx}\cosh x = \sinh x, \tag{74}$$

$$\frac{d}{dx}\sinh x = \cosh x. \tag{75}$$

In the same way as four other trigonometric functions are defined in terms of sine and cosine, so it is convenient to introduce four additional hyperbolic functions defined in terms of $\cosh x$ and $\sinh x$, namely

$$\tanh x = \frac{\sinh x}{\cosh x}, \quad \coth x = \frac{\cosh x}{\sinh x} = \frac{1}{\tanh x},$$

$$\operatorname{sech} x = \frac{1}{\cosh x}, \quad \operatorname{cosech} x = \frac{1}{\sinh x}.$$

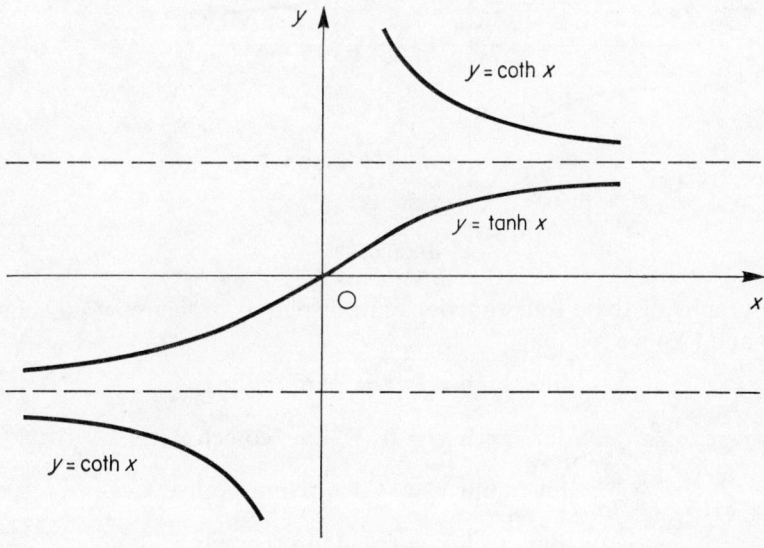

FIGURE 97

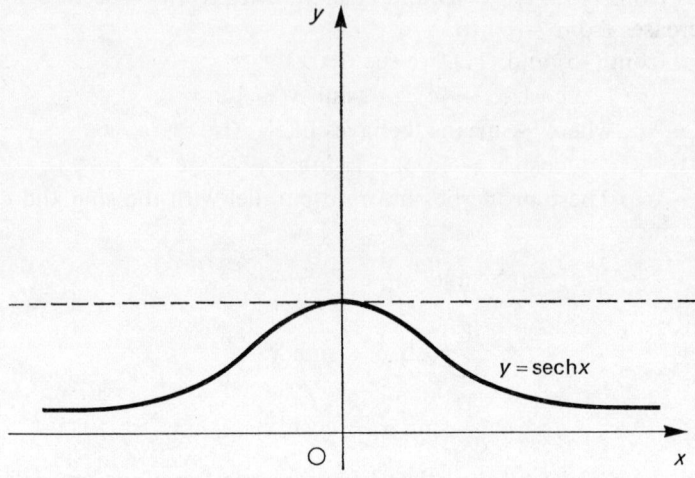

Figure 98

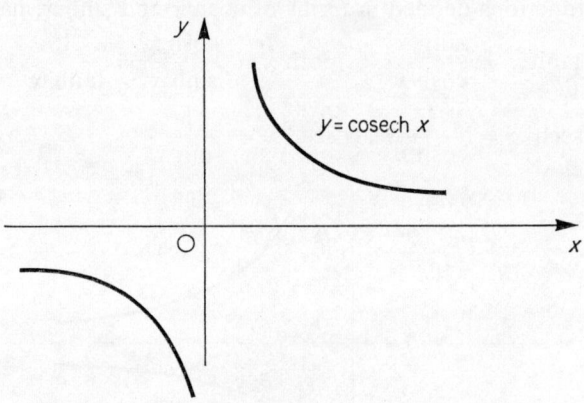

Figure 99

The graphs of these four functions are displayed in figures 97–99. From (72) and (73) we see that

$$\lim_{x \to +\infty} \tanh x = 1 = \lim_{x \to +\infty} \coth x,$$
$$\lim_{x \to +\infty} \operatorname{sech} x = 0 = \lim_{x \to +\infty} \operatorname{cosech} x,$$
$$\lim_{x \to -\infty} \tanh x = -1 = \lim_{x \to -\infty} \coth x,$$
$$\lim_{x \to -\infty} \operatorname{sech} x = 0 = \lim_{x \to -\infty} \operatorname{cosech} x.$$

Functions defined by series

The derivatives are calculated by using (74), (75) and standard rules. Thus

$$\frac{d}{dx} \operatorname{sech} x = \frac{d}{dx} \frac{1}{\cosh x}$$

$$= -\frac{1}{\cosh^2 x} \frac{d}{dx} \cosh x$$

$$= -\sinh x / \cosh^2 x$$

$$= -\operatorname{sech} x \tanh x.$$

The derivatives of the remaining functions are left as exercises.

Exercises

71. Establish the following formulae:
  (i) $\tanh(-x) = -\tanh x$,
  (ii) $\sinh 2x = 2 \sinh x \cosh x$,
  (iii) $\cosh 2x = \cosh^2 x + \sinh^2 x = 2 \cosh^2 x - 1 = 1 + 2 \sinh^2 x$,
  (iv) $1 - \tanh^2 x = \operatorname{sech}^2 x$,
  (v) $\coth^2 x - 1 = \operatorname{cosech}^2 x$,
  (vi) $\sinh(x+y) = \sinh x \cosh y + \cosh x \sinh y$,
  (vii) $\cosh(x+y) = \cosh x \cosh y + \sinh x \sinh y$,
  (viii) $\cosh x + \sinh x = e^x$, $\quad \cosh x - \sinh x = e^{-x}$,
  (ix) $(\cosh x + \sinh x)^n = \cosh nx + \sinh nx$ (n a positive integer),
  (x) $(1 + \tanh x)/(1 - \tanh x) = e^{2x}$,
  (xi) $\tanh 2x = \dfrac{2 \tanh x}{1 + \tanh^2 x}$,
  (xii) $\sinh x = \dfrac{2 \tanh \frac{1}{2}x}{1 - \tanh^2 \frac{1}{2}x}$.

72. Find $\tanh 2x$ when $\tanh x = \frac{1}{2}$.

73. Given the hyperbolic function below find the other five: (i) $\sinh x = 4/3$, (ii) $\cosh x = 37/35$ ($x > 0$), (iii) $\tanh x = -12/13$.

74. If $y = \ln \tan(\frac{1}{4}\pi + \frac{1}{2}x)$ for $-\frac{1}{2}\pi < x < \frac{1}{2}\pi$ prove that $\sinh y = \tan x$, $\cosh y = \sec x$, $\tanh y = \sin x$, $\tanh \frac{1}{2}y = \tan \frac{1}{2}x$.

75. The point P ($\cosh \theta, \sinh \theta$) with $\theta > 0$ lies on the hyperbola $x^2 - y^2 = 1$. Show that the line joining the origin to P intersects, at $(1, \tanh \theta)$, the tangent to the hyperbola at the vertex $(1, 0)$.

76. Prove that

(i) $(d/dx) \tanh x = \operatorname{sech}^2 x$, (ii) $(d/dx) \coth x = -\operatorname{cosech}^2 x$,
(iii) $(d/dx) \operatorname{cosech} x = -\operatorname{cosech} x \coth x$.

77. Evaluate (i) $\lim_{x \to 0} (\sinh x)/x$, (ii) $\lim_{x \to 0} (\tanh x)/x$, (iii) $\lim_{x \to 0} (1 - \cosh x)/x$
(iv) $\lim_{x \to +\infty} (\sinh x)^{2/x}$.

78. Examine the following series for convergence
   (i) $\operatorname{sech} 1 + \operatorname{sech} 2 + \ldots + \operatorname{sech} n + \ldots$,
   (ii) $\tanh 1 + \tanh 2 + \ldots + \tanh n + \ldots$.

79. Find the derivative of (i) $\sinh 4x$, (ii) $\cosh^2 2x$, (iii) $\tanh 5x$,
(iv) $\ln \sinh x$ $(x > 0)$, (v) $\tanh (1/x)$, (vi) $x \operatorname{cosech} x^2$, (vii) $\operatorname{sech}^3 x$,
(viii) $\ln (\coth 2x)^{\frac{1}{2}}$ $(x > 0)$.

80. Show that

$$\sin x \cosh x = x + \frac{1}{2}x^3 - \frac{1}{30}x^5 + \ldots,$$

$$\cos x \sinh x = x - \frac{1}{3}x^3 - \frac{1}{30}x^5 + \ldots,$$

$$\ln (1 + \sinh x) = x - \frac{1}{2}x^2 + \frac{1}{2}x^3 + \ldots.$$

81. If $ay = \cosh ax$ prove that

$$\frac{d^2y}{dx^2} = a\left[1 + \left(\frac{dy}{dx}\right)^2\right]^{\frac{1}{2}}.$$

82. If $y = A \cosh ax + B \cos ax + C \sinh ax + D \sin ax$, where A, B, C and D are constants, show that

$$\frac{d^4y}{dx^4} = a^4 y.$$

83. If $\sinh x = \tan y$ with $-\frac{1}{2}\pi < y < \frac{1}{2}\pi$ prove that

$$\frac{dy}{dx} = \operatorname{sech} x.$$

Functions defined by series

84. Examine $y = \sinh x$, $y = \cosh x$, $y = \tanh x$ for minima and points of inflection.

85. Find the smallest possible root of $\cos x \cosh x = -1$. What is the next largest?

10.7 Leibnitz's theorem

When calculating the higher derivatives of functions there is a theorem which is often of value and which arises as follows. It has been shown in Theorem 5.6 that

$$\frac{d}{dx}(uv) = \frac{du}{dx}v + u\frac{dv}{dx}.$$

Therefore

$$\frac{d^2}{dx^2}(uv) = \left(\frac{d^2u}{dx^2}v + \frac{du}{dx}\frac{dv}{dx}\right) + \left(\frac{du}{dx}\frac{dv}{dx} + u\frac{d^2v}{dx^2}\right)$$

$$= \frac{d^2u}{dx^2}v + 2\frac{du}{dx}\frac{dv}{dx} + u\frac{d^2v}{dx^2}.$$

Another derivatives with respect to x gives

$$\frac{d^3}{dx^3}(uv) = \left(\frac{d^3u}{dx^3}v + \frac{d^2u}{dx^2}\frac{dv}{dx}\right) + 2\left(\frac{d^2u}{dx^2}\frac{dv}{dx} + \frac{du}{dx}\frac{d^2v}{dx^2}\right) +$$

$$\left(\frac{du}{dx}\frac{d^2v}{dx^2} + u\frac{d^3v}{dx^3}\right)$$

$$= \frac{d^3u}{dx^3}v + 3\frac{d^2u}{dx^2}\frac{dv}{dx} + 3\frac{du}{dx}\frac{d^2v}{dx^2} + u\frac{d^3v}{dx^3}.$$

If these results are compared with

$$(a + b)^2 = a^2 + 2ab + b^2,$$
$$(a + b)^3 = a^3 + 3a^2b + 3ab^2 + b^3$$

it will be observed that the coefficients are the same, namely 1, 2, 1 for $(a + b)^2$ and the second derivative, and 1, 3, 3, 1 for $(a + b)^3$ and the third derivative. This suggests that the coefficients in the nth derivative will be the same as those in $(a + b)^n$. Therefore, the proposal is

THEOREM 10.1 (LEIBNITZ'S THEOREM)

$$\frac{d^n}{dx^n}(uv) = \frac{d^nu}{dx^n}v + n\frac{d^{n-1}u}{dx^{n-1}}\frac{dv}{dx} + \frac{n(n-1)}{2!}\frac{d^{n-2}u}{dx^{n-2}}\frac{d^2v}{dx^2} + \ldots + u\frac{d^nv}{dx^n}.$$

316   Introductory Analysis

Proof. The proof is by induction. Assume that the result is true for some value of n. Then

$$\frac{d^{n+1}}{dx^{n+1}}(uv) = \left(\frac{d^{n+1}u}{dx^{n+1}}v + \frac{d^n u}{dx^n}\frac{dv}{dx}\right) + n\left(\frac{d^n u}{dx^n}\frac{dv}{dx} + \frac{d^{n-1}u}{dx^{n-1}}\frac{d^2 v}{dx^2}\right) + \cdots$$

$$+ \frac{n(n-1)\cdots(n-r+1)}{r!}\left(\frac{d^{n-r+1}u}{dx^{n-r+1}}\frac{d^r v}{dx^r} + \frac{d^{n-r}u}{dx^{n-r}}\frac{d^{r+1}v}{dx^{r+1}}\right) + \cdots$$

$$+ \left(\frac{du}{dx}\frac{d^n v}{dx^n} + u\frac{d^{n+1}v}{dx^{n+1}}\right)$$

$$= \frac{d^{n+1}u}{dx^{n+1}}v + (1+n)\frac{d^n u}{dx^n}\frac{dv}{dx} + \cdots +$$

$$\left[\frac{n(n-1)\cdots(n-r+2)}{(r-1)!} + \frac{n(n-1)\cdots(n-r+1)}{r!}\right] \times$$

$$\frac{d^{n-r+1}u}{dx^{n-r+1}}\frac{d^r v}{dx^r} + \cdots + u\frac{d^{n+1}v}{dx^{n+1}}$$

$$= \frac{d^{n+1}u}{dx^{n+1}}v + (1+n)\frac{d^n u}{dx^n}\frac{dv}{dx} + \cdots +$$

$$\frac{(n+1)n\cdots(n-r+2)}{r!}\frac{d^{n-r+1}u}{dx^{n-r+1}}\frac{d^r v}{dx^r} + \cdots + u\frac{d^{n+1}v}{dx^{n+1}}$$

since

$$\frac{n(n-1)\cdots(n-r+2)}{(r-1)!} + \frac{n(n-1)\cdots(n-r+1)}{r!}$$

$$= \frac{n(n-1)\cdots(n-r+2)}{r!}(r+n-r+1)$$

$$= \frac{(n+1)n\cdots(n-r+2)}{r!}.$$

Hence, if the result holds for some value of n, it must be valid for the value $n+1$. Since it is known to be true for $n = 2$, it must be true for $n = 3$, and therefore when $n = 4$, and so for any positive integer n.

Example 6 Find the nth derivative of $x \ln x$.

Take $u = \ln x$, $v = x$. Then

$$\frac{du}{dx} = \frac{1}{x}, \qquad \frac{d^2 u}{dx^2} = -\frac{1}{x^2}, \ldots, \qquad \frac{d^n u}{dx^n} = \frac{(n-1)!(-1)^{n-1}}{x^n},$$

$$\frac{dv}{dx} = 1, \qquad \frac{d^2 v}{dx^2} = 0, \ldots, \qquad \frac{d^n v}{dx^n} = 0 \ (n \geq 2).$$

Hence Leibnitz's theorem gives

$$\frac{d^n}{dx^n}(x \ln x) = \frac{(n-1)!(-1)^{n-1}}{x^n} \cdot x + n \cdot \frac{(n-2)!(-1)^{n-2}}{x^{n-1}} \cdot 1$$

$$= \frac{(n-2)!(-1)^{n-1}}{x^{n-1}}(n-1-n) = \frac{(n-2)!(-1)^n}{x^{n-1}}.$$

Example 7 If $x = \cos[(1/a) \ln y]$ for $y > 0$, show that

$$(1-x^2)\frac{d^2y}{dx^2} - x\frac{dy}{dx} - a^2y = 0.$$

Taking a derivative with respect to x, we have

$$1 = -\sin\left(\frac{1}{a}\ln y\right)\frac{d}{dx}\left(\frac{1}{a}\ln y\right) = -\frac{1}{ay}\frac{dy}{dx}\sin\left(\frac{1}{a}\ln y\right)$$

or

$$-ay = \frac{dy}{dx}\sin\left(\frac{1}{a}\ln y\right). \tag{76}$$

After another derivative with respect to x we obtain

$$-a\frac{dy}{dx} = \frac{d^2y}{dx^2}\sin\left(\frac{1}{a}\ln y\right) + \frac{1}{ay}\left(\frac{dy}{dx}\right)^2 \cos\left(\frac{1}{a}\ln y\right).$$

Multiply both sides by $\sin\left(\frac{1}{a}\ln y\right)$ and use (76). Then

$$a^2y = \frac{d^2y}{dx^2}\sin^2\left(\frac{1}{a}\ln y\right) - \frac{dy}{dx}\cos\left(\frac{1}{a}\ln y\right).$$

Since $\sin^2\left(\frac{1}{a}\ln y\right) = 1 - \cos^2\left(\frac{1}{a}\ln y\right) = 1 - x^2$ it follows that

$$(1-x^2)\frac{d^2y}{dx^2} - x\frac{dy}{dx} - a^2y = 0.$$

The nth derivative of this is, according to Leibnitz's theorem,

$$\left[(1-x^2)\frac{d^{n+2}y}{dx^{n+2}} + n(-2x)\frac{d^{n+1}y}{dx^{n+1}} + \frac{n(n-1)}{2}(-2)\frac{d^ny}{dx^n}\right]$$

$$- \left[x\frac{d^{n+1}y}{dx^{n+1}} + n \cdot 1 \cdot \frac{d^ny}{dx^n}\right] - a^2\frac{d^ny}{dx^n} = 0$$

or

$$(1-x^2)\frac{d^{n+2}y}{dx^{n+2}} - (2n+1)x\frac{d^{n+1}y}{dx^{n+1}} - (n^2+a^2)\frac{d^ny}{dx^n} = 0.$$

Exercises

86. Find the nth derivative of (i) $x^2 \, e^x$, (ii) $x^3 \, e^{ax}$, (iii) $x^2 \ln x$, (iv) $e^{-at} \sin at$, (v) $x^2(1+x)^n$, (vi) $x^2 y$,

(vii) $$x^2 \frac{d^2 y}{dx^2} + x \frac{dy}{dx} = 0.$$

87. Prove that
$$\frac{d^n}{dx^n}(x^2 \ln x) = 2 \frac{d^{n-2}}{dx^{n-2}} \ln x.$$

88. Prove that
$$\sin x \cosh x = x + \tfrac{1}{3} x^3 - \ldots + (-1)^{n-1} \frac{2^{n-\frac{1}{2}} x^{2n-1}}{(2n-1)!} \cos (2n-1) \frac{\pi}{4} + \ldots .$$

10.8 Inverse functions

The new functions that have been introduced in this chapter are studied because they arise often in practical situations. Any function expresses the relation between one variable and another. For example, we might be given that the distance travelled by a person at time t is
$$y = 30t.$$
From this we could deduce that, if we wanted the time to reach a certain distance y, we could calculate it by writing
$$t = y/30.$$
In doing this we have expressed t as a function of y, and we say that $t = y/30$ is the *inverse function* of $y = 30t$.

On the other hand, we might be told that the height of the cable of a suspension bridge above its lowest point is
$$y = x^2/100$$
when a horizontal distance x has been travelled. Now, if we are given the height y and ask for the horizontal distance gone we find
$$x = \pm 10\sqrt{y},$$
i.e. x is a double-valued function of y. The reason is quite simple: we might step out from the lowest point in either direction along the bridge (figure 100). Of course, if the cable supports were not symmetrically placed one of the values of x might have to be rejected because it was beyond a support.

Functions defined by series

As far as the inverse function of $y = x^2/100$ is concerned, note that, when $x > 0$, we must take $x = 10\sqrt{y}$. This gives the full line on figure 101. But, when $x < 0$, we must take $x = -10\sqrt{y}$ and follow the dashed line in figure 101. These two curves are called the *branches* of the inverse function, one branch being used when $x > 0$ and the other when $x < 0$.

The two branches join at $x = 0$ which corresponds in figure 100 to the

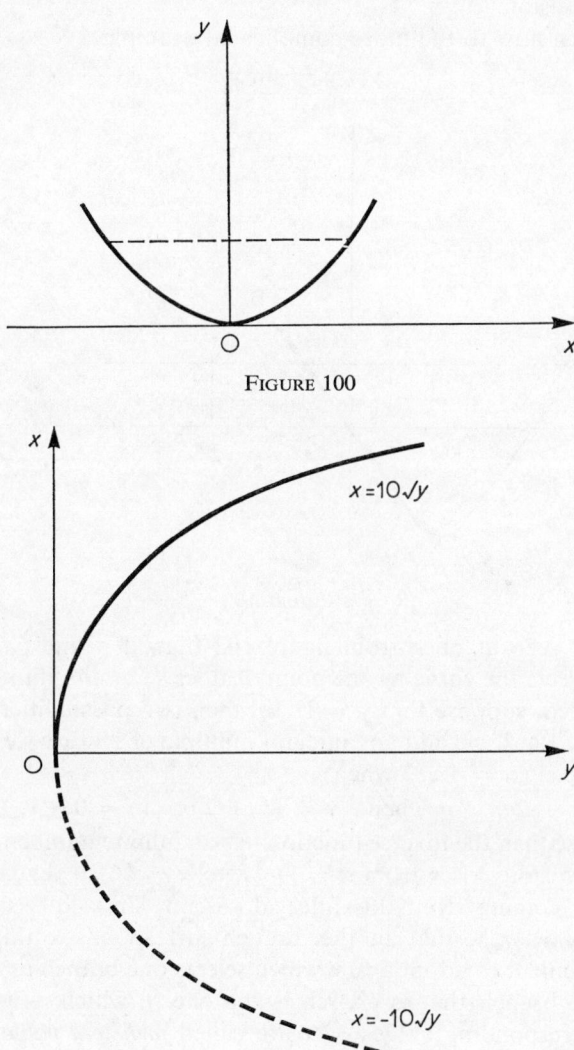

FIGURE 100

FIGURE 101

point where the graph changes from falling to rising. This suggests that so long as we are on a part of a curve which is steadily rising or steadily falling we shall stay on one branch of the inverse function and shall not suddenly switch to another branch.

Sometimes physical conditions tell us what branch or branches we must choose and sometimes we have to lay down a rule for ourselves just to prevent confusion.

Let us turn now to the more complicated example
$$y = \sin x.$$

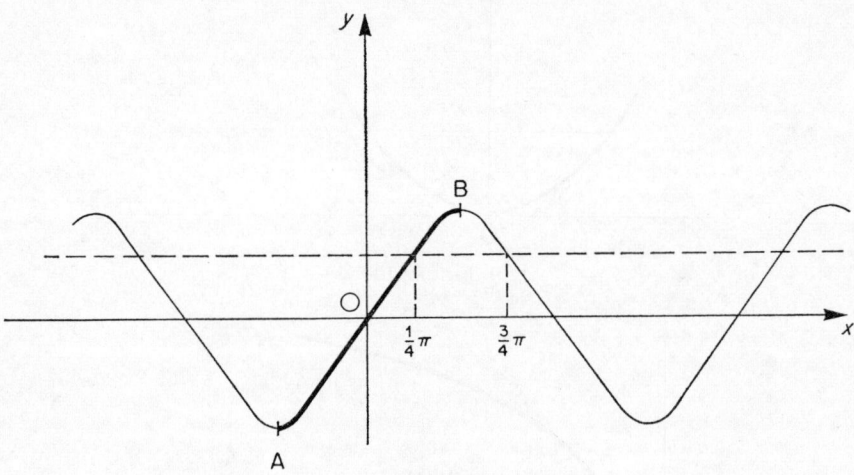

FIGURE 102

It can be seen at once from figure 102 that, if a line parallel to the x-axis intersects the curve at one point, it does so at an infinite number of points. Indeed, suppose that $y = 1/\sqrt{2}$; then two possibilities are $x = \tfrac{1}{4}\pi$ and $x = \tfrac{3}{4}\pi$. But if we add any integral multiple of 2π to x we do not alter $\sin x$. Hence $\sin x = 1/\sqrt{2}$ when

$$x = \tfrac{1}{4}\pi + 2n\pi \quad \text{or when} \quad x = \tfrac{3}{4}\pi + 2n\pi \quad (n = 0, \pm 1, \pm 2, \ldots).$$

In this case then the inverse function has an infinite number of branches. Now $\sin x$ steadily rises from $x = -\tfrac{1}{2}\pi$ to $x = \tfrac{1}{2}\pi$ so that there will be one branch coming from this interval. Again, $\sin x$ falls steadily from $x = \tfrac{1}{2}\pi$ to $x = \tfrac{3}{2}\pi$ so that another branch corresponds to this interval.

It is convenient to adopt a rule which selects one branch as the *principal branch*. The branch that is chosen is the one in which $-\tfrac{1}{2}\pi \leqslant x \leqslant \tfrac{1}{2}\pi$ and the corresponding values of x are called *principal values*. Thus, the possible points on the curve $y = \sin x$ are restricted to those lying on the solid curve AOB of figure 102. Then, given any value of y from -1 to 1,

Functions defined by series

there will be one and only one value of x on this portion of the curve. Hence the equation
$$y = \sin x$$
with the restriction $-\tfrac{1}{2}\pi \leqslant x \leqslant \tfrac{1}{2}\pi$ gives a single-valued function
$$x = \sin^{-1} y \quad (-\tfrac{1}{2}\pi \leqslant x \leqslant \tfrac{1}{2}\pi) \tag{77}$$
called the 'inverse sine of y'; it is defined for $-1 \leqslant y \leqslant 1$ and takes values from $-\tfrac{1}{2}\pi$ to $\tfrac{1}{2}\pi$.

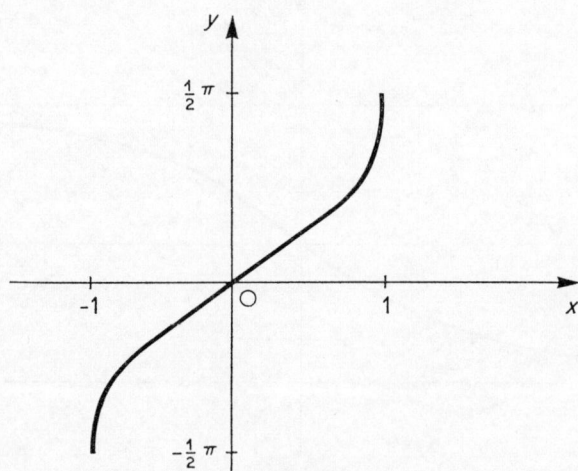

FIGURE 103

For instance, in our previous example where $y = 1/\sqrt{2}$ there is only one x which satisfies the conditions, namely $x = \tfrac{1}{4}\pi$. Therefore
$$\sin^{-1}(1/\sqrt{2}) = \tfrac{1}{4}\pi.$$
The function (77) is the inverse function of $y = \sin x$. Therefore, when we want the inverse function of $x = \sin y$ we write
$$y = \sin^{-1} x \quad (-\tfrac{1}{2}\pi \leqslant y \leqslant \tfrac{1}{2}\pi).$$
A graph showing the principal value of $y = \sin^{-1} x$ is shown in figure 103.

It should be observed that the notation $\sin^{-1} x$ does *not* mean $1/\sin x$. In fact, $\sin^{-1} x$ is a way of describing a new function which is defined implicitly by $x = \sin y$. The notation $y = f^{-1}(x)$ is often used for the function defined implicitly by $x = f(y)$.

An alternative notation for $\sin^{-1} x$ is arcsin x, but this is not employed so much now because it is more lengthy and does not fit in with the general usage of $f^{-1}(x)$ to denote an inverse function.

322   *Introductory Analysis*

To make sure that $1/\sin x$ is not confused with $\sin^{-1} x$ it should be written

$$\frac{1}{\sin x} \text{ or } (\sin x)^{-1} \text{ or cosec } x.$$

Each of the trigonometric functions possesses an inverse and in each case some restriction is necessary to obtain an inverse function which is single-valued.

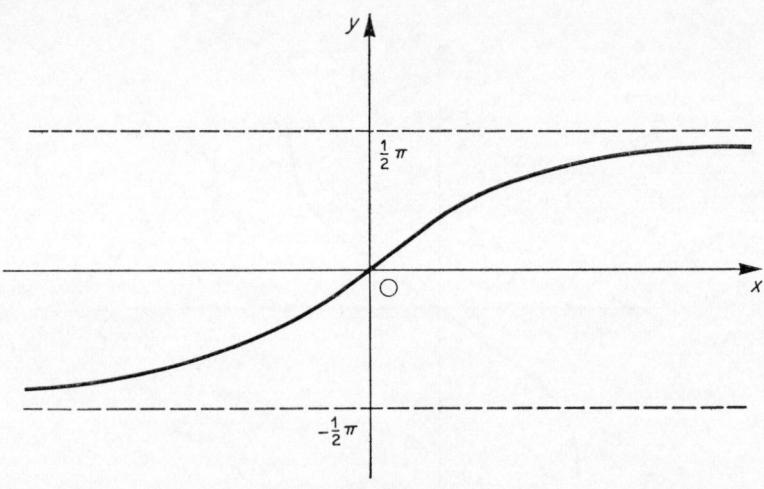

FIGURE 104

Suppose we have the equation $x = \tan y$. Given x there is an infinite number of solutions of this equation, all differing by integer multiples of π. By turning figure 69 on its side, we see that we can choose a single branch by specifying that $-\tfrac{1}{2}\pi < y < \tfrac{1}{2}\pi$. This is called the principal branch and we write

$$y = \tan^{-1} x \quad (-\tfrac{1}{2}\pi < y < \tfrac{1}{2}\pi).$$

The values $y = \pm \tfrac{1}{2}\pi$ are excluded because $\tan y$ takes infinite values at these points. Actually the lines $y = \pm \tfrac{1}{2}\pi$ are asymptotes of the curve $y = \tan^{-1} x$, as can be seen from the graph of this function in figure 104.

Similarly, the principal branch of the inverse cosine comes from $x = \cos y$ and is defined as

$$y = \cos^{-1} x \quad (0 \leqslant y \leqslant \pi).$$

Its graph is shown in figure 105.

The inverse cotangent solves $x = \cot y$ and the principal branch is given by

$$y = \cot^{-1} x \quad (0 < y < \pi).$$

Functions defined by series 323

The definition of the inverse function of $x = \sec y$ is less simple. Some of the branches of this function are shown in figure 106.

Since $|\sec y| \geq 1$ for all y, $\sec^{-1}$ cannot be defined unless $|x| \geq 1$. There is no way of choosing a principal branch such that, as x increases

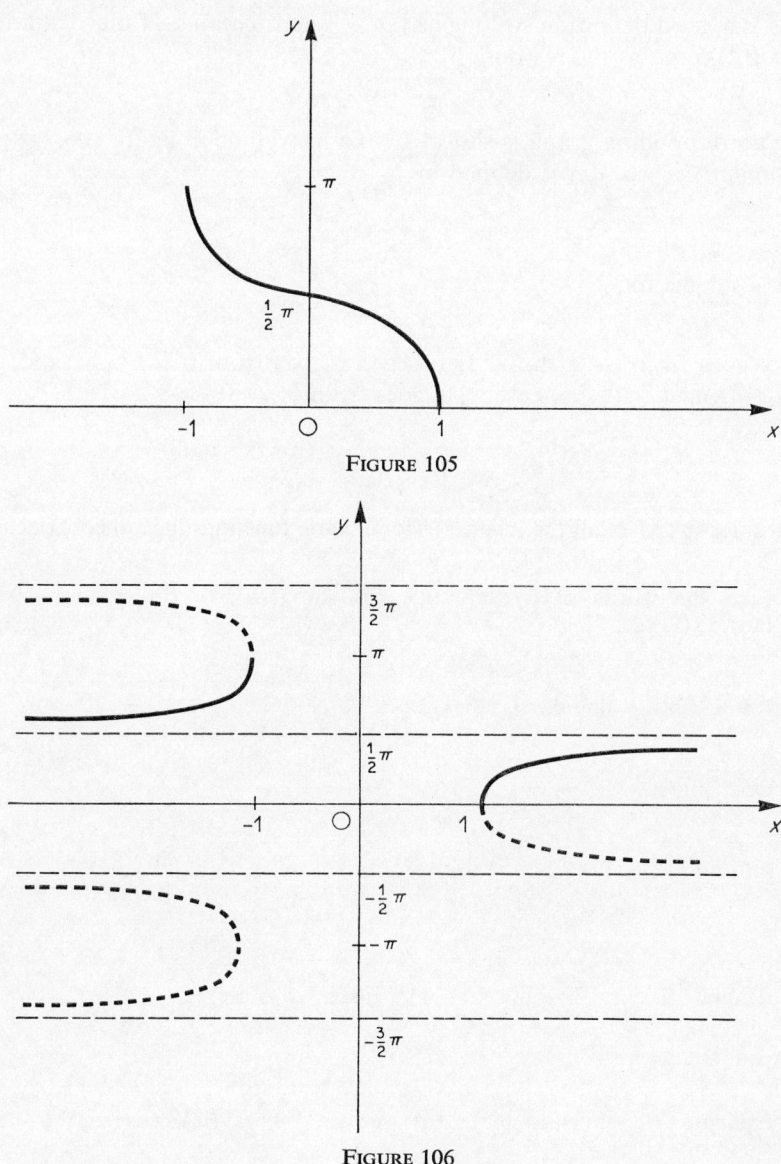

FIGURE 105

FIGURE 106

from $-\infty$ to $+\infty$, y steadily increases or steadily decreases. Therefore, since $\cos y = 1/x$, we select

$$\sec^{-1} x = \cos^{-1} \frac{1}{x}.$$

This is a good definition so long as $|x| \geq 1$ and, because of our definition of $\cos^{-1}$, imposes the restriction

$$0 \leq \sec^{-1} x \leq \pi.$$

The corresponding graph is shown by the heavy line in figure 106.

Similarly, $\operatorname{cosec}^{-1} x$ is defined by

$$\operatorname{cosec}^{-1} x = \sin^{-1} \frac{1}{x} \tag{78}$$

and is subject to

$$-\tfrac{1}{2}\pi \leq \operatorname{cosec}^{-1} x \leq \tfrac{1}{2}\pi.$$

As x goes from $-\infty$ to -1, $\operatorname{cosec}^{-1} x$ goes from 0 to $-\tfrac{1}{2}\pi$ and, as x goes from 1 to $+\infty$, $\operatorname{cosec}^{-1} x$ goes from $\tfrac{1}{2}\pi$ to 0.

Exercises

In these exercises all the inverse trigonometric functions have their principal values.

89. Find the values of (i) $\sin^{-1} 0{\cdot}5$, (ii) $\sin^{-1} (-0{\cdot}5)$, (iii) $\cos^{-1} (-0{\cdot}5)$, (iv) $\tan^{-1} 1$, (v) $\sec^{-1} 2$.

90. If $\theta = \sin^{-1} \tfrac{1}{2}$, find $\cos \theta$, $\tan \theta$, $\sec \theta$.

91. Does $\sin (\sin^{-1} x) = x$?

92. Prove that $\cos^{-1} x = \tfrac{1}{2}\pi - \sin^{-1} x$, $\sin^{-1} (-x) = -\sin^{-1} x$,

$$\cot^{-1} x = \tfrac{1}{2}\pi - \tan^{-1} x.$$

93. Evaluate (i) $\tan^{-1} 1 - \tan^{-1} (-1)$, (ii) $\sec^{-1} 2 - \sec^{-1} (-2)$, (iii) $\lim_{x \to +\infty} \tan^{-1} x$.

94. Evaluate (i) $\sin (\sin^{-1} 0{\cdot}5)$, (ii) $\sec (\sin^{-1} 0{\cdot}5)$, (iii) $\cos (\sec^{-1} (-2))$, (iv) $\sin (2 \sin^{-1} 0{\cdot}5)$.

Functions defined by series

The derivatives of inverse trigonometric functions can be found from the derivatives of the trigonometric functions themselves. Suppose that
$$y = \sin^{-1} x$$
with $-\tfrac{1}{2}\pi \leqslant y \leqslant \tfrac{1}{2}\pi$. Then
$$\sin y = x$$
and
$$\cos y \frac{dy}{dx} = 1.$$
Hence
$$\frac{dy}{dx} = \frac{1}{\cos y}.$$
Now $\cos^2 y + \sin^2 y = 1$ implies that
$$\cos y = \pm (1 - \sin^2 y)^{\frac{1}{2}} = \pm (1 - x^2)^{\frac{1}{2}}.$$
But, when $-\tfrac{1}{2}\pi \leqslant y \leqslant \tfrac{1}{2}\pi$, $\cos y$ cannot be negative and therefore
$$\cos y = (1 - x^2)^{\frac{1}{2}}.$$
Consequently,
$$\frac{dy}{dx} = \frac{1}{(1 - x^2)^{\frac{1}{2}}}.$$
It follows from Theorem 5.8 that
$$\frac{d}{dx} \sin^{-1} u = \frac{1}{(1 - u^2)^{\frac{1}{2}}} \frac{du}{dx}.$$
Similarly, we may prove that
$$\frac{d}{dx} \cos^{-1} u = - \frac{1}{(1 - u^2)^{\frac{1}{2}}} \frac{du}{dx}. \tag{79}$$
This also follows from (79) and Exercise 92.

The same method may be employed to obtain the derivative of $\operatorname{cosec}^{-1}$ or we can proceed as follows. Put $u = 1/x$ in (79). Then, from (78),
$$\frac{d}{dx} \operatorname{cosec}^{-1} x = \frac{d}{dx} \sin^{-1} \frac{1}{x} = \frac{1}{(1 - 1/x^2)^{\frac{1}{2}}} \left(-\frac{1}{x^2}\right).$$
It is now tempting to write
$$\left(1 - \frac{1}{x^2}\right)^{\frac{1}{2}} = \left(\frac{1}{x^2}\right)^{\frac{1}{2}} (x^2 - 1)^{\frac{1}{2}} = \frac{1}{x}(x^2 - 1)^{\frac{1}{2}}.$$
But we must remember that the left-hand side is positive because of the derivation of (79). Therefore the right-hand side must be positive and, with $(x^2 - 1)^{\frac{1}{2}}$ positive, this means that the square root of x^2 must be

chosen positive. Hence it is necessary to use x when $x > 0$ and $-x$ when $x < 0$. Briefly, this can be denoted by $|x|$ and so

$$\frac{d}{dx}\operatorname{cosec}^{-1} x = \frac{|x|}{(x^2-1)^{\frac{1}{2}}}\left(-\frac{1}{x^2}\right) = -\frac{1}{|x|(x^2-1)^{\frac{1}{2}}}$$

since $x^2 = |x|^2$.

The calculation of the derivatives of the remaining inverse trigonometric functions is left as an exercise for the reader.

Exercises

95. Prove that

$$\frac{d}{dx}\operatorname{cosec}^{-1} u = -\frac{1}{|u|(u^2-1)^{\frac{1}{2}}}\frac{du}{dx},$$

$$\frac{d}{dx}\tan^{-1} u = \frac{1}{1+u^2}\frac{du}{dx},$$

$$\frac{d}{dx}\sec^{-1} u = \frac{1}{|u|(u^2-1)^{\frac{1}{2}}}\frac{du}{dx},$$

$$\frac{d}{dx}\cot^{-1} u = -\frac{1}{1+u^2}\frac{du}{dx}.$$

96. Find the derivative of (i) $\sin^{-1}\frac{1}{4}x$, (ii) $\sin^{-1} x^{\frac{1}{2}}$ ($1 \geqslant x \geqslant 0$), (iii) $\cos^{-1} x^2$, (iv) $\cos^{-1}(\sin x)$, (v) $\tan^{-1} 2x/(1-x^2)$, (vi) $(\cos^{-1} x)^2$, (vii) $(1-x^2)^{\frac{1}{2}}\sin^{-1} x$, (viii) $\operatorname{cosec}^{-1}(1-x^2)^{-\frac{1}{2}}$, (ix) $\cos^{-1}[(1-x^2)/(1+x^2)]$.

97. By taking a derivative prove that

$$\tan^{-1} x + \tan^{-1}\frac{1}{x} = \tfrac{1}{2}\pi$$

when $x > 0$. What is the value when $x < 0$?

98. If $uv < 1$, prove that

$$\tan^{-1} u + \tan^{-1} v = \tan^{-1}\frac{u+v}{1-uv}.$$

99. Show that, when $x > 0$,

$$\tan^{-1} x > x - \tfrac{1}{3}x^3.$$

100. Find the local maxima and minima of $\tan^{-1} x - \tfrac{1}{2}\ln(1+x^2)$.

101. Show that $y = \tan^{-1} x$ is concave downward for $x < 0$.

102. Evaluate

(i) $\lim\limits_{x \to 0} \dfrac{\sin^{-1} 2x - 2\sin^{-1} x}{x^3}$,   (ii) $\lim\limits_{x \to 0} \left(\dfrac{\sin^{-1} x}{x}\right)^{2/x^2}$.

103. By using the mean-value theorem prove that
(i) $|\tan^{-1} u - \tan^{-1} v| \leqslant |u - v|$,   (ii) $x(1 + x^2)^{-1} < \tan^{-1} x < x$ $(x > 0)$.

104. If $y = \sin(m \sin^{-1} x)$ prove that
$$(1 - x^2)\dfrac{d^2 y}{dx^2} - x\dfrac{dy}{dx} + m^2 y = 0.$$

105. Obtain the Maclaurin expansion
$$\sin^{-1} x = x + \dfrac{1}{2} \cdot \dfrac{x^3}{3} + \dfrac{1 \cdot 3}{2 \cdot 4} \cdot \dfrac{x^5}{5} + \dfrac{1 \cdot 3 \cdot 5}{2 \cdot 4 \cdot 6} \cdot \dfrac{x^7}{7} + \ldots.$$

The series converges for $|x| < 1$.

106. Obtain the expansion
$$\tan^{-1} x = x - \dfrac{x^3}{3} + \dfrac{x^5}{5} - \dfrac{x^7}{7} + \ldots.$$

The series converges for $|x| \leqslant 1$.
For what values of x can $\tan^{-1} x$ be approximated correct to 2 decimal places by (i) x, (ii) $x - \tfrac{1}{3}x^3$? By using Exercise 98 indicate how $\tfrac{1}{4}\pi$ could be computed fairly rapidly.

107. Obtain the Maclaurin expansions
$$e^{\sin^{-1} x} = 1 + x + \dfrac{x^2}{2!} + \dfrac{(1 + 1^2)}{3!}x^3 + \dfrac{(1 + 2^2)}{4!}x^4 + \ldots,$$
$$\tan^{-1}(1 + x) = \tfrac{1}{4}\pi + \tfrac{1}{2}x - \tfrac{1}{4}x^2 + \tfrac{1}{12}x^3 - \ldots,$$
$$(\sin^{-1} x)^2 = \dfrac{2}{2!}x^2 + \dfrac{2^3}{4!}x^4 + \dfrac{2^3 \cdot 4^2}{6!}x^6 + \ldots,$$
$$x\tan^{-1} x - \tfrac{1}{2}\ln(1 + x^2) = \dfrac{x^2}{1 \cdot 2} - \dfrac{x^4}{3 \cdot 4} + \dfrac{x^6}{5 \cdot 6} - \ldots.$$

108. Prove that $\tan^{-1}(e^x) - \tan^{-1}(\tanh \tfrac{1}{2}x) = \tfrac{1}{4}\pi$.

109. Draw a flow chart for an iteration procedure to calculate $\sin^{-1} x$ when $|x| < 1$. It is of interest to compare on a graph plotter $\sin^{-1} x$, x and $x + \tfrac{1}{6}x^3$.

110. Solve $x = \tan^{-1}(\pi + x)$.

The numerical values of the inverse trigonometric functions can be determined from the tables of the trigonometric functions in the same way that antilogarithms are obtained from a table of logarithms. For example, suppose that $\sin^{-1} 0\cdot 57$ is required. If $x = \sin^{-1} 0\cdot 57$

$$\sin x = 0\cdot 57.$$

From a table of sines we find

x	$\sin x$
35° 40′	0·5688
35° 50′	0·5712

The exact value 0·57 does not appear in the right-hand column but the appropriate value of x can be found by linear interpolation (section 1.4). The two numbers on the right differ by 0·0024. The upper differs from 0·57 by 0·0012 and the ratio is 12/24 or $\frac{1}{2}$. Therefore the required value of x lies halfway between the two on the left and is 35° 45′. Actually we ought to work in radians and obtain 0·6065 as the value of $\sin^{-1} 0\cdot 57$ in radians.

10.9 Inverse hyperbolic functions

The hyperbolic functions have inverse functions just as the trigonometric functions do. Let

$$x = \sinh y. \tag{80}$$

As y increases steadily from $-\infty$ to ∞ so does x. Therefore if we choose a particular value of y and draw a line through it parallel to the x-axis it will meet the curve in one and only one point. Conversely, a line through a given value of x will meet the curve in one and only one point. Therefore we can write

$$y = \sinh^{-1} x \tag{81}$$

without any difficulty about branches because y and x are in one-to-one correspondence. Therefore equations (80) and (81) mean the same.

When we turn to the equation

$$x = \cosh y$$

we find that y is a double-valued function of x because $\cosh y = \cosh(-y)$. Also, since $\cosh$ is never less than unity, a real y cannot be found unless $x \geqslant 1$. As principal branch we select that one on which y takes positive values, i.e.

$$y = \cosh^{-1} x \quad (x \geqslant 1, y \geqslant 0). \tag{82}$$

The only other inverse hyperbolic function which is double-valued is the secant and this is defined by

$$\operatorname{sech}^{-1} x = \cosh^{-1} \frac{1}{x} \quad (0 < x \leqslant 1). \tag{83}$$

Functions defined by series

The remaining inverse hyperbolic functions present no problem and

$$\operatorname{cosech}^{-1} x = \sinh^{-1} \frac{1}{x} \qquad (x \neq 0), \tag{84}$$

$$x = \tanh y, \quad y = \tanh^{-1} x, \quad (|x| < 1), \tag{85}$$
$$x = \coth y, \quad y = \coth^{-1} x, \quad (|x| > 1). \tag{86}$$

The inverse hyperbolic functions can be expressed as natural logarithms. If $y = \sinh^{-1} x$ then

$$x = \sinh y = \tfrac{1}{2}(e^y - e^{-y}).$$

Let $e^y = z$; then $e^{-y} = 1/z$ and

$$2x = z - \frac{1}{z}$$

or

$$z^2 - 2xz - 1 = 0.$$

Hence

$$z = x \pm (x^2 + 1)^{\tfrac{1}{2}}.$$

The lower sign is not permissible because that makes z negative which contradicts the fact that e^y must be positive for real y. Hence

$$e^y = z = x + (x^2 + 1)^{\tfrac{1}{2}}$$

and so

$$y = \ln [x + (x^2 + 1)^{\tfrac{1}{2}}].$$

Therefore

$$\sinh^{-1} x = \ln [x + (x^2 + 1)^{\tfrac{1}{2}}] \tag{87}$$

for all real x.

One can show in a similar manner that

$$\cosh^{-1} x = \ln [x + (x^2 - 1)^{\tfrac{1}{2}}] \qquad (x \geqslant 1), \tag{88}$$

$$\tanh^{-1} x = \tfrac{1}{2} \ln \frac{1 + x}{1 - x} \qquad (|x| < 1), \tag{89}$$

$$\operatorname{cosech}^{-1} x = \ln \left[\frac{1}{x} + \frac{(1 + x^2)^{\tfrac{1}{2}}}{|x|} \right] \qquad (x \neq 0), \tag{90}$$

$$\operatorname{sech}^{-1} x = \ln \left[\frac{1 + (1 - x^2)^{\tfrac{1}{2}}}{x} \right] \qquad (0 < x \leqslant 1), \tag{91}$$

$$\coth^{-1} x = \tfrac{1}{2} \ln \frac{x + 1}{x - 1} \qquad (|x| > 1). \tag{92}$$

The derivatives of the inverse hyperbolic functions can be obtained from (87)–(92) by using the formula for the derivative of a logarithm.

Alternatively, one can use the following method, which is illustrated for $\cosh^{-1} x$. If $y = \cosh^{-1} x$,

$$\cosh y = x$$

and

$$\sinh y \frac{dy}{dx} = 1.$$

Now

$$\cosh^2 y - \sinh^2 y = 1$$

so that

$$\sinh y = \pm (x^2 - 1)^{\frac{1}{2}}.$$

However, $y \geqslant 0$ on the principal branch of $\cosh^{-1}$ and so $\sinh y \geqslant 0$. Hence the positive sign of the square root must be taken and

$$\frac{dy}{dx} = \frac{1}{(x^2 - 1)^{\frac{1}{2}}}.$$

One may show similarly that

$$\frac{d}{dx} \sinh^{-1} u = \frac{1}{(1 + u^2)^{\frac{1}{2}}} \frac{du}{dx}, \tag{93}$$

$$\frac{d}{dx} \tanh^{-1} u = \frac{1}{1 - u^2} \frac{du}{dx} \quad (|u| < 1), \tag{94}$$

$$\frac{d}{dx} \operatorname{cosech}^{-1} u = - \frac{1}{|u|(1 + u^2)^{\frac{1}{2}}} \frac{du}{dx} \quad u \neq 0), \tag{95}$$

$$\frac{d}{dx} \operatorname{sech}^{-1} u = - \frac{1}{u(1 - u^2)^{\frac{1}{2}}} \frac{du}{dx} \quad (0 < u < 1), \tag{96}$$

$$\frac{d}{dx} \coth^{-1} u = \frac{1}{1 - u^2} \frac{du}{dx} \quad (|u| > 1). \tag{97}$$

Exercises

111. Establish (88)–(92).

112. Establish (93)–(97).

113. If $|x| > 1$ prove that $\coth^{-1} x = \tanh^{-1} (1/x)$.

114. Find the derivative of (i) $\sinh^{-1} \frac{1}{2}x$, (ii) $\cosh^{-1} (x^2/9)$, (iii) $\tanh^{-1} (\sin x)$, (iv) $\cosh^{-1} e^x$, (v) $\coth^{-1} (\sec x)$, (vi) $\tanh^{-1} (\tan \frac{1}{2}x)$, (vii) $\frac{1}{2} \operatorname{sech}^{-1} (\sin 2x)$.

115. Prove that
$$\tanh^{-1} x = x + \frac{x^3}{3} + \frac{x^5}{5} + \ldots$$
for $|x| < 1$. Obtain the Maclaurin expansion
$$\sinh^{-1} x = x - \frac{1}{2} \cdot \frac{x^3}{3} + \frac{1 \cdot 3}{2 \cdot 4} \frac{x^5}{5} - \ldots .$$

116. Find the two smallest positive roots of
$$x = \cosh^{-1} (\sec x).$$

117. Find a root near 2 of
$$x + \tanh^{-1} (\tan x) = 0.$$

11
Elementary integration methods

11.1 The inverse of differentiation

The reader knows already how to find the derivatives of functions compounded from the 'elementary functions' of analysis: powers, trigonometric functions, exponentials and inverse functions. We shall now consider the inverse problem: to find a function whose derivative is given.

Suppose, for example, that we require a function F such that $dF/dx = x^2$ in the interval $-\infty < x < \infty$. $F(x) = \tfrac{1}{3}x^3$ is clearly such a function. Notice, however, that it is not the only one; for instance $F(x) = \tfrac{1}{3}x^3 + 6$ also satisfies the conditions.

A general form of the question would read: *find a function F such that*

$$\frac{dF(x)}{dx} = f(x) \tag{1}$$

where f is a given function.

(1) is an *equation* for the unknown *function F*, and it is called a *differential equation*. If f is a function of a sufficiently simple type an answer can sometimes be obtained with the help of a table of derivatives and some guesswork. Suppose, for example, that $f(x) = x\,e^x$, so that the equation becomes

$$\frac{dF(x)}{dx} = x\,e^x. \tag{2}$$

We know that

$$\frac{d}{dx}(x\,e^x) = x\,e^x + e^x, \tag{3}$$

so the 'trial function' $F(x) = x\,e^x$ gives the required $x\,e^x$ plus a term e^x on the right-hand side. This unwanted term can be eliminated by noticing that $e^x = (d/dx)(e^x)$, and we can rearrange equation (3) to give:

$$x\,e^x = \frac{d}{dx}(x\,e^x) - e^x = \frac{d}{dx}(x\,e^x) - \frac{d}{dx}(e^x)$$

$$= \frac{d}{dx}(x\,e^x - e^x).$$

Elementary integration methods 333

Therefore, $F(x) = x\,\mathrm{e}^x - \mathrm{e}^x$ satisfies equation (2). Notice that if we add to this function *any constant* the resulting function still satisfies (2).

The student should work through the following exercises, trial and error methods being entirely appropriate. Each answer should be differentiated to confirm its correctness.

Exercises

Find F when f has the form shown in the following questions

1. (i) x; (ii) $3x$; (iii) x^2; (iv) $\tfrac{1}{2}x^2$; (v) $x^{-\tfrac{1}{2}}$; (vi) x^n, n integer $\neq -1$; (vii) 2; (viii) 0; (ix) ax^n, n integer $\neq -1$; (x) x^α, $\alpha \neq -1$, $x > 0$; (xi) $1/x$, $x > 0$; (xii) $(-x)^\alpha$, $\alpha \neq -1$, $x < 0$: (xiii) $(x^{\tfrac{3}{2}} + x^2)/\sqrt{x}$; (xiv) $x^3 + x^2 + x + 1 + (1/x) + (1/x^2) + (1/x^3) + (1/x^4)$; (xv) $1/x$, $x < 0$.

2. (i) $\cos x$; (ii) $\sin x$; (iii) $1/\cos^2 x$; (iv) $1/\sin^2 x$; (v) $3 \sin x$; (vi) $\sin 2x$; (vii) $\cos 3x$.

3. (i) e^x; (ii) e^{-x}; (iii) e^{2x}; (iv) e^{-3x}; (v) $A\,\mathrm{e}^{-ax}$.

4. (i) $\sec^2 ax$; (ii) $\operatorname{cosec}^2 ax$.

5. (i) $1/\sqrt{(1-x^2)}$, $-1 < x < 1$; (ii) $1/\sqrt{(x^2-1)}$, $|x| > 1$.

6. (i) $x\,\mathrm{e}^{-x}$; (ii) $x^2\,\mathrm{e}^x$; (iii) $x \cos x$; (iv) $x \sin x$; (v) $2x\,\mathrm{e}^{x^2}$.

7. (i) $(2x+1)^2$; (ii) $(x+1)^4$; (iii) $(2x+3)^4$; (iv) $\sin(2x+3)$; (v) $\cos(1-x)$; (vi) $\cos(1-2x)$; (vii) $1/(x+1)$. $x > -1$; (viii) $2/(2x+3)$, $x > -\tfrac{3}{2}$.

8. (i) $\sin x \cos x$; (ii) $\sin^2 x$; (iii) $\cos^2 x$; (iv) $\cos^2 3x$; (v) $x \sin x - \sin x$.

11.2 The integrals of a function

We shall for the present restrict consideration to cases where f is a continuous function. We define the *integral* of f as follows.

Let f be a continuous function on $a < x < b$; then any function F, differentiable and satisfying the equation

$$\frac{\mathrm{d}F(x)}{\mathrm{d}x} = f(x)$$

in $a < x < b$ is said to be an integral* of the function f in the interval.†

* The terms *anti-derivative* or *indefinite integral* are also frequently used.
† We shall also consider cases where the interval is a closed one, i.e. $a \leqslant x \leqslant b$.

We have already noted in particular cases that *the solution to our problem is not unique. If $F^*(x)$ is one solution, then so is $F^*(x) + C$, where C is any constant.* This is so because

$$\frac{d}{dx}[F^*(x) + C] = \frac{dF^*(x)}{dx} = f(x),$$

as required.

The reasonableness of this fact can be illustrated geometrically. We can display our first solution $y = F^*(x)$ on a graph. The slope of the graph is equal to $f(x)$ at every point (provided that the vertical and horizontal

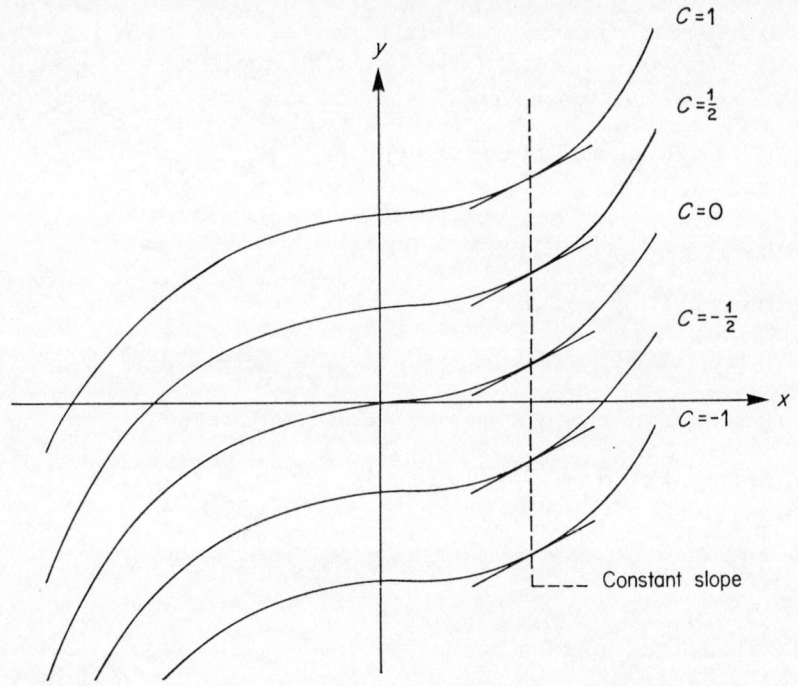

Parallel curves $y = \frac{1}{3} x^3 + C$

FIGURE 107

scales are the same). By shifting (or *displacing*) the curve up and down by various distances we produce a *family* of similar curves, each displaced curve being a member of the family. Clearly, at any value of x the slope of every curve is the same, so that each one satisfies the original condition $dy/dx = f(x)$. But the equations of these curves are simply $y = F^*(x) + C$,

Elementary integration methods 335

where C takes a different constant value on each curve. In figure 107 the situation is illustrated with $f(x) = x^2$ and $F^*(x) = \tfrac{1}{3}x^3$.

The question now arises as to whether the family of functions $F^*(x) + C$, where F^* is any one particular function satisfying $\mathrm{d}F^*(x)/\mathrm{d}x = f(x)$ and C is an *arbitrary constant* (that is to say, any value may be assigned to it) includes *all* the integrals of the function f; or whether there are perhaps further types of solution so far undetected.

There are in fact no more. Suppose that $F(x) = F_1(x)$, $F(x) = F_2(x)$ are any two solutions of the equation $\mathrm{d}F(x)/\mathrm{d}x = f(x)$ in $a < x < b$. We shall show that these can only differ by a *constant* (though we cannot say what its value is). Put
$$G(x) = F_1(x) - F_2(x).$$
Then
$$\frac{\mathrm{d}G(x)}{\mathrm{d}x} = \frac{\mathrm{d}}{\mathrm{d}x}[F_1(x) - F_2(x)] = \frac{\mathrm{d}F_1(x)}{\mathrm{d}x} - \frac{\mathrm{d}F_2(x)}{\mathrm{d}x}$$
$$= f(x) - f(x) = 0, \qquad a < x < b.$$

The only differentiable function whose derivative is zero throughout an interval is a constant, as has been proved in Corollary 7.2. Therefore $G(x) = C$, where C is a constant, and so
$$F_1(x) - F_2(x) = C,$$
which is what had to be proved.

The practical conclusion of this section is that if, by some means or other, we can find one integral of a function, we have in effect found them all. If $F^*(x)$ is one integral, the whole family of integrals is given by $F^*(x) + C$, where C is an arbitrary constant. When the complete family of integrals is required we shall ask for *the integral* of the given function.

Example 1 Find the integral of the function $\sin x \sin 3x$.

To make this recognizable as the derivative of something it must be manipulated into a suitable form. We can write
$$2 \sin x \sin 3x = \cos 2x - \cos 4x$$
using the usual trigonometrical formulae. This form suggests derivatives of sines. In fact $(\mathrm{d}/\mathrm{d}x)(\sin 2x) = 2 \cos 2x$ and $(\mathrm{d}/\mathrm{d}x)(\sin 4x) = 4 \cos 4x$. Therefore, $\cos 2x = (\mathrm{d}/\mathrm{d}x)(\tfrac{1}{2} \sin 2x)$ and $\cos 4x = (\mathrm{d}/\mathrm{d}x)(\tfrac{1}{4} \sin 4x)$. Finally we have
$$\sin x \sin 3x = \frac{\mathrm{d}}{\mathrm{d}x}(\tfrac{1}{4} \sin 2x - \tfrac{1}{8} \sin 4x)$$
so that the integral of $\sin x \sin 3x$ is $\tfrac{1}{4} \sin 2x - \tfrac{1}{8} \sin 4x + C$, where C is an arbitrary constant.

The student should make a habit of writing in the arbitrary constant as in this example; its presence is absolutely necessary in most applications. The condition that a function should be the integral of a given function is not alone sufficient to establish a definite value for the constant—any numerical value which is given to the constant produces a function which is eligible as an integral. In applications, however, there are usually further conditions which serve to fix the value that the constant must take. The following example is a simple illustration.

Example 2 A point moves in a straight line with constant *acceleration a. It begins its motion with velocity u. Find its velocity $v(t)$ at all later times (t represents time elapsed).*

The *definition* of acceleration as the rate of change of velocity:

$$\frac{dv(t)}{dt} = a$$

gives the basic equation connecting the unknown function $v(t)$ with the given function a (constant). Here the variable t takes the place of the variable x of the preceding theory and we see that the function v must be some integral of a. All integrals of a are included in the family $at + C$, where C is an arbitrary constant (verify by differentiation, noticing that this is not true unless a is *constant*). Thus there must be *some* value of C for which the equation

$$v(t) = at + C$$

is true.

To fix C we use the rest of the given information: that $v = u$ when the motion starts at $t = 0$ (this information is called the *initial condition*). Then

$$u = a \cdot 0 + C.$$

Thus, $C = u$ and therefore

$$v(t) = at + u,$$

a well-known formula in mechanics.

Exercises

9. Find which integrals of the given functions f satisfy the following conditions:

   (a) $f(x) = 2x^2$; the integral to take the value 1 at $x = 0$.
   (b) $f(x) = (1/x^2) + (1/x^3)$; the integral to take the value 1 at $x = 1$.
   (c) $f(x) = (1/x) + (1/x^2)$; the integral to take the value -1 at $x = -1$.
   (d) $f(x) = (1/x^2)\,e^{(1/x)}$; the integral to tend to zero as $x \to \infty$.

(Sketch the functions and state the intervals on which the results hold.)

10. Find and sketch the integral of the function $1/(x-1)^2$ which
    (a) takes the value 1 at $x = 0$,
    (b) tends to 1 as $x \to \infty$,
    (c) tends to 1 as $x \to -\infty$.

11. Show that the following results are not in conflict with the result of section 11.2.
    (a) $\frac{1}{6}(2x+3)^3$ and $\frac{4}{3}x^3 + 6x^2 + 9x$ are integrals of $(2x+3)^2$.
    (b) $-\frac{1}{2}\cos 2x$, $\sin^2 x$ and $-\cos^2 x$ are all integrals of $\sin 2x$.
    (c) $\ln[\sec^2(\frac{1}{2}x + \frac{1}{4}\pi)]$ and $-2\ln|\cos \frac{1}{2}x - \sin \frac{1}{2}x|$ are both integrals of $\tan(\frac{1}{2}x + \frac{1}{4}\pi)$.
    (d) $\sin^{-1} x$ and $(-\cos^{-1} x)$ are integrals of $(1-x^2)^{-\frac{1}{2}}$.

12. Show that if a body moves in a straight line with *constant* acceleration f then the distance $s(t)$ that it travels in time t is given by $s(t) = \frac{1}{2}ft^2 + at + b$ where a and b are constants.

    A stone falls *from rest* at time $t = 0$. (a) Assuming its acceleration under gravity is constant at 32 ft/sec² find how far it falls in 3 sec. (b) Find its speed in terms of the *distance* it has fallen.

    $$\text{velocity } v(t) = \mathrm{d}s(t)/\mathrm{d}t, \quad \text{acceleration} = \mathrm{d}v(t)/\mathrm{d}t$$

13. Show that if y is a function satisfying $\mathrm{d}^3y/\mathrm{d}x^3 = 0$ in $-\infty < x < \infty$, it must have the form $y = ax^2 + bx + c$, where a,b,c are constants. (Note that

    $$\frac{\mathrm{d}^3y}{\mathrm{d}x^3} = \frac{\mathrm{d}}{\mathrm{d}x}\left(\frac{\mathrm{d}^2y}{\mathrm{d}x^2}\right),$$

    etc.)

14. (To show that functions which are not 'smooth' may have integrals.) Sketch the function $f(x) = |x|$, and show that the function F given by

    $$F(x) \begin{aligned} &= \tfrac{1}{2}x^2, & x \geq 0 \\ &= -\tfrac{1}{2}x^2, & x < 0 \end{aligned}$$

    is an integral of f in $-\infty < x < \infty$, according to the definition on page 333.

15. (a) Show that the function f defined by

    $$f(x) \begin{aligned} &= 0, & -\infty < x \leq 0, \\ &= 1, & x > 0 \end{aligned}$$

    has no integral valid over the whole interval $-\infty < x < \infty$.
    (b) Construct a function $F^*(x)$ which is an integral on $-\infty < x < 0$ and $x > 0$ separately, and is not differentiable at $x = 0$ but is *continuous* there.
    (c) Show that all functions of the form $F^*(x) + C$, where C is any constant, satisfy the conditions in (b), and conversely that every function satisfying the conditions must be of the form $F^*(x) + C$.

(In other words, we can define a family of functions $F^*(x) + C$ which depart from the definition of integral given on page 333 only in not being differentiable at the point where f is discontinuous. We shall later extend the definition of integral to include cases such as this.)

16. Show that if F is an integral of f in $x_1 < x < x_2$ then

(a) $-F(-x)$ is an integral of $f(-x)$ in $-x_2 < x < -x_1$;
(b) $(1/a)F(ax + b)$ is an integral of $f(ax + b)$, where a and b are constants, $a \neq 0$. In what interval is this true?
(c) $\frac{1}{2}F(x^2)$ is an integral of $xf(x^2)$ in some interval, if $x_2 > 0$. What is the interval?
(d) Construct the integral of $x^2 f(x^3)$, and attempt to generalize further.

17. Use the results of question 16 to evaluate the integrals of the following functions: (a) $(-x)^{\frac{1}{2}}$, $x < 0$; (b) x^{-1}, $x < 0$; (c) $(ax+b)^n$, $a \neq 0$, n an integer, $n \neq -1$; (d) $(ax+b)^{\frac{1}{2}}$, $a \neq 0$, $x > -b/a$; (e) $\sqrt{(1-x)}$, $x < 1$; (f) $x \sin x^2$, $-\infty < x < \infty$; (g) $x(1+x^2)^{-1}$, $-\infty < x < \infty$; (h) $x(1-x^2)^{-\frac{1}{2}}$, $-1 < x < 1$.

11.3 The symbol for integral

Suppose that a function F is any integral of a function f. Let it be written as a function of some variable x. Then we shall denote $F(x)$ by a new symbol:

$$F(x) \equiv \int f(x)\,\mathrm{d}x. \qquad (4)$$

Likewise we shall write $F(t) \equiv \int f(t)\,\mathrm{d}t$ and so on. Equation (4) is to be read: *'an integral of f with respect to the variable x'*, or 'integral eff of ex dee ex.'

The operational part of the new symbolism is

$$\int \ldots \mathrm{d} \ldots$$

into which the name of a function (the *integrand*) and the name of a variable, the *variable of integration*, remain to be inserted. Thus

$$\int x^2\,\mathrm{d}x = \tfrac{1}{3}x^3 + C, \qquad -\infty < x < \infty$$

$$\int \sin \omega t\,\mathrm{d}t = -\frac{1}{\omega}\cos \omega t + D, \qquad -\infty < x < \infty$$

and so on, where C and D are constants.

It is always assumed that, in the course of any particular problem, $\int f(x)\,\mathrm{d}x$ retains the same meaning throughout, as with other mathematical symbols. Thus although $\sin^{-1} x$ and $-\cos^{-1} x$ are both integrals of $(1-x^2)^{-\frac{1}{2}}$, we must not allow $\int (1-x^2)^{-\frac{1}{2}}\,\mathrm{d}x$ to mean $\sin^{-1} x$ in one part of an argument and $-\cos^{-1} x$ in another, because these two functions are not the same.

The great practical convenience of this notation will appear as we go along.*

The following results are really only reformulations of results we already know, in the new notation.

I. *If* $\mathrm{d}F(x)/\mathrm{d}x = f(x)$ *then* $\int f(x)\,\mathrm{d}x = F(x) + C$, *where C is a constant.*

II. $(\mathrm{d}/\mathrm{d}x)\int f(x)\,\mathrm{d}x = f(x)$.

(I and II merely restate the definition of an integral.)

III. *If, in an interval,*
$$\int f(x)\,\mathrm{d}x = \int g(x)\,\mathrm{d}x,$$
then $f(x) = g(x)$ in the interval.

This follows from II, for $(\mathrm{d}/\mathrm{d}x)\int f(x)\,\mathrm{d}x = (\mathrm{d}/\mathrm{d}x)\int g(x)\,\mathrm{d}x$ in the interval, so that $f(x) = g(x)$.

IV. $\int [\mathrm{d}F(x)/\mathrm{d}x]\,\mathrm{d}x = F(x) + C$, *where C is a constant.*

This is to be read: *given F, any integral of $\mathrm{d}F(x)/\mathrm{d}x$ differs from $F(x)$ only by a constant.*

For $\int [\mathrm{d}F(x)/\mathrm{d}x]\,\mathrm{d}x$ is some integral of the function $\mathrm{d}F(x)/\mathrm{d}x$, and $F(x)$ is another (by the definition of integral). Therefore they differ by at most a constant.

V. $\int [af(x) + bg(x)]\,\mathrm{d}x = a\int f(x)\,\mathrm{d}x + b\int g(x)\,\mathrm{d}x + C$, *where C is a constant, and a and b are any given constants.*

This is easily verified by differentiation. A similar result is true for the sum of any finite number of functions, not just for two.

---

* A student who has studied differentials can for the present interpret the symbol $\int f(x)\,\mathrm{d}x$ as meaning 'some function $F(x)$ of which $f(x)\,\mathrm{d}x$ is the differential'. Thus
$$x^2\,\mathrm{d}x = \mathrm{d}(\tfrac{1}{3}x^3 + C), \qquad -\infty < x < \infty$$
and
$$(1-x^2)^{-\frac{1}{2}}\,\mathrm{d}x = \mathrm{d}(\sin^{-1} x + D), \qquad |x| < 1,$$
for any value of the constants C and D.

These results only have any meaning when the various functions, f, g, dF/dx appearing under the integration signs actually have integrals. We have not proved, and in fact it is not true, that every function has an integral. For the present in our examples we confine ourselves to functions of which we can actually write down the integrals so that we know they exist.

11.4 A table of basic integrals

We need to form a table of integrals of the simpler functions which occur in analysis which can be used as a basis for handling more complicated functions. Such a table is a table of derivatives in reverse, and most of the entries can simply be read off from a table of derivatives.

There will still be some gaps, for example, $\int \ln x \, dx$ is not immediately obtainable from a table of derivatives. We shall deal with this function and some others which may cause difficulty first.

I $\int \ln x \, dx, \quad x > 0.$

We make a guess, testing how near we can get by using the 'trial function' $x \ln x$. We have

$$\frac{d}{dx}(x \ln x) = \ln x + 1,$$

which can be rearranged to give

$$\ln x = \frac{d}{dx}(x \ln x) - 1 = \frac{d}{dx}(x \ln x - x).$$

Therefore, for $x > 0$, $\int \ln x \, dx = x \ln x - x + C$, where C is a constant. We have only to consider $x > 0$, since $\ln x$ is real only for $x > 0$.

II $\int x^{-1} \, dx, \quad x > 0 \ \text{and} \ x < 0.$

We know that $(d/dx) \ln x = x^{-1}$. Also $\ln x$ exists for $x > 0$, so that $\ln x + C$ is the integral of x^{-1} for $x > 0$. However, x^{-1} *is defined* (i.e. it has a meaning) also for $x < 0$, and the integral is required for this case too. For $x < 0$, $\ln(-x)$ is defined, and

$$\frac{d}{dx} \ln(-x) = \frac{1}{(-x)} \cdot (-1) = \frac{1}{x}.$$

Thus $\ln(-x)$ is an integral of x^{-1} for x negative. The results are therefore:

$$\int x^{-1} \, dx = \ln x + A, \qquad A \text{ arbitrary, for } x > 0$$

$$\int x^{-1} \, dx = \ln(-x) + B, \qquad B \text{ arbitrary, for } x < 0.$$

Elementary integration methods

The two results are often written as one:

$$\int x^{-1} \, dx = \ln |x| + C,$$

where C is arbitrary. This contains a convenient reminder that there will be occasions when x is negative, but the reader should interpret this form as implying that there are two distinct and independent families of integrals, holding on the intervals $x < 0$ and $x > 0$, respectively. There is no presumption in any

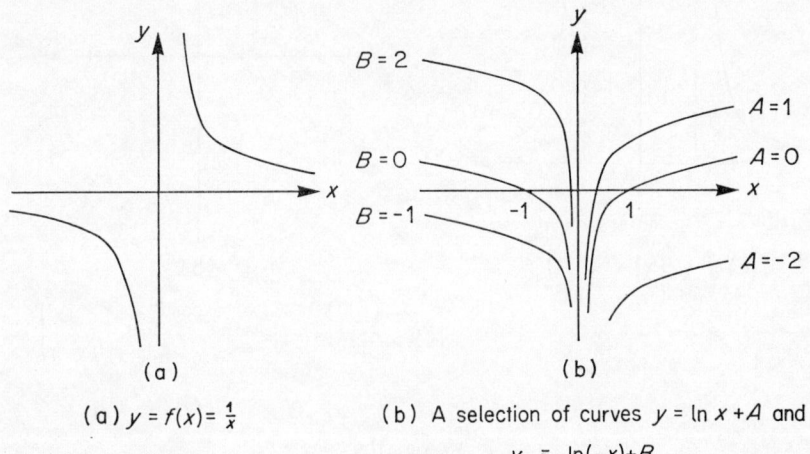

(a) $y = f(x) = \frac{1}{x}$

(b) A selection of curves $y = \ln x + A$ and
$y = \ln(-x) + B$

FIGURE 108

practical context that the arbitrary constant C must take the same values on each. A graph of the function x^{-1} and some of its integrals is given in figure 108.

Note that equivalent forms for the integrals are $\ln |Ax|$, where A is an arbitrary constant, and $\ln B|x|$ where B is an arbitrary *positive* constant.

III $\qquad \displaystyle\int \frac{dx}{\sqrt{(x^2 - 1)}}, \qquad x > 1 \text{ and } x < 1.$

A similar difficulty occurs here. We know that

$$\frac{d}{dx}(\cosh^{-1} x) = \frac{1}{\sqrt{(x^2 - 1)}}$$

where the positive sign is understood for the square root. $\cosh^{-1} x$ is therefore an integral, but although the right-hand side is real for $|x| > 1$, $\cosh^{-1} x$ is real only for $x > 1$. Therefore the integral has been found only for $x > 1$.

To find the integral in $x < -1$, put $y = -x$, so that $y > 1$; then

$$\frac{1}{\sqrt{(x^2 - 1)}} = \frac{1}{\sqrt{(y^2 - 1)}} = \frac{d}{dy}(\cosh^{-1} y) = -\frac{d}{dx}[\cosh^{-1}(-x)].$$

Thus for $x < -1$ the integral is $-\cosh^{-1}(-x) + C$. (The reader should confirm this by differentiation.)

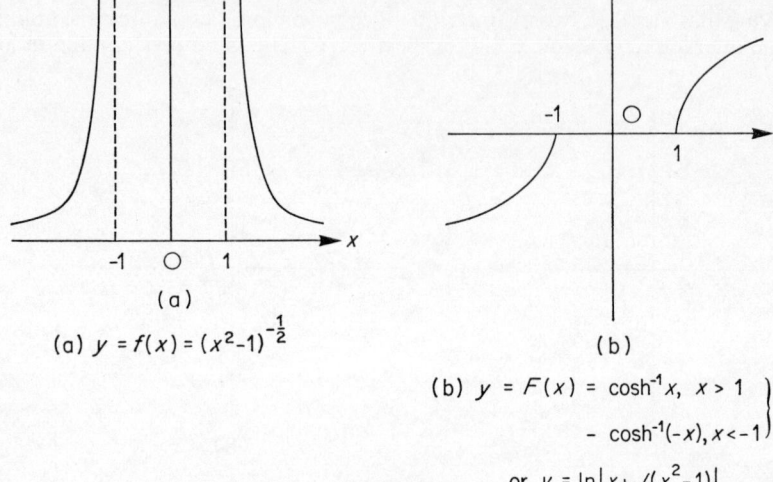

(a) $y = f(x) = (x^2-1)^{-\frac{1}{2}}$

(b) $y = F(x) = \cosh^{-1}x, \ x > 1$
$\quad\quad\quad\quad - \cosh^{-1}(-x), x < -1$

or $y = \ln|x + \sqrt{(x^2-1)}|$

FIGURE 109

IV $\displaystyle\int \frac{dx}{\sqrt{(x^2-1)}},$ (*an alternative result*).

For $x > 1$, $\cosh^{-1} x = \ln[x + \sqrt{(x^2-1)}] = F(x)$, say. For $x < -1$ the integral obtained in III is $-F(-x)$. We have

$$-F(-x) = -\ln[-x + \sqrt{(x^2-1)}]$$
$$= \ln\{1/[-x + \sqrt{(x^2-1)}]\}$$
$$= \ln[-x - \sqrt{(x^2-1)}] = \ln\{-[x + \sqrt{(x^2-1)}]\}.$$

Therefore

$$\int \frac{dx}{\sqrt{(x^2-1)}} = \ln|x + \sqrt{(x^2-1)}| + C, \text{ for } x > 1 \text{ and } x < -1.$$

V $\displaystyle\int \sqrt{(1-x^2)}\,dx, \ -1 < x < 1.$

We use $x\sqrt{(1-x^2)}$ as a trial function:

$$\frac{d}{dx}x\sqrt{(1-x^2)} = \sqrt{(1-x^2)} + \frac{(-x^2)}{\sqrt{(1-x^2)}}.$$

But
$$\frac{-x^2}{\sqrt{(1-x^2)}} = \frac{(1-x^2)-1}{\sqrt{(1-x^2)}} = \sqrt{(1-x^2)} - \frac{1}{\sqrt{(1-x^2)}}.$$

Therefore,
$$\frac{d}{dx}x\sqrt{(1-x^2)} = 2\sqrt{(1-x^2)} - \frac{1}{\sqrt{(1-x^2)}} = 2\sqrt{(1-x^2)} - \frac{d}{dx}(\sin^{-1}x).$$

Thus
$$\sqrt{(1-x^2)} = \frac{d}{dx}\tfrac{1}{2}[x\sqrt{(1-x^2)} + \sin^{-1}x]$$

and
$$\int \sqrt{(1-x^2)}\,dx = \tfrac{1}{2}[x\sqrt{(1-x^2)} + \sin^{-1}x] + C.$$

VI $\int \sqrt{(1+x^2)}\,dx$, $-\infty < x < \infty$.

Proceeding as in V, the integral is
$$\tfrac{1}{2}[x\sqrt{(1+x^2)} + \sinh^{-1}x] + C.$$

VII $\int \sqrt{(x^2-1)}\,dx$, $x > 1$ or $x < -1$.

The integration is performed as in V to give the result in the table below.

VIII $\int \dfrac{dx}{x^2-1}$.

We may write
$$\frac{1}{x^2-1} = \frac{1}{2}\left(\frac{1}{x-1} - \frac{1}{x+1}\right). \tag{5}$$

It can be confirmed by differentiation that one integral of $1/(x-1)$ is $\ln|x-1|$ and that one integral of $1/(x+1)$ is $\ln|x+1|$ (compare I above). Therefore
$$\int \frac{dx}{x^2-1} = \tfrac{1}{2}(\ln|x-1| - \ln|x+1|) + C$$
$$= \tfrac{1}{2}\ln\left|\frac{x-1}{x+1}\right| + C$$

(See figure 110), where C is a constant.

The methods used to evaluate some of the above integrals may seem to be rather artificial, and in Volume 2 we describe some general techniques which apply to the present integrals as special cases. However, the use of *ad hoc* methods enables us to have available a substantial list of basic integrals at an early stage. Such a table is given on page 475. It will be a considerable help to the reader to commit most of the entries to memory.

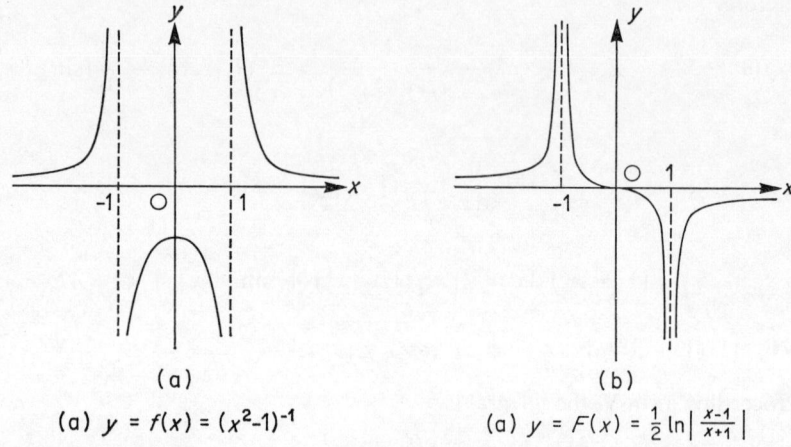

(a) $y = f(x) = (x^2-1)^{-1}$

(a) $y = F(x) = \frac{1}{2} \ln \left| \frac{x-1}{x+1} \right|$

FIGURE 110

11.5 Functions of $ax + b$

If $\int f(x)\,dx$ is known, $\int f(ax + b)\,dx$ can be written down.

Suppose that $f(x)$ has the integral $F(x)$ in $x_1 < x < x_2$. Then

$$\frac{dF(y)}{dy} = f(y), \qquad x_1 < y < x_2.$$

(It does not matter what we call the variable—we do not have to call it x.) Now put $y = ax + b$ into this statement and use the rule for changing the variable in a derivative. We obtain, for $a \neq 0$,

$$\frac{1}{a}\frac{d}{dx}F(ax + b) = f(ax + b), \qquad x_1 < ax + b < x_2.$$

Therefore, if $F(x) = \int f(x)\,dx$,

$$\int f(ax + b)\,dx = \frac{1}{a}F(ax + b) + C, \qquad x_1 < ax + b < x_2, \quad (6)$$

which is the required result. The new interval is equivalent to $(x_1 - b/a) < x < (x_2 - b)/a$ if $a > 0$ and to $(x_2 - b)/a < x < (x_1 - b)/a$ if $a < 0$.

Be prepared to recognize the special cases

$$\int f(x+b)\,dx = F(x+b) + C, \qquad \int f(ax)\,dx = \frac{1}{a}F(ax) + C,$$

$$\int f(-x)\,dx = -F(-x) + C, \qquad \int f(b-x)\,dx = -F(b-x) + C.$$

Examination of the new *interval* ensures that we do not write nonsense—we cannot say that because $\int x^{\frac{1}{2}}\,dx = \frac{2}{3}x^{\frac{3}{2}}$, $x > 0$, therefore, $\int (-x)^{\frac{1}{2}}\,dx = -\frac{2}{3}(-x)^{\frac{3}{2}}$, $x > 0$. This second relation is true, however, in the new interval $x < 0$.

Example 3 Find (a) $\int \sin(2x+1)\,dx$, (b) $\int \ln(1-x)\,dx$, (c) $\int dx/(1-x)$.

(a) * $\int \sin y\,dy = -\cos y$, so $\int \sin(2x+1)\,dx = -\frac{1}{2}\cos(2x+1)$, $-\infty < x < \infty$. (Here $a = 2$, $b = 1$.)

(b) $\int \ln y\,dy = y\ln y - y$, so

$$\int \ln(1-x)\,dx = -[(1-x)\ln(1-x) - (1-x)], \quad x < 1.$$

(Here $a = -1$, $b = 1$.)

(c) $\int dy/y = \ln|y|$, $\quad y > 0$ and $y < 0$, so

$$\int dx/(1-x) = \ln|1-x|, \quad x < 1 \text{ and } x > 1.$$

(Here $a = -1$, $b = 1$.)

* Notice the following mistaken procedure. To evaluate $\int \sin(2x+1)\,dx$:

$$\int \sin y\,dy = -\cos y;$$

therefore, putting $y = 2x + 1$,

$$\int \sin(2x+1)\,dx = -\cos(2x+1) + C$$

which is the *wrong* answer, the correct answer being as in (a). The reader will be helped not to go wrong if he notices that, for example in (a), under the integral sign, it seems reasonable to put

$$dy = d(2x+1) = 2\,dx,$$

and not dx. The student who has studied differentials formally and read the note in section 11.3 will see that this is perfectly correct. Others will regard it as part of the commonsense of the integral notation that it is possible to do this kind of thing.

Example 4

$$\int \frac{dx}{\sqrt{(4x^2 - 9)}}.$$

We can write

$$\int \frac{dx}{\sqrt{(4x^2 - 9)}} = \frac{1}{3} \int \frac{dx}{\sqrt{[(\tfrac{2}{3}x)^2 - 1]}}.$$

Now an integral of $1/\sqrt{(y^2 - 1)}$ is given by

$$\int \frac{dy}{\sqrt{(y^2 - 1)}} = \ln|y + \sqrt{(y^2 - 1)}|, \quad \text{in } y > 1 \text{ and } y < -1.$$

Therefore, since in this example $a = \tfrac{2}{3}$, $b = 0$, we have

$$\frac{1}{3} \int \frac{dx}{\sqrt{[(\tfrac{2}{3}x)^2 - 1]}} = \tfrac{1}{3} \cdot \tfrac{3}{2} \ln |\tfrac{2}{3}x + \sqrt{[(\tfrac{2}{3}x)^2 - 1]}|$$

$$\text{in } \tfrac{2}{3}x > 1 \text{ and } \tfrac{2}{3}x < -1$$

or, ignoring constant terms,

$$\int \frac{dx}{\sqrt{(4x^2 - 9)}} = \tfrac{1}{2} \ln |2x + \sqrt{(4x^2 - 9)}| \quad \text{in } |x| > \tfrac{3}{2}.$$

Example 5

$$\int \frac{dx}{(a^2x^2 - c^2)^{\frac{1}{2}}}.$$

Example 2 suggests that we put

$$\frac{1}{\sqrt{(a^2x^2 - c^2)}} = \frac{1}{c\sqrt{[(ax/c)^2 - 1]}}. \tag{7}$$

which leads in the same way to

$$\int \frac{dx}{\sqrt{(a^2x^2 - c^2)}} = \frac{1}{c}\frac{c}{a} \ln \left|\frac{ax}{c} + \sqrt{\left[\left(\frac{ax}{c}\right)^2 - 1\right]}\right|$$

$$= \frac{1}{a} \ln |ax + \sqrt{(a^2x^2 - c^2)}| - \frac{1}{a} \ln c, \tag{8}$$

this holding in the intervals $ax/c < -1$ and $ax/c > 1$, that is, in $|ax/c| > 1$. The second term in equation (8) is constant and so may be absorbed into the arbitrary constant, which we have not written in.

Equation (7) is correct only when c is positive (since '$\sqrt{}$' implies the positive square root), and the correctness of (8) might seem to be dependent on this.

Elementary integration methods

However, the derivative of (8) is evidently equal to the original function in the given intervals regardless of the sign of c, as is to be expected from the fact that the given function involved only the square of c. Therefore

$$\int \frac{dx}{\sqrt{(a^2x^2 - c^2)}} = \frac{1}{a} \ln |ax + \sqrt{(a^2x^2 - c^2)}|, \qquad |x| > \left|\frac{c}{a}\right|.$$

Notice that despite appearances, this result is independent also of the sign of a, for

$$\frac{1}{-a} \ln |-ax + \sqrt{(a^2x^2 - c^2)}| = \frac{1}{a} \ln |ax + \sqrt{(a^2x^2 - c^2)}|$$

(see IV, section 11.4).

Example 6

$$\int \frac{dx}{4x^2 + 4x + 5}.$$

Take the terms $4x^2 + 4x$ and 'complete the square', writing

$$4x^2 + 4x = (2x + 1)^2 - 1.$$

Then

$$\frac{1}{4x^2 + 4x + 5} = \frac{1}{(2x + 1)^2 + 4} = \frac{1}{4} \frac{1}{(x + \tfrac{1}{2})^2 + 1}.$$

Now

$$\int \frac{dy}{y^2 + 1} = \tan^{-1} y + \text{a constant,}$$

and therefore

$$\frac{1}{4} \int \frac{dx}{(x + \tfrac{1}{2})^2 + 1} = \tfrac{1}{4} \tan^{-1} (x + \tfrac{1}{2}) + \text{a constant.}$$

Example 7

$$\int \frac{dx}{x^2 - 2x - 3}.$$

Complete the square as in the last example: we obtain

$$\frac{1}{x^2 - 2x - 3} = \frac{1}{(x - 1)^2 - 4} = \frac{1}{4} \cdot \frac{1}{(\tfrac{1}{2}x - \tfrac{1}{2})^2 - 1}.$$

From the table of integrals we find that

$$\int \frac{dy}{y^2 - 1} = \tfrac{1}{2} \ln \left| \frac{y-1}{y+1} \right|; \qquad y < -1, \; -1 < y < 1, \; y > 1,$$

and, therefore,

$$\int \frac{dx}{(x-1)^2 - 4} = 2 \cdot \tfrac{1}{4} \cdot \tfrac{1}{2} \ln \left| \frac{\tfrac{1}{2}(x-1) - 1}{\tfrac{1}{2}(x-1) + 1} \right| = \tfrac{1}{4} \ln \left| \frac{x-3}{x+1} \right|;$$

$$x < -1, \; -1 < x < 3, \; x > 3.$$

Example 8

$$\int \frac{dx}{\sqrt{(3 + 2x - x^2)}}.$$

By writing $x^2 - 2x = (x-1)^2 - 1$, we obtain

$$\frac{1}{\sqrt{(3 + 2x - x^2)}} = \frac{1}{\sqrt{[4 - (x-1)^2]}} = \frac{1}{2\sqrt{\{1 - [\tfrac{1}{2}(x-1)]^2\}}}.$$

Now

$$\int \frac{dy}{\sqrt{(1 - y^2)}} = \sin^{-1} y, \qquad |y| < 1,$$

and therefore

$$\int \frac{dx}{2\sqrt{[1 - (\tfrac{1}{2}x - \tfrac{1}{2})^2]}} = \frac{1}{2} \frac{1}{(1/2)} \sin^{-1}(\tfrac{1}{2}x - \tfrac{1}{2})$$

$$= \sin^{-1}[\tfrac{1}{2}(x-1)],$$

in $|\tfrac{1}{2}(x-1)| < 1$; that is in $-1 < x < 3$.

Integrals of the form $\int \sqrt{(px^2 + qx + r)} \, dx$ are to be treated similarly.

Exercises

Find $\int f(x) \, dx$ where f is given by the following functions (Exercises 18 to 47):

18. $x^3, x^2, x, 1, 0, 1/x, 1/x^2, 1/x^3$.

19. $x^{\frac{1}{2}}(1 + x), (x+1)/x^{\frac{1}{3}}, 2x^2 + 3x + 1, \tfrac{1}{2}x^{-\frac{1}{2}} + 1 + \tfrac{3}{2}x^{\frac{1}{2}}, (x+1)^2/\sqrt{x}$.

20. $(2x^2 + 1)^2, \sqrt{x}\,[2\sqrt{x} - 1]^2, \sum_{r=1}^{n} ax^{r-1}$.

Elementary integration methods 349

21. e^x, e^{-x}, $\cosh x = \frac{1}{2}(e^x + e^{-x})$, $\sinh x = \frac{1}{2}(e^x - e^{-x})$.

22. $\sec^2 x$, $\cosec^2 x$, $\tan^2 x$, $\cot^2 x$ (express $\tan^2 x$ and $\cot^2 x$ in terms of $\sec^2 x$ and $\cosec^2 x$).

23. $\sin 2x$, $\cos 2x$, $\sin^2 x$, $\cos^2 x$ (note $\cos 2x = 2\cos^2 x - 1 = 1 - 2\sin^2 x$).

24. e^{ax}, e^{-ax}, a^x $(a > 0)$.

25. $\sin ax$, $\cos ax$, $\sinh ax$, $\cosh ax$.

26. $\sec^2 3x$, $\cosec^2 5x$, $\sin^2 2x$, $\cos^2 3x$.

27. $\sin^4 x$, $\cos^4 x$ (express in terms of $\cos 2x$).

28. $\sin x \cos x$, $\sin 2x \cos 2x$, $\sin x \cos^3 x$, $\cos x \sin^3 x$.

29. $\ln(2x)$, $\ln(2x^{\frac{1}{2}})$.

30. $\sinh^2 x$, $\cosh^2 x$, $\sinh 2x$, $\cosh 2x$.

31. $(3x + 4)^{10}$, $(1 - x)^5$, $(2 - 3x)^4$, $1/(4x + 3)^2$.

32. $\sqrt{(x-1)}, \sqrt{(1-2x)}, (1-x)^{\frac{1}{3}}, (1-3x)^{-\frac{1}{3}}, 1/\sqrt{(2-x)}$, (state the intervals).

33. $1/x$, $1/(x+1)$, $1/(2x+1)$, $1/(1-x)$, $1/(3-2x)$, (state the intervals).

34. $1/(a - x)^n$, $1/(c - dx)^n$, (n positive integer).

35. $\sin mx \cos nx$, $\sin mx \sin nx$, $\cos mx \cos nx$ when $n \neq m$ (express as the sum of two trigonometric functions).

36. $(x + 1)/(x - 1)$ (divide numerator by denominator), $(x^2 + 3x + 1)/(x + 1)$, $(x^n - 1)/(x - 1)$ (n positive integer).

37. $(x + 1)/\sqrt{(x-1)}$ $(x + 1 = x - 1 + 1)$, $x/(2x + 1)^n$ (n positive integer).

38. $(2x + 1)/(3x - 1)^2$, $(x^2 + 2x - 1)/(x - 1)^3$.

39. $\ln(1 + 2x)$, $\ln(1 - x)$, $\ln(3 - 4x)$, (state the intervals).

40. $\ln(x^2 - 4)$, $\ln[(x + 1)/(x - 1)]$, $\ln[(1 - x)/(1 + x)]$, (state the intervals).

41. $1/[(x + 1)^2 + 1]$, $1/[(2x + 1)^2 + 1]$, $1/[(2x - 1)^2 + 9]$.

42. $1/(x^2 + 4x + 5)$, $1/(4x^2 + 12x + 13)$ (write in the form of question 41).

43. $1/\sqrt{(x^2 - 2x)}$ $[x^2 - 2x = (x - 1)^2 - 1]$, $1/\sqrt{(2x - x^2)}$.

44. $1/(x^2 - 1)$, $1/[(2x + 1)^2 - 1]$.

45. $1/(x^2 - 2x)$, $1/(4x^2 - 4x - 3)$.

46. $1/(x^2 + 3x)$, $1/(x^2 - 6x + 8)$.

47. $1/\sqrt{(4x^2 - 4x - 3)}$, $1/\sqrt{(5 - 4x - x^2)}$, $1/\sqrt{(-x^2 - 4x - 3)}$, (state the intervals).

48. Write down $\int dx/\sqrt{(x^2 - c^2)}$, $\int dx/\sqrt{(c^2 - x^2)}$, $\int dx/(x^2 + c^2)$, $\int \sqrt{(c^2 - x^2)}\, dx$ (state intervals).

49. Find $\int dx/(a^2x^2 - c^2)$.

50. Find $\int dx/[\sqrt{x} + \sqrt{(1 + x)}]$ (rationalize the denominator).

51. Find $\int dx/[\sqrt{(a - x)} - \sqrt{(b - x)}]$.

52. Prove that $\int dx/(ax^2 + bx + c)$ can be expressed (a) as an inverse tangent when the equation $ax^2 + bx + c = 0$ has no real roots, (b) as a logarithm when the roots are real and different. (c) Consider the case when the roots are coincident.

53. Find the integral $\int x\,dx/(x-1)$ which (a) has the value 0 when $x = 2$, (b) has the value 0 when $x = -1$. Sketch the integral functions.

54. By differentiating the given series with respect to x show that
$$x + \frac{n}{2!}x^2 + \frac{n(n-1)}{3!}x^3 + \ldots + \frac{n!}{(n+1)!}x^{n+1} = \frac{(1+x)^{n+1} - 1}{n+1}.$$

11.6 Factorable integrals

Let $g(u)$ be a given function of a variable u, and suppose that we can find a function $G(u)$ such that
$$\frac{dG(u)}{du} = g(u). \tag{9}$$

Then G is some integral of g:
$$G(u) = \int g(u)\,du. \tag{10}$$

Now let $t(x)$ be a function of x. According to the chain rule:
$$\frac{dG[t(x)]}{dx} = \frac{dG}{dt}\frac{dt}{dx}$$
$$= g(t)\frac{dt}{dx} \tag{11}$$

by (9) above. (11) implies that $G[t(x)]$ *is an integral of the function* $g[t(x)]dt(x)/dx$ *with respect to* x; *and G is obtainable from* (10), *by integrating g*.

Consider an application of this result.

Example 9
$$\int x \sin x^2\,dx$$

We observe that the integral can be written in the form
$$x \sin x^2 = [\tfrac{1}{2}\sin(x^2)]\frac{d(x^2)}{dx}$$
$$= \tfrac{1}{2}\sin t\frac{dt}{dx}$$

say, if we write $t(x) = x^2$. This is now in the form (11) with $g(t) = \tfrac{1}{2}\sin t$. Its integral is $G[t(x)]$, where
$$G(u) = \int g(u)\,du = \int \tfrac{1}{2}\sin u\,du = -\tfrac{1}{2}\cos u + C.$$
Therefore
$$G[t(x)] = G(x^2) = -\tfrac{1}{2}\cos x^2 + C.$$

The result contained in (9), (10) and (11) can be expressed in the following, more memorable, way:

Suppose $f(x)$ can be written in the form

$$f(x) = g(t)\frac{dt}{dx}$$

where t is some function of x. Then

$$\int f(x)\, dx = \int g(t)\frac{dt}{dx}\, dx = \int g(t)\, dt + C.$$

This result is easy to remember because the apparent cancellation of the symbol dx on the left leads to the expression on the right.*†

Another way to express the rule is to say that *if an integral of $g(x)$ is known to be $G(x)$, then an integral of $g[t(x)]\, dt/dx$ is $G[t(x)]$.*

The use of the rule leads to the evaluation of integrals which, unlike Example 9, are not at all easy.

Example 10

$$\int \frac{2x+3}{x^2+3x+3}\, dx.$$

It is necessary to notice that

$$f(x) = \frac{2x+3}{x^2+3x+3} = \frac{1}{x^2+3x+3}\frac{d}{dx}(x^2+3x+3)$$

$$= \frac{1}{t}\frac{dt}{dx}$$

where $t = x^2 + 3x + 3$. This is of the form $g(t)(dt/dx)$, where $g(t) = 1/t$. Therefore

$$\int \frac{2x+3}{x^2+3x+3}\, dx = \int \frac{1}{t}\, dt + C = \ln|t| + C = \ln(x^2+3x+3) + C.$$

---

* We achieve this neat form at the cost of some ambiguity: the letter t has a mixture of meanings. It appears on the right as a variable of integration, then *after* integration it is understood to be a function of x. On the left it must be understood as a function of x *before* integration.

† The student familiar with differentials will see that if t is a function,

$$g(t)\frac{dt}{dx}\, dx = g(t)\, dt$$

so that the anti-differential forms $\int g(t)(dt/dx)\, dx$ and $\int g(t)\, dt$ differ only by a constant (see this interpretation of '$\int$' in section 11.3).

Example 11 Find
$$\int \frac{x \, dx}{\sqrt{(1-x^2)}}.$$

$$f(x) = \frac{x}{\sqrt{(1-x^2)}} = -\frac{\frac{1}{2}}{\sqrt{(1-x^2)}} \frac{d}{dx}(1-x^2).$$

Putting
$$t = 1 - x^2$$
we obtain
$$f(x) = \frac{-\frac{1}{2}}{t^{\frac{1}{2}}} \frac{dt}{dx}, \quad (g(t) = -\tfrac{1}{2}/t^{\frac{1}{2}} \text{ in the theory}).$$

Then
$$\int f(x) \, dx = \int \frac{-\frac{1}{2}}{t^{\frac{1}{2}}} \frac{dt}{dx} dx = \int \frac{-\frac{1}{2}}{t^{\frac{1}{2}}} \, dt + \text{constant}$$
$$= -\tfrac{1}{2}(2t^{\frac{1}{2}}) + \text{constant}$$
$$= -\sqrt{(1-x^2)} + \text{constant}.$$

Example 12 Find
$$\int \cos^3 x \, dx.$$

$$\cos^3 x = \cos^2 x \cdot \cos x = (1 - \sin^2 x) \frac{d(\sin x)}{dx}.$$

Putting $\sin x = t$, we have $g(t) = 1 - t^2$ and therefore
$$\int \cos^3 x \, dx = \int (1 - t^2) \frac{dt}{dx} dx = \int (1 - t^2) \, dt + \text{constant}$$
$$= t - \tfrac{1}{3}t^3 + \text{constant}.$$

To use this method to its best advantage it is necessary to be able to *see* that f may be 'factorized' into the form $g(t)(dt/dx)$ for some choice of function t. The student should go carefully through the following table which consists of various factorizations worked through. It will gradually become possible for him to see quickly when a simple factorization is available. When the factorization has been observed, it is still necessary to be able to integrate g if the method is to be of any use.

$f(x)$	t	$\dfrac{dt}{dx}$	Factorization	$g(t)$	$G(t)$ $[G'(t) = g(t)]$	$F(x)$ $[F'(x) = f(x)]$
$x\sqrt{(x^2+1)}$	x^2	$2x$	$[\tfrac{1}{2}\sqrt{(x^2+1)}]\dfrac{d(x^2)}{dx}$	$\tfrac{1}{2}\sqrt{(t+1)}$	$\tfrac{1}{3}(t+1)^{\frac{3}{2}}$	$\tfrac{1}{3}(x^2+1)^{\frac{3}{2}}$
or $x\sqrt{(x^2+1)}$	x^2+1	$2x$	$[\tfrac{1}{2}\sqrt{(x^2+1)}]\dfrac{d(x^2+1)}{dx}$	$\tfrac{1}{2}\sqrt{t}$	$\tfrac{1}{3}t^{\frac{3}{2}}$	$\tfrac{1}{3}(x^2+1)^{\frac{3}{2}}$
$\dfrac{2x+1}{(x^2+x-1)^{\frac{3}{2}}}$	x^2+x-1	$2x+1$	$\dfrac{1}{(x^2+x-1)^{\frac{3}{2}}}\dfrac{d(x^2+x-1)}{dx}$	$\dfrac{1}{t^{\frac{3}{2}}}$	$-\dfrac{2}{t^{\frac{1}{2}}}$	$-\dfrac{2}{(x^2+x-1)^{\frac{1}{2}}}$
$\sin x \cos^3 x$	$\cos x$	$-\sin x$	$(-\cos^3 x)\dfrac{d(\cos x)}{dx}$	$-t^3$	$-\tfrac{1}{4}t^4$	$-\tfrac{1}{4}\cos^4 x$
$\tan x = \dfrac{\sin x}{\cos x}$	$\cos x$	$-\sin x$	$\left(-\dfrac{1}{\cos x}\right)\dfrac{d(\cos x)}{dx}$	$-\dfrac{1}{t}$	$-\ln|t|$	$-\ln|\cos x|$ $= \ln|\sec x|$
$\tanh x = \dfrac{e^x - e^{-x}}{e^x + e^{-x}}$	$e^x + e^{-x}$	$e^x - e^{-x}$	$\dfrac{1}{e^x + e^{-x}}\dfrac{d(e^x + e^{-x})}{dx}$	$\dfrac{1}{t}$	$\ln|t|$	$\ln(e^x + e^{-x})$
$x\,e^{x^2}$	x^2	$2x$	$(\tfrac{1}{2}e^{x^2})\dfrac{d(x^2)}{dx}$	$\tfrac{1}{2}e^t$	$\tfrac{1}{2}e^t$	$\tfrac{1}{2}e^{x^2}$
$\dfrac{\ln x}{x}$	$\ln x$	$\dfrac{1}{x}$	$(\ln x)\dfrac{d(\ln x)}{dx}$	t	$\tfrac{1}{2}t^2$	$\tfrac{1}{2}\ln^2 x$

Elementary integration methods

The factorization $f(x) = g(t)(dt/dx)$ is *not unique*. We may choose t to be some quite arbitrary and unsuitable function of x and still construct the factorization. For example, suppose $f(x) = x \sin(x^2)$. We may decide to put, instead of $t = x^2$, $t = \sin(x^2)$. Then $(dt/dx) = 2x \cos x$ and therefore

$$f(x) = \frac{x \sin(x^2)}{2x \cos x} \frac{dt}{dx} = \frac{1}{2} \frac{t}{\sqrt{(1-t^2)}} \frac{dt}{dx}.$$

In this case $g(t) = \frac{1}{2}[t/\sqrt{(1-t^2)}]$ and we are no better off than before.

Exercises

Obtain the integrals of the following functions.

55. $x(x^2 + 1)^{\frac{3}{2}}$, $x(1 - x^2)^{-\frac{1}{2}}$, $x \sec^2(x^2)$, $(1/x^2) e^{1/x}$, $x/(x^2 + 1)$.

56. $x \ln x \;(= \frac{1}{2} x \ln x^2)$, $x/(x^2 - a^2)^n$, $x^2(2 + 3x^3)^{\frac{1}{2}}$.

57. $x^{\frac{3}{2}}/(1 + x^{\frac{5}{2}})^2$, $1/[\sqrt{x}\sqrt{(1 - \sqrt{x})}]$.

58. $(2ax + b)/(ax^2 + bx + c)^\alpha \;(\alpha \neq -1)$, $(2ax + b)/(ax^2 + bx + c)$.

59. $(x + 1)/(x^2 + 1)$, $(x^2 + 1)/(x^2 + 2x + 3)$ (express as the sum of suitable terms).

60. $(8x + 4)/\sqrt{(4x^2 + 4x + 5)}$, $(8x + 5)/\sqrt{(4x^2 + 4x + 5)}$ (express as the sum of two suitable terms).

61. $(18x + 6)/\sqrt{(9x^2 + 6x + 5)}$, $(x + 1)/\sqrt{(9x^2 + 6x + 5)}$ (relate $x + 1$ to $18x + 6$, and so express as the sum of two suitable terms).

62. $x^3(1 + x^2)^{\frac{1}{2}}$, $x^2/\sqrt{(1 - x^6)}$ $[x^6 = (x^3)^2]$, $x/(x^4 + 3)$.

63. $1/[\sqrt{x}(1 - x)]$, $x^3/(1 + x^2)^3$ {consider $(d/dx)[1/(1 + x^2)]$}.

64. $x/(1 - x^4)$, $(1/x)[1/(1 - x^4)]$.

65. $x^n \ln x \;(n \neq -1)$ (see question 56), $(\ln x)^n/x \;(n \neq -1)$, $(x \ln x)^{-1}$.

1—13

66. $f'(x)f''(x)$.

67. $\sin x \cos x$, $\sin x \cos^{\frac{1}{2}} x$, $\sin x/(a + b\cos x)$, $\cos x/(a + b\sin x)$.

68. $\tan x$, $\cot x$.

69. $\tan x \sec^2 x$, $\cot x \csc^2 x$.

70. $\tan^3 x$, $\tan^4 x$, $(\sec^2 x = 1 + \tan^2 x)$.

71. $\sin^2 x/\cos^6 x$, $1/(2\cos^2 x + 3\sin^2 x)$, $1/[\sin^2 x (1 + 2\cos^2 x)]$, $(\sec^2 x = 1 + \tan^2 x, \csc^2 x = 1 + \cot^2 x)$.

72. $\cos x/(\cos^2 x + 3\sin^2 x)$, $\cos x/(2\cos^2 x + \sin^2 x)$.

73. $(3\cos x + \sin x)/(\cos x + \sin x)$, $(2\cos x + \sin x)/(\cos x - \sin x)$.
[Hint: $3\cos x + \sin x = (-\sin x + \cos x) + 2(\cos x + \sin x)$].

74. $1/\sinh x$, $1/\cosh x$.

11.7 The integral of rational functions with a quadratic denominator and related types

Consider the integration of functions of the form
$$\frac{Ax + B}{px^2 + qx + r}. \tag{12}$$
The integrals of such functions can always be written down. We notice that
$$\frac{d}{dx}(px^2 + qx + r) = 2px + q,$$
and that $Ax + B$ can always be written in the form
$$Ax + B = \lambda(2px + q) + \mu$$
where
$$\lambda = A/2p, \quad \mu = B - Aq/2p.$$
Therefore
$$\frac{Ax + B}{px^2 + qx + r} \equiv \lambda\frac{2px + q}{px^2 + qx + r} + \mu\frac{1}{px^2 + qx + r}. \tag{13}$$

Elementary integration methods

The first term on the right is of the type considered in section 11.6 whose integral is $\lambda \ln |px^2 + qx + r|$. The second term on the right can always be integrated by the methods of section 11.5 by 'completing the square' in $px^2 + qx$, and so writing

$$\mu \frac{1}{px^2 + qx + r} = \mu \frac{1}{(ax + b)^2 \pm c^2}. \tag{14}$$

where a, b and c are constants. The integral can be evaluated by referring to the table on page 475: the integral of (14) is

$$\frac{\mu}{ac} \tan^{-1} \left(\frac{a}{c} x + \frac{b}{c} \right)$$

when there is a '+' sign, and

$$\frac{\mu}{ac} \ln \left| \left(\frac{a}{c} x + \frac{b}{c} - 1 \right) \middle/ \left(\frac{a}{c} x + \frac{b}{c} + 1 \right) \right|$$

when there is a '−' sign.* If $px^2 + qx + r$ is already a perfect square, $c = 0$ and the integral is $- \mu/a(ax + b)$.

Integrals of the form

$$\int (Ax + B)\sqrt{(px^2 + qx + r)}\, dx \quad \text{and} \quad \int \frac{(Ax + B)\, dx}{\sqrt{(px^2 + qx + r)}}$$

are to be treated similarly.

Example 13

$$\int \frac{(4x + 1)\, dx}{4x^2 + 4x + 3}.$$

We can write

$$\frac{4x + 1}{4x^2 + 4x + 3} = \frac{1}{2} \frac{8x + 4}{4x^2 + 4x + 3} - \frac{1}{4x^2 + 4x + 3}$$

$$= \frac{1}{2} \frac{8x + 4}{4x^2 + 4x + 3} - \frac{1}{(2x + 1)^2 + 2}.$$

Therefore

$$\int \frac{(4x + 1)\, dx}{4x^2 + 4x + 3} = \tfrac{1}{2} \ln (4x^2 + 4x + 3) - \frac{1}{2\sqrt{2}} \tan^{-1} \left(\frac{2x + 1}{\sqrt{2}} \right),$$

omitting the arbitrary constant.

The modulus sign has not been written under the logarithm since $4x^2 + 4x + 3$ is always greater than zero.

* These results are independent of the sign of c.

Example 14

$$\int \frac{x \, dx}{3x^2 - x - 2}.$$

$$\frac{x}{3x^2 - x - 2} = \frac{1}{6} \frac{6x - 1}{3x^2 - x - 2} + \frac{1}{6} \frac{1}{\left(\sqrt{3}\, x - \frac{1}{2\sqrt{3}}\right)^2 - \frac{25}{12}},$$

and therefore

$$\int \frac{x \, dx}{3x^2 - x - 2} = \tfrac{1}{6} \ln |3x^2 - x - 2| + \tfrac{1}{30} \ln \left|\frac{\tfrac{6}{5}x - \tfrac{1}{5} - 1}{\tfrac{6}{5}x - \tfrac{1}{5} + 1}\right|$$

$$= \tfrac{1}{6} \ln |3x^2 - x - 2| + \tfrac{1}{30} \ln \left|\frac{x - 1}{3x + 2}\right| + \tfrac{1}{30} \ln 3$$

$$= \tfrac{1}{6} \ln |(x - 1)(3x + 2)| + \tfrac{1}{30} \ln \left|\frac{x - 1}{3x + 2}\right| + \tfrac{1}{30} \ln 3$$

$$= \tfrac{1}{6} \ln |x - 1| + \tfrac{2}{15} \ln |3x + 2| + \tfrac{1}{30} \ln 3.$$

The term $\tfrac{1}{30} \ln 3$, being constant, may be omitted if it is merely required to produce a particular integral of the given function. We remind the student that an arbitrary constant is normally required to be written in and it has been omitted here merely for brevity.

In this last example the roots of the equation are both real. Then the sign before c^2 in equation (14) is negative. But also, when the roots are real, $px^2 + qx + r$ factorizes, and in such a case it is far easier to treat the integral by the method of 'partial fractions' as described in the next section.

Example 15

$$\int \frac{(x + 1) \, dx}{\sqrt{(2 + x - x^2)}}.$$

Proceed as in the previous examples:

$$\frac{x + 1}{\sqrt{(2 + x - x^2)}} = -\frac{1}{2} \frac{-2x + 1}{\sqrt{(2 + x - x^2)}} + \frac{3}{2} \frac{1}{\sqrt{(2 + x - x^2)}}.$$

Therefore

$$\int \frac{(x + 1) \, dx}{\sqrt{(2 + x - x^2)}} = -\sqrt{(2 + x - x^2)} + \tfrac{3}{2} \sin^{-1}\left(\frac{2x - 1}{3}\right),$$

the arbitrary constant being omitted.

Example 16

$$\int \frac{3x^3 + 2x^2 + x + 1}{x^2 + 2x + 1}\,dx.$$

By division of the denominator into the numerator this example can be reduced to the type considered:

$$\frac{3x^3 + 2x^2 + x + 1}{x^2 + 2x + 1} = 3x - 4 + \frac{6x + 5}{x^2 + 2x + 1}$$

$$= 3x - 4 + 3\frac{2x + 2}{x^2 + 2x - 1} - \frac{1}{(x + 1)^2}.$$

Therefore

$$\int \frac{3x^3 + 2x^2 + x + 1}{x^2 + 2x + 1}\,dx = \tfrac{3}{2}x^2 - 4x + 3\ln(x^2 + 2x + 1) + \frac{1}{x + 1},$$

arbitrary constant omitted.

Exercises

Integrate the following functions:

75. $x/(x^2 + 2x + 2)$.

76. $(2x - 1)/(4x^2 + 4x + 3)$.

77. $(2x + 3)/(4x^2 + 4x + 1)$.

78. $x/\sqrt{(2x^2 + 2x + 1)}$.

79. $1/\sqrt{(x^2 + x)}$.

80. $(x - 1)/\sqrt{(1 - x^2)}$.

81. $(2x + 1)/\sqrt{(1 + 2x - x^2)}$.

82. $(x + 1)\sqrt{(2x - x^2)}$.

83. $x\sqrt{(x^2 - 2x - 1)}$.

11.8 Integration of more general rational functions by resolution into partial fractions

In Chapter 3 is described a method for expressing a rational function as the sum of a number of simpler terms called 'partial fractions', in the case when the denominator factorizes into a number of linear and quadratic terms. In this event, the function will appear as the sum of terms of the form

$$\frac{A}{(ax+b)^n} \quad \text{and} \quad \frac{Cx+D}{(cx^2+dx+e)^m}$$

where A, a, b, C, D, c, d, e are constants and n and m are positive integers. We are already in a position to work out the integrals of any terms of this form, and therefore the integral of any rational function whose denominator factorizes suitably can be obtained.

Example 17   Find

$$\int \frac{(x+2)\,dx}{(x+1)(x^2+2x+2)}$$

By the methods of Chapter 3, we can find that

$$\frac{x+2}{(x+1)(x^2+2x+2)} = \frac{1}{(x+1)} - \frac{x}{(x^2+2x+2)}.$$

Therefore the integral is equal to

$$\ln|x+1| - \tfrac{1}{2}\ln(x^2+2x+2) + \tan^{-1}(x+1).$$

(See question 75.)

Exercises

Integrate the following functions:

84. $1/[x(x-1)]$, $1/[(x+1)(2x+1)]$, $1/[(2-3x)(x-2)]$, $1/[x(x^2-1)]$.

85. $1/[x(x-1)(x-2)\ldots(x-n)]$, n a positive integer.

86. $1/[(x^2-a^2)(x^2-b^2)]$. Is your answer independent of the sign of a and b?

87. $(x-1)/[x(x+1)]$, $[(x-1)(x-2)]/[(x-3)(x-4)]$, $(x^3+x^2+x+1)/(2x-1)^2$.

Elementary integration methods 361

88. $1/[x^2(x-1)]$, $1/[x(x-1)^2]$, $x/[(x^2-1)(x-1)^2]$, $x^2/(x-1)^3$.

89. $(x^2+1)/(x^2-1)^3$, $(x^4+1)/[x(x+1)^2]$.

90. $1/[x(x^2-2x+2)$, $(x+1)/[(x-1)(x^2+x+1)]$, $1/(x^3+1)$.

91. $1/[x(x+1)(2x^2-x-1)]$, $1/[(x^2-1)(x^2+1)]$ (form partial fractions in terms of x^2 instead of x), $1/[(x^2+1)(x^2+4)]$.

12

The definite integral and some applications

12.1 The application of integration to area problems

Many problems in geometry and in other applications of mathematics can be formulated in such a way as to produce an equation of the form

$$\frac{dF(x)}{dx} = f(x) \tag{1}$$

where f is a given function and F is related to the quantity required. In other words, if F cannot be found directly it may be possible to state its rate of change. The problem is thus reduced to finding an integral of f. Example 1 of Chapter 11 is such a case, the equation $ds(t)/dt = v(t)$ being of the form (1). There it was simply a *definition* written in mathematical terms, but in other cases the corresponding equation has to be constructed.

We shall first consider the problem of finding the area of a region bounded by the x-axis, two ordinates $x = a$ and $x = b$, and the graph of a continuous function f, $y = f(x)$, $a \leqslant x \leqslant b$, as shown in figure 111. This quantity is called *the area under the portion of the curve*, or *the area between it and the x-axis*. Frequently, quantities of importance in physical and other applications can be represented by the area of such a region, where the axes refer to suitable physical variables.

The area of figures bounded by curved lines is not a subject treated in elementary geometry, and it is unlikely that the student has heard of any formal definition of area as applied to such figures. We shall simply *assume* that the regions with which we shall be concerned have areas: *positive* numbers which serve as a measure of the space they occupy. These 'areas' are required to be *additive*, meaning in effect that the area of

The definite integral and some applications

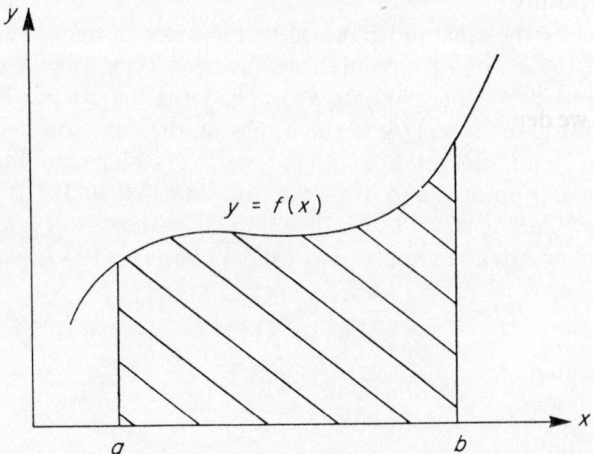

FIGURE 111

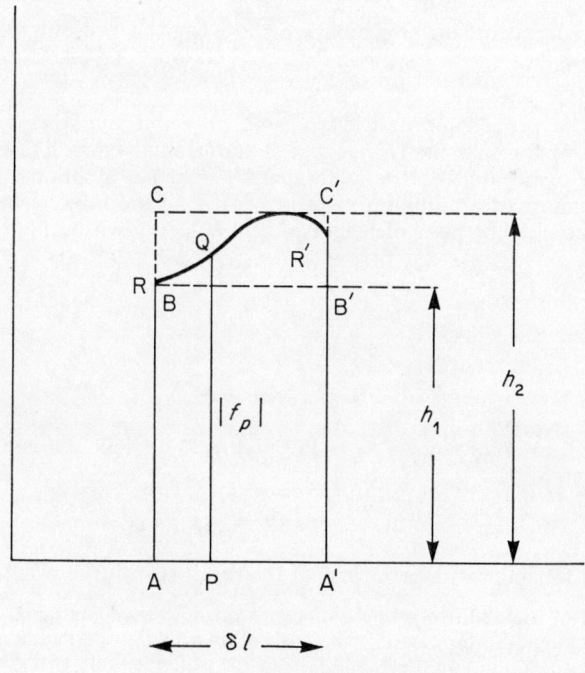

FIGURE 112

the whole is equal to the sum of the areas of any parts into which it may be divided.*

Now consider the area under the curve indicated in figure 112, which is a portion of the graph of a continuous function $f(x)$. Take a point P on the x-axis and draw the ordinate PQ. Then the length PQ is given by $PQ = |f_p|$ (positive) where f_p is the value of the function at the point. Now let AA′ be an interval of length δl (positive) which contains P (P may be one of its end-points), and draw the ordinates AR and A′R′. The area of AA′R′R we denote by δA (a positive number). Then, if δl is quite small, this region is nearly a rectangle, and we may write

$$\delta A \approx AA' \cdot PQ = \delta l |f_p|$$

or

$$\frac{\delta A}{\delta l} \approx |f_p|$$

where $\approx$ means 'is nearly equal to'.

As $\delta l \to 0$ in any way while continuing to contain the point P, we expect that in the limit

$$\lim_{\delta l \to 0} \frac{\delta A}{\delta l} = |f_p|; \tag{2}$$

which is the equation upon which we shall for the present base our discussion of area.

This may be proved more strictly, using only our hypotheses about area discussed above. Suppose firstly that $f_p \neq 0$. If δl is small enough f_p will be of one sign, positive or negative, in the interval AA′, since f is a continuous function. Let the maximum and minimum values of $|f(x)|$ in the interval be h_2 and h_1. Construct rectangles on AA′ of height h_2 and h_1 as shown on figure 112. Then

$$\text{area AA'B'B} \leq \delta A \leq \text{area AA'C'C}$$

or

$$h_1 \delta l \leq \delta A \leq h_2 \delta l$$

or

$$h_1 \leq \frac{\delta A}{\delta l} \leq h_2.$$

Now if $\delta l \to 0$ in any manner, $h_1 \to PQ$ and $h_2 \to PQ$, since f is a continuous function. Therefore

$$\lim_{\delta l \to 0} \frac{\delta A}{\delta l} = PQ = |f_p|$$

as predicted. It can easily be shown that the result is still true when $f_p = 0$.

* The truth of the additive property depends on how we define area. If we *defined* the area of rectangles to be length^2 × breadth^2, this would give a measure of the space occupied, but we should find that when a rectangle is divided into two other rectangles their areas would in general not add up to that of the original rectangle.

The definite integral and some applications

Equation (2) can be turned into a *differential equation* of the form (1) by introducing the coordinate x. Let the point P have coordinate x, and define the interval AA' to be $(x, x + \delta x)$, as in figure 113, where δx may be positive or negative. The length, δl, of the interval is positive. Therefore

$$\delta l = (x + \delta x) - x = \delta x \quad (\delta x \text{ positive}),$$
$$= -\delta x \quad (\delta x \text{ negative}).$$

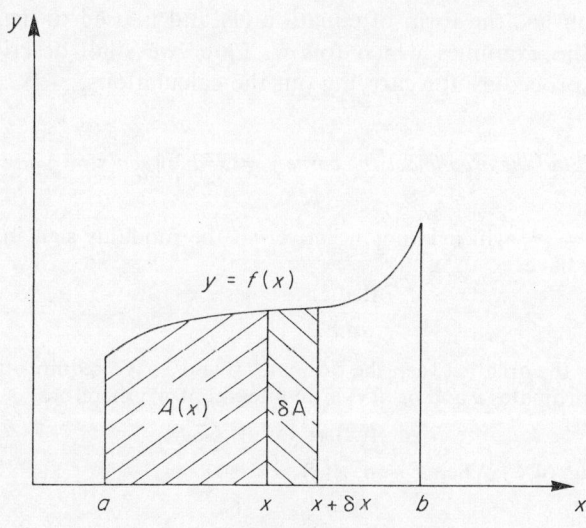

FIGURE 113

Now introduce a function $A(x)$ by:

$$A(x) = \text{area between } a \text{ and } x$$

(shown in figure 113). Then, since δA is positive,

$$\delta A = A(x + \delta x) - A(x) \quad (\delta x \text{ positive}),$$
$$= A(x) - A(x + \delta x) \quad (\delta x \text{ negative}),$$

from the additive property of areas. In either case

$$\frac{\delta A}{\delta l} = \frac{A(x + \delta x) - A(x)}{\delta x}.$$

Now, from equation (2),

$$\lim_{\delta l \to 0} \frac{\delta A}{\delta l} = |f_p| = |f(x)|.$$

Also
$$\lim_{\delta l \to 0} \frac{\delta A}{\delta l} = \lim_{\delta x \to 0} \frac{A(x + \delta x) - A(x)}{\delta x} = \frac{dA(x)}{dx}$$
(for the derivative to exist, we had to allow δx to be positive or negative). Thus
$$\frac{dA(x)}{dx} = |f(x)|. \tag{3}$$

This equation has the form of equation (1) and is used to determine the area as in the examples which follow. Later we shall describe a more streamlined procedure for carrying out the calculations.

Example 1   Find the area under the curve $y = x^2$ between $x = a$ and $x = b$.

Here $f(x) = x^2$, which is not negative, so the modulus sign in (3) can be dropped. We have
$$\frac{dA(x)}{dx} = x^2,$$
where $A(x)$ is the area between the ordinates 0 and x. We require $A(b)$, the area between the ordinates a and b. $A(x)$ is an integral of x^2; therefore,
$$A(x) = \tfrac{1}{3}x^3 + C$$
for some value of C. When $x = a$, $A(x) = 0$, so
$$0 = \tfrac{1}{3}a^3 + C$$
and $C = -\tfrac{1}{3}a^3$. Therefore,
$$A(b) = \tfrac{1}{3}b^3 - \tfrac{1}{3}a^3.$$

Example 2   Find the area under the curve $y = \sin x$, $0 \leqslant x \leqslant 2\pi$.

Letting $A(x)$ be the area from 0 to x,
$$\frac{dA(x)}{dx} = |\sin x|. \tag{4}$$
$\sin x$ is positive between $x = 0$ and $x = \pi$, and negative between $x = \pi$ and $x = 2\pi$. Therefore, to find the area, we could split the interval 0 to 2π into two parts and obtain their areas separately,
$$\frac{dA(x)}{dx} = \sin x, \qquad 0 \leqslant x \leqslant \pi, \tag{5}$$
$$\frac{dA(x)}{dx} = -\sin x, \qquad \pi \leqslant x \leqslant 2\pi.$$

The definite integral and some applications

However, the symmetry of the graph of sin x about $x = \pi$, (figure 114) shows that the area in $\pi \leqslant x \leqslant 2\pi$ is equal to the area in $0 \leqslant x \leqslant \pi$. In $0 \leqslant x \leqslant \pi$, from (5),

$$A(x) = -\cos x + C, \tag{6}$$

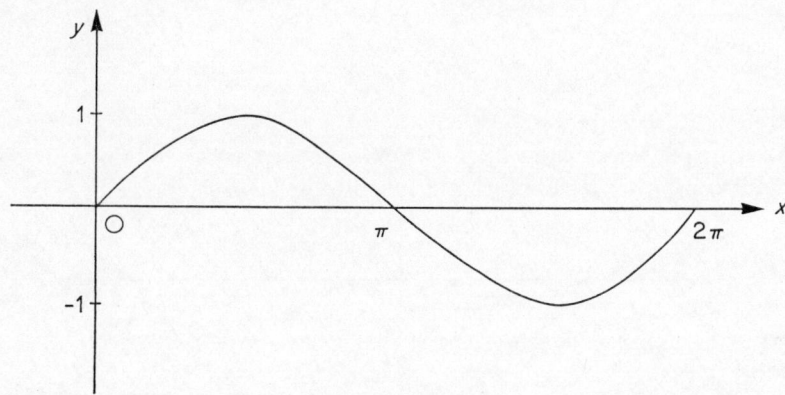

$y = \sin x,\ 0 \leq x \leq 2\pi$

FIGURE 114

for some value of C. $A(0) = 0$, therefore

$$0 = -\cos 0 + C = -1 + C$$

so that $C = 1$. Then from (6),

$$A(\pi) = -\cos \pi + 1 = 2.$$

The total area is twice this, equalling 4.

Example 3 Find the area of the ellipse

$$\frac{x^2}{a^2} + \frac{y^2}{b^2} = 1.$$

The ellipse is a closed curve, and a possible procedure would be to calculate the area above the x-axis, bounded by

$$y = b\sqrt{\left(1 - \frac{x^2}{a^2}\right)}, \qquad -a \leqslant x \leqslant a$$

and the area below the x-axis, bounded by

$$y = -b\sqrt{\left(1 - \frac{x^2}{a^2}\right)}, \qquad -a \leqslant x \leqslant a$$

separately and add them. However, the symmetry of the ellipse is such that it is sufficient to obtain the area of the shaded quadrant in figure 115, which is $\frac{1}{4}$ of the whole. Let $A(x)$ represent the area between 0 and x. Then

$$\frac{dA(x)}{dx} = b\sqrt{\left(1 - \frac{x^2}{a^2}\right)},$$

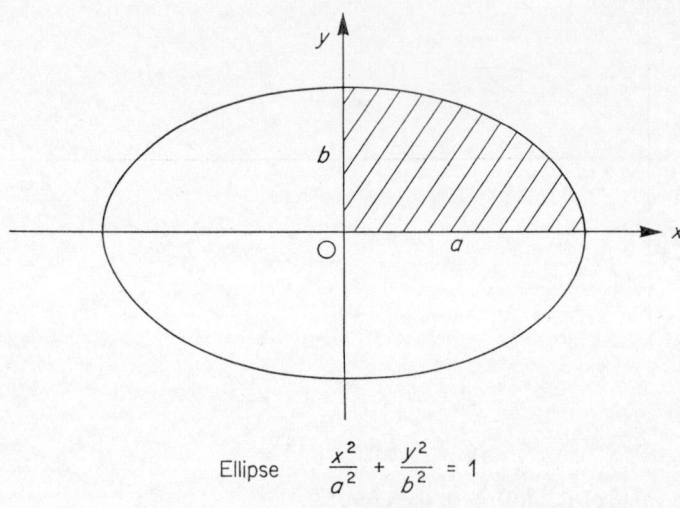

FIGURE 115

and therefore

$$A(x) = \tfrac{1}{2}ab\left[\frac{x}{a}\sqrt{\left(1 - \frac{x^2}{a^2}\right)} + \sin^{-1}\frac{x}{a}\right] + C,$$

for some constant C. Since $A(0) = 0$, C is zero. Therefore

$$\text{area} = 4A(a) = 4 \cdot \tfrac{1}{2}ab \sin^{-1} 1 = 4 \cdot \tfrac{1}{4}\pi ab = \pi ab.$$

In the following sections we show how the process of finding areas can be reduced to a simpler routine.

12.2 Further applications of the definite integral

The problem of area serves as a pattern for a great variety of physical and geometrical problems, some of which we shall treat in the sections which follow. They all lead to the following problem: given a function $f(x)$ find a function $F^*(x)$ such that

$$\frac{dF^*(x)}{dx} = f(x) \tag{7}$$

The definite integral and some applications

on an interval $a \leqslant x \leqslant b$ (which is an ordinary integration problem); with the extra condition which settles the value of the arbitrary constant,

$$F^*(a) = 0. \tag{8}$$

(In the area problem $F^*(x)$ represented the area between a and x.)

In fact $F^*(x)$ is not usually required, but only the value of $F^*(x)$ at $x = b$. (In the case of area, the area between the ordinates at a and b.)

Suppose that $F(x)$ is *any* integral of $f(x)$ on an interval $a \leqslant x \leqslant b$, then since $F^*(x)$ is also an integral (from (7)), we must have

$$F^*(x) = F(x) + C, \tag{9}$$

where C is a constant. To determine C, we use (8) which gives

$$F^*(a) = 0 = F(a) + C,$$

so that

$$C = -F(a)$$

and finally from (9):

$$F^*(b) = F(b) - F(a), \tag{10}$$

which is often written

$$F^*(b) = [F(x)]_a^b.$$

If we denote $F(x)$ by $\int f(x)\,dx$, we have

$$F^*(b) = \left[\int f(x)\,dx\right]_a^b$$

which is given the special notation:

$$F^*(b) = \int_a^b f(x)\,dx. \tag{11}$$

This is to be read: *the integral of the function f between the* lower limit *a and the* upper limit b.

The form (11) is called a *definite integral*. It represents a definite *number* depending on a, b and f which is independent of what particular *indefinite integral*, $F(x)$, is used in evaluating it, since all such functions differ only by constants and the constant vanishes in forming the difference (10). It is important to understand that no 'arbitrary constant' appears in definite integrals. Thus

$$\int_{-1}^{3} x^2\,dx = [\tfrac{1}{3}x^3]_{-1}^{3} = \tfrac{1}{3}\cdot 27 - \tfrac{1}{3}\cdot(-1) = \frac{28}{3}.$$

In evaluating this we naturally chose the simplest form for $\int x^2\,dx$, namely $\tfrac{1}{3}x^3$, but if we had chosen $\int x^2\,dx = \tfrac{1}{3}x^3 + 10$ we should get the same answer.

It must be emphasized that *the value of a definite integral is unaffected by the letter used to signify the variable of integration*. Thus

$$\int_a^b f(x)\,dx = \left[\int f(x)\,dx\right]_{x=b} - \left[\int f(x)\,dx\right]_{x=a},$$

$$\int_a^b f(t)\,dt = \left[\int f(t)\,dt\right]_{t=b} - \left[\int f(t)\,dt\right]_{t=a},$$

and so on, all these expressions having the same value, a number depending on a and b. In practice, the choice of letter may not be entirely a matter of indifference—sometimes it is convenient to think in terms of x as in the area problems, but in the following example it is more likely that we should choose to work in terms of t.

Example 4 Find the distance travelled by a particle moving with speed $v(t)$ (where t is the time) between $t = t_1$ and $t = t_2$.

We have

$$\frac{ds(t)}{dt} = v(t),$$

where v is the speed and s is the distance travelled from time t_1. Also $s(t_1) = 0$, and we require $s(t_2)$. This is exactly the form of problem considered first in this section, with s for F^*, v for f, t for x, t_1 for a and t_2 for b. Therefore,

$$s(t_2) = \int_{t_1}^{t_2} v(t)\,dt. \tag{12}$$

Now suppose instead that we wanted to plot a graph of distance travelled against time, starting at $t = t_1$. The function required is $s(t)$, that is to say t replaces t_2:

$$s(t) = \int_{t_1}^{t} v(t)\,dt. \tag{13}$$

However, there is some confusion of function in (13) between the t which defines the interval considered, the upper limit of the integral, and the t which signifies the variable of integration, and this is a situation which could lead to actual mistakes. We might then prefer to write (13) as

$$s(t) = \int_{t_1}^{t} v(u)\,du \quad \text{or} \quad \int_{t_1}^{t} v(t')\,dt',$$

using as variable of integration some letter not committed to another purpose.

The definite integral and some applications

In area problems we have for determining the area A under $y = f(x)$, $a \leqslant x \leqslant b$, the conditions

$$\frac{\mathrm{d}A(x)}{\mathrm{d}x} = |f(x)|, \qquad a \leqslant x \leqslant b,$$

(section 12.1), with $A(x)$ defined as the area between $x = a$ and $x = b$. Also $A(a) = 0$. From the argument at the beginning of this section,

$$A = A(b) = \int_a^b |f(x)|\, \mathrm{d}x. \tag{14}$$

We shall rework Example 3 in this way.

Example 5   Find the area of a quadrant of the ellipse

$$\frac{x^2}{a^2} + \frac{y^2}{b^2} = 1.$$

Consider the quadrant $y \geqslant 0$, $0 \leqslant x \leqslant a$. Then

$$\frac{\mathrm{d}A(x)}{\mathrm{d}x} = b\sqrt{\left(1 - \frac{x^2}{a^2}\right)}$$

with $A(0) = 0$. The area is given by

$$A = A(a) = \int_0^a b\sqrt{\left(1 - \frac{x^2}{a^2}\right)}\mathrm{d}x$$

$$= \tfrac{1}{2}ab\left[\frac{x}{a}\sqrt{\left(1 - \frac{x^2}{a^2}\right)} + \sin^{-1}\frac{x}{a}\right]_0^a$$

$$= \tfrac{1}{2}ab \sin^{-1} 1 = \tfrac{1}{4}\pi ab.$$

This formulation saves a good deal of writing since no constant C has explicitly to be determined.

Exercises

Prove the following results:

1. (i) $\int_0^1 x^2\, \mathrm{d}x = \tfrac{1}{2}$; (ii) $\int_0^{2\pi} \sin x\, \mathrm{d}x = 0$;

   (iii) $\int_0^{\frac{1}{2}\pi} \cos x\, \mathrm{d}x = 1$; (iv) $\int_0^{\frac{1}{2}\pi} \sin x\, \mathrm{d}x = 1$;

(v) $\int_{-\frac{1}{2}\pi}^{\frac{1}{2}\pi} x \cos x \, dx = 0$; (vi) $\int_{\pi}^{2\pi} \sin x \, dx = -2$;

(vii) $\int_a^b e^{-cx} \, dx = \frac{1}{c}(e^{-ac} - e^{-bc})$.

2. (i) $\int_0^2 (x-1)^3 \, dx = 0$; (ii) $\int_{-2}^{-1} (x+1)^5 \, dx = -\frac{1}{6}$;

(iii) $\int_0^1 \frac{dx}{\sqrt{(1-x^2)}} = \frac{1}{2}\pi$; (iv) $\int_{-2}^{-1} \frac{dx}{\sqrt{(x^2-1)}} = \ln(2+\sqrt{3})$;

(v) $\int_{-1}^1 (1+x^2)^{\frac{1}{2}} \, dx = \sqrt{2} + \ln(1+\sqrt{2})$; (vi) $\int_0^{\frac{1}{4}\pi} \sec^2 \theta \, d\theta = 1$.

3. (i) $\int_1^2 \ln u \, du = 2 \ln 2 - 1$; (ii) $\int_0^1 t \ln t \, dt = -\frac{1}{4}$.

4. (i) $\int_{-2}^{-1} x^{-1} \, dx = -\ln 2$;

(ii) $\int_{a^{-1}}^1 x^{-1} \, dx = \int_1^a x^{-1} \, dx = \ln a, \quad a > 0$.

5. $\int_a^b \frac{dt}{1+t^2} = \int_{-b}^{-a} \frac{dt}{1+t^2} = \tan^{-1} b - \tan^{-1} a$.

6. $\int \frac{dx}{x^3} = -\frac{1}{2x^2} + C$, therefore $\int_{-1}^1 \frac{dx}{x^3} = 0$. Is this legitimate?

7. $\int_{-3}^{-2} \frac{dx}{\sqrt{(x^2-1)}} = \ln \frac{2-\sqrt{3}}{3-\sqrt{8}}$.

8. Find the area under the curve $y = (x^2 + a^2)^{-1}$ between $x = 0$ and $x = a$.

9. Find the area between the x-axis and that part of the curve $y = a^2 - b^2 x^2$ lying above it.

10. Find the area of the closed region between the curve $y = 2/x$ and the lines $x = 1, y = 1$.

11. A hyperbola has the equation $x^2/a^2 - y^2/b^2 = 1$. From a point (x,y) on the branch of the curve in the first quadrant a perpendicular is dropped on to the x-axis. Show that the area of the closed region formed by the x-axis, the perpendicular and the curve for $x \geqslant 0$ is given by $\frac{1}{2}[xy - ab \ln (x/a + y/b)]$.

12. Find the area of the closed region contained between the two parabolas $y = x^2$ and $y = 12 - 2x^2$ (sketch the two curves).

13. Find the area of the closed region bounded by the lines $y = 1 + x^{\frac{1}{2}}$, $y = 1 - x^{\frac{1}{2}}, x = 1$.

14. Find the area between the curve $y = x \ln x$ and the x-axis, $0 \leqslant x \leqslant 2$.

15. Find the area of the circular cap formed between the shorter segment of the circumference of a circle of radius a and a chord distant b, $b < a$, from the centre.

16. Find the area of the lens-shaped region common to two circles of radius a, the circumference of one passing through the centre of the other.

17. Traffic flows past a check-point into an otherwise empty stretch of one-way road at a rate $At^2 e^{-at}$ cars/hr, $(t \geqslant 0)$. Assuming the stretch is l miles long and that the traffic moves at a constant speed v m/hr find the number of cars upon it at all times $t > 0$.

18. A ray of light passes through a block of coloured glass, its intensity at a point distant x from the surface at which it enters being $I(x)$. Its loss of intensity, $-\delta I$, in passing a small distance δx through the glass is governed by the law 'fractional loss in intensity per unit distance' = a constant a:

$$\frac{-\delta I}{I}\bigg/ \delta x \approx a.$$

Obtain a differential equation connecting I and x.
If the intensity of the light on entering the block is I_0, show that its intensity at the point $x = l$ is $I_0 e^{-al}$.

19. Prove that the area under the parabola $y = px^2 + qx + r, = f(x)$, say, $d \leqslant x \leqslant d + h$, is equal to $\frac{1}{6}h[f(d) + 4f(d + \frac{1}{2}h) + f(d + h)]$.

(Assume for simplicity that $p > 0$ and $q^2 - 4pr \leqslant 0$ so that $f(x) \geqslant 0$ in the interval. This result is the basis of Simpson's rule, section 12.9.)

20. Find the area enclosed between the parabola $y = 8 - x^2$ and the line $y = 2x$. (Consider the figure below, taking the shaded element as having area δA and let AA′ have length δl. Then, as in section 12.1, $\lim_{\delta l \to 0} \delta A/\delta x = $ QQ′ leading to $dA(x)/dx = $ QQ′, where $A(x)$ is the area between the curves.)

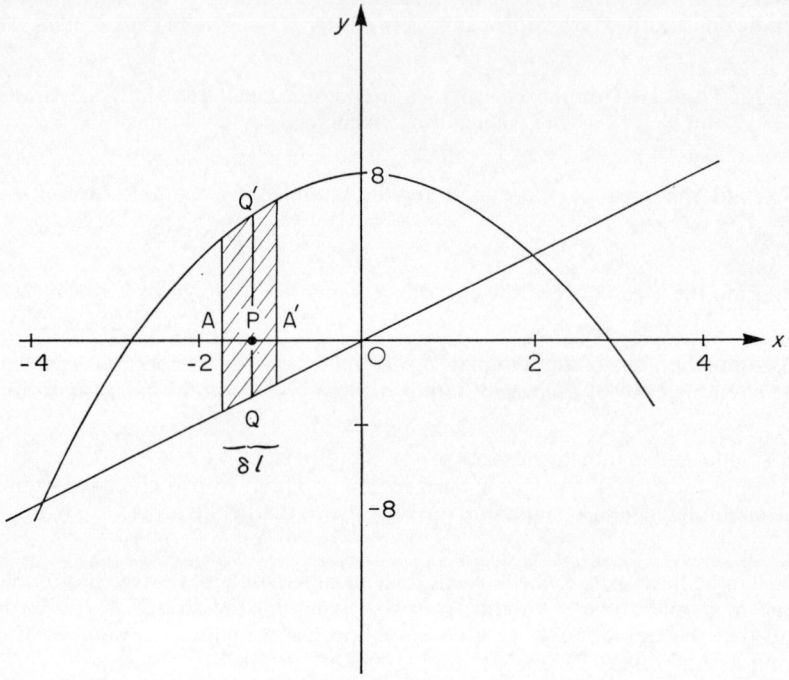

21. Find the area enclosed between the curves $y = x$ and $y = \sin x$, $0 \leqslant x \leqslant 2\pi$ by the method of question 20.

22. Find the area of the closed region formed by the curve $y = \cos^{-1} x$, the line $x = 0$ and the line $y = \frac{1}{3}\pi$. (Straightforwardly, this would involve integrating $\cos^{-1} x$ from 0 to $\frac{1}{2}$ and subtracting the area of STUO. We have as yet given no method for integrating $\cos^{-1} x$. However, if the roles of the axes are reversed and strips taken parallel to the x-axis, the width being δl and area δA, we obtain as in section 12.1 $\lim_{\delta l \to 0} \delta A/\delta l = $ PQ and therefore $dA(y)/dy = \cos y$, so the problem is reduced to a simpler integration. Concealed here is also a method for integrating $\cos^{-1} x$: let the upper limit be x instead of $\frac{1}{2}$, and complete the argument as above.)

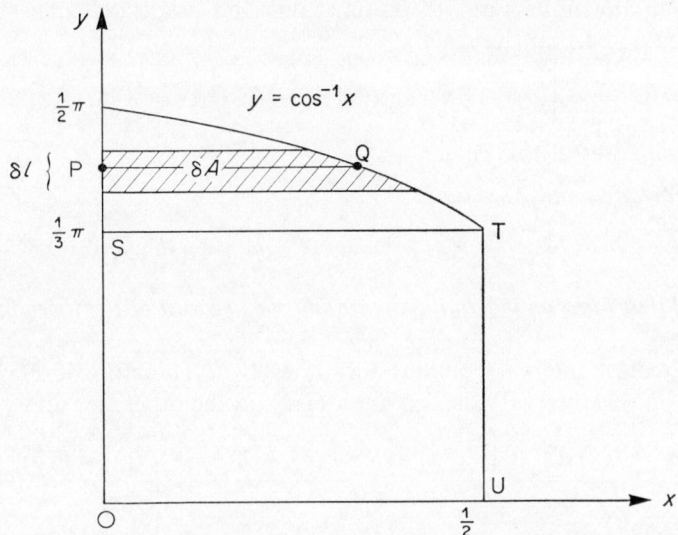

12.3 Some important properties of the definite integral

In our applications we shall not encounter only integrals $\int_a^b f(x)\,dx$ in which $b > a$, and the following results are applicable whether or not $b > a$, provided the function has an integral on the interval concerned.

The definition
$$\int_a^b f(x)\,dx = \left[\int f(x)\,dx\right]_a^b$$
is equally applicable when $b \leqslant a$, leading to

I. $$\int_a^a f(x)\,dx = 0 \qquad (15)$$

and

$$\int_a^b f(x)\,dx = -\int_b^a f(x)\,dx. \qquad (16)$$

II. *If f and g have integrals on $a \leqslant x \leqslant b$, then*

$$\int_a^b [Af(x) + Bg(x)]\,dx = A\int_a^b f(x)\,dx + B\int_a^b g(x)\,dx,$$

where A and B are any constants.

This follows immediately from V of section 11.3. The result extends to sums over any finite number of functions and associated constants. It should be noted that the results

$$\int_a^b Af(x)\,dx = A\int_a^b f(x)\,dx \quad \text{and} \quad \int_a^b [-f(x)]\,dx = -\int_a^b f(x)\,dx \quad (17)$$

are special cases of this rule.

III. *For any numbers a,b,c;*

$$\int_a^b f(x)\,dx + \int_b^c f(x)\,dx = \int_a^c f(x)\,dx, \quad (18)$$

provided that f has an indefinite integral on the greatest of the three intervals concerned.*

The greatest interval contains the other two. Therefore if $F(x)$ is an integral on this interval it is also an integral on the other two and we have

$$\int_a^b f(x)\,dx + \int_b^c f(x)\,dx = F(b) - F(a) + F(c) - F(b)$$

$$= F(c) - F(a) = \int_a^c f(x)\,dx.$$

12.4 The area analogy for definite integrals

The use of the definite integral is not confined to obtaining the areas under curves which, indeed, is a rather minor application. However, any definite integral can be *interpreted* in terms of the area under a certain curve, and this provides a very useful way of visualizing the mathematical properties of integrals and of suggesting results which are true but which may be hard to prove analytically.

If f is a continuous function and $f(x) \geqslant 0$ on $a \leqslant x \leqslant b$, then $\int_a^b f(x)\,dx$ is equal to the area under the curve $y = f(x)$ between a and b. We can make this true if $f(x)$ is negative on all or part of its range if we modify our definition of area by introducing 'negative areas'. We attribute a sign to the area between the graph of $f(x)$ and the x-axis according to the following rule: where the graph lies above the x-axis (i.e. $f(x) \geqslant 0$), the associated area is given a positive sign, and where it lies below, it is given a negative sign. Thus, in figure 116, let the areas of the sections between the zeros of the function be $A_1, A_2, \ldots$, (all positive). Then we assert that

$$\int_a^b f(x)\,dx$$

---

* We cannot say that $a \leqslant x \leqslant c$ is automatically the greatest interval, because c is not necessarily greater than b.

is equal to the 'signed area' between the curve and the x-axis; that is, to
$$A_1 - A_2 + A_3 - A_4.$$
This is true because $\int_a^{c_1} f(x)\,dx$ is positive, $\int_{c_1}^{c_2} f(x)\,dx$ is negative, and so on.

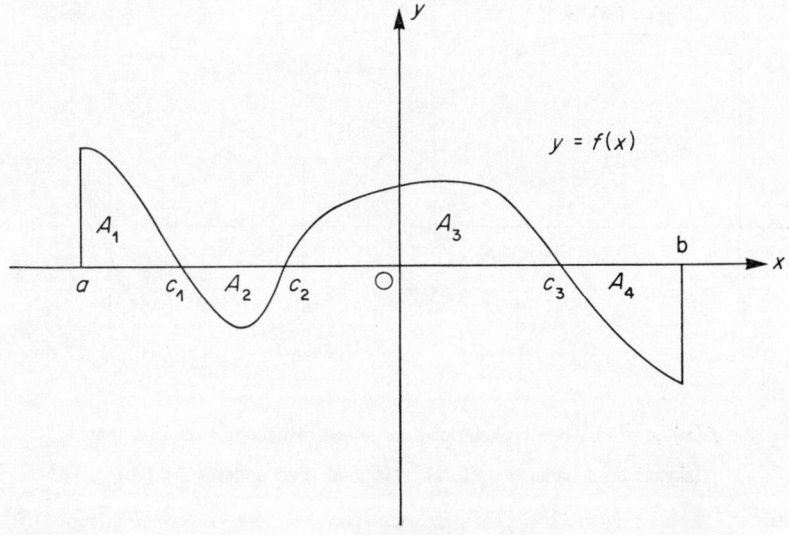

FIGURE 116

But these are, respectively, equal in absolute value to $A_1, A_2, \ldots$, etc. In figure 116
$$\int_a^b f(x)\,dx = \int_a^{c_1} f(x)\,dx + \int_{c_1}^{c_2} f(x)\,dx + \int_{c_2}^{c_3} f(x)\,dx + \int_{c_3}^b f(x)\,dx$$
(by III, section 12.3)
$$= \int_a^{c_1} |f(x)|\,dx - \int_{c_1}^{c_2} |f(x)|\,dx + \int_{c_2}^{c_3} |f(x)|\,dx - \int_{c_3}^b |f(x)|\,dx$$
(by equation (17) of section 12.3, since on the interval where $f(x)$ is negative, $f(x) = -|f(x)|$),
$$= A_1 - A_2 + A_3 - A_4.$$
Thus we may say that $\int_a^b f(x)\,dx$ *is equal to the area between the curve* $y = f(x)$ *and the x-axis, provided that we count the areas corresponding to sections where the function is positive as being positive, and count them as negative where the function is negative.* For an example, the reader might reconsider Example 2 of section 12.1.

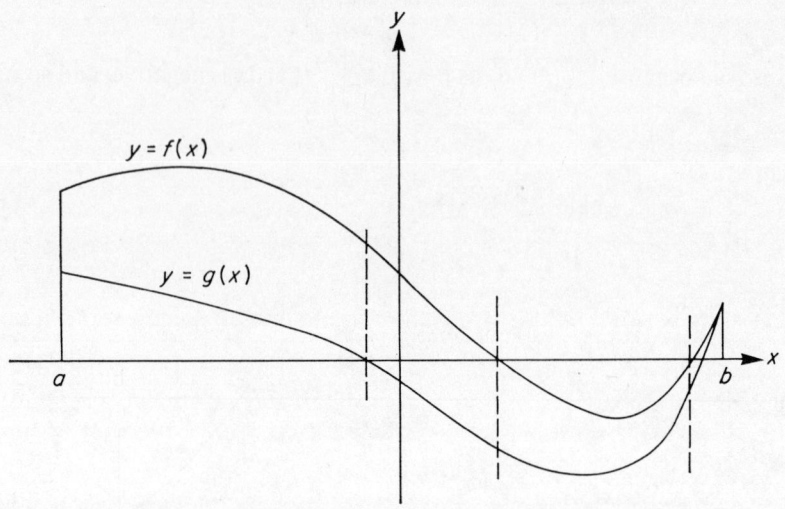

$f(x) \geq g(x)$ When broken down as shown it becomes obvious that
(signed area under $y = f(x)$) ≥ (signed area under $y = g(x)$)

FIGURE 117

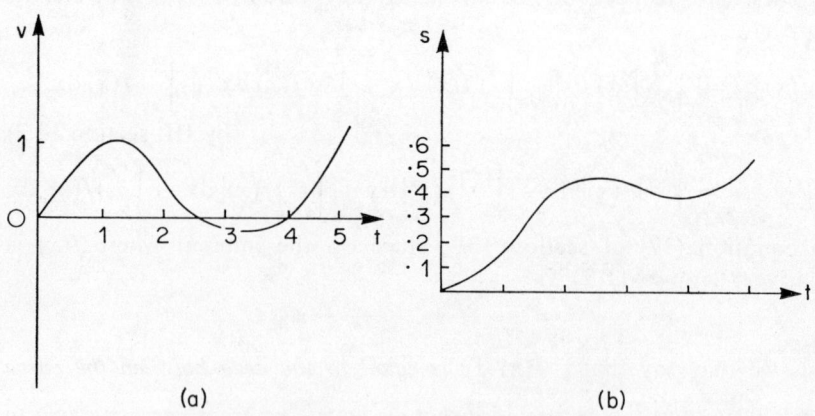

(a)   (b)
(a) velocity v plotted against time   (b) distance s sketched as a function of time from (a)

FIGURE 118

As an instance of its use in deriving properties of integrals, a result proved in section 12.11, that if

$$f(x) \geqslant g(x) \quad \text{on} \quad a \leqslant x \leqslant b$$

then

$$\int_a^b f(x)\,dx \geqslant \int_a^b g(x)\,dx$$

is convincingly suggested by a sketch such as that shown in figure 117.

The idea is often helpful in thinking about physical situations in graphical terms. If the velocity of a car is $v(t)$, the distance travelled between times t_1 and t has been shown to be given by $s(t) = \int_{t_1}^t v(u)\,du$ (Example 4, section 12.2). Therefore, $s(t)$ is equal to the area (suitably signed) under the graph of v against t between t_1 and t, as in figure 118(a). From this figure, the function $s(t)$ has been roughly sketched (figure 118(b)) by estimating the progressive development of the signed area as t increases.

Results suggested by the area analogy cannot be regarded as complete proofs, since our treatment of area has not been a strict one. Intuitions about area break down when a function is considered such as $x \sin(1/x)$, which changes sign infinitely often in every interval containing the origin.

Exercises

23. Use the area analogy to show that

(i) $\int_{-a}^a x\,dx = 0;$ (ii) $\int_{-a}^a x^3\,dx = 0;$

(iii) $\int_{-a}^a (x^3 - 2x)\,dx = 0;$ (iv) $\int_{b-a}^{b+a} (x-b)^3\,dx = 0;$

(v) $\int_{-a}^a x^2\,dx = 2\int_0^a x^2\,dx;$ (vi) $\int_0^\pi \cos x\,dx = 0;$

(vii) $\int_{-1}^1 \frac{dx}{1+x^2} = 2\int_0^1 \frac{dx}{1+x^2};$ (viii) $\int_0^{\frac{1}{2}\pi} \cos 2x\,dx = 0;$

(ix) $\int_0^{2n\pi} \sin mx\,dx = 0;$ m,n integers; (x) $\int_{-a}^a \sin^{-1} x\,dx = 0;$

(xi) $\int_{-a}^a [f(x) - f(-x)]\,dx = 0;$

(xii) $\left|\int_a^b f(x)\,dx\right| \leqslant \int_a^b |f(x)|\,dx.$

24. Use arguments based on area considerations to show that

(i) $\displaystyle\int_a^b f(x+c)\,dx = \int_{a+c}^{b+c} f(u)\,du;$

(ii) $\displaystyle\int_a^b f(-x)\,dx = \int_{-b}^{-a} f(u)\,du;$

(iii) $\displaystyle\int_a^b f(cx)\,dx = \frac{1}{c}\int_{ac}^{bc} f(u)\,du.$

25. Draw a figure incorporating the graphs of e^{-x} and $\sin x^2$ for $x > 0$.

(i) Use the area analogy to show that if
$$u_n = \int_{\sqrt{(n\pi)}}^{\sqrt{[(n+1)\pi]}} e^{-x} \sin x^2\,dx,$$
n an integer $\geqslant 0$ then $\{u_n\}$ is an alternating sequence and that $|u_n|$ decreases and tends to zero as n increases.

(ii) Prove that $\displaystyle\int_0^X e^{-x} \sin x^2\,dx$ tends to a limit as X tends to infinity.

26. Consider a function h, such that $h(x_2) > h(x_1)$ for $a \leqslant x_1 < x_2 \leqslant b$. It has an inverse, H, say, in this interval, so if $y = h(x)$, $a \leqslant x \leqslant b$, we have $x = H(y)$. Let $H(a) = A$, $H(b) = B$.

(i) By interpreting suitable figures in terms of area, show that
$$\int_a^b h(x)\,dx = Bb - Aa - \int_A^B H(y)\,dy;$$

(ii) Hence evaluate
$$\int_0^c \sin^{-1} y\,dy \qquad (0 < c \leqslant \tfrac{1}{2}\pi);$$

(iii) Also evaluate
$$\int_{\frac{1}{2}}^1 \frac{\sqrt{(1-y^2)}}{y}\,dy.$$

12.5 The definite integral as a function of its upper limit

The value of $\displaystyle\int_a^b f(x)\,dx$ depends on the two numbers a and b—it is in fact a *function* of a and b. Suppose that we wish to study how its value varies as b varies, a being a fixed number. We then look upon b as a variable

The definite integral and some applications

and to stress this interest we adopt, instead of b, some letter traditionally assigned to variables, say x. Then

$$\int_a^x f(u)\,du = F(x) - F(a), \tag{19}$$

where F is any integral of f. We have in (19) changed the name of the *variable of integration*, formerly x, to u to avoid confusing the uses of x as upper limit and as variable of integration, as discussed in section 12.2.

From (19),

$$\frac{d}{dx}\int_a^x f(u)\,du = \frac{d}{dx}[F(x) - F(a)] = f(x),$$

and therefore, $\int_a^x f(u)\,du$ is an indefinite integral of $f(x)$.

We may now look at our area-type problem of section 12.2 in yet another way. We have to solve

$$\frac{dF^*(x)}{dx} = f(x)$$

in $a \leqslant x \leqslant b$, where also $F^*(a) = 0$.

Evidently

$$F^*(x) = \int_a^x f(u)\,du$$

is just the function required since it is an integral of $f(x)$, and is zero when $x = a$ (by I of section 12.3). Therefore,

$$F^*(b) = \int_a^b f(u)\,du,$$

which we had before.

12.6 Areas of sectorial regions expressed in terms of polar coordinates

The equations of certain types of closed curve are expressible most easily in polar coordinates (figure 119) and the area enclosed can be expressed as an integral whose integrand is written in terms of polar coordinates.

Suppose the equation of the curve in figure 119 is of the form

$$r = g(\theta) \tag{20}$$

where g is a continuous function. Generally there is some restriction on the range of θ; in the figure the sector being considered is defined by (20) in $\alpha \leqslant \theta \leqslant \beta$.

382 *Introductory Analysis*

We shall also suppose that any radius in this range cuts the curve only once, that is to say $g(\theta)$ is defined to be a single-valued function of θ.

Take a point P on the curve, and two points A,A' on it such that the interval AA' contains P, and AA' subtends an angle $\delta\theta$ (> 0) at the

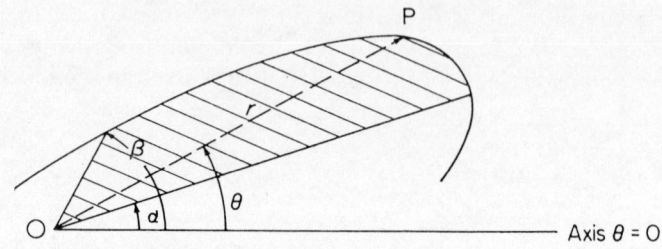

Sector of $r = g(\theta)$, $\alpha \leq \theta \leq \beta$

FIGURE 119

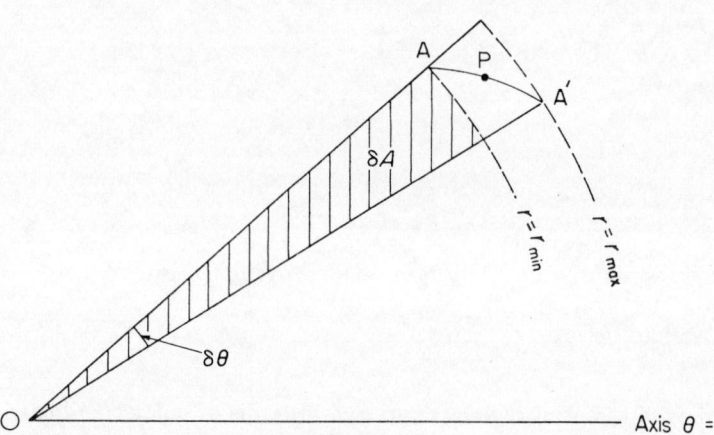

Sectorial Element of area, δA

FIGURE 120

origin. Within the sector (figure 120) the radial coordinate r takes on a minimum value $r_{\min}$ and a maximum value $r_{\max}$. If the area of the region concerned is δA, then using the hypotheses about area employed earlier:

$$\tfrac{1}{2}r_{\min}^2 \, \delta\theta \leq \delta A \leq \tfrac{1}{2}r_{\max}^2 \, \delta\theta,$$

The definite integral and some applications

and as A and A′ tend to the point P, $r_{\min}$ and $r_{\max}$ tend to OP, so that in the limit we have

$$\lim_{\delta\theta\to 0} \frac{\delta A}{\delta\theta} = \tfrac{1}{2}\mathrm{OP}^2 = \tfrac{1}{2}g_p^2$$

(compare equation (2) of Section 12.1), where $\mathrm{OP} = g_p$.

Now let $A(\theta)$ be the area of that sector of the curve contained between the radii at angles α and θ. Following the argument of section 12.1, we have

$$\frac{dA(\theta)}{d\theta} = \tfrac{1}{2}g^2(\theta)$$

and $A(\alpha) = 0$. Therefore,

$$\text{area of sector} = A(\beta) = \int_\alpha^\beta \tfrac{1}{2}g^2(\theta)\,d\theta.$$

This may also be remembered as

$$\text{area of sector} = \int_\alpha^\beta \tfrac{1}{2}r^2\,d\theta.$$

Example 6 Find the area of one loop of the curve $r = a\cos\theta$.

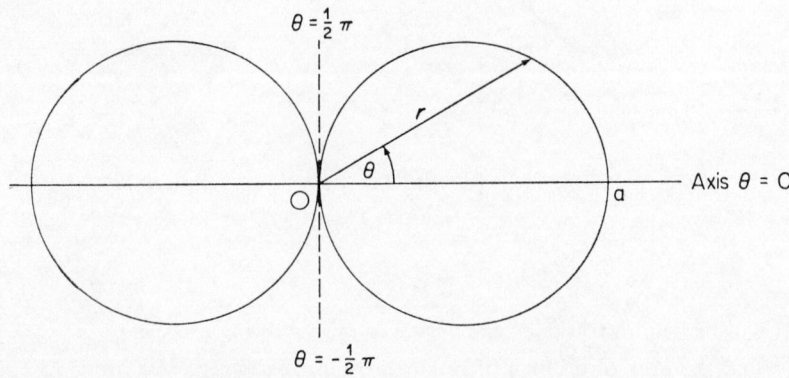

$r = a\cos\theta$

FIGURE 121

The curve $r = a\cos\theta$, $-\pi < \theta \leqslant \pi$ is as shown. We shall take the loop $-\tfrac{1}{2}\pi \leqslant \theta \leqslant \tfrac{1}{2}\pi$. The area is

$$\int_{-\frac{1}{2}\pi}^{\frac{1}{2}\pi} \tfrac{1}{2}(a\cos\theta)^2\,d\theta = \tfrac{1}{2}a^2[\tfrac{1}{2}\theta + \tfrac{1}{4}\sin 2\theta]_{-\frac{1}{2}\pi}^{\frac{1}{2}\pi}$$
$$= \tfrac{1}{2}a^2[(\tfrac{1}{4}\pi + \tfrac{1}{4}\sin\pi) - (-\tfrac{1}{4}\pi + \tfrac{1}{4}\sin(-\pi))]$$
$$= \tfrac{1}{4}a^2\pi.$$

384 *Introductory Analysis*

Example 7 *Find the area of one loop of the curve $r^2 = a^2 \sin 2\theta$.*

The shape of the curve is as depicted in figure 122.
The loop in the first quadrant corresponds to $0 \leqslant \theta \leqslant \tfrac{1}{2}\pi$. The area is
$$\int_0^{\frac{1}{2}\pi} \tfrac{1}{2}a^2 \sin 2\theta \, d\theta = \tfrac{1}{2}a^2.$$

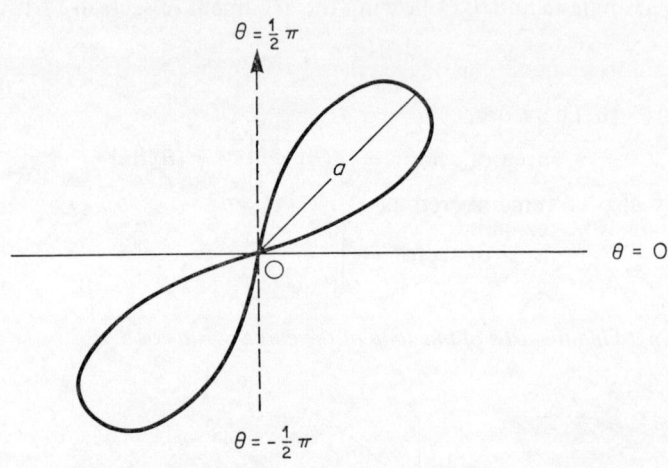

$r^2 = a^2 \sin 2\theta$

FIGURE 122

Exercises

It will be helpful to sketch the curves in the following exercises.

27. Find the area of a circle of radius a by using the representations in polar coordinates: (i) $r = a$; (ii) $r = 2a \sin \theta$.

28. Trace the curve and find the area of one loop of the 'lemniscate'
$$r^2 = c^2 \cos 2\theta.$$

29. Find the area enclosed by the 'cardioid' $r = c(1 + \cos \theta)$.

30. Find the area of the 'limaçon' $r = a + b \cos \theta$, $a > b$.

31. Sketch the curve $r^2 = a^2 \cos^2 \theta + b^2 \sin^2 \theta$ and find its area.

The definite integral and some applications

32. Find the area enclosed by one loop of the spiral $r = a\,e^\theta$, $a \leqslant \theta \leqslant \alpha + 2\pi$.

33. Show that the polar equation for the curve for which the tangent at each point makes a constant angle α with the radius vector $\boldsymbol{r}$ can be obtained from the relation $\mathrm{d}r/\mathrm{d}\theta = r \cot \alpha$.

If the point $\theta = 0$, $r = a$ lies on the curve, find the area of the sector $0 \leqslant \theta \leqslant 2\pi$.

34. By transforming to polar coordinates prove that the area of the ellipse
$$ax^2 + 2hxy + by^2 = 1$$
is $\pi(ab - h^2)^{-\frac{1}{2}}$.

12.7 Volumes of revolution

If the graph of a function f, $y = f(x)$, $a \leqslant x \leqslant b$, is rotated about the x-axis, a body of circular but variable cross-section is formed, the radius of the section through the point x being $|f(x)|$. Some possibilities are shown in figure 123.

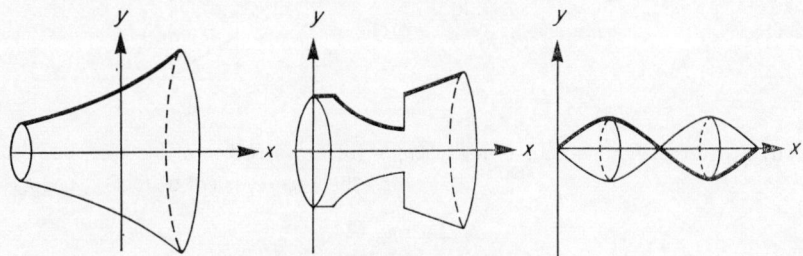

Solids of revolution obtained by rotating various graphs $y = f(x)$ (heavy lines) about the, x-axis

FIGURE 123

Such solids are said to be *axially symmetrical* (symmetrical about an axis) and many projectiles, vessels, machine parts and so on are axially symmetrical. We shall construct a differential relation analogous to (1) of section 12.1 for area, but concerning the volume of such solids, and we shall make geometrical assumptions about the concept 'volume' similar to those we made about area in that section.

Take a point P on the axis, and a slice of the body containing P bounded by two sections perpendicular to the x-axis a distance $\delta l (> 0)$ apart (figure 124). Let the volume of this slice be denoted by $\delta V (> 0)$. If δl is small, the slice approximates to a cylinder of height δl and radius $|f_p|$. Its volume is, therefore, approximately given by

$$\delta V \approx \pi f_p^2 \delta l.$$

Thus we expect that $\lim_{\delta l \to 0} (\delta V / \delta l) = \pi f_p^2$, the modulus signs being dropped since $|f_p|^2 = f_p^2$ whether f_p is positive or negative.

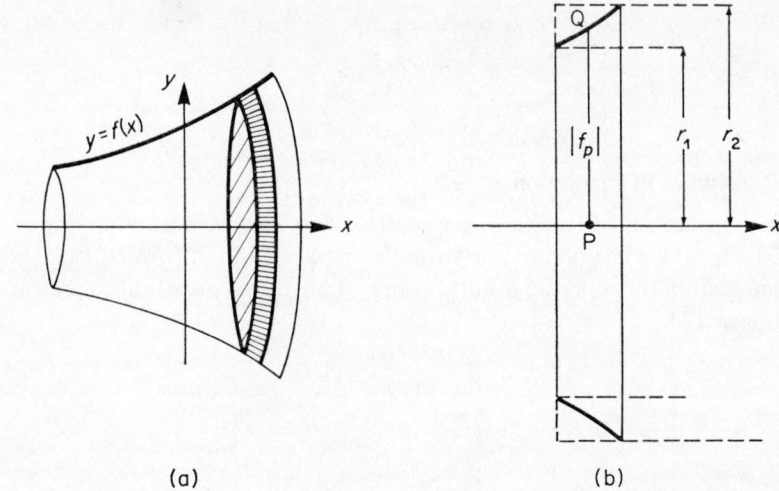

(a)   (b)

(a) Slice taken from a solid of revolution;   (b) Cross-section of the slice, bounded between cylinders of radii r_1 and r_2

FIGURE 124

We can prove this, assuming certain properties of volume, as follows.

If f is continuous at P, the slice in the figure has a maximum radius r_2 and a minimum radius r_1, and so can be enclosed between two cylinders having these radii (see figure 124). Thus

$$\pi r_1^2 \delta l \leqslant \delta V \leqslant \pi r_2^2 \delta l,$$

or

$$\pi r_1^2 \leqslant \frac{\delta V}{\delta l} \leqslant \pi r_2^2.$$

f is continuous at P and therefore, as $\delta l \to 0$, $r_1 \to |f_p|$ and $r_2 \to |f_p|$. Therefore

$$\lim_{\delta l \to 0} \frac{\delta V}{\delta l} = \pi f_p^2 \tag{21}$$

as expected. We shall not always go through this argument in full in future cases.

The definite integral and some applications

Now let $V(x)$ represent the volume between the planes through a and x. Then, proceeding exactly as with area by considering the interval $[x, x + \delta x]$, we have

$$\frac{dV(x)}{dx} = \pi f^2(x) \tag{22}$$

with $V(a) = 0$. Therefore,

$$\text{volume between } a \text{ and } b = \int_a^b \pi f^2(x)\, dx. \tag{23}*$$

This result is easier to remember in the form

$$V_{ab} = \int_a^b \pi y^2\, dx \tag{24}$$

where $y = f(x)$.

Example 8 Find the volume of the right circular cone of height h with half angle at the vertex α.

The cone is generated by rotating the line $y/x = \tan \alpha$, i.e. $y = x \tan \alpha$, $0 \leqslant x \leqslant h$, about the x-axis (figure 125). If $V(x)$ represents the volume between 0 and x, equation (22) gives

$$\frac{dV(x)}{dx} = \pi x^2 \tan^2 \alpha.$$

$V(0) = 0$ and $V(h)$ is the volume of the cone. From the preceding theory,

$$\text{volume} = V(h) = \int_0^h \pi x^2 \tan^2 \alpha\, dx = \pi \tan^2 \alpha [\tfrac{1}{3}x^3]_0^h = \tfrac{1}{3}\pi h^3 \tan^2 \alpha.$$

If the radius of the base is a, then $a = h \tan \alpha$ and the result takes the simpler form $\tfrac{1}{3}\pi a^2 h$.

Example 9 Find the volume of a sphere of radius a.

The sphere is obtained by rotating the semi-circle $y = (a^2 - x^2)^{\frac{1}{2}}$, $-a \leqslant x \leqslant a$, about the x-axis (figure 126). If $V(x)$ represents the volume between $-a$ and x, then

$$\frac{dV(x)}{dx} = \pi f^2(x) = \pi(a^2 - x^2).$$

---

* This is equivalent to another *area* problem: the *volume* of the body between $x = a$ and $x = b$ is numerically equal to the *area* under the graph $y = \pi f^2(x)$ between $x = a$ and $x = b$.

1—14

388 Introductory Analysis

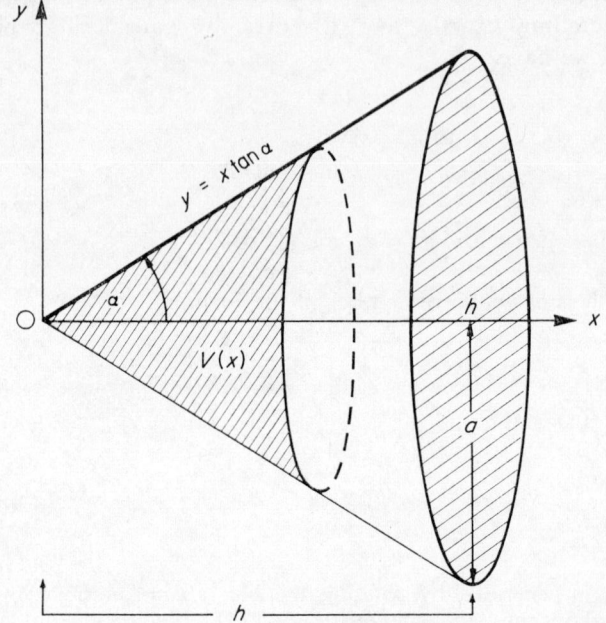

FIGURE 125

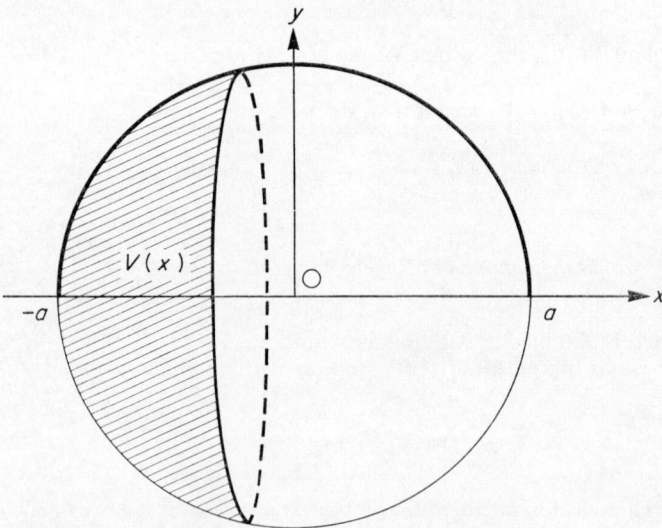

FIGURE 126

$V(-a) = 0$, and we require $V(a)$. $V(x)$ satisfying the requirements is
$$V(x) = \int_{-a}^{x} \pi(a^2 - u^2)\,du.$$
Therefore,
$$V(a) = \int_{-a}^{a} \pi(a^2 - u^2)\,du = \pi[a^2 u - \tfrac{1}{3}u^3]_{-a}^{a} = \tfrac{4}{3}\pi a^3.$$

Sometimes it will be convenient to *formulate* the problem in a different way, as in the following examples.

Example 10 Find the volume of the torus formed by rotating the circle
$$x^2 + (y-a)^2 = c^2; \quad a > 0,\; c > 0,\; a \geqslant c$$
about the x-axis.

The circle is a closed curve and the previous method is not, therefore, applicable immediately (figure 127). There are several ways to adapt the method to this type of curve. For example, if a diameter DD' is drawn parallel to the x-axis, and perpendiculars DN and DN' are dropped, we have (figure 127(a)),

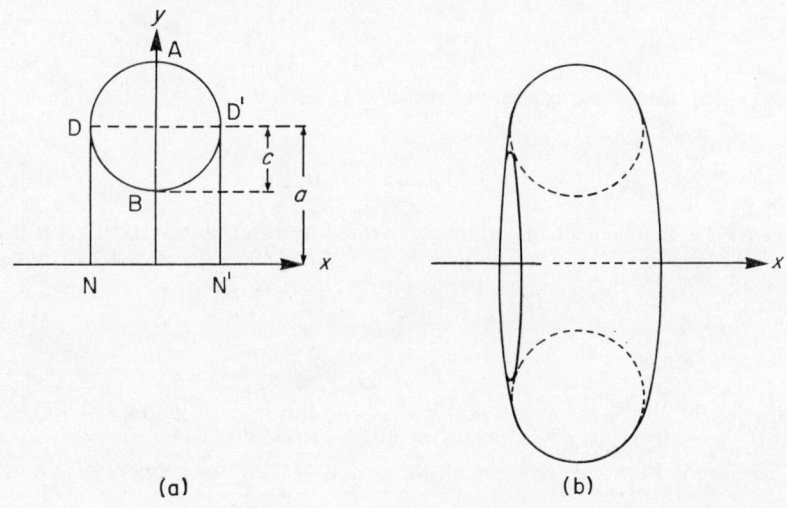

Torus (b) formed by rotating the circle (a) about the x - axis

FIGURE 127

(*volume V of the torus*) = (*volume V_1 generated by rotating the figure* DAD'N'N)
− (*volume V_2 generated by rotating the figure* DBD'N'N).

The volumes V_1 and V_2 are of the right type, the equations of the corresponding generating curves being

The upper semi-circle DAD': $y = a + \sqrt{(c^2 - x^2)}$
The lower semi-circle DBD': $y = a - \sqrt{(c^2 - x^2)}$.

This leads to an unnecessarily lengthy calculation which can be shortened as follows. Take a point P on the axis and an interval NN' of length δl containing P (figure 128).

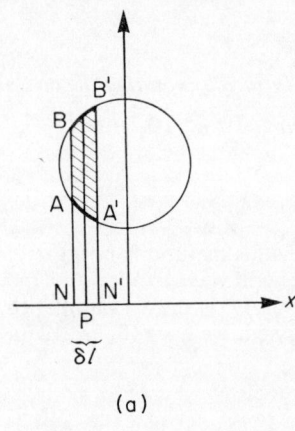
(a)

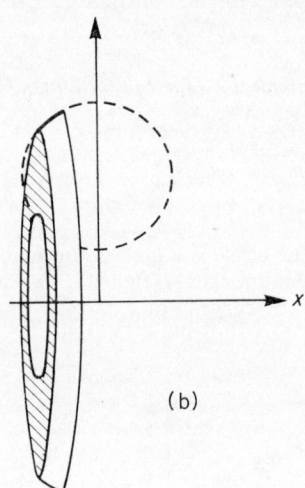
(b)

Slice (b) formed by rotating the section AA'B'B of the circle (a) about the x - axis

FIGURE 128

Let δV be the element of volume generated by rotating the slice AA'B'B of the circle between N and N' about the x-axis, δV_1 that generated by rotating NN'B'B, and δV_2 that generated by rotating NN'A'A. Then

$$\delta V = \delta V_1 - \delta V_2$$

and

$$\lim_{\delta l \to 0} \frac{\delta V}{\delta l} = \lim_{\delta l \to 0} \frac{\delta V_1 - \delta V_2}{\delta l} = \lim_{\delta l \to 0} \frac{\delta V_1}{\delta l} - \lim_{\delta l \to 0} \frac{\delta V_2}{\delta l} = \pi(\mathrm{PQ'}^2 - \mathrm{PQ}^2).$$

Therefore if $V(x)$ is the volume of the section of the torus between $-c$ and x

$$\frac{dV(x)}{dx} = \pi\{[a + \sqrt{(c^2 - x^2)}]^2 - [a - \sqrt{(c^2 - x^2)}]^2\}$$

$$= 4\pi a\sqrt{(c^2 - x^2)}.$$

The volume of the torus is given by

$$V(c) = \int_{-c}^{c} [4\pi a \sqrt{(c^2 - x^2)}] \, dx$$

$$= 4\pi a \left\{ \tfrac{1}{2} c^2 \left[\frac{x}{c} \sqrt{\left(1 - \frac{x^2}{c^2}\right)} + \sin^{-1} \frac{x}{c} \right]_{-c}^{c} \right\}$$

$$= 4\pi a \cdot \tfrac{1}{2} c^2 \cdot 2 \sin^{-1} 1$$

$$= 2\pi^2 a c^2.$$

Example 11 Find the volume of the solid of revolution obtained by rotating the figure $y = \cos^{-1} x$, $0 \leqslant x \leqslant \tfrac{1}{2}$, about the x-axis.

If this is done straightforwardly it will involve the integration of the function $(\cos^{-1} x)^2$; in fact (see figure 129(a)) the required volume is given by

volume generated by LMK = volume generated by OKLMT
                                        − volume generated by OKMT
                                        (a cylinder)

$$= \int_0^{\frac{1}{2}} \pi (\cos^{-1} x)^2 \, dx - [\pi (\tfrac{1}{3}\pi)^2] \cdot \tfrac{1}{2}.$$

The following alternative method leads to a simpler integral (see figure 129). Choose a point P on the y-axis between L and K and an interval NN′ of length δl containing P. Now let the element NAA′N′ be rotated about the x-axis, generating a volume δV. The shape generated is that of a thin cylindrical 'shell' of thickness δl, radius approximately OP, and height PQ. Therefore, since $PQ = \cos y_P$, where y_P is the value of y at P,

$$\delta V \approx 2\pi \, \text{OP} \cdot \delta l \cdot \text{PQ} = 2\pi y_P \delta l \cos y_P.$$

Therefore, omitting details,

$$\lim_{\delta l \to 0} \frac{\delta V}{\delta l} = 2\pi y_P \cos y_P,$$

leading to

$$\frac{dV(y)}{dy} = 2\pi y \cos y, \qquad \tfrac{1}{3}\pi \leqslant y \leqslant \tfrac{1}{2}\pi,$$

where $V(y)$ is the volume between the radius $\tfrac{1}{3}\pi$ and the radius y. Thus, $V(\tfrac{1}{3}\pi) = 0$. The volume required is $V(\tfrac{1}{2}\pi)$ and

$$V(\tfrac{1}{2}\pi) = \int_{\frac{1}{3}\pi}^{\frac{1}{2}\pi} 2\pi y \cos y \, dy = 2\pi [(y \sin y + \cos y)]_{\frac{1}{3}\pi}^{\frac{1}{2}\pi}$$

$$= 2\pi [(\tfrac{1}{2}\pi + 0) - (\tfrac{1}{3}\pi \cdot \tfrac{1}{2}\sqrt{3} + \tfrac{1}{2})]$$

$$= \pi^2 \left(1 - \frac{1}{\sqrt{3}}\right) - \pi.$$

392 Introductory Analysis

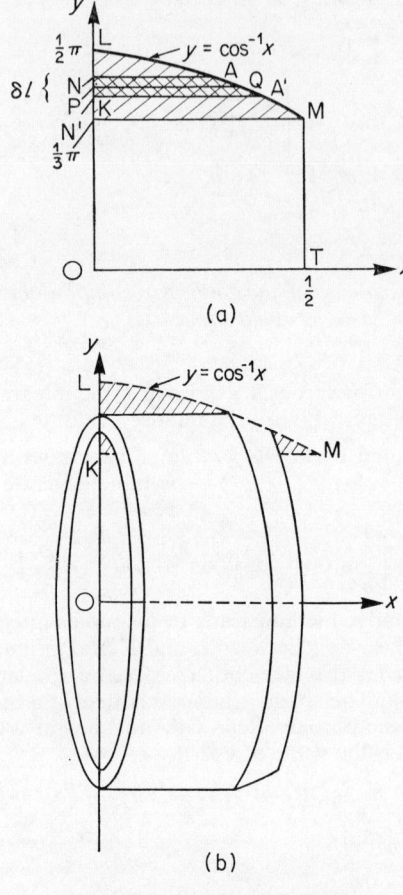

Cylindrical shell (b) formed by
rotating the portion NAA'N' of
(a) about the x-axis

FIGURE 129

The general formula resulting from this method of subdivision may be written

$$V = \int_{y_1}^{y_2} 2\pi x y \, dy \qquad (25)$$

where V is the volume of revolution produced by rotating about the x-axis the region bounded by the lines $x = 0$, $y = y_1$, $y = y_2$, and the curve itself.

Exercises

35. Find the volume of the axially symmetrical solid obtained by rotating the curve $y^2 = 4ax$, $b \leqslant x \leqslant c$ about the x-axis.

36. Find the volume of the solid of revolution formed by rotating the curve $y = x^{\frac{1}{2}}(l - x)^{\frac{1}{2}}$, $0 \leqslant x \leqslant l$ about the x-axis.

37. Find the volume of the region obtained by rotating one loop of the curve $y^2 = x^2(1 - x)/(1 + x)$ about the x-axis.

38. Find the volume of the 'ellipsoid of revolution' formed by rotating the elliptical figure $x^2/a^2 + y^2/b^2 = 1$ about the x-axis.

39. Find the volume of the spherical cap cut off from a sphere of radius a by a plane distance b ($b < a$) from the centre.

40. Find the volume of the body formed by rotating about the x-axis that part of the circle $x^2 + (y + c)^2 = a^2$, $a > c > 0$ which lies above the x-axis.

41. The parabola $y^2 = 4ax$, $0 \leqslant x \leqslant a$ rotates about the line $x = a$. Find the volume of the solid so formed. (Define elements of volume either by shells centred on $x = a$ or by discs centred on $x = a$.)

42. Find the volume of the blunt-nosed projectile obtained by rotating the curve $y = x^{\frac{1}{2}}/(1 + x^{\frac{1}{2}})$, $0 \leqslant x \leqslant 1$ about the x-axis (consider the method exemplified by Example 11).

43. Find the volume of the torus formed by rotating the ellipse
$$\frac{x^2}{a^2} + \frac{(y - c)^2}{b^2} = 1, \qquad c \geqslant a \geqslant b,$$
about the x-axis.

44. (a) A body is bounded by the planes $x = a$ and $x = b$, and the area of its cross-section perpendicular to the direction of the x-axis at a point x is $A(x)$. Give an argument to show that its volume is
$$\int_a^b A(x) \, dx.$$

(b) Find the volume of a trapezoid of height h based on a rectangle of sides a and b.

(c) A solid has a circular base of radius a. Every section perpendicular to the base is a square. Find its volume.

12.8 Problems on density

The density ρ of a uniform body (one having the same properties all the way through) is defined by

$$\rho = \frac{\text{Mass of the body}}{\text{Volume of the body}}.$$

Because of its supposed uniformity, the same value for this ratio will be obtained if any part of the body is considered instead of the whole. If the body is not uniform but its consistency varies smoothly from point to point, a density at each point may be defined in a common-sense way by means of limits. Let P be the point at which the density is to be read. Take a portion of the body of volume δV, containing P, and suppose that its mass δM is measured. $\delta M/\delta V$ will be close to anything we could reasonably call 'density at the point P' if δV is very small, though we should get a slightly different answer depending on precisely what *element of volume* δV was being considered. We say that provided $\delta M/\delta V$ tends to a limit, and always the same limit, as the dimensions of the element tend to zero in any way whatever (the element still containing the point P), then *this limit defines the density at* P, and we write

$$\rho = \lim_{\delta V \to 0} \frac{\delta M}{\delta V}.$$

The result has dimensions of mass per unit volume:

This limit is different from limits of the type $x \to a$, or $n \to \infty$, which the reader is familiar with because it involves consideration, in theory at any rate, of all the possible ways in which the dimensions of the element can tend to zero.*

---

* There are many density-like quantities involved in physics. For example, we define the heat content H of a quantity of substance of mass m at a *uniform* temperature T by

$$H = mcT$$

where c is a constant characteristic of the substance called its specific heat. If the body is *not* at a uniform temperature we can define its 'heat density' at any point by

$$\lim_{\delta V \to 0} \frac{\delta H}{\delta V}$$

having dimensions Heat/Volume. It is equal to $\rho c T$, where T is the temperature at the point and ρ is its density there.

The definite integral and some applications

In the following problems an assumption is made about bodies of variable density: that *the mass of such a body, M, lies between $V\rho_{min}$ and $V\rho_{max}$, where V is its volume and ρ_{min}, ρ_{max}, are the minimum and maximum densities respectively.*

We cannot prove this with the mathematics at our disposal, and we shall simply regard it as a hypothesis. We made very similar hypotheses of a geometrical character in our treatment of area and volume problems.

Example 12 Obtain in the form of an integral the mass of a prismatic column of uniform cross-section A and of length l, whose density is $\rho(x)$, where x is the distance from one end.

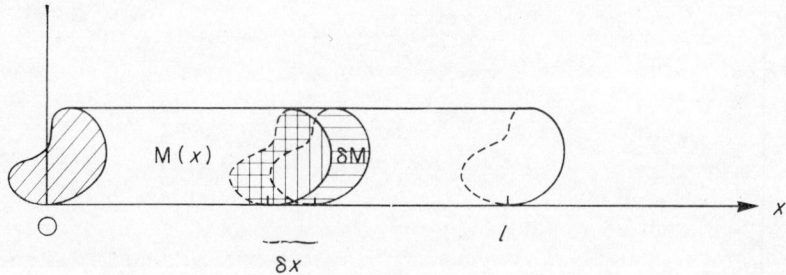

FIGURE 130

Introduce a function $M(x)$, the mass of the column between sections through O and x, and take an interval $(x, x + \delta x)$ (figure 130) the mass in this interval being δM, so that (assuming $\delta x > 0$)

$$\delta M = M(x + \delta x) - M(x).$$

Then

$$\delta M \approx \rho(x) \times \text{volume of the section} = A\rho(x)\,\delta x.$$

Thus

$$\frac{\delta M}{\delta x} \approx A\rho(x),$$

leading in the limit $\delta x \to 0$ to

$$\frac{dM(x)}{dx} = A\rho(x). \qquad (26)$$

Since $M(0) = 0$, we have as usual

$$M(x) = \int_0^x A\rho(u)\,du$$

and

$$\text{mass of column} = M(l) = \int_0^l A\rho(u)\,du.$$

It would be quite normal to formulate the argument in the above way, and we have several times provided a similar abbreviated treatment in connection with areas and volumes. It can be made more precise by using explicitly the hypotheses preceding this example in the following way. If in the section $(x, x + \delta x)$ the minimum density is ρ_{max} and the minimum density is ρ_{min}, then

$$A\rho_{min} \delta x \leq \delta M \leq A\rho_{max} \delta x,$$

or

$$A\rho_{min} \leq \frac{\delta M}{\delta x} \leq A\rho_{max}.$$

If ρ is assumed to be a continuous function, then as $\delta x \to 0$, $\rho_{min} \to \rho(x)$ and $\rho_{max} \to \rho(x)$, and equation (26) is obtained again.

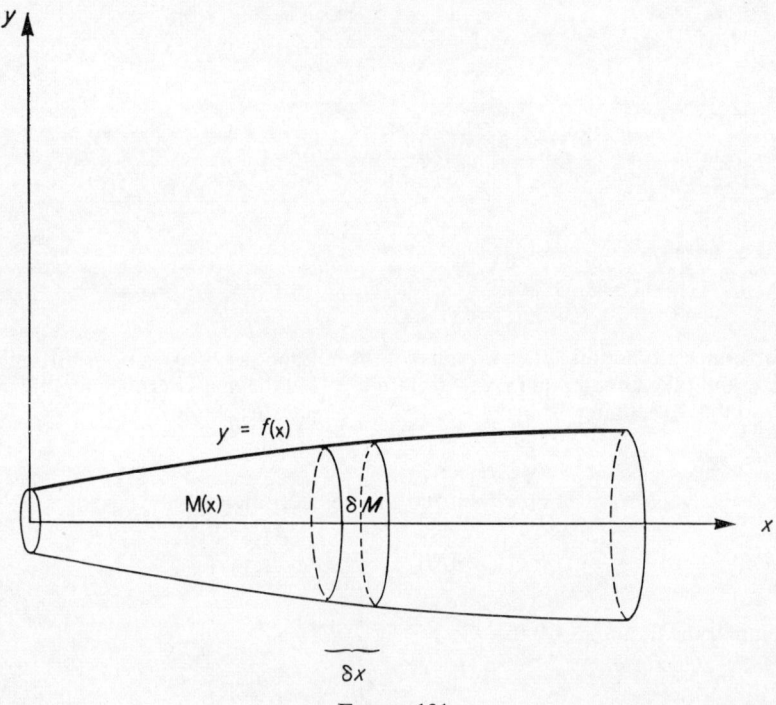

FIGURE 131

Example 13 *A beam of varying circular cross-section is formed by rotating the curve $y = f(x)$, $0 \leq x \leq l$, about the x-axis. Its density is given by $\rho(x)$. Find its mass.*

The beam is a solid of revolution, with the density uniform on each cross-section (figure 131). Let $M(x)$ be the mass of the part of the beam between the

sections at O and x. Consider the slice lying in the interval $(x, x + \delta x)$. Its mass is δM, given by
$$\delta M = M(x + \delta x) - M(x),$$
and we denote its volume by δV. Then
$$\delta M \approx \rho(x)\, \delta V, \tag{27}$$
and
$$\delta V \approx \pi f^2(x)\, \delta x. \tag{28}$$
Therefore,
$$\delta M \approx \pi f^2(x) \rho(x)\, \delta x,$$
and we expect that, therefore,
$$\frac{dM(x)}{dx} = \lim_{\delta x \to 0} \frac{\delta M}{\delta x} = \pi f^2(x)\rho(x).$$
Since $M(0) = 0$, we obtain finally
$$\text{mass} = M(l) = \int_0^l \pi f^2(x) \rho(x)\, dx.$$

The student should make this argument complete by applying the appropriate inequalities in place of the approximations (27) and (28).

Example 14 *The density of a sphere of radius c is given in terms of distance from the centre by $\rho(r) = a - br$, where a and b are constants. Find the total mass.**

Let $M(r)$ be the mass inside a sphere of centre O and radius r ($r < c$) (figure 132). Now consider an interval in the variable $r : (r, r + \delta r)$. This corresponds to a spherical shell, having volume δV, say, and mass δM.

We have
$$\delta V = \delta(\tfrac{4}{3}\pi r^3) \approx \delta r \frac{d}{dr}(\tfrac{4}{3}\pi r^3) = 4\pi r^2\, \delta r \tag{29}$$
and
$$\delta M \approx \rho(r)\, \delta V. \tag{30}$$
Thus
$$\delta M \approx 4\pi r^2 \rho(r)\, \delta r$$
from which we obtain as in previous cases
$$\frac{dM(r)}{dr} = 4\pi r^2 \rho(r). \tag{31}$$

---

* The density is said to have a spherically symmetrical distribution, since its value depends only on the distance from the centre. The Earth is a body having very nearly spherical symmetry.

Finally, after inserting the given density function, we have
$$\text{mass} = \int_0^c 4\pi r^2(a - br)\,dr = \tfrac{4}{3}\pi c^3(a - \tfrac{3}{4}bc).$$

Again, the approximate equations (29) and (30) may be replaced by inequalities to give a precise derivation.

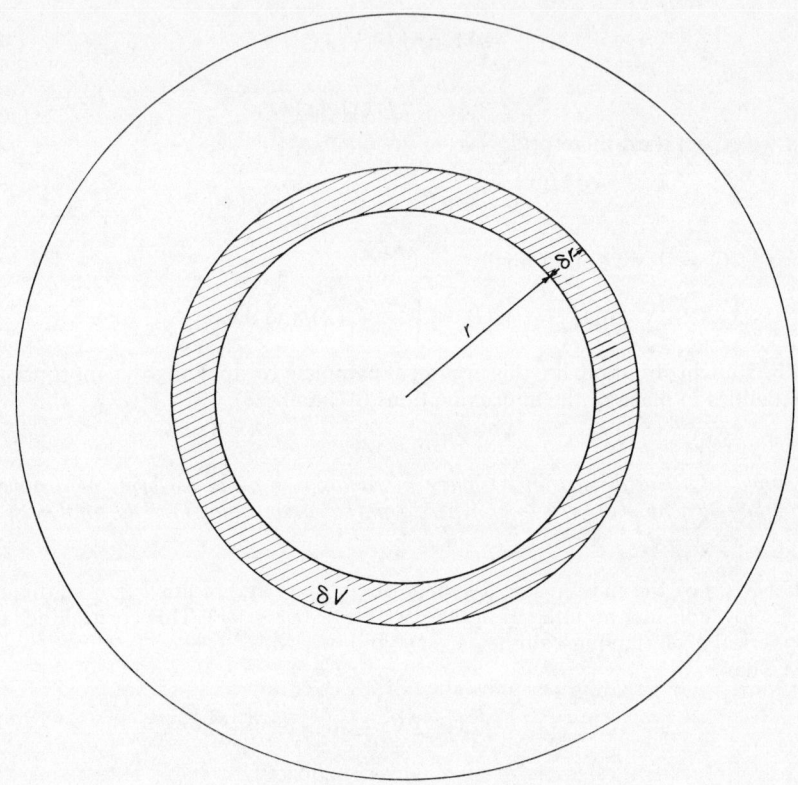

Spherical shell, radius r, thickness δr,
mass δm, volume δV in a sphere of radius c

FIGURE 132

Exercises

45. Find the mass of a solid circular cylinder of length l and radius a, the density of whose substance is given by $\rho(x) = \rho_0 + \varepsilon x$, where ρ_0 and ε are constants and x is the distance from one end. Prove that the mean density (mass/volume) is equal to the density at the mid-point of the rod.

The definite integral and some applications

46. A rod is in the form of a right circular truncated cone of height 10 cm, the radii of whose ends are 0·1 and 0·2 cm, respectively. Its density is given by $\rho(x) = 2 \cdot 0 - 0 \cdot 1 x$ gm/cm^3, where x is the distance from the narrow end. Find its mass.

47. The density of water in a lake varies with depth due to compression of the water at deeper levels. The density $\rho(x)$ at distance x below the surface is given by $\rho(x) = \rho_0(1 + \rho_0 x/\kappa)$, where ρ_0 is density at the surface and κ is a constant. If $\rho_0 = 1$ gm/cm^2 and $\kappa = 2 \cdot 5 \times 10^7$ gm/cm find the mass of water in a vertical column of cross-section 1 cm^2 extending to a depth of 1 km below the surface, and hence calculate the pressure (force/unit area) at this depth.

48. Find the heat content of a section of a right cylindrical pipe of length l, internal and external radii r_0 and r_1, respectively, made of a substance of density ρ and specific heat c, when the temperature at a radial distance r is given by $T(r) = a + b \ln r$, a and b being constants.

49. Find the heat content of a hollow sphere of internal and external radii R_0 and R_1, respectively, the density being ρ and the specific heat c, whose temperature distribution is given $T(R) = A + B/R$, R being the distance from the centre.

50. When a viscous liquid is driven along a narrow tube of circular cross-section, of radius a, a velocity distribution is set up given by

$$v(r) = A(a^2 - r^2)$$

where r is distance from the centre, $v(r)$ is the velocity of the liquid in the direction of the tube at a distance r from the centre, and A is a constant. Find an expression for the amount of liquid per unit time flowing over any cross-section.

51. A porous hemispherical bowl of radius a is filled with water. The leakage per unit area, per unit time, varies with depth below the surface, being equal to αx where α is a constant and x is depth below the surface. Find the total rate at which water begins to leak out while the bowl is still full. (Consider a circular strip of bowl surface cut out by two nearby horizontal planes and use polar coordinates.)

52. A cloud formation with a very long straight front moves across the country perpendicularly to its front at a speed $v = \beta t$ m.p.h., where β is a constant. It discharges rain at a steady rate $Ax^{\frac{1}{2}} e^{-\alpha x}$ gallons per hour per square mile of cloud at a distance x miles back from its front (where A and α are constants), starting at $t = 0$. Find an expression for the total amount of rain in gallons per square mile which by time t has fallen on a point on the ground initially under the cloud front.

12.9 The numerical estimation of definite integrals

In Volume 2 we shall describe some of the most important further available methods for obtaining integrals of commonly appearing types of function. However, the range of functions which can be integrated explicitly (in the sense that the integrals can be written as combinations of the elementary functions x^n, $\sin x$, e^x and so on) is quite small, and not nearly large enough to serve all practical purposes. This deficiency does not arise principally because sufficiently effective *methods* are not available but rather because our repertory of functions is not extensive enough to represent the integrals which we wish to evaluate. For example,

$$\int e^{-x^2} \, dx, \quad \int \frac{e^x}{x} \, dx, \quad \int \frac{dx}{\sqrt{[(a^2 - x^2)(b^2 - x^2)]}}$$

and many others are simply *not expressible* in terms of finite combinations of the elementary functions. We have nothing to express them with: the integrals define entirely new functions.

If the integral concerned has sufficiently wide application its properties and general character will be studied, tables of its numerical values will be constructed and it may be given a special name. It then enters the category of 'known' functions on the same footing as the elementary functions.

This situation is quite unlike that which exists in the case of differentiation. Any finite combination of the elementary functions can be differentiated to give an explicit result in the form of a similar finite combination of elementary functions by using a few simple rules. This is impossible in the case of integration.

The user of mathematics in any field of application will encounter integrals which he cannot evaluate. This may be because the integral defines a new function as described or simply because the integral is too complicated; whatever the reason, he need seldom be completely at a loss since the *numerical* value of any definite integral can in principle be determined to any required degree of accuracy by computational processes. The subject of 'numerical analysis'* contains a sophisticated study of such processes. Only the simplest methods can be treated here and in a very informal way.†

---

* See, for example J. M. Rushforth, Computers and Computing (Wiley).

† In order to discuss definite integrals we have up to the present point required that somehow we should know that there exists an indefinite integral, which is a question independent of whether we can actually write it down. This question is treated in the next chapter, but for the present we shall assume (correctly) that continuous functions have integrals.

The definite integral and some applications

The methods described will be based on the fact that a definite integral is equal numerically to the area (suitably interpreted; see section 12.4) between the graph of the integrand and the axis. However, instead of seeking an *approximation* to the area under the *true* graph of the function (as in the well known method of 'counting squares') we shall find the *true* area under another curve which *approximates* to the graph of the given function.

The simplest method is the following. Suppose we require an approximation to

$$\int_a^b f(x)\,dx$$

where f is a given function, whose graph is sketched in figure 133.

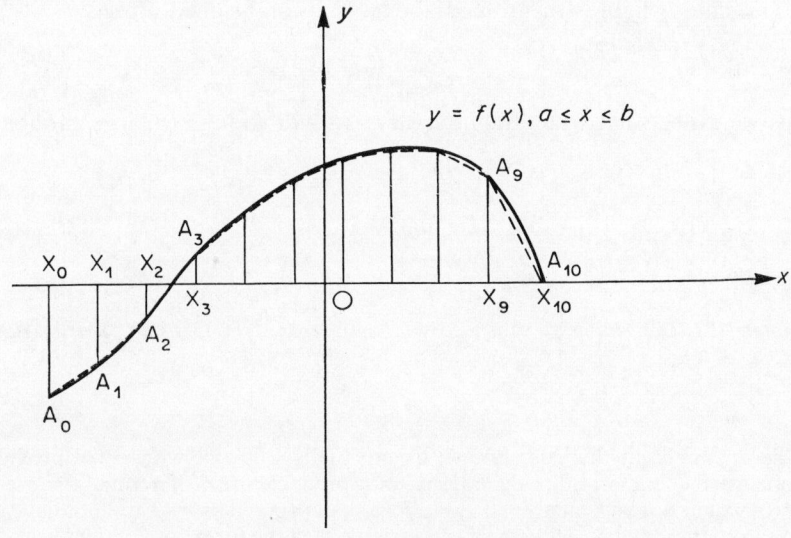

Approximation to a graph by a polygonal line

FIGURE 133

Divide the interval $[a,b]$ into a certain number n (10 in the figure) of equal intervals, each of length $h = (b - a)/n$. If we erect perpendiculars upon the points of subdivision to intersect the curve at the points $A_0, A_1, \ldots, A_{n-1}, A_n$, then the polygonal 'curve' $A_0 A_1, \ldots, A_{n-1} A_n$, constructed by joining the points by straight lines, approximates more or less closely to the original curve (it is assumed that the interval h is chosen small enough to make the approximation convincing). The *signed* area under the polygonal curve will be an approximation to the *signed* area under the curve $y = f(x)$, and this signed area is the sum of the areas of the trapezia

$$X_0X_1A_1A_0, \quad X_1X_2A_2A_1, \ldots, \quad X_{n-1}X_nA_nA_{n-1},$$

counted as positive when the trapezium lies entirely above the x-axis, negative when it lies below. The 'trapezia' $X_2X_3A_3A_2$ and $X_9A_{10}A_{10}A_9$ in the figure appear anomalous, and this point is referred to below.

If the points of subdivision $X_0, X_1, \ldots, X_n$ have abscissae $x_0 (= a)$, $x_1 = a + h$, $x_2 = a + 2h$, ..., $x_n = a + nh \,(= b)$, the ordinates at these points are $f(x_0), f(x_1), f(x_2), \ldots, f(x_n)$ which for brevity we shall call $f_0, f_1, f_2, \ldots, f_n$.

The signed area of the trapezium $X_rX_{r+1}A_{r+1}A_r$ is $\tfrac{1}{2}h(f_r + f_{r+1})$ and the reader will have no difficulty in proving that this formulae is true also for the anomalous figures mentioned above. The total signed area is

$$\tfrac{1}{2}h(f_0 + f_1) + \tfrac{1}{2}h(f_1 + f_2) + \ldots + \tfrac{1}{2}h(f_{n-1} + f_n) =$$
$$\tfrac{1}{2}h(f_0 + 2f_1 + 2f_2 + \ldots + 2f_{n-1} + f_n).$$

Thus we expect that when the interval h is small enough (n large enough),

$$\int_a^b f(x)\,dx \approx \tfrac{1}{2}h(f_0 + 2f_1 + 2f_2 + \ldots + 2f_{n-1} + f_n). \tag{32}$$

This result is called the *trapezium rule*.

Example 15 Determine $\int_0^{\frac{1}{2}\pi} \cos x \, dx$ *within an error of* 0·005, *using the trapezium rule.*

The answer is in this case known exactly: it is 1·000 ..., which will provide a conclusive check on our calculations. In a practical case, of course, the exact answer would not be available to us, and we shall proceed as if we did not have it.

We note that the trapezium rule can be written in the form:

$$\int_a^b f(x)\,dx \approx \tfrac{1}{2} \times \text{(interval)} \times [\text{sum of 1st and last ordinates} +$$
$$2 \times \text{(sum of intermediate ordinates)}]. \tag{33}$$

The accuracy we obtain will be estimated in the following way. We shall carry out the calculation firstly for 5 intervals and obtain an approximation. We shall then repeat the calculation, using twice the number of intervals. This second approximation will differ from the first by some amount. Suppose that the *difference* is substantially less than the permitted error (0·005), then we have some grounds for expecting that further subdivisions will involve further changes in the approximation only very much smaller than that already incurred, and consequently that our second approximation will itself be within the permitted error. If the difference is not small enough, the interval will be halved again and the calculation repeated.

This argument is based on the supposition that by doubling, quadrupling, ..., the number of intervals (and consequently halving, quartering, ..., the length of each interval) we obtain *a sequence of approximations which converges to the true value of the integral* sufficiently quickly.

We shall carry 4 decimal places in the working, since rounding-off errors in the last decimal place will not then accumulate, under the numerous additions involved, to a magnitude comparable with the permitted error.

$$\text{First ordinate } f_0 = \cos 0.0000 \ldots = 1.0000$$
$$\text{Last ordinate } f_n = \cos \tfrac{1}{2}\pi = 0.0000$$

Sum of first and last ordinates $(f_0 + f_n) = 1.0000$

First Estimate No. of intervals $n = 5$. Length of each interval 0·3142.

Table of intermediate ordinates:

x:	0·3142	0·6284	0·9426	1·2568
$\cos x$:	0·9510	0·8090	0·5877	0·3089
	(f_1)	(f_2)	(f_3)	(f_4)

Therefore,

$$2 \times (\text{sum of intermediate ordinate}) = 2 \times 2.2656 = 5.3132.$$

Thus (see equation (33)),

$$\text{first approximation} = \tfrac{1}{2} \times 0.3142 \times (1.0000 + 5.3132) = 0.9918.$$

This is in fact quite close to the true result, 1·0000, but we could not have known this in a practical case. We therefore test to see whether this result changes appreciably when the interval is halved.

Second Estimate We now double the number of intervals (halving the length of each) and increase the number of ordinates brought into consideration. However, there is no need to recalculate all the ordinates—many of our new points of subdivision coincide with previously calculated values, which is a good reason for *halving* the interval.

No. of intervals $n = 10$.   Length of intervals $= 0.1571$.

Table of *new* intermediate ordinates (those not occurring in the first estimate):

x:	0·1571	0·4713	0·7855	1·0997	1·4139
$\cos x$:	0·9877	0·8910	0·7070	0·4539	0·1563

$$2 \times (\text{sum of } all \text{ intermediate ordinates}) = 2 \times 5.8525 = 11.7050.$$

Thus,

$$\text{second approximation} = \tfrac{1}{2} \times 0.1571 \times (1.0000 + 11.7050) = 0.9980.$$

The difference between the first and second estimates is 0·0062 which is not yet within our permitted error, 0·005. We therefore halve the interval again.

Third Estimate

No. of intervals $n = 20$.   Length of each interval $= 0.0786$.

Table of *new* intermediate ordinates (those not already computed in the first or second estimates):

x:	0·0786	0·2357	0·3928	0·5499	0·7070
$\cos x$:	0·9969	0·9724	0·9239	0·8525	0·7603
x:	0·8641	1·0212	1·1783	1·3354	1·4925
$\cos x$:	0·6494	0·5223	0·3825	0·2332	0·0782

$2 \times$ (sum of *all* intermediate ordinates) $= 2 \times 12 \cdot 2241 = 24 \cdot 4482$.

Thus,

third approximation $= \frac{1}{2} \times 0 \cdot 0786 \times (1 \cdot 0000 + 24 \cdot 4482) = 0 \cdot 9996$.

The difference between the last two estimates is 0·0016 which seems comfortably within the permitted error, 0·005, and we may be inclined to stop here. The *true* error is in fact only 0·0004 or thereabouts (the last figure in the estimate will not be reliable, due to rounding errors). The true error in the second estimate was 0·0020.

To carry out such a process without at least a mechanical calculating machine is scarcely possible; using a calculating machine the time-consuming part is in looking up tables of the function. A digital computer permits vast calculations of this kind to be carried out in a very short time.

Shorn of the numerical detail, the above method is based on approximating to the given curve, $y = f(x)$, by selecting a set of points of subdivision $x_0, x_1, x_2, \ldots, x_n$ and calculating a corresponding set of ordinate values, $f_0, f_1, f_2, \ldots, f_n$ which carry all the explicit information we use about the function. In the above method we allowed the original curve to be approximated by a broken line coinciding with the curve at the points where the ordinates are calculated. We shall now describe another method where better use is made of the information carried by the number $f_0, f_1, \ldots, f_n$.

The points will be taken in groups of three and each group of three points connected by a parabolic curve. On general grounds we should expect a better 'fit' to the original curve than that obtained by the polygonal approximation which was used to derive the trapezium rule.*

Let us suppose firstly that we wish to connect three points on a graph; $(-h, y_1), (0, y_2)$ and (h, y_3) (figure 134) by means of a parabolic curve of the form

$$y = a\left(\frac{x}{h}\right)^2 + b\left(\frac{x}{h}\right) + c,$$

---

* There is no guarantee that this will give better results: the original curve $y = f(x)$ might, indeed, be exactly a polygonal line joining the given points, in which case our first method might be exact and anything else must be worse. In general, however, intuition leads one to expect that a smooth curve will usually give a better fit than two straight lines.

where a, b and c are constants, these points being chosen and the equation being written in this form in order to simplify the determination of the unknown constants a,b,c. By putting $x = -h, 0, h$ in turn we obtain:

$$y_1 = a - b + c,$$
$$y_2 = c,$$
$$y_3 = a + b + c.$$

The constants a, b and c are therefore given by

$$a = \tfrac{1}{2}(y_1 - 2y_2 + y_3), \qquad b = \tfrac{1}{2}(y_3 - y_1), \qquad c = y_2.$$

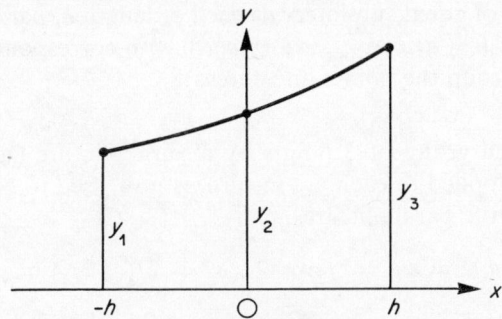

FIGURE 134

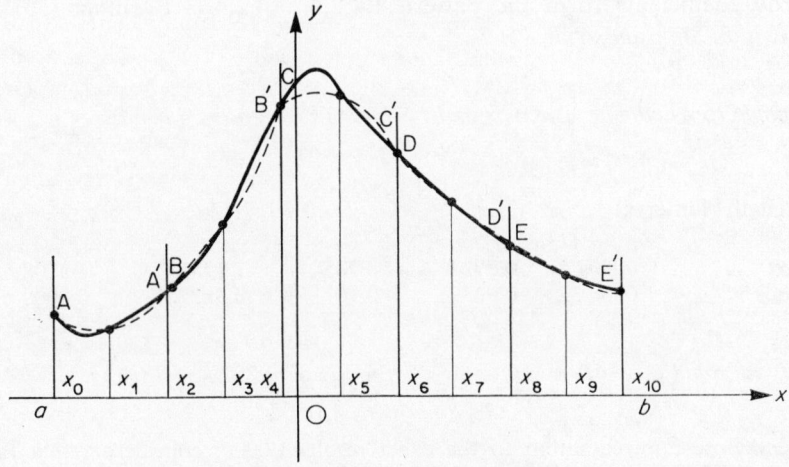

Illustrating the fit of a function (heavy line) by a sequence $AA', BB', ..$ of parabolic segments (broken lines)

FIGURE 135

The area under this curve is given by

$$\int_{-h}^{h} y \, dx = \int_{-h}^{h} [a(x/h)^2 + b(x/h) + c] \, dx$$
$$= 2h(\tfrac{1}{3}a + c) = \tfrac{1}{3}h(y_1 + 4y_2 + y_3). \tag{34}$$

It is clear that formula (34) for the area has the same *form* regardless of the position of the figure, the only requirement being that the three abscissae, having ordinates y_1, y_2, y_3, should be separated by intervals h.

We shall now approximate to $\int_a^b f(x) \, dx$. Divide the interval $[a,b]$ into an *even* number of equal sub-intervals each of length h, naming the points of subdivision $x_0(=a), x_1, x_2, \ldots, x_n$, ($n$ odd), with corresponding ordinates $f_0, f_1, f_2, \ldots, f_n$. Group the points into threes:

$$(x_0, x_1, x_2), (x_2, x_3, x_4), \ldots, (x_{n-2}, x_{n-1}, x_n)$$

and connect each group of 3 points by a parabolic arc (see figure 135). Equation (34), applied to each arc in turn, is now used to give the sum of the areas under the parabolic arcs,

$$\int_a^b f(x) \, dx \approx \tfrac{1}{3}h(f_0 + 4f_1 + f_2) + \tfrac{1}{3}h(f_2 + 4f_3 + f_4) + \ldots$$
$$+ \tfrac{1}{3}h(f_{n-2} + 4f_{n-1} + f_n)$$
$$= \tfrac{1}{3}h(f_0 + 4f_1 + 2f_2 + 4f_3 + \ldots + 2f_{n-2} + 4f_{n-1} + f_n). \tag{35}$$

The coefficients form the pattern 1,4,2,4,...,4,2,4,1. Formula (35) is known as *Simpson's rule*.

Example 16 Estimate $\int_0^{\frac{1}{2}\pi} \cos x \, dx$ by Simpson's rule, using 5 points.

Length of interval $\qquad h = \dfrac{3 \cdot 1416}{2 \times 4} = 0 \cdot 3927.$

x:	0·0000	0·3927	0·7854	1·1781	1·5708
$\cos x$:	1·0000	0·9239	0·7071	0·3827	0·0000
	(f_0)	(f_1)	(f_2)	(f_3)	(f_4)

Thus,
$$\tfrac{1}{3}h(f_0 + 4f_1 + 2f_2 + 4f_3 + f_4) = \tfrac{1}{3} \times 0 \cdot 3927 \times 7 \cdot 6406$$
$$= 1 \cdot 0002$$

a very close approximation to the exact result, 1·0000, considering that the amount of information about the function explicitly used was its values at only five points. (We obtained an error of 0·0004 using the trapezium rule with 21 points.) Such an economical attainment of an accurate result is not guaranteed, however, and it is always advisable to repeat the calculation using more points, or (eventually) to make a proper study of numerical techniques.

The definite integral and some applications

Exercises

53. Use the trapezium rule to calculate $\int_0^1 x^2 \, dx$ to 2 decimal places.

54. Obtain a value for ln 2 to 3 decimal places by using Simpson's rule to calculate $\int_1^2 dx/x$.

55. Use the trapezium rule to calculate $\int_0^1 \frac{x \, dx}{1 + x^2}$ to 2 decimal places (compare the exact answer, $\frac{1}{2} \ln 2$).

56. Obtain a value for π correct to 3 decimal places by calculating

$$\tfrac{1}{4}\pi = \int_0^1 \frac{dx}{1 + x^2}$$

using (a) the trapezium rule, (b) Simpson's rule.

57. Sketch an example to indicate that halving the interval in the trapezium rule may sometimes lead to a worse approximation (but convince yourself that repeated division of the interval will ultimately give a better approximation).

50. Use Simpson's rule to obtain a 2-decimal approximation to the area of the sectorial region given by $r = 2 \cos^2 \theta$, $-\tfrac{1}{4}\pi \leq \theta \leq \tfrac{1}{4}\pi$.

59. A rod of length 10 units lying in $-5 \leq x \leq 5$ is in the form of a solid of revolution formed by rotating the curve $y = 1 - 0.001x^2$ about the x-axis. Its density is given by $\rho(x) = 2 \, e^{-0.001 x^2}$ in suitable units. Find its mass to 2 decimal places.

60. Write a flow diagram for evaluating $\int_a^b f(x) \, dx$ by the trapezium rule, using n intervals each of length h. Use this as a unit in a flow diagram which causes the interval to be halved repeatedly until two successive estimates differ by less than ε.

Write an alternative program which requires each function value to be computed only once.

12.10 Integration of power series

In the earlier part of this book new functions, such as e^x, were explored by expressing them as power series. Integrals may also define new functions, as explained at the beginning of the last section. For example,

$$F(x) = \int_0^x e^{-t^2} \, dt$$

is a new function. We may hope to represent $F(x)$ as a power series by integrating over the known power series for the integrand

$$F(x) = \int_0^x \left(1 - t^2 + \frac{1}{2!}t^4 - \ldots\right) dt$$

$$= x - \tfrac{1}{3}x^3 + \frac{1}{2!5}x^5 - \ldots .$$

The general process is as follows.

Consider the function f defined by the power series

$$f(x) = \sum_{r=0}^{\infty} c_r x^r \tag{36}$$

within its interval of convergence: $-R < x < R$. The integral from 0 to x where $-R < x < R$ is given by

$$\int_0^x f(t) \, dt = \int_0^x \left(\sum_{r=0}^{\infty} c_r t^r\right) dt. \tag{37}$$

If we interchange the order of integration and summation on the right-hand side of (37) (which needs its legitimacy confirming, for it is not always right to interchange the order of application of two limiting processes), we obtain

$$\sum_{r=0}^{\infty} c_r \left(\int_0^x t^r \, dx\right)$$

which is equal to

$$\sum_{r=0}^{\infty} \frac{c_r}{r+1} x^{r+1}. \tag{38}$$

This series is the result of integrating the terms of (36) one at a time, and we shall show that it is always the correct series representation of the integral (37).

THEOREM *Power series may be integrated term by term over any interval within their interval of convergence.*

We first confirm that the derivative of (38) is equal to (36) for $-R < x < R$. We have

$$\left|\frac{c_r}{r+1} x^{r+1}\right| \leqslant |c_r x^r| \, |x| \leqslant R |c_r x^r|.$$

The definite integral and some applications 409

Therefore, by the Comparison Test for absolute convergence, (38) converges whenever (36) converges absolutely, i.e. in $-R < x < R$. Therefore put

$$F(x) = \sum_{r=0}^{\infty} \frac{c_r}{r+1} x^{r+1}, \quad -R < x < R. \tag{39}$$

By Theorem 9.13,

$$F'(x) = \sum_{r=0}^{\infty} \frac{d}{dx}\left(\frac{c_r}{r+1} x^{r+1}\right) = \sum_{r=0}^{\infty} c_r x^r = f(x).$$

Therefore F is an integral of f:

$$F(x) = \int_0^x f(t)\,dt + \text{constant},$$

and the constant is zero since $F(0) = 0$, from (39).

Integrals of the form

$$\int_a^b f(t)\,dt$$

where $|a|$ and $|b|$ are less than R can be written

$$\int_a^b f(t)\,dt = \int_0^b f(t)\,dt - \int_0^a f(t)\,dt = \sum_{r=0}^{\infty} \frac{c_r}{r+1}(b^{r+1} - a^{r+1}),$$

combining the series by Theorem 9.6, and this completes the theorem.

The result can be used to give numerical estimates, as in the following example.

Example 17 Find an approximation to 3 decimal places of the integral

$$\int_0^{0.1} \frac{\cos t}{1+t}\,dt.$$

$$\frac{\cos t}{1+t} = (1 - \tfrac{1}{2}t^2 + \tfrac{1}{24}t^4 - \ldots)(1 - t + t^2 - t^3 + t^4 - \ldots)$$

$$= 1 - t + \tfrac{1}{2}t^2 - \tfrac{1}{2}t^3 + \tfrac{13}{24}t^4 + \ldots,$$

with interval of convergence $-1 < t < 1$. Since $0.1 < 1$,

$$\int_0^{0.1} \frac{\cos t}{1+t}\,dt = \int_0^{0.1} (1 - t + \tfrac{1}{2}t^2 - \tfrac{1}{2}t^3 + \ldots)\,dt$$

$$= [t - \tfrac{1}{2}t^2 + \tfrac{1}{6}t^3 + \tfrac{1}{8}t^4 + \ldots]_0^{0.1}$$

$$= 0.095$$

to 3 decimals.

Exercises

61. Show that power series of the more general form $\sum_{r=0}^{\infty} c_r(x-a)^r$ may be integrated within their interval of convergence.

Find $\int_{0\cdot 5}^{0\cdot 7} \frac{e^t}{t} \, dt$ to 3 places of decimals.

62. By expanding $(1-t^2)^{-\frac{1}{2}}$ by the Binomial Theorem and integrating term-by-term prove that
$$\sin^{-1} x = x + \frac{1}{2}\frac{x^3}{3} + \frac{1.3}{2.4}\frac{x^5}{5} + \ldots$$
(why not $-\cos^{-1} x$?). Where is this series valid?

63. Find the first few terms of the power series for $\tan^{-1} x$. What is its radius of convergence? (Compare question 54.)

64. Obtain an approximation to the integral
$$\int_1^4 \frac{dx}{(1+x^4)^{\frac{1}{2}}}$$
in the form of the first few terms of a power series.

65. Prove that
$$\int_0^x e^{-t^2} \, dt = \sum_{r=0}^{\infty} \frac{(-1)^r x^{2r+1}}{r!(2r+1)}.$$

12.11 Some inequalities involving integrals

It is often important to be able to estimate the magnitude of definite integrals without necessarily evaluating them. The results below are of frequent use in this connection. They are subject to the assumption that somehow we know that the functions concerned actually *have* indefinite integrals (see Chapter 13).

I If $f(x) \geqslant 0$ on $a \leqslant x \leqslant b$, then $\int_a^b f(x) \, dx \geqslant 0$.

Put
$$F(x) = \int f(x) \, dx.$$

Then
$$\int_a^b f(x)\,dx = F(b) - F(a)$$
$$= (b-a)F'(\xi),$$
(for some number ξ, $a < \xi < b$, by the Mean Value Theorem (section 7.2)),
$$= (b-a)f(\xi).$$
But $b \geqslant a$ and $f(\xi) \geqslant 0$, therefore the result follows.

II *If $f(x) \geqslant g(x)$ on $a \leqslant x \leqslant b$, then*
$$\int_a^b f(x)\,dx \geqslant \int_a^b g(x)\,dx.$$
$f(x) - g(x) \geqslant 0$; so by I,
$$\int_a^b [f(x) - g(x)]\,dx \geqslant 0, \quad \text{or} \quad \int_a^b f(x)\,dx - \int_a^b g(x)\,dx \geqslant 0,$$
and the result follows.

III *If $m \leqslant f(x) \leqslant M$ on $a \leqslant x \leqslant b$, where m and M are two numbers, then*
$$m(b-a) \leqslant \int_a^b f(x)\,dx \leqslant M(b-a).$$
The proof, using II, is left to the reader.

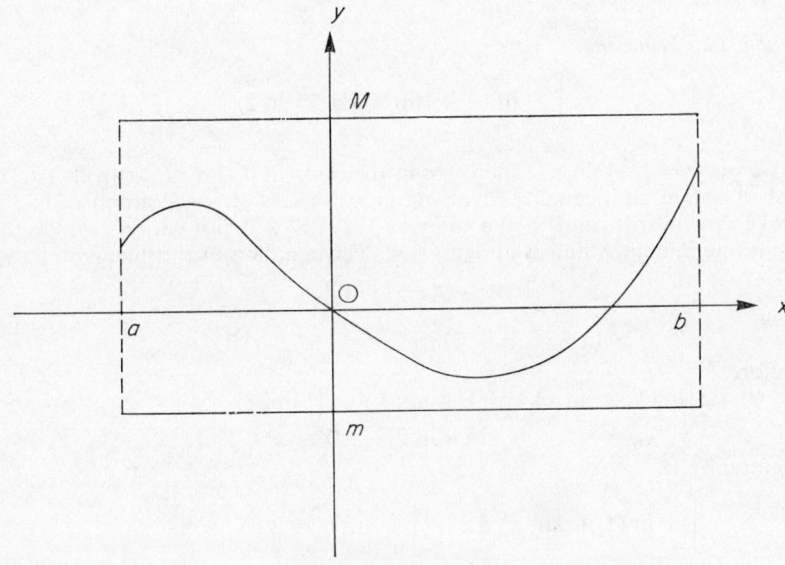

FIGURE 136

These three results are all strongly motivated by considering their meaning in terms of the area analogy (section 1.4). I says that a positive function gives rise to a positive area, II is interpreted already in section 12.6, and III is suggested by figure 136. In the figure we have made M positive and m negative. The result states that the total signed area between the curve and the x-axis is not greater than the area $M(b - a)$ of the upper rectangle, nor less than the signed area $m(b - a)$ (m a negative number) of the lower rectangle.

Example 18 *Prove that*
$$0{\cdot}500 \leqslant \int_0^{\frac{1}{2}} \frac{\mathrm{d}x}{\sqrt{(1 - x^4)}} \leqslant 0{\cdot}524.$$

We have
$$\sqrt{(1 - x^2)} \leqslant \sqrt{(1 - x^4)} \leqslant 1 \quad \text{on } 0 \leqslant x \leqslant \tfrac{1}{2}.$$
Therefore, using II,
$$\int_0^{\frac{1}{2}} \mathrm{d}x \leqslant \int_0^{\frac{1}{2}} \frac{\mathrm{d}x}{\sqrt{(1 - x^4)}} \leqslant \int_0^{\frac{1}{2}} \frac{\mathrm{d}x}{\sqrt{(1 - x^2)}}.$$
The last integral is equal to $[\sin^{-1} x]_0^{\frac{1}{2}} = \tfrac{1}{6}\pi = 0{\cdot}5235\ldots < 0{\cdot}524$. The first is equal to $\tfrac{1}{2} = 0{\cdot}500$. The student should try sketching the curves $y = 1$, $y = (1 - x^2)^{-\frac{1}{2}}$, $y = (1 - x^4)^{-\frac{1}{2}}$ to obtain a visual appreciation of the approximations involved.

Example 19 *Prove that*
$$\int_0^{\frac{1}{2}\pi} \ln(1 + \sin x)\, \mathrm{d}x \geqslant \ln 2.$$

If we put $z = 1 + \sin x$, z increases in the interval $0 \leqslant x \leqslant \tfrac{1}{2}\pi$ from 1 to 2. We shall obtain an inequality involving $\ln z$, $1 \leqslant z \leqslant 2$. The graph of $\ln z$ is concave downwards, and in the interval $1 \leqslant z \leqslant 2$ it lies entirely above the straight line joining A and B in figure 137. The equation of this line is
$$y = (z - 1) \ln 2,$$
so that
$$\ln z \geqslant (z - 1) \ln 2, \quad 1 \leqslant z \leqslant 2.$$
Therefore,
$$\ln(1 + \sin x) \geqslant [(1 + \sin x) - 1] \ln 2$$
$$= \sin x \ln 2, \quad 0 \leqslant x \leqslant \tfrac{1}{2}\pi.$$
Therefore, by II,
$$\int_0^{\frac{1}{2}\pi} \ln(1 + \sin x)\, \mathrm{d}x \geqslant \int_0^{\frac{1}{2}\pi} \sin x \ln 2\, \mathrm{d}x = \ln 2,$$
as required.

The definite integral and some applications

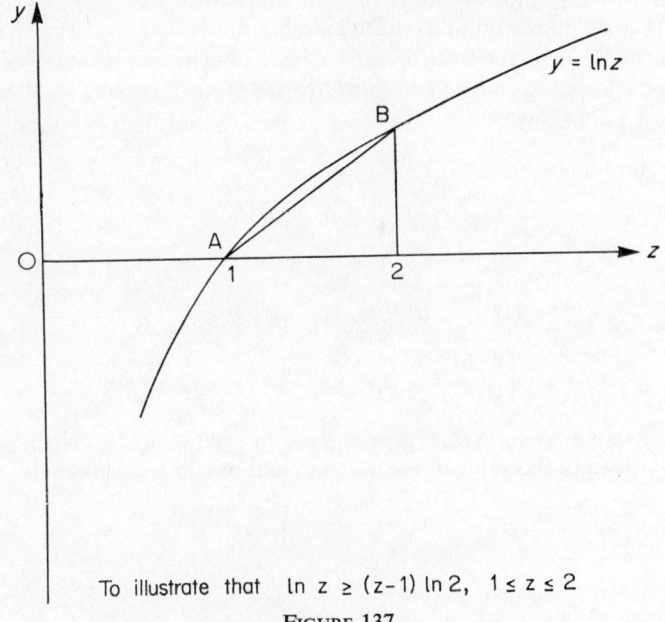

To illustrate that $\ln z \geq (z-1)\ln 2$, $1 \leq z \leq 2$

FIGURE 137

Exercises

66. Find what bounds on the integral $I = \int_0^1 \dfrac{e^x \, dx}{1 + x}$ are obtained by using the following alternative inequalities in $0 \leq x \leq 1$:
    (a) $\tfrac{1}{2} \leq (1 + x)^{-1} \leq 1$;
    (b) $1 - x \leq (1 + x)^{-1} \leq 1 - \tfrac{1}{2}x$;
    (c) $1 \leq e^x \leq e$;
    (d) $1 + x \leq e^x \leq e$;
    (e) $(1 + x) \leq e^x \leq 1 + (e - 1)x$.
[the exact result is $I = 1\cdot 125...$].

67. In the process of numerical integration of $\int_0^1 x^2 \sin\dfrac{1}{x}\,dx$ the computer wishes to stop short of the point $x = 0$ to avoid the rapid oscillations there, and so calculates $\int_\varepsilon^1 x^2 \sin\dfrac{1}{x}\,dx$. Find an ε small enough to ensure that the numerical value of the error incurred is less than $0\cdot 001$.

68. Prove that
$$\frac{1}{2} \leq \int_0^1 \frac{dx}{\sqrt{(4 - x^2 + x^4)}} \leq \frac{1}{6}\pi.$$

69. (a) Show (by drawing a figure, if necessary) that $2x/\pi \leqslant \sin x \leqslant x$, $0 \leqslant x \leqslant \tfrac{1}{2}\pi$. By integrating this result suitably, show that
$$1 - \tfrac{1}{2}x^2 \leqslant \cos x \leqslant 1 - \frac{x^2}{\pi}$$
in $0 \leqslant x \leqslant \tfrac{1}{2}\pi$.

(b) Obtain bounds for
$$I = \int_0^1 \frac{\cos x \, dx}{1 + x^2}.$$

70. Use of the inequality $-1 \leqslant \sin 20x \leqslant 1$ gives bounds for the integral
$$I = \int_0^1 \frac{\sin 20x}{1 + x^2} \, dx \quad \text{as} \quad -\tfrac{1}{4}\pi \leqslant I \leqslant \tfrac{1}{4}\pi.$$
Use the area analogy to explain why the use of the inequality on $\sin 20x$ gives such a poor idea of the value of the integral, and obtain better bounds.

13

Another view of integration: the integral as a sum

13.1 The need for a revised idea of integral

We have defined 'definite integral' as a number constructed from an indefinite integral. This approach is to some extent inadequate.

For example, we are not in a position to make any very strict statement about integrals which we cannot evaluate, that is to say those for which we cannot demonstrate in some way the existence of an indefinite integral from which definite integrals can be deduced. It would seem that we have strong reasons for expecting that there will be a definite integral in straightforward cases from the area analogy, for in considering $\int_a^b f(x) \, dx$ we could argue that $f(x)$ has a graph associated with it, and that there is an area under it between a and x, given by $\int_a^x f(u) \, du$, which is then an indefinite integral of $f(x)$. However, the geometrical conception of area has not really been examined, and the idea of graphs and areas in connection with functions such as that given in the note at the end of section 12.4 is quite obscure.

Secondly, and of more practical importance, it is possible to develop an alternative approach which makes the formulation of many types of physical problem more direct than the methods we used in Chapter 12, and which in a sense more closely reflects the fundamentals of the situation.

Thirdly, the new approach can be extended to problems in more dimensions—in two or three space dimensions, in several space dimensions and a time dimension, or many more dimensions in certain types of probability and physical problems.

In devising an alternative definition we have to be sure that it does not conflict with any results we have obtained previously.

416   *Introductory Analysis*

13.2 Integration as a summation process

One of our most valuable supporting intuitions about integrals is that they are closely related to area. In sections 12.1 and 12.4 this idea is explored, and in section 12.9 the idea of obtaining the numerical value of an integral by an approximating process is justified by intuitions about area. In section 12.9, Example 15, it was suggested that any numerical estimations of an integral could be regarded as one stage in an infinite sequence of such processes, in which the intervals used tend to zero length. This, if pursued, would give a sequence of approximations converging to the exact value of the integral. We shall proceed to *define* the value of a definite integral in some such way as this. We shall allow the area analogy to suggest an appropriate limiting process but will avoid using geometrical assumptions about area in deriving results from the definition.

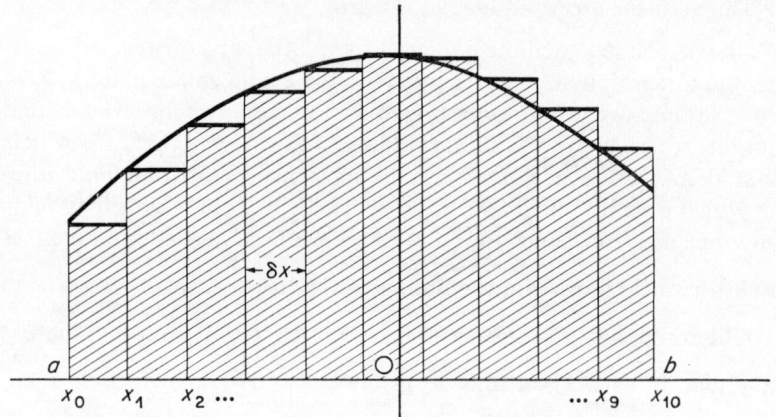

Area under the step function shaded
FIGURE 138

We need a system for approximating to regions under a curve which is capable of the most simple algebraic expression. The polygonal line fit leads to the trapezium rule for area (section 12.9), which is very simple but an even simpler one is possible. Since we are interested only in the limit of a sequence of progressively improving approximations arrived at by taking smaller and smaller intervals, we need not worry about getting the best possible approximation to the area at each stage. We will, therefore, approximate to the curve $y = f(x)$ by a *step function** as shown in

---

* A step function is one which is piecewise constant i.e. takes a constant value, say C_k, on $x_{k-1} < x < x_k$ for $k = 1, 2, \ldots, n$.

figure 138. Divide the interval, $a \leqslant x \leqslant b$, into n equal sub-intervals ($n = 11$ in figure 138), the coordinates of the points of subdivision being $x_0(=a), x_1, x_2, \ldots, x_{n-1}, x_n(=b)$. The ith interval is

$$x_{i-1} \leqslant x \leqslant x_i.$$

On the ith interval erect a rectangle of height equal to the function value at the beginning of the interval, $f(x_{i-1})$ as in figure 138. The area under the profile of the tops of the rectangles is an approximation to the area under the curve (though we can, of course, get better approximations). We call this approximating area A_n, and its value is equal to the sum of the areas of the rectangles,

$$A_n = \sum_{i=1}^{n} f(x_{i-1})(x_i - x_{i-1}) \tag{1}$$

$$= \sum_{i=1}^{n} f(x_{i-1}) \, \delta x, \tag{2}$$

say, where

$$\delta x = x_i - x_{i-1} = (b-a)/n = \text{constant}. \tag{3}$$

We expect that as the number of intervals n increases, the length δx of each interval consequently diminishing, the approximation will gradually improve, and that the area A under the true curve is given ultimately by

$$\int_a^b f(x) \, \mathrm{d}x = A = \lim_{n \to \infty} A_n = \lim_{n \to \infty} \left[\sum_{i=1}^{n} f(x_{i-1}) \, \delta x \right]. \tag{4}$$

We can now turn the argument upside down, and attempt to *define* the integral $\int_a^b f(x) \, \mathrm{d}x$ by

$$\int_a^b f(x) \, \mathrm{d}x = \lim_{n \to \infty} \left[\sum_{i=1}^{n} f(x_{i-1})(x_i - x_{i-1}) \right] = \lim_{n \to \infty} \left[\sum_{i=1}^{n} f(x_{i-1}) \, \delta x \right]. \tag{5}$$

This involves no mention of area.*

However, the provisional definition (5) is too wide. In fact, if *all* we require of an integral is that it is equal to the limit (5) (when this exists), it can be shown that we should have to accept as integrals $\int_a^x f(u) \, \mathrm{d}u$, some of which have no derivatives or which do not obey the indispensible rules given in section 12.3. For the present, however, we shall show that the program implied by (5) can actually be carried out in some cases to deliver the results we expect from our earlier theory.

---

* If some such definition proves to be consistent with that given in earlier chapters we can reinterpret the integral sign: we can regard it as an elongated S (for 'sum'), under which sign the elements δx are replaced by '$\mathrm{d}x$', and $f(x_{i-1})$ by a non-committal '$f(x)$'.

Example 1 $\int_a^b x^2 \, dx$, $b > a$, by summation.

Divide the interval $a \leqslant x \leqslant b$ into n equal intervals, each of length $\delta x = (b - a)/n$ (figure 139).

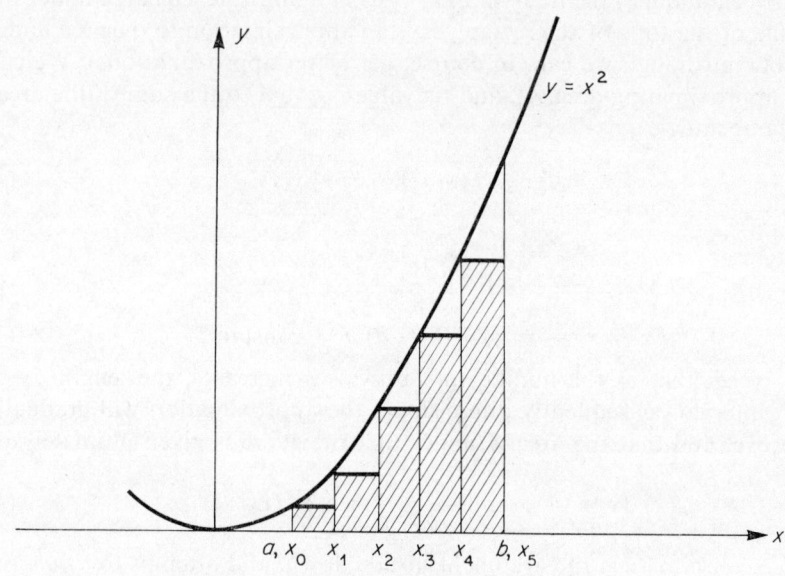

FIGURE 139

Then, $x_i = a + [(b - a)/n]i$, $\quad i = 0,1,2,...,n$.
The approximating area is

$$A_n = \sum_{i=1}^{n} (x_{i-1})^2 \, \delta x$$

$$= \sum_{i=1}^{n} \left[a + \frac{b-a}{n}(i-1) \right]^2 \frac{b-a}{n}$$

$$= \frac{b-a}{n} \sum_{i=1}^{n} \left[\left(a - \frac{b-a}{n} \right)^2 + 2\left(a - \frac{b-a}{n} \right) \frac{b-a}{n} i + \frac{b-a}{n} i^2 \right].$$

Since

$$\sum_{i=1}^{n} 1 = n, \qquad \sum_{i=1}^{n} i = \tfrac{1}{2} n(n+1), \qquad \sum_{i=1}^{n} i^2 = \tfrac{1}{6} n(n+1)(2n+1),$$

we have
$$A_n = \frac{b-a}{n}\left(a - \frac{b-a}{n}\right)^2 n + 2\left(a - \frac{b-a}{n}\right)\left(\frac{b-a}{n}\right)^2 \cdot \frac{1}{2}n(n+1)$$
$$+ \left(\frac{b-a}{n}\right)^3 \cdot \frac{1}{6}n(n+1)(2n+1).$$

The area A is given by
$$A = \lim_{n\to\infty} A_n = (b-a)a^2 + 2a(b-a)^2 \cdot \tfrac{1}{2} + (b-a)^3 \cdot \tfrac{2}{6} = \tfrac{1}{3}b^3 - \tfrac{1}{3}a^3$$
which agrees with the result obtained by our usual methods.

Clearly such a process is possible only in a few cases, since normally the required summations cannot be carried out. Moreover, we already have (Chapter 12) methods for calculating areas and integrals in general which appear to be far more powerful and adaptable. Nevertheless, the present approach, in addition to providing (potentially) a proper definition of integrals of functions regardless of whether or not we can find a *formula* for them, gives a new method for thinking about certain types of physical problem where our earlier methods appear rather artificial. The following examples are similar to those considered in sections 12.7 and 12.8, but regarded in this new light.

Example 2 *A particle moves in a straight line with velocity $v(t)$. Find the distance travelled between times $t = 0$ and $t = T$.*

Notice that $v(t)$ may be positive or negative. We shall also suppose that v is a continuous function. Divide the interval $0 \leqslant t \leqslant T$ into n equal intervals of length $\delta t = T/n$ and name the instants of subdivision $t_0 = 0, t_1, t_2,...,t_n = T$. Then in the ith interval $t_{i-1} \leqslant t \leqslant t_i$ the distance δs_i travelled is approximately equal to the length of the interval multiplied by some number representative of the velocity on the interval. Anticipating a limiting process we can take a suitable representative velocity as being the velocity at the beginning of the interval, $v(t_{i-1})$, and put
$$\delta s_i \approx v(t_{i-1})\,\delta t,$$
the approximation being more appropriate the smaller the interval length δt. Then the total distance s is given by
$$s = \sum_{i=1}^{n} \delta s_i \approx \sum_{i=1}^{n} v(t_{i-1})\,\delta t,$$
and in the limit
$$s = \lim_{n\to\infty} \sum_{i=1}^{n} v(t_{i-1})\,\delta t,$$
which we provisionally denote by
$$s = \int_0^T v(t)\,\mathrm{d}t$$

whether we know the indefinite integral of v or not. Of course, when $\int v(t)\,\mathrm{d}t$ is known, this is the expected answer which we obtained in section 11.2 by a different method.

Example 3 A ray of light travels along a tube containing an inhomogeneous substance in which the velocity of light is $v(x)$, where x is the distance from one end. Express the time taken to travel a distance s as an integral.

Divide the interval $0 \leqslant x \leqslant s$ into n equal intervals of length δx. The time taken, δt_i, to travel the section x_{i-1} to $x_{i-1} + \delta x$ is given approximately by

$$\delta t_i \approx \frac{\delta x}{v(x_{i-1})}.$$

The total time to travel between 0 and s, $t(s)$, is then

$$t(s) = \lim_{n\to\infty} \sum_{i=1}^{n} \delta t_i = \lim_{n\to\infty} \sum_{i=1}^{n} \frac{\delta x}{v(x_{i-1})},$$

which can be written as

$$\int_0^s \frac{\mathrm{d}x}{v(x)}.$$

Example 4 Find the heat content of a uniform bar of length l, cross-sectional area A, and density ρ whose temperature is given by $T(x)$, where x is the distance from one end.

This is very like the problems in density of section 12.7. In physics we define the heat content H of a quantity of a substance of mass m at *uniform* temperature T by

$$H = mcT,$$

where c is a constant characteristic of the substance called the specific heat. (Note that T may be negative, so that H can be negative—this is as insignificant a fact as that temperature may be negative.) The bar is not at a uniform temperature and it may be that no section, even the smallest, is at a uniform temperature. However, we expect that, if it is meaningful to speak of *the* heat content of a substance not at uniform temperature, an approximation to it would be obtained by splitting the bar into a large number of parts, estimating the heat content of each on the supposition that its temperature is nearly uniform at some representative value, and adding the results. The more sections we take the closer the approximation is likely to be, and this suggests a limiting process.

Divide the interval $0 \leqslant x \leqslant l$ into n equal parts, each of length $\delta x = l/n$, labelling the points of division $x_0 = 0$, $x_1, x_2, \ldots, x_n\ (= l)$. The mass of each section is $A\rho\delta x$, and if T_i^* is the 'representative temperature' of the ith section its heat content δH_i is given by

$$\delta H_i = T_i^* A\rho c \delta x.$$

Another view of integration: the integral as a sum

What set of temperatures T_i^* do we take as being representative? If we were merely aiming at a good approximation using a fixed finite number of sections we should choose some average temperature between the maximum and minimum temperature of each section. However, since it is the limit which concerns us we may simply take

$$T_i^* = T(x_{i-1}),$$

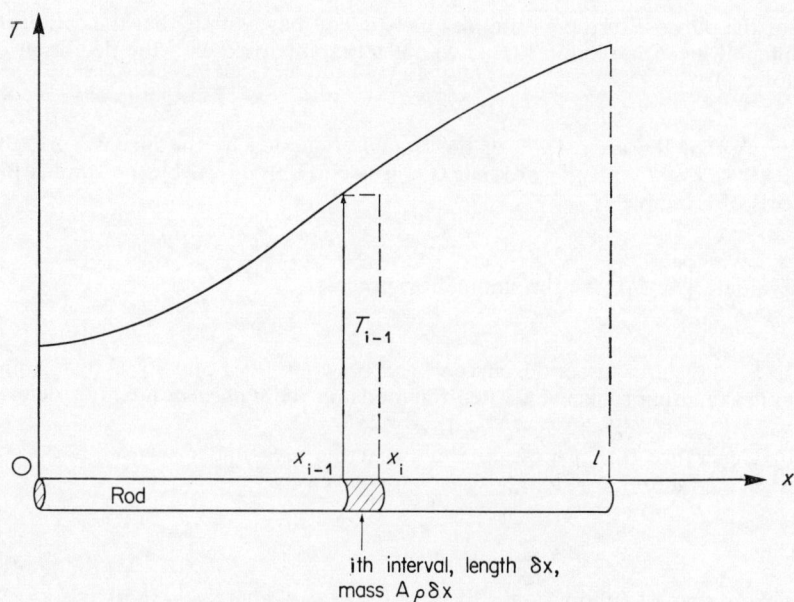

Temperature distribution in a rod

FIGURE 140

the temperature at the beginning of the section. Then the total heat content H is given approximately by

$$H = \sum_{i=1}^{n} \delta H_i \approx \sum_{i=1}^{n} A\rho c T(x_{i-1}) \, \delta x;$$

and finally

$$H = \lim_{n \to \infty} \sum_{i=1}^{n} A\rho c T(x_{i-1}) \, \delta x,$$

which we may write as

$$H = \int_0^l A\rho c T(x) \, dx.$$

Each of the above problems is capable of being formulated as the problems of sections 12.1, 12.7 and 12.8 were. We obtain the differential equations $dA(x)/dx = f(x)$ in Example 1, $dt/dx = v(x)$ in Example 3 and $dH(x)/dx = A\rho cT(x)$ in Example 4.

Exercises

For the purpose of these examples the student may assume that the definition of integral given in section 13.2 is completely compatible with the definition of Chapter 11.

1. Verify that the area, $\frac{1}{2}al^2$, of the triangle bounded by the line $y = ax$, the x-axis $0 \leqslant x \leqslant l$, and the ordinate $x = l$ is correctly given by the summation process of Example 1.

2. Evaluate $\int_a^b e^{cx}\,dx$ by the summation process.

3. (a) Find the area under the curve $y = x^5$ between $x = 1$ and $x = 2$ by forming an approximating sequence of step functions on the n *unequal* intervals defined by $x_0 = 1, x_1 = 2^{1/n}, x_2 = 2^{2/n}, ..., x_n = 2^{n/n} = 2$.

(b) Generalize this process to show that the area under $y = x^\alpha$, $\alpha \neq -1$, between $x = a$ and $x = b$, $a > 0$, $b > 0$ is given by

$$\frac{1}{\alpha + 1}(b^{\alpha+1} - a^{\alpha+1}).$$

(Let the points of subdivision x_i be $a, \varepsilon a, \varepsilon^2 a, ..., \varepsilon^n a$, where $\varepsilon = (b/a)^{1/n}$.)

4. In Example 1 $f(x_{i-1})$ is the minimum value of $f(x)$ in the ith interval $x_{i-1} \leqslant x \leqslant x_i$, since $f(x) = x^2$ is increasing with x. The area of the region (making the usual geometrical assumptions about area) is therefore always greater than the area under the step-functions. If $f(x_i)$ is taken always in place of $f(x_{i-1})$, the area of the region is always less than the areas under the corresponding step-functions (draw a figure). $\sum_{i=1}^{n} f(x_i)\,\delta x$ and $\sum_{i=1}^{n} f(x_{i-1})\,\delta x$, therefore, always bound the area from above and below respectively.

Use these bounds to prove rigorously that $\sum_{i=1}^{n} f(x_{i-1})\,\delta x$ tends to the value of the area, when $f(x) = x^2$.

5. Prove that $\lim_{n \to \infty} [(1/n) + 1/(n+1) + ... + (1/mn)] = \ln m$, where m is a positive integer, by considering $\int_1^m dx/x$ as the limit of a certain sum.

Another view of integration: the integral as a sum

6. Express the area of a sector of a curve given in polar coordinates as an integral by using the idea of summation.

7. Express the volume of a solid of revolution as an integral by using the idea of summation.

8. Express as an integral the mass of a rod of cross-sectional area $A(x)$ and density $\rho(x)$, x being distance from one end of the rod, by using the idea of summation.

9. The wind is blowing squarely against one rectangular side of a tall building of width a and height h. The wind speed v as a function of height x above the ground is given by $v(x) = Bx^\beta$. If the pressure exerted by a wind of speed v is Av^α per unit area under the prevailing conditions, express the total force on the building as an integral. Find the moment tending to topple the building over.

10. The work done by a *constant* force F acting on a particle which moves a distance d in the direction of the force is Fd, and $(-Fd)$ if the motion is opposite to the direction of the force. Now suppose that the force is variable, being a function only of the distance x from some fixed point on the line of motion.

    Construct an argument to show that the work, W_{ab}, done by the force when the particle moves from $x = a$ to $x = b$ ($b > a$) is given by

$$W_{ab} = \lim_{n \to \infty} \sum_{i=1}^{n} F(x_{i-1})(x_i - x_{i-1}) = \int_a^b F(x)\,dx,$$

where $x_0, x_1, \ldots, x_n$ are points of subdivision of the interval $a \leqslant x \leqslant b$, and $x_0 < x_1 < x_2 \ldots < x_n$. (Consider first the case where the particle moves steadily in a forward direction from a to b and then cases where it is permitted to move sometimes forward, sometimes backwards, but eventually arriving at $x = b$.)

11. The force on a unit electric charge at a point P due to a charge e at the point Q is equal to $-e/PQ^2$ (the strength of the field at P due to the charge at Q). A straight rod AB of length l is charged uniformly along its length with a total electric charge q. Show that the strength of the field at a point on the continuation of AB at a distance d from the end is equal to

$$-\frac{q}{l}\left(\frac{1}{d} - \frac{1}{d+l}\right).$$

12. Prove that

$$\lim_{n \to \infty} \sum_{k=1}^{n} \frac{1}{n+k} = \ln 2.$$

13.3 The Riemann Sum

In the examples of the last section an arbitrary and rigid scheme was imposed, which it would have been possible to relax considerably without affecting the plausibility of the arguments. For example, the condition that the sub-intervals should be equal does not seem to be essential, provided that in the limit *all* the sub-intervals tend to zero. Again we took as being representative of the function in each sub-interval its value at the beginning of the interval, a completely arbitrary choice. For example, if the direction of the bar in Example 4 were

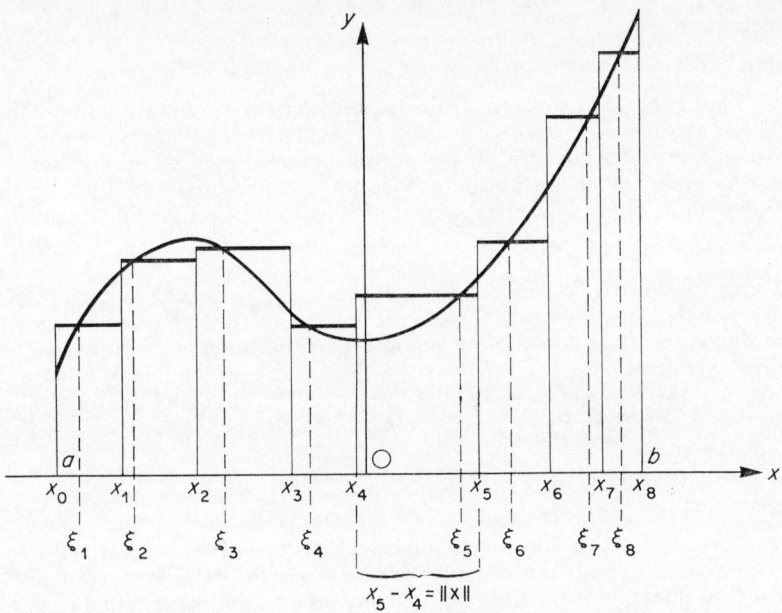

A function $y = f(x)$, $a \leq x \leq b$, with a partition and a set of representative points. The norm is equal to $x_5 - x_4$, the longest interval in this case

FIGURE 141

reversed and the calculation repeated, selection of temperatures at the *beginnings* of the new intervals would be equivalent to selecting temperatures at the *ends* of the old ones. *We shall now consider what kinds of function have associated with them sums which* always *tend to the* same *limit independently of the earlier restrictions.* Only such functions are likely to have a direct importance in applications, such as in finding areas, volumes, heat content and so on.

Let x be the variable and f the function, and let $f(x)$ be defined for all $a \leqslant x \leqslant b$ (regardless of whether we are talking about speed as a function of time, force as a function of distance, geometrical problems, or whatever the context may be). Note that $f(x)$ need not be of one sign in the interval, nor do we require it to be continuous.

Another view of integration: the integral as a sum

A *partition* on the interval [a,b] corresponds to any particular division into sub-intervals. Formally it is *any set of numbers, $x_0, x_1, x_2, \ldots, x_n$, which set we shall denote by* **x**, *such that* $x_0 = a$, $x_n = b$, *and* $x_0 < x_1 < \ldots < x_n$. The corresponding intervals are not necessarily equal.

The *norm of any partition* on the interval is *the length of its largest sub-interval*: the greatest of the numbers $x_i - x_{i-1}$, $i = 1, 2, \ldots, n$. The norm of the partition **x** is denoted by $\|\mathbf{x}\|$.

Given a partition **x** we may choose on each sub-interval an arbitrary *representative point*. We denote the point chosen on the ith interval by ξ_i,

$$x_{i-1} \leqslant \xi_i \leqslant x_i, \quad i = 1, 2, \ldots, n,$$

and the whole set of such points by $\boldsymbol{\xi}$. ($\boldsymbol{\xi}$ is thus restricted by the choice of **x**, without being wholly determined by it.)

Corresponding to a particular partition **x** and a set of representative points $\boldsymbol{\xi}$ we can construct the *step function equal to $f(\xi_i)$ in the interval $x_{i-1} < x \leqslant x_i$*. Pictorially the position is as shown in figure 141, for a particular function f in $a \leqslant x \leqslant b$, a particular choice of partition **x** and set of representative points $\boldsymbol{\xi}$.

The area under such a step function we shall indicate by $S_f(\mathbf{x}, \boldsymbol{\xi})$ (since it depends on the function f, and on the choice of **x** and $\boldsymbol{\xi}$). Then

$$S_f(\mathbf{x}, \boldsymbol{\xi}) = \sum_{x=a}^{x=b} f(\xi_i)(x_i - x_{i-1}). \tag{6}$$

Instead of $\sum_{i=1}^{n}$ we have written $\sum_{x=a}^{x=b}$ (sum over i from $x = a$ to $x = b$), since we do not want to drag in yet another index, n, which will have no importance in expressing what we want to say. We shall call the right-hand side of (6) the *Riemann Sum* corresponding to the function f, the partition **x** and the representative points $\boldsymbol{\xi}$.

We can now consider an *infinite sequence of partitions and their associated sets of representative points*, such that *the corresponding sequence of norms tends to zero*. This means that the subdivision of the interval [a,b] becomes finer and finer as we run through the sequence of partitions. The sums we formed in the last section involved such sequences of partitions and representative points, but using the inflexible scheme of equal intervals, with the first point of each interval as its representative point. The norm $\|\mathbf{x}\|$ was equal to $(b-a)/n$ in the nth partition, and $\xi_i = x_{i-1}$. To illustrate the start of a more haphazard, but perfectly admissible sequence we offer the following example. The interval is $0 \leqslant x \leqslant 1$.

1st Partition
 x: $\{0\cdot 0,\ 0\cdot 5,\ 1\cdot 0\}$
 $\boldsymbol{\xi}$: $\{0\cdot 25,\ 0\cdot 25\}$
 $\|\mathbf{x}\|$: $0\cdot 50$.

2nd Partition
 x: $\{0\cdot 0,\ 0\cdot 34,\ 0\cdot 51,\ 0\cdot 70,\ 1\cdot 0\}$
 $\boldsymbol{\xi}$: $\{0\cdot 30,\ 0\cdot 50,\ 0\cdot 65,\ 0\cdot 80\}$
 $\|\mathbf{x}\|$: $0\cdot 34$.

3rd Partition
  x: $\{0\cdot 0, 0\cdot 15, 0\cdot 45, 0\cdot 46, 0\cdot 70, 1\cdot 00\}$
  ξ: $\{0\cdot 10, 0\cdot 42, 0\cdot 46, 0\cdot 58, 0\cdot 85\}$
  $\|x\|$: $0\cdot 30$.

4th Partition
  x: $\{0\cdot 0, 0\cdot 20, 0\cdot 40, 0\cdot 60, 0\cdot 80, 1\cdot 00\}$
  ξ: $\{0\cdot 20, 0\cdot 40, 0\cdot 60, 0\cdot 80\}$
  $\|x\|$: $0\cdot 20$.

5th Partition
  x: $\{0\cdot 0, 0\cdot 12, 0\cdot 20, 0\cdot 30, 0\cdot 45, 0\cdot 60, 0\cdot 72, 0\cdot 80, 0\cdot 90, 1\cdot 00\}$
  ξ: $\{0\cdot 10, 0\cdot 18, 0\cdot 25, 0\cdot 34, 0\cdot 56, 0\cdot 68, 0\cdot 75, 0\cdot 82, 0\cdot 96\}$
  $\|x\|$: $0\cdot 15$.

The progress is slow and the choice of x and ξ deliberately erratic but the norm is decreasing and the student may try to imagine the sequence continued in any way he pleases so that the norm eventually tends to zero. There is, of course, no requirement that the norm should tend steadily to zero.

We can now ask the question: Does the corresponding sequence of numbers $S_f(x,\xi)$ tend to a limit? The answer depends on the function itself and on the partitions chosen. We can invent peculiar functions for which the limit does not exist. However, we want even more than this. *We require that the sequence of Riemann Sums $S_f(x,\xi)$ tend to a limit and the same limit, for every sequence of partitions and every corresponding choice of representative points, subject to the sole condition that the norm associated with each sequence tends to zero.*

This is necessary in physically meaningful cases since intuition gives no convincing criterion for preferring one such sequence of partitions to another. If, in an example such as 4, section 12.2, we obtained two different 'distances' by using two different sequences of partitions we could not account for this except by assuming that the whole idea is at fault, at any rate as applied to the particular velocity function being considered.

We shall write

$$S = \operatorname*{Lim}_{\|x\| \to 0} S_f(x,\xi) = \operatorname*{Lim}_{\|x\| \to 0} \sum_{x=a}^{x=b} f(\xi_i)(x_i - x_{i-1}) \tag{7}$$

for the limit in question, using 'Lim' instead of 'lim' to emphasize that it is not a simple limit that is implied, but that we imply that the *same* limit should be arrived at in an infinite variety of different ways: *through every possible sequence of partitions whose norm tends to zero and for every choice of representative points within these partitions.*

The proof of the existence or otherwise of the limit (7) for various kinds of function f is difficult and we shall simply state without proof the form of result which we shall use.

I. *If $f(x)$ is piecewise continuous on the interval $a \leqslant x \leqslant b$ the limit (as defined above) of the Riemann Sum on this interval*

$$\operatorname*{Lim}_{\|x\| \to 0} \sum_{x=a}^{x=b} f(\xi_i)(x_i - x_{i-1}) \tag{8}$$

exists.

Another view of integration: the integral as a sum

By 'piecewise continuous' is meant here that the function is continuous everywhere except possibly at a finite number of points, at which it is nevertheless finite. For example, step-functions are piecewise continuous, and so is a function with a continuous graph except for a finite number of points which lie 'off the line'. Such functions cover most physical requirements.

We shall denote expression (8) by

$$\int_a^b f(x)\,dx,$$

and use this as our *definition of definite integral* where the limit exists, but shall reserve the right to make statements like '$\int_0^1 \frac{dx}{x}$ does not exist', meaning that the limit in (8) does not exist. In the rest of this chapter we shall restrict consideration to *piecewise continuous functions* and all our proofs refer to this class of functions, though the results are true for a wider class.

Comparing (8) with (1) of section 13.2, we see that provided we require f to be piecewise continuous, (1) of section 13.2 gives correctly the value of the integral, since all partitions give the same result.

Exercises

13. Which of the following integrals does the definition on page 426 guarantee to exist?

(a) $\int_1^2 \dfrac{dx}{\sqrt{(1-x^2)}}$;

(b) $\int_0^1 x \ln x \, dx$, where $x \ln x$ is defined to be zero at $x = 0$;

(c) $\int_0^1 x^{-1} \, dx$;

(d) $\int_0^1 e^{-1/x} \, dx$, where $e^{-1/x}$ is defined to be zero at $x = 0$;

(e) $\int_{-1}^1 x \ln x \, dx$ (see (b) above);

(f) $\int_1^b \dfrac{dx}{\sqrt{[x(b-x)]}}$, $b > 0$;

(g) $\int_1^2 x^{-1} \, dx$;

(h) $\int_{\frac{1}{4}\pi}^{\frac{3}{4}\pi} \dfrac{dx}{\sin x}$;

(i) $\int_0^{2\pi} \dfrac{dx}{a + b \cos x}$ (consider all values of a and b);

(j) $\int_0^1 x \sin \dfrac{1}{x} \, dx$, where $x \sin \dfrac{1}{x}$ is defined to be zero at $x = 0$;

(k) $\int_{-\pi}^{\pi} |\sin x| \, dx$;

(l) $\int_0^3 f(x) \, dx$; $f(x) = 1, 0 \leqslant x \leqslant 1$; $f(x) = -1, 1 < x \leqslant 3$.

13.4 Some fundamental properties of the definite integral as newly defined

In deducing the various properties of the integral in this section we repeat that *we consider only functions which are piecewise continuous on the intervals*. All the integrals mentioned, then, exist by I of the last section, and this makes the proofs comparatively simple. These results simply reproduce those of sections 12.3 and 12.1, with the exception of IV which is new.

I. DEFINITIONS

$$\int_a^a f(x) \, dx = 0 \text{ for every } a. \tag{9}$$

In our new terms, this is a genuine definition—it does not come strictly from our theory since we cannot partition the interval $[a,a]$. If this definition is to be useful we also need $\lim\limits_{b \to a^+} \int_a^b f(x) \, dx = 0$, which is proved in V below.

$$\int_a^b f(x) \, dx = - \int_b^a f(x) \, dx. \tag{10}$$

All our theory related to 'upper limit > lower limit', so that hitherto $\int_b^a f(x) \, dx$ has had no meaning for $b > a$. We need this definition in order to make our theorems completely general.

On our earlier definition of integral these two results followed immediately from the definition: see section 12.3.

II. $\int_a^b f(x) \, dx + \int_b^c f(x) \, dx = \int_a^c f(x) \, dx$ for every a,b,c.

We shall prove the result only for $a < b < c$. Set up partitions on the intervals $[a,b]$ and $[b,c]$ both of whose norms are equal to the same number δ. Then in doing this we have also constructed a partition on $[a,c]$ whose norm is δ, and which has $x = b$ as a point of subdivision (see figure 142).

Another view of integration: the integral as a sum

For every choice of representative points ξ we have

$$\sum_{x=a}^{x=b} f(\xi_i)(x_i - x_{i-1}) + \sum_{x=b}^{x=c} f(\xi_i)(x_i - x_{i-1}) = \sum_{x=a}^{x=c} f(\xi_i)(x_i - x_{i-1}).$$

Now consider a sequence of such partitions on which $\delta \to 0$. Since f, if piecewise continuous on the intervals, is integrable, taking the limit as $\delta \to 0$ on this particular sequence of partitions we have

$$\int_a^b f(x)\,dx + \int_b^c f(x)\,dx = \int_a^c f(x)\,dx,$$

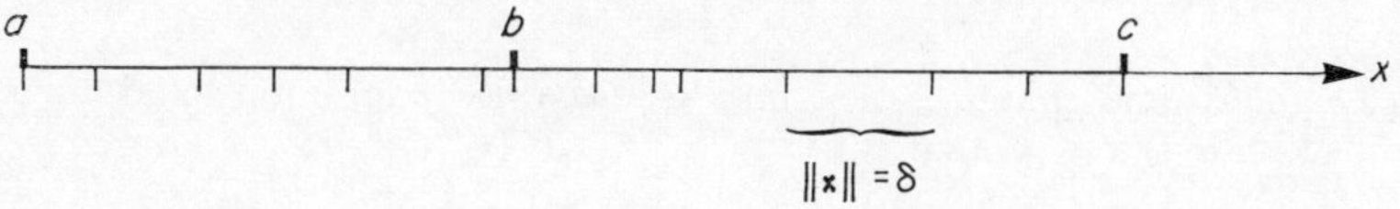

Interval $[a, c]$ partitioned with norm δ and b a point of subdivision

FIGURE 142

as required. Note how we have avoided having to consider *every* possible partition on $[a,c]$—we considered only partitions which had $x = b$ as a point of subdivision. We were able to do this from our knowledge that f was piecewise continuous on $[a,c]$, so that the choice of partition sequence is a matter of indifference.

III. *If f is piecewise continuous and $m \leq f(x) \leq M$ on $a \leq x \leq b$, where m and M are fixed numbers, then*

$$m(b - a) \leq \int_a^b f(x)\,dx \leq M(b - a).$$

For any partition on the interval, since $m \leq f(\xi_i) \leq M$, we have

$$m\sum_{x=a}^{x=b}(x_i - x_{i-1}) \leq \sum_{x=a}^{x=b} f(\xi_i)(x_i - x_{i-1}) \leq M\sum_{x=a}^{x=b}(x_i - x_{i-1}).$$

But

$$\sum_{x=a}^{x=b}(x_i - x_{i-1}) = x_1 - x_0 + x_2 - x_1 + \ldots$$
$$+ x_n - x_{n-1} = -x_0 + x_n = b - a.$$

Therefore,

$$m(b - a) \leq \sum_{x=a}^{x=b} f(\xi_i)(x_i - x_{i-1}) \leq M(b - a).$$

This is true independently of the choice of partition. It is therefore true in the limit for a sequence of partitions whose norm tends to zero; that is to say, for the integral, which is the result required.

IV. $\left| \int_a^b f(x)\,dx \right| \leq \int_a^b |f(x)|\,dx.$

This is a useful general-purpose result. It is proved by noticing that it is true for the Riemann Sums leading to the integral, and is therefore true for the limit of a sequence of Riemann Sums whose norm tends to zero. It can also be proved as in Exercise 18 below.

V. $\lim\limits_{h \to 0^+} \int_a^{a+h} f(x)\,dx = 0$ *if f is piecewise continuous on some interval* $a \leq x \leq b$.

We can assume that $a + h < b$. f is piecewise continuous and therefore bounded on $a \leq x \leq b$, that is to say

$$|f(x)| \leq M, \text{ say, } a \leq x \leq b.$$

By IV we have

$$0 \leq \left| \int_a^{a+h} f(x)\,dx \right| \leq \int_a^{a+h} |f(x)|\,dx$$

$$\leq M \int_a^{a+h} dx \quad \text{(by III)},$$

$$= Mh.$$

Therefore, $\lim\limits_{h \to 0^+} \left| \int_a^{a+h} f(x)\,dx \right| = 0$ and this is equivalent to the required result.

This makes the definition $\int_a^a f(x)\,dx = 0$ an obviously reasonable one.

Exercises

Assume that the given functions are piecewise continuous on the intervals concerned.

14. Interpret results II and III graphically.

15. Interpret result IV graphically.

16. Prove II for cases other than $a < b < c$.

17. Prove that

$$\int_a^b [Af(x) + Bg(x)]\,dx = A \int_a^b f(x)\,dx + B \int_a^b g(x)\,dx.$$

18. (a) Prove that if $f(x) \geq 0$, $a \leq x \leq b$, then
$$\int_a^b f(x)\,dx \geq 0;$$

(b) Prove that if $f_1(x) \leq f(x) \leq f_2(x)$, $a \leq x \leq b$, then
$$\int_a^b f_1(x)\,dx \leq \int_a^b f(x)\,dx \leq \int_a^b f_2(x)\,dx;$$

(c) Use (b) to prove result IV.

19. Write out the details of the method of proof of IV suggested in the text.

13.5 The integral as a continuous function of its upper limit

We are in a position to consider the integrals of *discontinuous* (but piecewise continuous) functions, which are of extensive practical use.

Consider the function (figure 143) defined by
$$f(x) = 0, \quad x < a,$$
$$= c, \quad a \leq x \leq b,$$
$$= 0, \quad x > b.$$

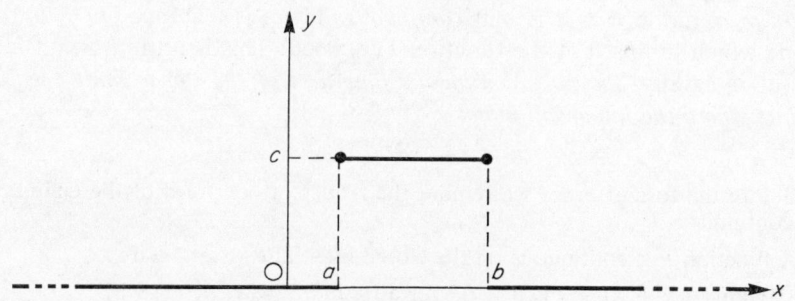

FIGURE 143

This is a type of function frequently encountered in physical problems and it has exactly the same validity as a 'smooth' function such as e^x, although we cannot represent it by a single formula throughout its range. For example, if x represents time, the function might represent the voltage applied to an electrical circuit when a battery is switched on at time a and off at time b. Again, if x were distance along a main road the function could represent the density of traffic at some moment in a queue caused by an obstruction at $x = b$.

Let us integrate this (piecewise continuous) function from the point $x = 0$, or try to calculate

$$F(x) = \int_0^x f(u)\,du.$$

Whatever the graph really stood for it is helpful to think in terms of area: for $0 \leqslant x < a$ the area is zero; in $a \leqslant x < b$ it climbs linearly (increases at a constant rate) to area $(b - a)c$ at $x = b$. Beyond $x = b$ it does not vary at all. The graph of $F(x)$ is shown in figure 144.

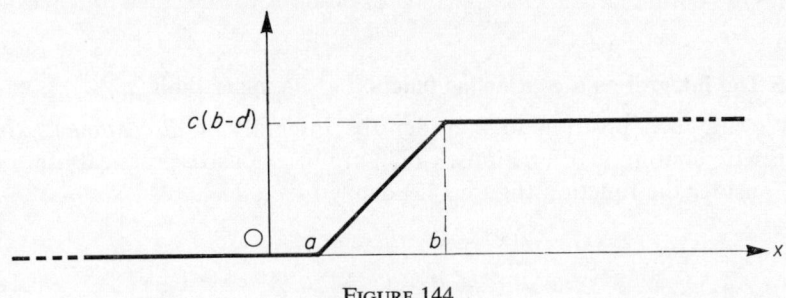

FIGURE 144

At all points except $x = a$ and $x = b$ the function is perfectly smooth. At $x = a$ and $x = b$ it is still *continuous*. There is a jump, but it is the slope which jumps, not the function $F(x)$ itself. This is an important fact about integrals—*they are continuous functions of the upper limit even at points where the integrand jumps.*

Before the formal proof we remind the reader of one form of the definition of continuity:

A function F is continuous on the closed interval $a \leqslant x \leqslant b$ if

(i) $\lim\limits_{h \to 0} [F(x + h) - F(x)] = 0$ for all x in $a < x < b$;

(ii) at $x = a$, $\lim\limits_{h \to 0^+} [F(a + h) - F(a)] = 0$ ($h \to 0$ through positive values);

(iii) at $x = b$, $\lim\limits_{h \to 0^-} [F(b + h) - F(b)] = 0$ ($h \to 0$ through negative values).

I *If f is piecewise continuous on $a \leqslant x \leqslant b$ then $\int_a^x f(u)\,du$ is continuous on this interval.*

Proof.

Put
$$\int_a^x f(u)\,du = F(x).$$

Then
$$F(x+h) - F(x) = \int_a^{x+h} f(u)\,du - \int_a^x f(u)\,du$$
$$= \int_x^{x+h} f(u)\,du,$$

by result II, section 13.4. By V of section 13.4, $\lim_{h \to 0} \int_x^{x+h} f(u)\,du = 0$ if $a < x < b$. Likewise $\lim_{h \to 0^+} \int_a^{a+h} f(u)\,du = 0$ and $\lim_{h \to 0^-} \int_b^{b+h} f(u)\,du = 0$. Therefore, $F(x)$ is continuous on $a \leqslant x \leqslant b$.

It is easy to show that $\int_x^b f(u)\,du$ is likewise a continuous function of x.

13.6 Comparison of the old definition of integration with the new

We shall prove that if f is piecewise continuous on $a \leqslant x \leqslant b$ then

$$\frac{d}{dx}\left(\int_a^x f(u)\,du\right) = f(x) \tag{11}$$

at every point x, $a < x < b$, where f is continuous (the integral being defined as in section 13.3).

This result essentially identifies the new definition (section 13.3) with the old (section 11.2). The earlier definition was concerned with functions *continuous* on an interval and what (11) says essentially is that if f is continuous, $\int_a^x f(u)\,du$ (defined as in section 13.3) is an indefinite integral in the earlier sense. Our earlier definition did not cover the case when f was permitted to have jumps, and the new definition is, therefore, seen as extending the old one to a wider class of functions.

We can also deduce that all continuous functions have integrals in the sense of Chapter 11.

We now give an analytical proof, but we need first another result.

I 'INTERMEDIATE VALUE THEOREM'. *If f is continuous on an interval $c \leq x \leq d$ then there is a number θ where $0 \leq \theta \leq 1$, such that*

$$\frac{1}{d-c}\int_c^d f(u)\,du = f[c + \theta(d-c)].$$

The number $c + \theta(d - c)$ lies between c and d. The result states that the function takes up, at some point on $[c,d]$, a value equal to its mean or average value over the interval. That this is likely to be true for a continuous function can be seen from a graph (figure 145).

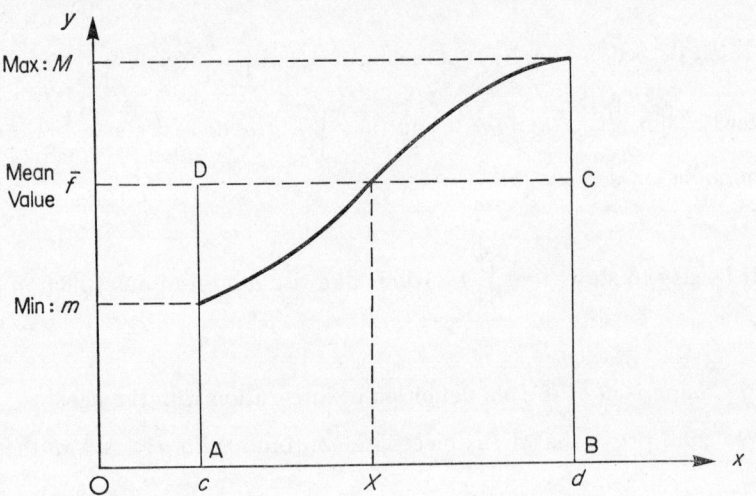

Area ABCD = area under the curve; $\bar{f}$ is the mean value of the function, and
$$\bar{f} = f(X)$$

FIGURE 145

Proof. Since f is continuous it is bounded and takes on, at some points in the interval, a maximum and a minimum value, M and m, respectively. Then by III of section 13.4, supposing that $d > c$,

$$m(d - c) \leq \int_c^d f(u)\,du \leq M(d - c). \tag{12}$$

But again, since f is continuous, it takes on every value between m and M at some point. Therefore $f(u)(d - c)$ takes up every value between $m(d - c)$ and $M(d - c)$ at some point. Since by (12) the integral has a value between

Another view of integration: the integral as a sum 435

$m(d-c)$ and $M(d-c)$, $f(u)(d-c)$ must take on *its* value at some point X (possibly more than one), where $c \leqslant X \leqslant d$. We may put $X = c + \theta(d-c)$, $0 \leqslant \theta \leqslant 1$ and we have the above theorem.

The proof for $d < c$ involves reversing the inequalities in (12). Notice that θ depends on c and d.

II (*The* 'FUNDAMENTAL THEOREM OF THE CALCULUS'). *If f is piecewise continuous and*
$$F(x) = \int_a^x f(u)\,du$$
then
$$\frac{dF(x)}{dx} = f(x)$$
at every point where f is continuous.

Let $x > a$ be a *fixed* point at which f is continuous. We shall show that
$$\frac{dF}{dx} = \lim_{h \to 0} \frac{F(x+h) - F(x)}{h}$$
exists and equals $f(x)$. Notice that h may take positive and negative values. In all cases, $h \neq 0$,
$$\frac{F(x+h) - F(x)}{h} = \frac{1}{h}\left(\int_a^{x+h} f(u)\,du - \int_a^x f(u)\,du\right)$$
$$= \frac{1}{h}\int_x^{x+h} f(u)\,du,$$
(by II of section 13.4). Then by the Intermediate Value Theorem, this is equal to
$$f(x + \theta h), \quad 0 \leqslant \theta \leqslant 1.$$
Now let $h \to 0$; since x is a point of continuity of f,
$$\lim_{h \to 0} f(x + \theta h) = f(x)$$
and the theorem is proved.

13.7 The indefinite integral

The definition of integral given in this chapter permits an extension of the concept of indefinite integral over that given in Chapter 11. We shall say that *F is an indefinite integral of f in $a \leqslant x \leqslant b$ if (i) F is continuous on $a \leqslant x \leqslant b$ and (ii) $dF(x)/dx = f(x)$ at every point in the interval where f is continuous.*

The extension lies in the fact that dF/dx may not exist at several points in the interval, whereas the indefinite integral was previously required to have a derivative at every point in the interval.

Example 14 Find $\int f(x)\,dx$, $-\infty < x < \infty$, where (*figure 146(a)*),

$$\begin{aligned} f(x) &= 0, & x < 0, \\ &= 1, & 0 \leqslant x \leqslant 1, \\ &= 2 - x, & x \geqslant 1. \end{aligned}$$

We can construct $\int f(x)\,dx$ by stringing together indefinite integrals appropriate to the three sections separately to form a *continuous* function.

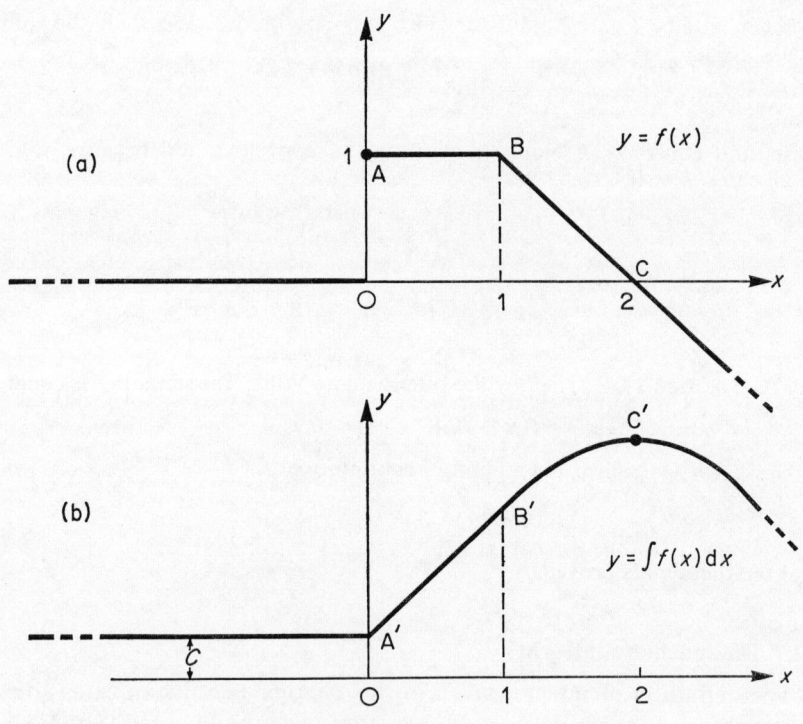

A discontinuous function and its integral

FIGURE 146

For $x < 0$, $\int f(x)\,dx = C$, an arbitrary constant.

For $0 < x < 1$, $\int f(x)\,dx = x + D$, where D is arbitrary.

Continuity at $x = 0$ requires $D = C$.

For $x > 1$, $\int f(x)\,dx = \int (2 - x)\,dx = 2x - \tfrac{1}{2}x^2 + E$, E arbitrary.

Continuity at $x = 1$ requires $2 - \tfrac{1}{2} + E = 1 + D$, so $E = D - \tfrac{1}{2} = C - \tfrac{1}{2}$.
Therefore (figure 146(b)).

$$\int f(x)\,dx = C, \qquad\qquad x < 0,$$
$$= x + C, \qquad\qquad 0 \leqslant x < 1,$$
$$= 2x - \tfrac{1}{2}x^2 - \tfrac{1}{2} + C, \quad x > 1,$$

where C is an arbitrary constant.

Note that $\int f(x)\,dx$ is differentiable at B′ since f is continuous at B, but is not differentiable at A′ since f is not continuous at A.

Exercises

20. Sketch the function $F(x) = \int_0^x f(u)\,du$ where f is given by the following:

(a) $f(u) = |\sin u|$; (b) $f(u) = |\cos u|$.

(c) The function shown in the following figure:

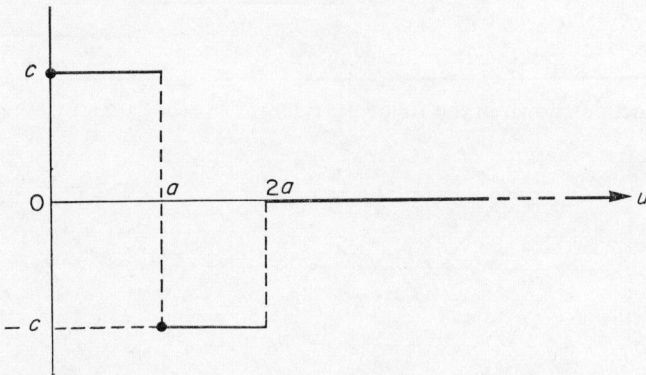

Consider the general way in which the shape of the graph of F changes when c increases and a decreases.

(d) The saw-tooth function shown in the following figure:

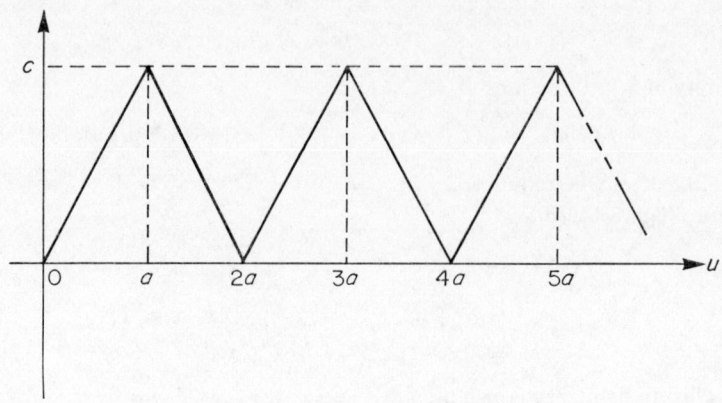

(e) The step-function which lies between the graphs of $y = 1/u$ and $y = 1/2u$ as shown in the following figure:

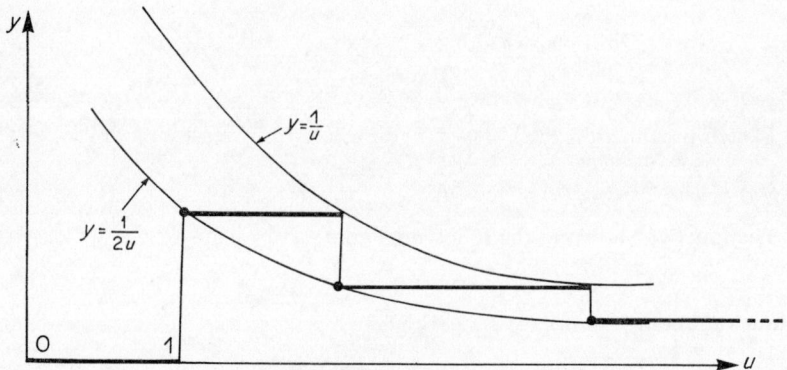

(f) The function shown in the following figure:

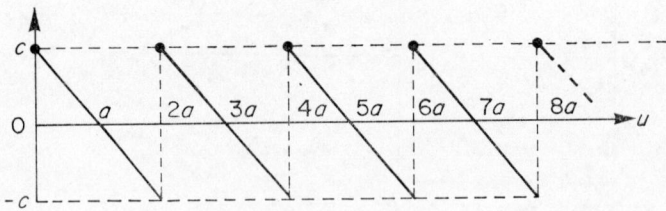

21. Prove that if f is piecewise continuous on $a \leqslant x \leqslant b$ any two indefinite integrals differ only by a constant. (This was proved in section 11.2 only for the

case where the indefinite integrals were differentiable at every point in the interval. They are now permitted to be non-differentiable at the isolated points where f is not continuous.)

22. Let $x = a$ be a point of discontinuity (a jump) in $f(x)$, a piecewise continuous function. Show that the jump in $f(x)$ at the point is equal to the jump in the slope of $F(x)$ there, $F(x)$ being any indefinite integral.

23. Prove that, given a function $\phi(x)$ with a piecewise continuous derivative on $a \leqslant x \leqslant b$,
$$\int_a^b \phi'(x) \, dx = \phi(b) - \phi(a).$$

24. The 'moving average' of a function over an interval of length h; $\bar{f}_h(x)$; is defined by
$$\bar{f}_h(x) = \frac{1}{h} \int_{x-\frac{1}{2}h}^{x+\frac{1}{2}h} f(u) \, du.$$
Prove that if $f(x)$ is piecewise continuous, $\bar{f}_h(x)$ is a continuous function of x.

25. Sketch the moving average (as defined in question 24) of the function given in (f) of question (20), when $h = \frac{1}{4}a$, $h = a$, $h = 2a$, $h = 3a$, respectively.

26. Prove that
$$\lim_{h \to 0} \frac{1}{h} \int_{a-\frac{1}{2}h}^{a+\frac{1}{2}h} f(u) \, du = f(a)$$
at points $x = a$, where $f(x)$ is continuous. (Assume f is piecewise continuous in an interval about $x = a$.)

27. (a) Use the Intermediate Value Theorem, I of section 13.7 to prove that
$$\int_0^1 (1 - u)^n g(u) \, du = (1 - \theta)^n g(\theta), \quad 0 \leqslant \theta \leqslant 1,$$
when $g(u)$ is continuous on $0 \leqslant u \leqslant 1$. (b) Let $f(x) \geqslant 0$ and be piecewise continuous on $a \leqslant x \leqslant b$ and let $g(x)$ be continuous on $a \leqslant x \leqslant b$. Prove by the method of I, section 13.7, that
$$\int_a^b f(u) \, g(u) \, du = g(a + \theta_1(b - a)) \int_a^b f(u) \, du, \quad 0 \leqslant \theta_1 \leqslant 1.$$
Hence, show that
$$\int_0^1 (1 - u)^n g(u) \, du = \frac{1}{n+1} g(\theta_1), \quad 0 \leqslant \theta_1 \leqslant 1,$$
if $g(u)$ is continuous on $0 \leqslant u \leqslant 1$.

Answers

Chapter 1

1. (i) I and IV; (ii) III and IV; (iii) IV.
3. (i) (0,2); (ii) (−2,0).
4. 45°.
5. (a) 60°; (b) 60°.
6. (i) Q(2,2),3,−4,5; (ii) Q(2,4),8,−15,17; (iii) Q(1,−5)3,2,$\sqrt{13}$; (iv) $d = 4$.
10. (i) (4,−1); (ii) (0,−5).
11. (i) 1; (ii) $-\frac{1}{2}$; (iii) $-\frac{1}{2}$; (iv) 0; (v) $-y/x$.
12. (i) Yes; (ii) No; (iii) Yes; (iv) Yes.
13. (i) Rectangle; (ii) Neither; (iii) Rectangle; (iv) Parallelogram.
14. 45°, 45°, 90°.
15. (i) $(\frac{4}{3},\frac{5}{3})$; (ii) $(0,-\frac{5}{4})$; (iii) (−4,4); (iv) $(-\frac{2}{3},\frac{8}{3})$.
17. $(\frac{4}{3},\frac{4}{3})$.
18. (8,−4).
19. 3.
20. (i) 0·506; (ii) 0·435; (iii) 6·875; (iv) 0·949.
21. (i) $y - 5x + 3 = 0$; (ii) $y + x + 1 = 0$; (iii) $3y + x - 5 = 0$; (iv) $4y - x + 8 = 0$.
22. (i) $x + 2y = 0$; (ii) $x = 0$; (iii) $3x - 5y - 1 = 0$; (iv) $y = -2$; (v) $3x + 5y - 1 = 0$.
23. (i) 4; (ii) $-\frac{3}{4}$; (iii) $-\frac{4}{3}$; (iv) $\frac{1}{5}$.
24. $4x - 3y = 15$.
25. (−2,1); $7x - 2y + 16 = 0$.
27. $p - x_1 \cos \alpha - y_1 \sin \alpha$.
28. (i) 1·7; (ii) 5.
29. $-0·204$; 492.
31. (i) $3X - 4Y - 4 = 0$; (ii) $3X - 4Y - 16 = 0$; (iii) $3X - 4Y = 0$; (iv) $3X - 4Y + 13 = 0$; (v) $3X - 4Y = 0$.
32. Unaltered.
33. (1,−1).
35. $\cos \alpha = \frac{4}{5}$, $\sin \alpha = \frac{3}{5}$; $\frac{1}{5}(4k - 3h)$.
36. $\cos \alpha = \frac{4}{5}$, $\sin \alpha = -\frac{3}{5}$; 1.
37. $x = 1 + (1/\sqrt{2})(X_1 - Y_1)$, $y = 2 + (1/\sqrt{2})(X_1 + Y_1)$. Choose new origin (h,k) where $Ah + Bk + C = 0$, $A_1 h + B_1 k + C_1 = 0$ and then rotate axes through angle α where $(AA_1 - BB_1) \tan 2\alpha = AB_1 + BA_1$.
41. 5.
42. $\frac{1}{2} r_1 r_2 \sin (\theta_2 - \theta_1)$.
44. $r \cos \theta = 1$.
45. $r \cos \theta + 1 = 0$.
46. $\theta = \frac{5}{6}\pi$.
47. $r \sin \theta = -\frac{1}{2}$.
48. $r \cos (\theta - \frac{1}{3}\pi) = \sqrt{3}$; $r = \sqrt{3}$; no positive r.

Chapter 2

1. (i) $x^2 + y^2 - 6x - 8y + 21 = 0$; (ii) $x^2 + y^2 + 8x - 6y = 0$; (iii) $x^2 + y^2 + 12x + 8y + 51 = 0$.
2. $x^2 + y^2 - 6x + 4y - 19 = 0$.
3. $x^2 + y^2 + 6y = 0$.
4. (i) Circle $C(1,0)$, $r = 2$; (ii) Circle $C(-\frac{3}{2}, 1)$, $r = \frac{3}{2}$; (iii) Circle $C(-\frac{3}{8}, -\frac{9}{8})$, $r = \sqrt{26}/8$; (iv) Point $(-12, 5)$; (v) No real solutions; (vi) Circle $C(-\frac{1}{2}, \frac{1}{2})$, $r = \sqrt{6}/2$.
5. $x^2 + y^2 - 6x + 4y - 12 = 0$.
6. $9x^2 + 9y^2 - 11x - 31y - 88 = 0$.
7. $x^2 + y^2 + 2x + 2y = 23$.
10. $2x^2 + 2y^2 - 2x - 6y + 7 = 0$.
11. $4x^2 + 4y^2 + 9x + y + 2 = 0$.
14. $8x^2 + 8y^2 - 6x + 3y = 0$.
15. Inside.
16. $r = a$.
17. (i) Vertex $(0,0)$, $F(2,0)$, $L: x = -2$; (ii) $(0,0)$, $(0, \frac{5}{4})$, $y = -\frac{5}{4}$; (iii) $(0,0)$, $(-\frac{3}{16}, 0)$, $x = \frac{3}{16}$; (iv) $(-2, 3)$, $(-1, 3)$, $x = -3$; (v) $(1, -2)$, $(\frac{1}{4}, -2)$, $x = 1\frac{3}{4}$; (vi) $(2, -6)$, $(2, -5\frac{1}{2})$, $y = -6\frac{1}{2}$.
18. Circle $x^2 + y^2 - 8y + 12 = 0$.
19. (i) $y^2 = 8x$; (ii) $x^2 = -12y$; (iii) $y^2 = 2x + 3y - 12$; (iv) $x^2 = 6y - 9$.
20. (i) $y^2 = 4x - 4$; (ii) $y^2 = 4x + 4y - 12$.
21. $x^2 = -16y + 16$.
22. $\frac{3}{4}$.
23. $y^2 = 3x + 7y - 3$.
24. $x^2 = 8x - 2y + 4$.
25. (a) 75 ft; (b) 3 m.
26. (a) 4 ft; (b) 1·225 m.
27. (a) 12·5 sec, 10,000 ft; (b) 7·1 sec, 1500 m.
28. (a) 20 ft; (b) 20 m.
29. 40 million km.
30. (i) $-1, 1$; (ii) $-\frac{4}{3}, \frac{3}{4}$; (iii) $\frac{7}{2}, -\frac{2}{7}$.
31. (i) $4x + 3y = 25$, $4y = 3x$; (ii) $x = 2$, $y = 0$; (iii) $4x + 3y = 19$, $4y - 3x + 8 = 0$.
34. $x^2 + y^2 + 6x - 2y + 5 = 0$, $x^2 + y^2 + 3x - 10y + 125 = 0$, $(-2, -1)$, $(-10, -5)$.
36. 4.
37. Point $(-2, 3)$; $x^2 + y^2 + 4x - 6y + 10 = 0$.
39. $x = y + 2$, $x + y = 6$.
40. $x - 2y + 2 = 0$.
47. $x^2/25 + y^2/9 = 1$, $\frac{4}{5}$.
48. 94,500,000 and 91,500,000 miles.
49. (i) 26, 24, $(\pm 5, 0)$, $\frac{5}{13}$; (ii) 10, 8, $(\pm 3, 0)$, $\frac{3}{5}$; (iii) $2\sqrt{10}$, $2\sqrt{5}$, $(\pm 2\sqrt{5}, 0)$, $1\sqrt{2}$.
50. $4x^2 + 25y^2 = 100$.
51. $3x^2 + 4y^2 = 24$.
53. $5x^2 + 9y^2 = 180$.
54. $3x^2 + y^2 = 7$.
55. $y - 2x = 3$, $3x + 6y = 8$.
59. $\sqrt{(5 - 2\cos^2\theta)}$.

63. $x^2/9 - y^2/7 = 1; \frac{4}{3}$.
64. (i) $(\pm\sqrt{34},0), \sqrt{34}/5, 3x = \pm 5y, (0,0)$; (ii) $(\pm\sqrt{15},0), \sqrt{(3/2)}, x = \pm\sqrt{2}y, (0,0)$; (iii) $(\pm\sqrt{15},0), \sqrt{3}, \sqrt{2}x = \pm y, (0,0)$.
65. The branch of $9x^2 - 16y^2 = 5$ in $x > 0$.
66. $4x^2 - y^2 = 20$.
67. $16x^2 - y^2 = 12$.
69. $8x^2 - y^2 = 8$.
72. $3x - y = 5, 9x - 7y = 5$.
80. $9x^2 - 12xy + 4y^2 - 12x - 18y + 3 = 0$.
81. $220x^2 - 144xy + 112y^2 - 76x - 152y + 1121 = 0, (\frac{1}{2},1)$.
82. $28x^2 - 144xy - 80y^2 + 116x + 232y + 319 = 0, (\frac{1}{2},1)$.
83. $7x^2 + 4xy + 7y^2 - 14x - 4y = 173$.
84. $56x^2 - 216xy + 119y^2 - 216x + 238y = 81$.
85. (i) Straight lines $3x - y + 2 = 0$, $x + 2y = 1$; $(-\frac{3}{7},\frac{5}{7})$; (ii) hyperbola, $(\frac{3}{7},\frac{5}{7})$; (iii) parabola; (iv) hyperbola; $(-\frac{3}{2},\frac{1}{5})$; (v) ellipse, $(0,-1)$; (vi) circle, $(-\frac{1}{2},\frac{1}{2})$.
86. (i) $X^2 + 2Y^2 = 1$, ellipse; (ii) $5X^2 = 2$, parallel lines; (iii) $X^2 - Y^2 = 4$, hyperbola; (iv) $X^2 + 3Y = 0$, parabola; (v) $X^2 + Y^2 + 4 = 0$, no real locus.
87. $\tan^{-1} \frac{52}{21}$.
90. $xy = a^2$.

Chapter 3

1. $0, 10, 4, 4, -200$.
2. $21, 5, 0, -0.884, (x + 2)(x + 6)/x^2$.
3. $0, 0, 0, 0$.
4. $ax^2 + (b + 2a)x + a + b + c, ax^2 + (b - 2a)x + a - b + c$, $2ahx + ah^2 + bh, (a + bx + cx^2)/x^2$.
8. (i) One, all real x; (ii) Two, $x \geqslant 0$; (iii) One, all real x; (iv) One for all real x (m odd): Two with $-a \leqslant x \leqslant a$ (m even); (v) Two, $-\sqrt{6} \leqslant x \leqslant \sqrt{6}$; (vi) Two, $0 \leqslant x \leqslant 2$; (vii) One, all x except $x = -1$; (viii) Two $-2 < x < 2$; (ix) Two, $0 \leqslant x < 2$.
9. (a) $y = x(250 - x); 0 < x < 250$; (b) $y = x(300 - x); 0 < x < 300$.
11. $40, 120, 240, 400$ ft; $20, 40, 60, 80$ ft.
16. (i) $-3 < x < 3$; (ii) $x \geqslant 3$ or $x \leqslant -3$; (iii) $-1 < x < 5$; (iv) $x \geqslant 5$ or $x \leqslant -1$; (v) $-2\frac{1}{2} < x < 2$ or $-4 < x < -3\frac{1}{2}$; (vi) $-2\frac{1}{2} < x \leqslant -2$ or $-4 \leqslant x < -3\frac{1}{2}$; (vii) $-5 < x \leqslant -3$.
17. (i) If $1 < x < 5$ then $-5 < f(x) < 5$; (ii) if $x \geqslant 5$ or if $x \leqslant 1$ then $-2 < f(x) < 6$; (iii) if $-1 \leqslant x \leqslant 1$ then either $-1\frac{1}{2} < f(x) < 1$ or $-3 < f(x) < -2\frac{1}{2}$.
18. See figure on page 443.
20. $x = \pm 1$.
25. -102.
26. $3a + b = 27$.
27. $a = \frac{2}{3}, b = 1, c = \frac{1}{3}, d = 0$.
28. $b^2 = 3ac, b^3 = 27a^2d$.
30. $y_1 P_1(x) + y_2 P_2(x) + \ldots + y_n P_n(x)$ with

$$P_k(x) = \frac{(x - x_1) \ldots (x - x_{k-1})(x - x_{k+1}) \ldots (x - x_n)}{(x_k - x_1) \ldots (x_k - x_{k-1})(x_k - x_{k+1}) \ldots (x_k - x_n)}.$$

31. (i) $\dfrac{2}{x-1}+\dfrac{3}{x-2}-\dfrac{4}{x-3}$; (ii) $\dfrac{4}{1-3x}-\dfrac{5}{1-2x}$;

(iii) $1+\dfrac{1}{x}-\dfrac{1}{5(x-1)}-\dfrac{8}{5(2x+3)}$;

(iv) $\dfrac{5}{(x-1)^4}-\dfrac{7}{(x-1)^3}+\dfrac{1}{(x-1)^2}+\dfrac{3}{x-1}$; (v) $\dfrac{3x}{x^2+2x-5}-\dfrac{1}{x-3}$;

(vi) $\dfrac{2}{(x-1)^3}+\dfrac{3}{(x-1)^2}+\dfrac{2}{x-1}$;

(vii) $\dfrac{1}{4a(x-a)}-\dfrac{1}{4a(x+a)}+\dfrac{1}{2(x^3+a^2)}$; (viii) $\dfrac{1}{(x^2+1)^2}+\dfrac{2}{x^2+1}$;

(ix) $\dfrac{1}{x^2+2x+2}+\dfrac{x+3}{(x^2+2x+2)^2}+\dfrac{2}{(x-1)^2}$.

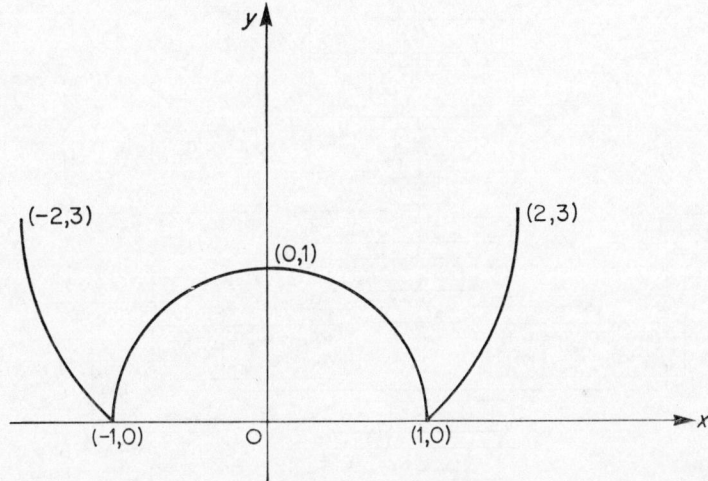

Chapter 4

1. (i) -2; (ii) -2; (iii) $-1/100$; (iv) 6; (v) $\tfrac{1}{4}$; (vi) 3; (vii) 3; (viii) -4; (ix) 0; (x) b/d; (xi) 2/7; (xii) p_2/q_2; (xiii) a^2/b^3; (xiv) 0; (xv) $3x^2$; (xvi) $\sqrt{2}$; (xvii) 1; (xviii) $-1/8\sqrt{a}$; (xix) 6; (xx) $7a^6$; (xxi) $1/2\sqrt{a}$; (xxii) $7a^2/5$; (xxiii) 2; (xxiv) a; (xxv) a/b; (xxvi) $\tfrac{1}{2}(b^2-a^2)$; (xxvii) a^2/b^2; (xxviii) 0; (xxix) a/b; (xxx) 0; (xxxi) $-3/5$.
3. (a) 1/750; (b) 1/7500; (c) $\varepsilon/7\cdot 5$.
5. (viii) 1; (ix) 1; (x) a/c; (xi) 4/3; (xiii) 0; (xiv) 2. Yes; (ix) becomes -1.
6. (i) 0; (ii) 3.
7. (i) Yes, 0; (ii) Yes, 0; (iii) Yes, 0; (iv) Yes, 1; (v) No; (vi) Yes, 0; (vii) Yes, 0; (viii) No; (ix) No; (x) 0.
8. (i) $x=0$, lims $+\infty, -\infty$; (ii) $x=1$, lims $+\infty, -\infty$; $x=-3$, lims $+\infty, -\infty$; (iii) $x=-q$, lims $\pm\infty$, $\mp\infty$ according as $q^2+qa+b \gtrless 0$; (iv) $x=3$, lims $+\infty, +\infty$; (v) $x=0$, lims $-\infty, +\infty$; $x=-2$, lims $+\infty, +\infty$.

444   *Introductory Analysis*

9. (i) 0,0; (ii) 0,0; (iii) $+\infty, -\infty$; (iv) 1,1; (v) 8,8.
10. (i) $+\infty$; (ii) $+\infty(n > m)$, $1(n = m)$, $0(n < m)$; (iii) $\frac{1}{2}$.
11. No.
14. $-\frac{2}{3}; 2; 4; n\pi; \pm\sqrt{2}, \pm 2; \frac{1}{4}(2n + 1)\pi; \frac{1}{6}(2n + 1)\pi; (2n + \frac{3}{2})\pi$; continuous.
15. Yes, take $0 < c \leqslant \varepsilon$. Implies $|x|$ continuous.
16. 48.
17. (i) Yes; (ii) No; (iii) Yes.
20. Max. 1, min. 0.
21. (i) Neither; (ii) Max. 1, no min.; (iii) Max. 1, min. 0.
25. (i) $1\cdot047_5$; (ii) $0\cdot68_5$; (iii) $1\cdot16_5$.
26.

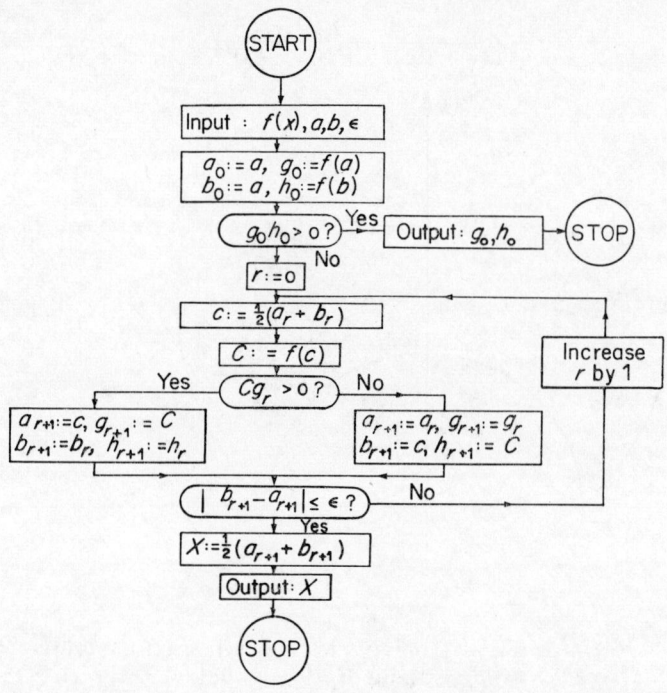

Chapter 5

1. (i) 91; (ii) 76·51; (iii) 75·1501; (iv) $75 + 15h + h^2$. Yes, 75. Same limit.
2. $3x^2$.
3. (i) 6, 10, 20 cm²/sec; (ii) 4 cm/sec.
4. (i) 27, 432 in³/sec; (ii) 1·73 in/sec.
5. (i) 36π m³/sec; (ii) 24π m²/sec.
6. $\frac{1}{3}$ ft/sec.
7. (i) 24 cm²/sec; (ii) 1 cm/sec.
8. (i) $4\frac{1}{2}$ in³/min; (ii) $1/8\pi$ in/min.
9. $16/9\pi$ m/min.

Answers

10. (i) Sphere; ratio $4\pi/3$; (ii) Sphere; ratio $2\pi/3$.
11. (i) $2x$; (ii) 3; (iii) a; (iv) $3x^2$; (v) $-3/x^4$; (vi) $6x - 2$; (vii) $2ax + b$; $-2/(2x + 1)^2$.
12. (i) $4/(x + 2)^2$; (ii) $-13/(x - 2)^2$; (iii) $(ad - bc)/(cx + d)^2$;
    (iv) $-(x^2 + 1)/(x^2 - 1)^2$; (v) $1/\sqrt{x}$; (vi) $-3/2x^{3/2}$; (vii) $-b/x^2$;
    (viii) $\frac{1}{2}a/\sqrt{(ax + b)}$; (ix) $x/\sqrt{(x^2 + 2)}$.
13. $x = 1$.
15. (i) h^2; (ii) $3xh^2 + h^3$; (iii) $6x^2h^2 + 4xh^3 + h^4$; (iv) $h^2/x^2(x + h)$.
16. (i) $3\cdot083$; (ii) $9\cdot95$; (iii) $144\cdot48$; (iv) $0\cdot00691$; (v) $5\cdot01204$.
17. (i) $1\cdot22054$; (ii) $0\cdot01255625$.
19. (i) $2x$; (ii) $x/2a$; (iii) $-3/x^4$; (iv) $2ax + b$.
20. Decreases.
22. $u + gt$.
23. (i) $2\cdot7$; (ii) $-4\cdot8$; (iii) 0; $-1\cdot7$.
24. $\frac{1}{30}$; increases; $\frac{1}{20}$.
25. (i) -20, (ii) 20, (iii) 0.
26. No if $x < \frac{1}{2}$, yes if $x > \frac{1}{2}$.
27. $0\cdot4722$, $0\cdot4822$, $0\cdot4770$. Correct value $0\cdot4767$.
28. $4x^3$, $7x^6$, $18x^{17}$, $56x^{55}$.
29. (i) $2x - 3$; (ii) $8x - 5$; (iii) $2ax + b$; (iv) $12x^2 - 16x + 9$; (v) $2x - 6$;
    (vi) $3a_1x^2 + 2a_2x + a_3$; (vii) $8x^3 - 2x$; (viii) $60x^9 - 21x^2$;
    (ix) $-3x^2 + 12x - 5$; (x) $3x^2 + 12x + 12$; (xi) $12x - 1$;
    (xii) $2x^5 - 2x^3 + 2x$; (xiii) $4x^3 - 12x$.
30. $f'(x) + g'(x)$.
31. 8.
32. $(-1,2), (1,-2)$.
33. $c = 2$, $(2,4)$.
34. $t = 2$; 64 ft/sec; 32 ft/sec.
35. (a) $x < 1$ and $x > 2$; (b) $1 < x < 2$.
36. $2\cdot65$.
37. $x_2 = \dfrac{1}{2}\left(x_1 + \dfrac{a}{x_1}\right)$;

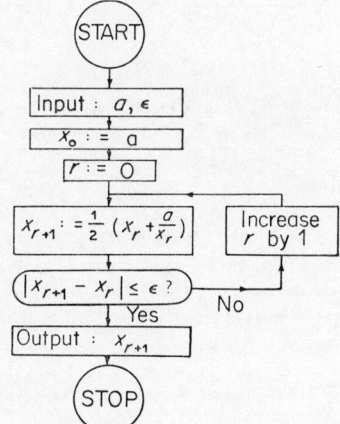

446 Introductory Analysis

38. $x_2 = \dfrac{1}{3}\left(2x_1 + \dfrac{a}{3x_1^2}\right)$; 1·44.

39. $x_2 = \dfrac{2(x_1^3 - x_1^2 - 5)}{3x_1^2 - 4x_1 - 5}$; 2·24.

41. $x_2 = x_1 - \dfrac{f(x_1)}{f'(x_1)}$.

42.

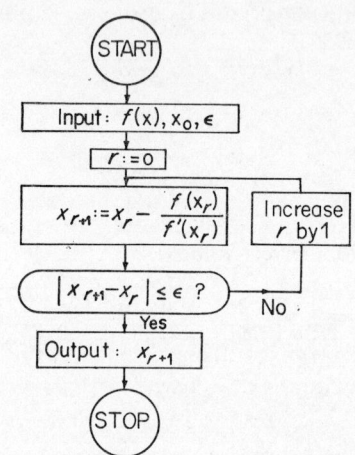

43. x_2 is on the other side of the root from x_1 due to bend in the curve; in fact, $x_2 = 1$, $x_3 = 1·25$, $x_4 = 1·31$, $x_5 = 1·317$.
44. $1/5x^{4/5}$, $\tfrac{7}{4}x^{3/4}$, $4/7x^{3/7}$, $1/qx^{1-1/q}$, $-3/x^4$, $-9/x^{10}$, $-p/x^{p+1}$, $-3/4x^{7/4}$, $1/qx^{1/q+1}$, $-5/8x^{13/8}$, $-1/qx^{1/q+1}$.
45. 10·05, 0·0995, 0·1005, 1·008, 3·9969, 6·0092.
46. $5x^4 + 6x^2 - 2x$, $4x^3 + 9x^{7/2}$, $3adx^2 + 2(ae + bd)x + be + cd$, $-1/x^{3/2} - 15/2x^{7/2}$, $6x^5 - 4x^3 + 3x^2 - 1$, $15(5x + 1)^2$, $2(4x + 1)(2x^2 + x + 1)$.
47. $nf'(x)[f(x)]^{n-1}$.
48. $f'(x)g(x) + f(x)g'(x)$.
49. $-5/(3x + 2)^2$, $14x/(x^2 + 8)^2$, $-(6x + 5)/(3x^2 + 5x + 3)^2$, $1/(\sqrt{x} - x)$, $-4(x + 1)(x + 2)/(x^2 - 2)^2$, $+2(x^2 + 3x + 3)/x^2(x + 2)^2$.
50. $-nf'(x)[f(x)]^{-n-1}$.
51. $[f'(x)g(x) - f(x)g'(x)]/[g(x)]^2$.
52. (i) $-x/y$; (ii) $-x^2/y^2$; (iii) x/y; (iv) $-1/y(x-1)^2$; (v) $-1/x^2$; (vi) $-y^{\frac{1}{2}}/x^{\frac{1}{2}}$; (vii) $(2x - 3y)/(3x - 2y)$; (viii) $(1/y)[x - (1/x^3)]$; (ix) $-y^{\frac{1}{3}}/x^{\frac{1}{3}}$; (x) $x(2x^2 - y^2)/y(x^2 - 2y^2)$; (xi) $-y(5x^3 + 2y^2)/x(2x^3 + 5y^3)$; (xii) $3(x + 1)^2/5y^4$; (xiii) $-y^3/x^3$; (xiv) $(2/3xy)(x^2 + 1)^{-\frac{1}{3}} - (y/x)$.
53. (i) $-3/13$; (ii) 12; (iii) $-\tfrac{1}{4}$; (iv) ∞, the tangent is parallel to the y-axis.
54. (i) $x + 2y = 5$, $y = 2x$; (ii) $7y - 3x = 10$, $7x + 3y + 4 = 0$; (iii) $y = x - 4$, $x + y = 0$; (iv) $y - 2x = 1$, $x + 2y = 2$.
56. (i) 90°; (ii) 15·25° at (2,3); (iii) they touch at (2,0); (iv) 90° at the four points $(\pm 1/\sqrt{3}, \pm\sqrt{8/3})$.

57. $51° 20'$ or $51·33°$.
60. (i) $10x(x^2 + 1)^4$; (ii) $x(x^2 - 1)^{-\frac{1}{2}}$; (iii) $-(3x + 1)(3x^2 + 2x + 1)^{-\frac{3}{2}}$;
    (iv) $5x^2/(4 - x^3)^{8/3}$; (v) $(1 - x^2)/[2x^{\frac{1}{2}}(x^2 + 1)^{\frac{3}{2}}]$; (vi) $(4x + 1)/[(3x + 1)^{\frac{2}{3}}]$;
    (vii) $-(3x^2 + 2)/[x^3(3x^2 + 1)^{\frac{1}{2}}]$; (viii) $8x/[3t(t^2 + 1)^2]$.
61. (i) $1/2x^{\frac{1}{2}}(1 - x^{\frac{1}{2}})^2$; (ii) $3t^2/2(t + 1)$; (iii) $x/(1 + x^2)^{\frac{1}{2}}$.
62. Yes.
64. $-\frac{1}{2}, \frac{1}{7}$.
65. $3/4t$.
66. (i) $2\pi r\, dr/dt$; (ii) $4\pi r^2\, dr/dt$.
68. $9/25\pi$ cm/sec.
69. $7\frac{1}{2}$ ft/sec, $2\frac{1}{2}$ ft/sec.
70. Anti-clockwise.
71. 2 min.
72. $5/72\pi$ cm/hr.
73. 1600 ft/sec.
74. (i) $4 \cos 4x$; (ii) $8 \cos(8x - \frac{1}{2}\pi) = 8 \sin 8x$; (iii) $a \sec^2 ax$; (iv) $b \sec bx \tan bx$;
    (v) $n \sin^{n-1} x \cos x$; (vi) $-\frac{1}{3} \sin x/(\cos x)^{\frac{4}{3}}$; (vii) $4 \sec^4 x \tan x$;
    (viii) $-3 \cot^2 x \csc^2 x$; (ix) $ma \tan^{m-1} ax \sec^2 ax$; (x) $-\frac{1}{2} \sin 2x$;
    (xi) $3x^2 \sin 4x + 4x^3 \cos 4x$; (xii) $(1/2x^{\frac{1}{2}}) \tan x + x^{\frac{1}{2}} \sec^2 x$;
    (xiii) $(3 \cos 3x)/x^2 - (2 \sin 3x)/x^3$; (xiv) $\cos 2x$; (xv) $-6(\sin 6x + \sin 4x)$;
    (xvi) $-\cot \frac{1}{2}x \csc^2 \frac{1}{2}x$; (xvii) $2x \sec^2 x^2$; (xviii) $-2 \sin x/(3 + 4 \cos x)^{\frac{1}{2}}$;
    (xix) $-\cos 2x/(\sin 2x)^{\frac{3}{2}}$; (xx) $2/(1 - \sin 2x)$;
    (xxi) $-3 \sec 2x \tan 2x/(\sec 2x - 1)^{\frac{5}{2}}$; (xxii) $3 \sin^2 x \sin 4x$.
75. (i) $-2 \cos 2x/3 \sin 3y$; (ii) $-(1/x)[y + \cos^2(xy)]$; (iii) $\sin 2x/\sin 2y$;
    (iv) $y/(x^2 + y^2 + x)$.
76. $1000\pi/3$ m/sec.
77. $\frac{1}{2}\pi$ ft^2/min.
78. (i) $y = x$, (ii) $y = -x$.
79. $0·74$.
80. (i) $119·26\pi/180$; (ii) $3·3422$.
81. (i) $0·8657$; (ii) $2·001$; (iii) $1·0006$; (iv) $0·2492$.
82. $3·73$ cm.
83. $1·992$ m.
84. 1 yd^2 too small.
85. (i) $a(a - b \cos C)/c^2$; (ii) $abC \sin C/c^2$.
86. (i) $5·05$; (ii) $-0·4$.
87. (i) $5·036$; (ii) $-0·56$.
88. $R_0(a + 2bT)t$.
91. $7·82$ m.
92. $1·7$.

Chapter 6

1. (i) $9x^8$, $72x^7$, $504x^6$, $9!x^{9-n}/(9-n)! (n < 9)$ $9! (n = 9)$ and $0(n > 9)$;
   (ii) $16x^3 + 6x^2 - 3$, $48x^2 + 12x$, $96x + 12$, $96 (n = 4)$ $0(n > 4)$; (iii) $-3/x^4$,
   $12/x^5$, $-60/x^6$, $(n + 2)!(-1)^n/2x^{n+3}$; (iv) $1/2x^{\frac{1}{2}}$, $-1/4x^{\frac{3}{2}}$, $3/8x^{\frac{5}{2}}$,
   $(-1)^{n-1} 1 . 3 . 5 ... (2n - 3)/2^n x^{n-\frac{1}{2}}$; (v) $-3/(3x + 2)^2$, $18/(3x + 2)^3$,
   $-162/(3x + 2)^4$, $n!(-3)^n/(3x + 2)^{n+1}$; (vi) $-\sin x$, $-\cos x$, $\sin x$,
   $\cos(\frac{1}{2}n\pi + x)$; (vii) $\sin 2x$, $2 \cos 2x$, $-4 \sin 2x$, $-2^{n-1} \cos(\frac{1}{2}n\pi + 2x)$.

2. (i) $-3x/(2-3x^2)^{\frac{1}{2}}$, $-6/(2-3x^2)^{\frac{3}{2}}$; (ii) $x\cos x + \sin x$, $-x\sin x + 2\cos x$; (iii) $x(2+x)/(1+x)^2$, $2/(1+x)^3$;
(iv) $\tan x + x\sec^2 x$, $2\sec^2 x(1 + x\tan x)$;
(v) $(x-2)/2(x-1)^{\frac{3}{2}}$, $(4-x)/4(x-1)^{\frac{5}{2}}$.
4. $(2 - 6t + 6t^2)/(1 - 2t)^3$.
7. 0.
9. (i) $x = -1$ (min.); (ii) $x = 1$ (max.); (iii) $x = -1$ (max.), $x = 1$ (min.); (iv) $x = 1$ (max.), $x = \frac{3}{2}$ (min.); (v) none; (vi) $x = \frac{1}{2}$ (not max. or min.); (vii) $x = 0$ (min.), $x = \frac{1}{2}$ (max.), $x = \frac{5}{2}$ (min.); (viii) $x = 0$ (not max. or min.), $x = \frac{1}{2}$ (max.), $x = \frac{3}{2}$ (min.); (ix) $x = \frac{3}{4}$ (min.); (x) $x = \frac{1}{2}$ (max.), $x = \frac{5}{6}$ (min.); (xi) $x = 0$ (max.); (xii) $x = -1$ (max.), $x = 1$ (min.); (xiii) $x = -2$ (min.), $x = 2$ (max.); (xiv) $x = \frac{4}{3}$ (max.); (xv) $x = (2n + \frac{1}{4})\pi$ (max.), $x = (2n + \frac{5}{4})\pi$ (min.), n integer; (xvi) $x = \frac{1}{2}\pi(n + \frac{1}{6})$ (max.), $x = \frac{1}{2}\pi(n - \frac{1}{6})$ (min.), n integer; (xvii) none; (xviii) $x = (n + \frac{1}{3})\pi$ (max.), $x = (n - \frac{1}{3})\pi$ (min.), $x = n\pi$ (not max. or min.), n integer; (xix) $x = (n + \frac{1}{4})\pi$ (min.), $x = (n + \frac{3}{4})\pi$ (max.), n integer; (xx) $x = (n + \frac{1}{6})\pi$ (min.), $x = (n - \frac{1}{6})\pi$ (max.), n integer.
10. If $a \leq 0$ none; if $a > 0$, min. $b - 2a^{\frac{3}{2}}$, max. $b + 2a^{\frac{3}{2}}$.
12. (i) 4 at $x = 0$, 0 at $x = 2$; (ii) 5 at $x = 0$, 4 at $x = \pm 3$; (iii) no greatest; least 0 at $x = 0$; (iv) 2 at $x = 6$, 0 at $x = 2$.
13. If $2b^2 \geq 1$, min. at $\omega = 0$; if $2b^2 < 1$, max. at $\omega = 0$ and min. at
$$\omega = \pm \Omega(1 - 2b^2)^{\frac{1}{2}}.$$
14. $x = \frac{1}{2}$.
17. -0.2520, -0.2510, -0.2500, -0.2480, -0.2000. Accuracy does not necessarily improve as h decreases.
18. 0·692.
19. $x = 1.5708$, max.
20. (i) Min. of 27 at $x = 3$; (ii) Min. of $\frac{1}{2}$ at $x = -1$, max. of $\frac{3}{2}$ at $x = 1$; (iii) Max. of 81/16 at $x = -\frac{1}{2}$, min. of 0 at $x = 1, -2$.
21. 2.
23. $v^2/64$.
24. Square.
25. Local min. of 2 at $x = 1$, local max. of -2 at $x = -1$.
26. 30 m.p.h.
29. (i) AC $= \frac{3}{4}$ km; (ii) C = B.
30. Length 2 ft, circumference 4 ft.
31. $2a/\sqrt{3}$, a.
33. $\frac{1}{3}$ height of glass.
34. $(3v/\pi)^{\frac{1}{3}}$.
35. 1,221,000.
36. $x = \frac{1}{2}l$.
37. $d/[1 + (s_2/s_1)^{\frac{1}{3}}]$ from s_1.
38. $\sqrt{2a}$.
39. $|(av - bu)\sin\theta|/(u^2 + v^2 - 2uv\cos\theta)^{\frac{1}{2}}$.
41. (i) (a) $a > 0$, (b) $a < 0$; (ii) (a) $x > 0$, (b) $x < 0$; (iii) (a) $x > 0$, (b) $x < 0$; (iv) (a) $y < 0$, (b) $y > 0$; (v) (a) $x > 2$ and $x < 1$, (b) $1 < x < 2$; (vi) (a) $x > -1$, (b) $x < -1$.
42. (i) $x = -\frac{1}{2}$; (ii) $x = n\pi$, n integer; (iii) $x = \frac{1}{3}(a + b + c)$; (iv) $x = n\pi$, n integer; (v) None.

43. (i) Max. $x = -1$, min. $x = 1$, infl. $x = 0$; (ii) Max. $x = \pm\sqrt{3}$, min. $x = 0$, infl. $x = \pm 1$; (iii) Max. $x = (2n - \frac{1}{4})\pi$, min. $x = (2n + \frac{3}{4})\pi$, infl. $x = (n + \frac{1}{4})\pi$, n integer; (iv) Max. $x = 2$, min. $x = -2$, infl. $x = 0$, $\pm 2\sqrt{3}$; (v) min. $x = -1$, infl. $x = -2$, infinite $x = 1$.
45. (i) $x = \pm 2$; (ii) $x = 3$.
47. Yes, $x = 1$.

Chapter 7

1. (i) $-\frac{3}{2}$; (ii) $\frac{1}{2}\pi$.
3. (i) Yes; (ii) no, f and f' not continuous at $x = 1$; (iii) no, f and f' not continuous at $x = \frac{1}{2}\pi$; (iv) yes; (v) no, f and f' not continuous at $x = 0$; (vi) no, f' does not exist at $x = 2$.
7. (i) $\frac{5}{2}$; (ii) $\frac{1}{2}(a + b)$; (iii) 3; (iv) 35/27.
8. (b) Yes.
12. (i) No; (ii) no; (iii) yes; (iv) no.
15. (i) $3\frac{1}{6}$, with error between 0 and -0.005; (ii) 4·004 with error between 0 and 0·000001; (iii) 0·001001 with error between 0 and 0·000 000 000 5.
18. 0·1.
20. (i) $h/8$ if $h > 0$; (ii) $\frac{1}{2}h \sin(1 + h)$ if $h > 0$.
22. (i) $\frac{2}{3}$; (ii) a/b; (iii) $1/2a$; (iv) 1; (v) $-\frac{1}{6}$; (vi) 2; (vii) $\frac{1}{6}$; (viii) $-\sin a$; (ix) $-\frac{1}{2}$; (x) $\frac{1}{12}$.
23. (a) Rule not applicable because numerator and denominator do not both vanish; (b) Same as (a) but at second step.
24. Rule says, if $\lim f'/g'$ exists, then so does $\lim f/g$ but does not assert the converse.
25. (i) $0 (a \neq 1)$, $-2 (a = 1)$; (ii) 16; (iii) $-\frac{1}{6}$.
26. (i) 1; (ii) $\frac{1}{3}$; (iii) $\frac{1}{2}$; (iv) -2.
27. (i) $\frac{1}{3}$; (ii) 1.
28. (i) 0; (ii) 0; (iii) $\frac{1}{6}$; (iv) $-\frac{1}{8}$; (v) ∞; (vi) 0; (vii) ∞; (viii) 1; (ix) 1; (x) $\frac{1}{3}$.
30. (i) $1 + \mu x + \dfrac{\mu(\mu - 1)}{2!}x^2 + \ldots + \mu(\mu - 1)\ldots(\mu - n + 2)\dfrac{x^{n-1}}{(n-1)!}$
    $+ \mu(\mu - 1)\ldots(\mu - n + 1)\dfrac{x^n}{n!}(1 + \theta x)^{\mu-n}$;
    (ii) 0, 3; $(1/\sqrt{2}) - 1$, $4\sqrt{2} - 1$; (iii) $\frac{4}{7}$, $\frac{1}{7}$.
32. (i) $\sin a + (x - a)\cos a - \dfrac{(x - a)^2}{2!}\sin a - \dfrac{(x - a)^3}{3!}\cos a$;
    (ii) $\cos a - (x - a)\sin a - \dfrac{(x-a)^2}{2!}\cos a + \dfrac{(x - a)^3}{3!}\sin a$;
    (iii) $\tan a + (x - a)\sec^2 a + (x - a)^2 \sec^2 a \tan a$.
33. (i) $x + \frac{1}{3}x^3 + \frac{2}{15}x^5$; (ii) $1 + \frac{1}{2}x^2 + \frac{5}{24}x^4$; (iii) $1 + \frac{1}{6}x^2 + \frac{7}{360}x^4$; (iv) $1 + 2x + 2x^2$.
34. $-\frac{1}{2}\pi < x < \frac{1}{2}\pi$.
37. $\theta = 25/256$; extra terms become infinite at $a = 0$.
39. $-8° 15'$ to $8° 15'$.
40. $-37\frac{1}{2}°$ to $37\frac{1}{2}°$.
41. 3.
42. Yes.

450   *Introductory Analysis*

43. $48 + 16(x - 4) - 2(x - 4)^2 - (x - 4)^3 - \frac{1}{16}(x - 4)^4$; 37·89.
44. (i) 0·685; (ii) 0·682; (iii) 1·036; (iv) 4·160; (v) 1·428; (vi) 0·977.
45. ± 0·63. Newton's method gives ± 0·635.
46. (i) $-\frac{1}{2}hf''(x + \theta h)$; (ii) $\frac{1}{2}hf''(x + \theta h)$; (iii) $-\frac{1}{6}h^2f'''(x + \theta h)$;
    (iv) $\frac{1}{30}h^4f^{(v)}(x + \theta h)$, $-$ 0·00016 and 0·000 0007.
47. $-h^2f^{(iv)}/12$.
49. $-(ax^5/240) + (ax^7/8064)$.
50. $\frac{1}{3}$.

Chapter 8

1. (i) 1, 2, 3, 4; (ii) $\frac{4}{5}$, 1, $\frac{6}{5}$, $\frac{7}{5}$; (iii) 0, $-$ 1, 0, 1; (iv) 0, 1, 0, $\frac{1}{2}$; (v) 1, $1/\sqrt{2}$, $1/\sqrt{3}$, $\frac{1}{2}$;
   (vi) $\frac{2}{3}$, $\frac{2}{5}$, $\frac{2}{7}$, $\frac{4}{9}$.
2. (i) $1/2^n$; (ii) $1 - (1/n)$; (iii) $1/[n(n + 1)]$; (iv) $1 + (1/2^n)$; (v) $(-1)^n/n!$.
3. (i) No; (ii) No; (iii) No; (iv) Yes, 0; (v) Yes, 0; (vi) Yes, $\frac{1}{2}$.
4. (i) $N \geqslant 101, 10001$; (ii) $N \geqslant 200, 20000$; (iii) $N \geqslant 101, 10001$;
   (iv) $N \geqslant 101, 10001$.
12. Bounded: (ii), (iii). Convergent: (ii).
13. (i) 0; (ii) 2; (iii) 2; (iv) 0.
17. $1·047_5$.
18. $0·68_5$.
20. (i) If $|u_1| < x_1$ converges to x; if $|u_1| > x_1$ diverges. (ii) Converges to x_1.
22. Sequence does not converge because roots not real.
24. $-1/f'(x)$.
27. $\frac{1}{4}$, $\frac{1}{8}$; $\frac{1}{128}$, $1/2^{127}$; 0·81, 0·73; 0·48, 0·000 0015.
30. (a) 842·2, 7·565, 5402, 402·0, 108·4, 69730. (b) 842·23, 7·56, 5402·04, 402·04, 108·38, 69732·12.
31. 0·48, 0·49.
32. $a = 3, \quad 2, \quad 0, \quad -2, \quad 1, \quad 0, \quad 4$
    $b = 0·332, 0·252, 0·106, -0·26, 0·1, -0·1, 0·5$
33. 0·5; sum may not be accurate to one decimal place.

Chapter 9

1. $\frac{3}{4}$; $\frac{1}{2}$.
2. (i) 3/2; (iii) 8/5; (iv) 50/9; (ii), (iv) divergent.
7. 16 ft.
8. 451/999, 41/22; Yes.
10. (i) No; (ii) no; (iii) yes; (iv) no.
13. (i) Divergent; (ii) convergent ($p > 1$), divergent ($p \leqslant 1$); (iii) convergent; (iv) convergent; (v) divergent; (vi) convergent ($p > 2$), divergent ($p \leqslant 2$); (vii) convergent; (viii) divergent; (ix) divergent; (x) convergent ($0 \leqslant a \leqslant 1$), divergent ($a > 1$); (xi) convergent ($0 \leqslant a \leqslant 1$), divergent ($a > 1$); (xii) convergent ($0 \leqslant a < 1$), divergent ($a \geqslant 1$).
15. (i) Convergent; (ii) convergent; (iii) divergent; (iv) convergent ($a^2 \leqslant 1$), divergent ($a^2 > 1$); (v) convergent; (vi) convergent; (vii) convergent ($a < 1$), divergent ($a > 1$), and, when $a = 1$, convergent ($\gamma > \alpha + \beta$), divergent ($\gamma < \alpha + \beta$).

17. (i) Convergent; (ii) divergent; (iii) convergent; (iv) convergent; (v) divergent; (vi) divergent; (vii) convergent; (viii) divergent; (ix) divergent; (x) divergent; (xi) convergent; (xii) divergent; (xiii) convergent.
18. (i) Convergent; (ii) convergent; (iii) convergent; (iv) divergent; (v) convergent; (vi) divergent; (vii) convergent; (viii) divergent; (ix) convergent.
19. (a) (i) $1/5^2$; (ii) $1/5!$; (iii) $1/27$; (v) $1/\sqrt{5}$; (vii) $5/4^4$; (ix) $1/\sqrt{30}$;
    (b) (i) $-1/8^2$; (ii) $-1/8!$; (iii) $-1/66$; (v) $1/\sqrt{8}$; (vii) $-8/4^7$; (ix) $-1/\sqrt{72}$.
22. No.
24. (i), (ii), (vii).
25. (i) Absolute; (ii) absolute; (iii) conditional; (iv) absolute; (v) absolute; (vi) absolute; (vii) absolute; (viii) absolute; (ix) absolute; (x) absolute; (xi) divergent; (xii) absolute.
28. $1 + \frac{1}{3} + \frac{1}{5} + \ldots; \frac{1}{2} + \frac{1}{4} + \ldots$.
35. 0.568.
36. 1.202.
37. (i) $-1 < x < 1, -1 < x < 1$; (ii) $-1 \leqslant x \leqslant 1, -1 \leqslant x \leqslant 1$;
    (iii) $-1 < x < 1, -1 < x < 1$; (iv) $-1 < x < 1, -1 < x < 1$;
    (v) all finite x, all finite x; (vi) $-1 < x < 1, -1 < x < 1$;
    (vii) $-1 < x < 1, -1 < x < 1$; (viii) $-1 \leqslant x \leqslant 1, -1 \leqslant x \leqslant 1$;
    (ix) all finite x, all finite x; (x) if $p \leqslant 1, -1/p \leqslant x < 1/p, -1/p < x < 1/p$;
    if $p > 1, -1/p \leqslant x \leqslant 1/p, -1/p \leqslant x \leqslant 1/p$; (xi) $|x| < 4$; (xii) $|x| < \frac{1}{4}$;
    (xiii) $x = 0$ only; (xiv) $-28 < x \leqslant 28, -28 < x < 28$; (xv) $-1 < x < 1, -1 < x < 1$.
38. (i) $2 \leqslant x < 4$; (ii) $\frac{1}{3} < x \leqslant 1$; (iii) all finite x; (iv) $x = -2$; (v) $-1 \leqslant x < 5$; (vi) $-1 < x \leqslant 0$; (vii) $1 \leqslant x \leqslant 3$; (viii) $-3 \leqslant x \leqslant -1$.
40. (i) $1 + 2x + \ldots + nx^{n-1} + \ldots, |x| < 1$;

    (iii) $-\frac{3}{2} + \frac{10x}{3} - \ldots + (-1)^n \frac{(2n-1)(n-1)}{n} x^{n-2} + \ldots, |x| < 1$;

    (v) $\frac{1!}{2!} - \frac{2!2}{4!}x + \ldots + (-1)^{n-1} \frac{n!nx^{n-1}}{(2n)!} + \ldots$, all finite x;

    (vii) $\frac{3}{5} + \ldots + \frac{2^n+1}{2^{n+1}+1} nx^{n-1} + \ldots, |x| < 1$;

    (ix) $-\frac{2}{1!} + \frac{3x}{1!} - \ldots + (-1)^n \frac{n+1}{(n+1)!} x^{n-1} + \ldots$, all finite x;

    (xi) $\frac{(1!)^2}{2!} + \frac{(2!)^2}{4!} 2x + \ldots + \frac{(n!)^2}{(2n)!} nx^{n-1} + \ldots, |x| < 4$;

    (xiii) Derivative does not exist.
45. (i) $1 + \frac{1}{2}x - \frac{1}{8}x^2, R = 1$; (ii) $1 - \frac{2}{5}x - \frac{3}{25}x^2, R = 1$; (iii) $1 - 2x^2 + 3x^4, R = 1$; (iv) $1 + 2x^2 + x^4, R = \infty$; (v) $1 - x + \frac{3}{2}x^2, R = \frac{1}{2}$;
    (vi) $1 - x + \frac{2}{3}x^2, R = 3$; (vii) $4(1 + x - \frac{1}{4}x^2), R = \frac{2}{3}$;
    (viii) $\frac{1}{2a^{\frac{1}{2}}}\left(1 + \frac{x}{a} + \frac{3x^2}{2a^2}\right), R = \frac{1}{2}a$.
46. (i) 10.010; (ii) 6.999; (iii) 0.198; (iv) 0.008.
47. (i) $1 - (23x/6)$; (ii) $1 - (5x/8)$; (iii) $1 - (343x/24)$.
52. (i) $1 - (23/6x)$; (ii) $1 - (5/8x)$.

452   *Introductory Analysis*

Chapter 10

1. 1.
3. $2[1 + x^2/2! + \ldots + x^{2n}/(2n)! + \ldots]$.
4. (i) $3e^{3x}$; (ii) $(2x + x^2)e^x$; (iii) $-e^{-x}$; (iv) $-(\sin x + \cos x)e^{-x}$; (v) $-2\sin 2x\, e^{\cos 2x}$; (vi) $4/(e^x + e^{-x})^2$; (vii) $-(1/x^2)e^{1/x}$; (viii) $3e^{3x}\sec^2 3x$; (ix) $-\tfrac{1}{2}[1 + a\,e^{ax}\,\mathrm{cosec}\,(x + 2y)]$; (x) $\left(-\dfrac{2}{x^3} + \dfrac{a}{x^2}\right)e^{ax}$; (xi) e^{x+e^x}.
6. $x = -1$, local minimum.
7. $x = \pm b/\sqrt{2}$.
8. Ratio $- e^{-\frac{1}{2}\pi}$.
12. $- ah/2T^2$.
13. e^{-2}.
14. (i) Converges to 0; (ii) divergent; (iii) converges to 0; (iv) converges to 1; (v) converges to 1.
15. (i) Convergent; (ii) divergent; (iii) convergent; (iv) convergent for all finite x.
16. (i) $\tfrac{1}{2}$; (ii) -1; (iii) $\tfrac{1}{3}$; (iv) e^3; (v) 6; (vi) $\tfrac{1}{3}$; (vii) 0; (viii) $\tfrac{1}{4}$; (ix) $\tfrac{1}{6}$; (x) $\tfrac{3}{2}$; (xi) 0; (xii) 0.
20. $C_0(1 - e^{-nt_0})/(e^{t_0} - 1)$.
21. 0·59.
22. 0·0001; $-0·1 < x < 0·1$.
23. 0·3679, 0·1353.
24. (i) 8; (ii) 6; $e = 2·7183$ correct to 4 decimal places.
25. 0·567.
27.

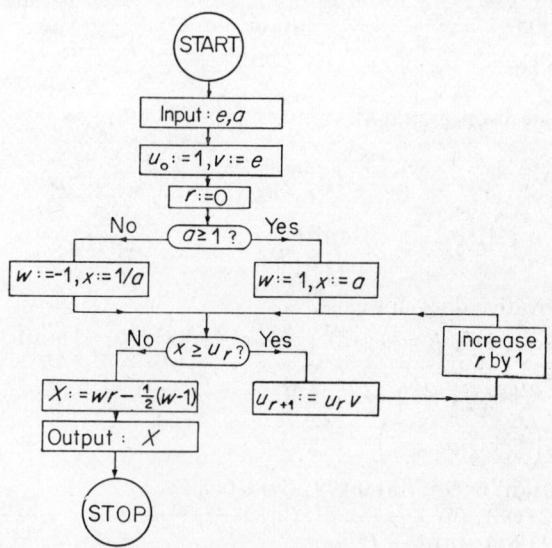

28. $x_{r+1} = x_r - 1 + a\,\mathrm{e}^{-x_r}$; x_0 might be taken as zero or located approximately by question 27.

32. (i) $\dfrac{3}{3x-1}$; (ii) $\dfrac{1}{x-3}$; (iii) $\dfrac{b}{a+bx}$; (iv) $\dfrac{2x}{x^2-1}$; (v) $\dfrac{b}{bx-a}$;

   (vi) $\dfrac{2x-3}{x^2-3x+2}$; (vii) $\dfrac{3x^2+1}{x^3+x}$; (viii) $\dfrac{5\cos x}{2+5\sin x}$; (ix) $2\operatorname{cosec} 2x$;

   (x) $\dfrac{b\sin 2x}{a-b\cos^2 x}$; (xi) $nx^{n-1}\ln(x^2-1) + \dfrac{2x^{n+1}}{x^2-1}$; (xii) $\dfrac{1}{x^2}(1-\ln x)$;

   (xiii) $\dfrac{a}{x^{\frac{1}{2}}(ax+b)} - \dfrac{\ln(ax+b)}{2x^{\frac{3}{2}}}$; (xiv) $\tfrac{1}{2}\cot x$;

   (xv) $\dfrac{1}{x+1} + \dfrac{1}{x-2} - \dfrac{2}{x-3}$;

   (xvi) $\dfrac{b\cos x}{a+b\sin x} + \dfrac{q\sin x}{p+q\cos x}$; (xvii) $\dfrac{1}{(x^2+1)^{\frac{1}{2}}}$; (xviii) $\dfrac{1}{\mathrm{e}^x+1}$;

   (xix) $\dfrac{\mathrm{e}^{2x}}{1+\mathrm{e}^{2x}}$; (xx) $\sin(\ln x) + \cos(\ln x)$; (xxi) $\dfrac{1}{x\ln x}$; (xxii) $\dfrac{2\operatorname{cosec} 2x}{\ln\tan x}$;

   (xxiii) $(\ln x)^2 + 2\ln x$.

33. (i) $-1/(x-2)^2$; (ii) $\mathrm{e}^{-x}(\ln x - 2/x - 1/x^2)$.
36. $1/\mathrm{e}$.
37. $-\tfrac{1}{2}$.
39. $x^2 - \tfrac{2}{3}(1+\tfrac{1}{2})x^3 + \tfrac{1}{4}(1+\tfrac{1}{2}+\tfrac{1}{3})x^4 - \dots$.
41. $[(-1)^{n-1}3^n + 2^n]x^n/n$.
43. (i) -1; (ii) $(a/b)^2$; (iii) 0; (iv) 1; (v) 3; (vi) $-\tfrac{1}{2}$.
44. Divergent.
45. 0.
46. $-\ln 2 - \sqrt{3}(x - \tfrac{1}{3}\pi) - 2(x - \tfrac{1}{3}\pi)^2$.
48. 0·223.
49. Error does not exceed $0·01|x|$.
50. 0·693.
54. $x = 2·552$.
55. 0·657.
56. No.
57. (i) $\dfrac{x+1}{[x(x+2)]^{\frac{1}{2}}}$; (ii) $x^2\mathrm{e}^{-3x}\sin 2x\,(2\cot 2x - 3 + 2/x)$;

   (iii) $\tfrac{1}{3}y\left[\dfrac{1}{x+2} + \dfrac{2x}{x^2-1} - \dfrac{2}{2x+5} - \dfrac{2x+2}{2x^2+2x+5}\right]$;

   (iv) $2x^{(\ln x - 1)}\ln x$; (v) $(\ln x)^x[\ln\ln x + 1/\ln x]$; (vi) $a^x a^{a^x}(\ln a)^2$;

   (vii) $x^x x^{x^x}[\ln x + (\ln x)^2 + 1/x]$; (viii) $x^{\mathrm{e}^{-x}}\mathrm{e}^{-x}(-\ln x + 1/x)$;

   (ix) $2^{\tan x}\sec^2 x \ln 2$; (x) $x^{\sin x}(\cos x \ln x + \sin x/x)$;

   (xi) $(\sec x)^{\sin x}(\cos x \ln\sec x + \sin x \tan x)$.

58. (i) e^{-6}; (ii) e^{-6}; (iii) e; (iv) 1; (v) 1; (vi) e^{-2}; (vii) e^6; (viii) 1; (ix) e^2; (x) $\mathrm{e}^{-2/\pi}$
   (xi) e; (xii) e^a; (xiii) 1.

454   *Introductory Analysis*

60. (i) 3; (ii) $\frac{3}{2}$; (iii) $\frac{7}{3}$; (iv) -3; (v) -4; (vi) $-\frac{4}{3}$; (vii) $\frac{2}{3}$; (viii) $-\frac{3}{2}$.
61. (i) $\dfrac{\ln 4}{\ln\left(\frac{5}{3}\right)}$; (ii) $\dfrac{5\ln 5}{2\ln 3 + 3\ln 2}$; (iii) 24.
65. 1·7709, 3·7709, $\bar{3}$·7709, $\bar{5}$·7709.
66. (i) 1·8062; (ii) 1·6460; (iii) 0·1781.
72. 4/5.
73. (i) $\cosh x = 5/3$, $\tanh x = 4/5$, $\coth x = 5/4$, $\text{sech } x = 3/5$, $\text{cosech } x = 3/4$;
    (ii) $\sinh x = 12/35$, $\tanh x = 12/37$, $\coth x = 37/12$, $\text{sech } x = 35/37$, $\text{cosech } x = 35/12$; (iii) $\sinh x = -12/5$, $\cosh x = 13/5$, $\coth x = -13/12$, $\text{sech } x = 5/13$, $\text{cosech } x = -5/12$.
77. (i) 1; (ii) 1; (iii) $-\frac{1}{2}$; (iv) e^2.
78. (i) Convergent; (ii) divergent.
79. (i) $4\cosh 4x$; (ii) $2\sinh 4x$; (iii) $5\,\text{sech}^2 5x$; (iv) $\coth x$;
    (v) $-(1/x^2)\,\text{sech}^2(1/x)$; (vi) $(1 - 2x^2\coth x^2)\text{cosech }x^2$;
    (vii) $-3\,\text{sech}^3 x \tanh x$; (viii) $-2\,\text{cosech }4x$.
84. $\sinh x$ and $\tanh x$ each have point of inflection at $x = 0$; $\cosh x$ has minimum at $x = 0$.
85. 1·8751, 4·6941.
86. (i) $(x^2 + 2nx + n^2 - n)e^x$;
    (ii) $a^{n-3}e^{ax}[a^3x^3 + 3na^2x^2 + 3n(n-1)ax + n(n-1)(n-2)]$;
    (iii) $(n-3)!2(-1)^{n-1}/x^{n-2}$; (iv) $(-a)^n 2^{\frac{1}{2}n} e^{-at}\sin(at - \frac{1}{4}n\pi)$;
    (v) $n![x^2 + 2nx(1+x) + \frac{1}{2}n(n-1)(1+x)^2]$;
    (vi) $x^2\dfrac{d^n y}{dx^n} + 2nx\dfrac{d^{n-1}y}{dx^{n-1}} + n(n-1)\dfrac{d^{n-2}y}{dx^{n-2}}$;
    (vii) $x^2\dfrac{d^{n+2}y}{dx^{n+2}} + (2n+1)x\dfrac{d^{n+1}y}{dx^{n+1}} + n^2\dfrac{d^n y}{dx^n} = 0$.
89. (i) $\pi/6$; (ii) $-\pi/6$; (iii) $2\pi/3$; (iv) $\pi/4$; (v) $\pi/3$.
90. $\sqrt{3}/2$, $1/\sqrt{3}$, $2/\sqrt{3}$.
91. Yes.
93. (i) $\frac{1}{2}\pi$; (ii) $-\pi/3$; (iii) $\frac{1}{2}\pi$.
94. (i) 0·5; (ii) $2/\sqrt{3}$; (iii) -0.5; (iv) $\sqrt{3}/2$.
96. (i) $1/(16 - x^2)^{\frac{1}{2}}$; (ii) $1/2x^{\frac{1}{2}}(1-x)^{\frac{1}{2}}$; (iii) $-2x/(1-x^4)^{\frac{1}{2}}$; (iv) $-1\,(\cos x > 0)$, $1\,(\cos x < 0)$; (v) $2/(1+x^2)$; (vi) $-2\cos^{-1}x/(1-x^2)^{\frac{1}{2}}$;
    (vii) $1 - x\sin^{-1}x/(1-x^2)^{\frac{1}{2}}$;
    (viii) $-(1-x^2)^{-\frac{1}{2}}\,(x > 0)$, $(1-x^2)^{-\frac{1}{2}}\,(x < 0)$;
    (ix) $2/(1+x^2)$; (x) $x/(1+x^2)$; (xi) $(\sin 2x)/(1 - \frac{1}{2}\sin^2 2x)$.
97. $-\frac{1}{2}\pi$.
100. Local maximum of $\frac{1}{4}\pi - \frac{1}{2}\ln 2$ at $x = 1$.
102. (i) 1; (ii) $e^{\frac{2}{3}}$.
106. (i) $|x| < 0.3$; (ii) $|x| < 0.54$. Put $u = \frac{1}{2}$, $v = \frac{1}{3}$ and calculate $\tan^{-1}$ by series.
110. 1·3515.
114. (i) $1/(x^2 + 4)^{\frac{1}{2}}$; (ii) $2x/(x^4 - 81)^{\frac{1}{2}}$; (iii) $\sec x$; (iv) $e^x/(e^{2x} - 1)^{\frac{1}{2}}$;
    (v) $-\text{cosec }x$; (vi) $\frac{1}{2}\sec x$; (vii) $-\text{cosec }2x\,(\cos 2x > 0)$, $\text{cosec }2x\,(\cos 2x < 0)$.
116. 4·7300, 7·8532.
117. 2·365.

Answers

Chapter 11

In Exercises 1–8, 17–51, and 55–91, any constant may be added to the answers given, the result being still correct.

1. (i) $\frac{1}{2}x^2$; (ii) $\frac{3}{2}x^2$; (iii) $\frac{1}{3}x^3$; (iv) $\frac{1}{6}x^3$; (v) $2x^{\frac{1}{2}}$; (vi) $x^{n+1}/(n+1)$; (vii) $2x$; (viii) constant; (ix) $ax^{n+1}/(n+1)$; (x) $x^{\alpha+1}/(\alpha+1)$; (xi) $\ln x$; (xii) $-(-x)^{\alpha+1}/(\alpha+1)$; (xiii) $\frac{1}{2}x^2 + \frac{2}{5}x^{\frac{5}{2}}$;

   (xiv) $\frac{1}{4}x^4 + \frac{1}{3}x^3 + \frac{1}{2}x^2 + x + \ln x - \frac{1}{x} - \frac{1}{2x^2} - \frac{1}{3x^3}$; (xv) $\ln(-x)$.

2. (i) $\sin x$; (ii) $-\cos x$; (iii) $\tan x$; (iv) $-\cot x$; (v) $-3\cos x$; (vi) $-\frac{1}{2}\cos 2x$; (vii) $\frac{1}{3}\sin 3x$.
3. (i) e^x; (ii) $-e^{-x}$; (iii) $\frac{1}{2}e^{2x}$; (iv) $-\frac{1}{3}e^{-3x}$; (v) $-(A/a)e^{-ax}$.
4. (i) $\tan ax/a$; (ii) $-\cot ax/a$.
5. (i) $\sin^{-1} x$ or $-\cos^{-1} x$; (ii) $\cosh^{-1} x$.
6. (i) $-xe^{-x} - e^{-x}$; (ii) $x^2 e^x - 2x e^x + 2 e^x$; (iii) $x \sin x + \cos x$; (iv) $-x \cos x + \sin x$; (v) e^{x^2}.
7. (i) $\frac{4}{3}x^3 + 2x^2 + x$ (or $\frac{1}{2 \cdot 3}(2x+1)^3$); (ii) $\frac{1}{5}(x+1)^5$; (iii) $\frac{1}{2 \cdot 5}(2x+3)^5$; (iv) $-\frac{1}{2}\cos(2x+3)$; (v) $-\sin(1-x)$; (vi) $-\frac{1}{2}\sin(1-2x)$; (vii) $\ln(x+1)$; (viii) $\ln(2x+3)$.
8. (i) $\frac{1}{2}\sin^2 x$; (ii) $\frac{1}{2}x - \frac{1}{4}\sin 2x$ ($\cos 2x = 1 - 2\sin^2 x$); (iii) $\frac{1}{2}x - \frac{1}{4}\sin 2x$ ($\cos 2x = 2\cos^2 x - 1$); (iv) $\frac{1}{4}\{\frac{1}{3}(3x) - \frac{1}{4}\sin[2(3x)]\}$; (v) $-x \cos x$.
9. (a) $\frac{2}{3}x^3 + 1$; (b) $-(1/x) - (1/2x^2) + \frac{5}{2}$, $x > 0$; (c) $\ln(-x) - (1/x) - 2$; (d) $-e^{1/x} + 1$, $x > 0$.
10. (a) $-1/(x-1)$, $x < 1$; (b) $1 - 1/(x-1)$, $x > 1$; (c) $1 - 1/(x-1)$, $x < 1$.
11. The functions given differ by constants.
12. (a) $\frac{1}{2} \cdot 32 \cdot 3^2$ ft/sec; (b) $v^2 = 64s$.
14. The point is that $F(x)$ is differentiable with zero derivative at $x = 0$.
15. One such function is $F^*(x) = 0, x \leq 0$; $= x, x > 0$. All integrals on $-\infty < x < 0$ are of the form $F^*(x) + C$, and for continuity at $x = 0$, we require $F(x) = x + C$ for $x > 0$.
16. (b) $x_1 < ax + b < x_2$; (c) $x_1 < x^2 < x_2$.
17. (a) $-\frac{2}{3}(-x)^{\frac{3}{2}}$;
    (c) $\frac{1}{a(n+1)}(ax+b)^{n+1}$, $x > -\frac{b}{a}$, $x < -\frac{b}{a}$ (assuming $a > 0$);
    (d) $(2/3a)(ax+b)^{\frac{3}{2}}$; (e) $-\frac{2}{3}(1-x)^{\frac{3}{2}}$; (f) $-\frac{1}{2}\cos(x^2)$;
    (g) $\frac{1}{2}\tan^{-1}(x^2)$; (h) $\frac{1}{2}\sin^{-1}(x^2)$.
18. $\frac{1}{4}x^4, \frac{1}{3}x^3, \frac{1}{2}x^2, x$, any constant, $\ln|x|, -1/x, -1/2x^2$.
19. $\frac{2}{3}x^{\frac{3}{2}} + \frac{3}{5}x^{\frac{5}{2}}, \frac{3}{5}x^{\frac{5}{3}} + \frac{3}{2}x^{\frac{4}{3}}, \frac{3}{2}x^3 + \frac{3}{2}x^2 + x, x^{\frac{1}{2}} + x + x^{\frac{3}{2}}, \frac{2}{5}x^{\frac{5}{2}} + \frac{4}{3}x^{\frac{3}{2}} + 2x^{\frac{1}{2}}$.
20. $\frac{4}{3}x^5 + \frac{4}{3}x^3 + x, \frac{8}{5}x^{\frac{5}{2}} - 2x^2 + \frac{2}{3}x^{\frac{3}{2}}, \sum_{r=1}^{n} \frac{a}{r} x^r$.
21. $e^x, -e^{-x}, \sinh x, \cosh x$.
22. $\tan x, -\cot x, \tan x - x, -\cot x - x$.
23. $-\frac{1}{2}\cos 2x, \frac{1}{2}\sin 2x, \frac{1}{2}x - \frac{1}{4}\sin 2x, \frac{1}{4}\sin 2x + \frac{1}{2}x$.
24. $\frac{1}{a}e^{ax}, -\frac{1}{a}e^{-ax}, \frac{1}{\ln a}a^x$.

25. $-\dfrac{1}{a}\cos ax, \dfrac{1}{a}\sin ax, \dfrac{1}{a}\cosh ax, \dfrac{1}{a}\sinh ax$.
26. $\tfrac{1}{3}\tan 3x, -\tfrac{1}{5}\cot 5x, \tfrac{1}{2}x - \tfrac{1}{8}\sin 4x, \tfrac{1}{12}\sin 6x + \tfrac{1}{2}x$.
27. $\tfrac{1}{8}(3x - 2\sin 2x + \tfrac{1}{4}\sin 4x), \tfrac{1}{8}(3x + \tfrac{1}{4}\sin 2x + \tfrac{1}{4}\sin 4x)$.
28. $\tfrac{1}{2}\sin^2 x$ or $-\tfrac{1}{2}\cos^2 x, \tfrac{1}{4}\sin^2 2x$ or $-\tfrac{1}{4}\cos^2 2x, -\tfrac{1}{4}\cos^4 x, \tfrac{1}{4}\sin^4 x$.
29. $x \ln (2x) - x, x \ln (2x^{\frac{1}{2}}) - \tfrac{1}{2}x$.
30. $\tfrac{1}{2}(\tfrac{1}{2}\sinh 2x - x), \tfrac{1}{2}(\tfrac{1}{2}\sinh 2x + x), \tfrac{1}{2}\cosh 2x, \tfrac{1}{2}\sinh 2x$.
31. $\tfrac{1}{33}(3x + 4)^{11}, -\tfrac{1}{6}(1 - x)^6, -\tfrac{1}{15}(2 - 3x)^5, -\tfrac{1}{4}(4x + 3)^{-1}$.
32. $\tfrac{2}{3}(x - 1)^{\frac{3}{2}}, -\tfrac{1}{3}(1 - 2x)^{\frac{3}{2}}, -\tfrac{3}{4}(1 - x)^{\frac{4}{3}}, -\tfrac{1}{2}(1 - 3x)^{\frac{3}{2}}, -2(2 - x)^{\frac{1}{2}}$.
33. $\ln |x|, \ln |x + 1|, \tfrac{1}{2}\ln |2x + 1|, -\ln |1 - x|, -\tfrac{1}{2}\ln |3 - 2x|$.
34. $\dfrac{1}{n-1}\cdot\dfrac{1}{(a-x)^{n-1}}, \dfrac{1}{d(n-1)}\dfrac{1}{(c-dx)^{n-1}}$.
35. $\dfrac{1}{2}\left(-\dfrac{\cos(m+n)x}{m+n} - \dfrac{\cos(m-n)x}{m-n}\right), \dfrac{1}{2}\left(\dfrac{\sin(m-n)x}{m-n} - \dfrac{\sin(m+n)x}{m+n}\right),$
$\dfrac{1}{2}\left(\dfrac{\sin(m-n)x}{m-n} + \dfrac{\sin(m+n)x}{m+n}\right), (m \neq \pm n)$.
36. $x + 2\ln|x - 1|, \tfrac{1}{2}x^2 + 2x - \ln|x + 1|, x + \tfrac{1}{2}x^2 + \tfrac{1}{3}x^3 + \ldots + \dfrac{1}{n}x^n$.
37. $\tfrac{2}{3}(x - 1)^{\frac{3}{2}} + 4(x - 1)^{\frac{1}{2}}, \dfrac{-1}{4(n-2)(2x+1)^{n-2}} + \dfrac{1}{4(n-1)(2x+1)^{n-1}}$.
38. $\tfrac{2}{9}\ln|3x - 1| - \dfrac{5}{9}\dfrac{1}{3x-1}, \ln|x - 1| - \dfrac{4}{x-1} - \dfrac{1}{(x-1)^2}$.
39. $\tfrac{1}{2}(1 + 2x)(\ln|1 + 2x| - 1), -(1 - x)(\ln|1 - x| - 1),$
$-\tfrac{1}{4}(3 - 4x)(\ln|3 - 4x| - 1)$.
40. $x \ln(x^2 - 4) + 2\ln\dfrac{x+2}{x-2} - 2x, \quad (x > 2, x < -2);$

$x \ln\dfrac{x+1}{x-1} + \ln(x^2 - 1) - 2, \quad (x > 1, x < -1);$

$-\ln(1 - x^2) + x\ln\dfrac{1-x}{1+x} + 2x, \quad (-1 < x < 1)$.
41. $\tan^{-1}(x + 1), \tfrac{1}{2}\tan^{-1}(2x + 1), \tfrac{1}{6}\tan^{-1}(\tfrac{2}{3}x - \tfrac{1}{3})$.
42. $\tan^{-1}(x + 2), \tfrac{1}{4}\tan^{-1}(x + \tfrac{3}{2})$.
43. $\cosh^{-1}(x - 1), (x > 2); \quad \sin^{-1}(x - 1), (0 < x < 2)$.
44. $\tfrac{1}{2}\ln\left|\dfrac{x-1}{x+1}\right|, (x < -1, -1 < x < 1, x > 1); \tfrac{1}{4}\ln\left|\dfrac{x}{x+1}\right|$.
45. $\tfrac{1}{2}\ln\left|1 - \dfrac{2}{x}\right|, \tfrac{1}{8}\ln\left|\dfrac{2x-3}{2x+1}\right|$.
46. $\tfrac{1}{3}\ln\left|\dfrac{x}{x+3}\right|, \tfrac{1}{2}\ln\left|\dfrac{x-4}{x-2}\right|$.
47. $\tfrac{1}{2}\cosh^{-1}(x - \tfrac{1}{2}), \sin^{-1}(\tfrac{1}{3}x + \tfrac{2}{3}), \sin^{-1}(x + 2)$.
48. Example 11, $\sin^{-1}\dfrac{x}{|c|}, \dfrac{1}{|c|}\tan^{-1}\dfrac{x}{|c|}, \dfrac{1}{2}\left[x\sqrt{(c^2 - x^2)} + c^2\sin^{-1}\dfrac{x}{|c|}\right]$.
49. $\dfrac{1}{2|ac|}\ln\left|\dfrac{x - |c/a|}{x + |c/a|}\right|$.

Answers 457

50. $\frac{2}{3}[(1 + x)^{\frac{3}{2}} - x^{\frac{3}{2}}]$.
51. $\dfrac{2}{3(b - a)}[(a - x)^{\frac{3}{2}} + (b - x)^{\frac{3}{2}}]$.
52. See section 11.7.
53. $x + \ln(x - 1) - 2$, $(x > 1)$; $x + \ln(1 - x) + 1 - \ln 2$, $(x < 1)$.
55. $\frac{1}{5}(x^2 + 1)^{\frac{5}{2}}$, $-(1 - x^2)^{\frac{1}{2}}$, $\frac{1}{2}\tan(x^2)$, $-e^{1/x}$, $\frac{1}{2}\ln(1 + x^2)$.
56. $\frac{1}{4}x^2(2\ln x - 1)$, $-\dfrac{1}{2(n - 1)}\dfrac{1}{(x^2 - a^2)^{n-1}}$, $\frac{2}{27}(2 + 3x^3)^{\frac{3}{2}}$.
57. $-\frac{2}{5}(1 + x^{\frac{5}{2}})^{-1}$, $-4\sqrt{(1 - \sqrt{x})}$.
58. $\dfrac{1}{1 - \alpha}(ax^2 + bx + c)^{-\alpha+1}$, $\ln|ax^2 + bx + c|$.
59. $\frac{1}{2}\ln(x^2 + 1) + \tan^{-1} x$, $1 - \ln(x^2 + 2x + 3)$.
60. $2(4x^2 + 4x + 5)^{\frac{1}{2}}$, $2(4x^2 + 4x + 5)^{\frac{1}{2}} + \frac{1}{2}\ln\{(x + \frac{1}{2}) + \sqrt{[(x + \frac{1}{2})^2 + 1]}\}$.
61. $2(9x^2 + 6x + 5)^{\frac{1}{2}}$, $\frac{1}{9}(9x^2 + 6x + 5)^{\frac{1}{2}} + \frac{2}{9}\ln\{(\frac{3}{2}x + \frac{1}{2}) + \sqrt{[(\frac{3}{2}x + \frac{1}{2})^2 + 1]}\}$.
62. $\frac{1}{5}(x^2 + 1)^{\frac{5}{2}} - \frac{1}{3}(x^2 + 1)^{\frac{3}{2}}$, $\frac{1}{3}\sin^{-1}(x^3)$, $\dfrac{1}{2\sqrt{3}}\tan^{-1}\dfrac{x^2}{\sqrt{3}}$.
63. $\ln\left|\dfrac{\sqrt{x + 1}}{\sqrt{x - 1}}\right|$, $\dfrac{1}{4}\dfrac{1}{(1 + x^2)^2} - \dfrac{1}{2}\dfrac{1}{1 + x^2}$.
64. $-\frac{1}{4}\ln\left|\dfrac{x^2 - 1}{x^2 + 1}\right|$, $\frac{1}{4}\ln\left|\dfrac{x^4}{x^4 - 1}\right|$.
65. $\dfrac{1}{n + 1}x^{n+1}\left(\ln|x| - \dfrac{1}{n + 1}\right)$, $\dfrac{1}{n + 1}(\ln x)^{n+1}$, $\ln|\ln|x||$.
66. $\ln|f(x)|$ when $n = -1$, $\dfrac{1}{n + 1}f^{n+1}(x)$ when $n \neq -1$.
67. $\frac{1}{2}\sin^2 x$, $-\frac{2}{3}\cos^{\frac{3}{2}} x$, $-(1/b)\ln|a + b\cos x|$, $(1/b)\ln|a + b\sin x|$.
68. $\ln|\sec x|$, $\ln|\sin x|$.
69. $\frac{1}{2}\tan^2 x$, $-\frac{1}{2}\cot^2 x$.
70. $\frac{1}{2}\tan^2 x - \ln|\sec x|$, $\frac{1}{3}\tan^3 x - \tan x + x$.
71. $\frac{1}{3}\tan^3 x + \frac{1}{5}\tan^5 x$, $\dfrac{1}{\sqrt{6}}\tan^{-1}(\sqrt{\frac{3}{2}}\tan x)$,
    $-\left[\frac{1}{3}\cot x + \dfrac{2}{3\sqrt{3}}\tan^{-1}(\sqrt{3}\cot x)\right]$.
72. $\dfrac{1}{\sqrt{2}}\tan^{-1}(\sqrt{2}\sin x)$, $\dfrac{1}{2\sqrt{2}}\ln\dfrac{\sqrt{2} + \sin x}{\sqrt{2} - \sin x}$.
73. $\ln|\cos x + \sin x| + 2x$, $\frac{1}{2}\ln|\cos x - \sin x| - \frac{3}{2}x$.
74. $\ln\left|\dfrac{e^x - 1}{e^x + 1}\right| = \ln|\tanh x|$, $2\tan^{-1} e^x$.
75. $\frac{1}{2}\ln(x^2 + 2x + 2) - \tan^{-1}(x + 1)$.
76. $\frac{1}{4}\ln(4x^2 + 4x + 3) - \dfrac{1}{\sqrt{2}}\tan^{-1}\dfrac{2x + 1}{\sqrt{2}}$.
77. $\frac{1}{4}\ln(4x^2 + 4x + 1) - \dfrac{1}{2x + 1}$.

78. $\frac{1}{4}\sqrt{(2x^2 + 2x + 1)} - \dfrac{1}{2\sqrt{2}} \sinh^{-1}(2x + 1)$.
79. $\ln|(x + \frac{1}{2}) + \sqrt{(x^2 + x)}|$.
80. $-\sqrt{(1 - x^2)} - \sin^{-1} x$.
81. $-2\sqrt{(1 + 2x - x^2)} + 3\sin^{-1}\left(\dfrac{x-1}{\sqrt{2}}\right)$.
82. $\frac{1}{3}(2x - x^2)^{\frac{3}{2}} + (x - 1)\sqrt{(2x - x^2)} + \sin^{-1}(x - 1)$
83. $\frac{1}{3}(x^2 - 2x - 1)^{\frac{3}{2}} + \frac{1}{4}(x - 1)\sqrt{(x^2 - 2x - 1)} -$
$\qquad \frac{1}{2}\ln|(x - 1) - \sqrt{(x^2 - 2x - 1)}|$.
84. $\ln\left|1 - \dfrac{1}{x}\right|$, $\ln\left|\dfrac{2x-1}{x+1}\right|$, $\frac{1}{4}\ln\left|\dfrac{2-3x}{x-2}\right|$, $\frac{1}{2}\ln\left|\dfrac{x-1}{x+1}\right| - \ln|x|$.
85. $\displaystyle\sum_{r=0}^{n} \dfrac{(-1)^{n-r}}{r!(n-r)!} \ln|x - r|$.
86. $\dfrac{1}{2(a^2 - b^2)}\left(\dfrac{1}{a}\ln\left|\dfrac{x-a}{x+a}\right| - \dfrac{1}{b}\ln\left|\dfrac{x-b}{x+b}\right|\right)$.
87. $2\ln|x + 1| - \ln|x|$, $6\ln|x - 4| - 2\ln|x - 3| + x$,
$\quad \frac{1}{8}x^2 + \frac{3}{16}\ln(2x + 1) - \dfrac{5}{16}\dfrac{1}{2x+1}$.
88. $\dfrac{1}{x} + \ln\left|1 - \dfrac{1}{x}\right|$, $-\dfrac{1}{x-1} - \ln\left|1 - \dfrac{1}{x}\right|$,
$\quad -\dfrac{1}{4}\dfrac{1}{(x-1)^2} - \dfrac{1}{4}\dfrac{1}{x-1} + \dfrac{1}{8}\ln\left|\dfrac{x+1}{x-1}\right|$,
$\quad -\dfrac{1}{2}\dfrac{1}{(x-1)^2} - 2\dfrac{1}{x-1} + \ln|x - 1|$.
89. $\dfrac{1}{8}\left[\dfrac{1}{(x+1)^2} + \dfrac{1}{x+1} - \dfrac{1}{(x-1)^2} + \dfrac{1}{x+1} + \ln\left|\dfrac{x-1}{x+1}\right|\right]$.
90. $\frac{1}{2}\ln|x| - \frac{1}{4}\ln(x^2 - 2x + 2) + \frac{1}{2}\tan^{-1}(x - 1)$,
$\quad \frac{2}{3}\ln|x - 1| - \frac{1}{3}\ln(x^2 + x + 1)$,
$\quad \frac{1}{3}\ln|x + 1| - \frac{1}{6}\ln(x^2 - x + 1) - \dfrac{4}{9\sqrt{3}}\tan^{-1}\left(\dfrac{2x-1}{\sqrt{3}}\right)$.
91. $-\ln|x| - \frac{1}{2}\ln|x + 1| + \frac{4}{3}\ln|2x + 1| + \frac{1}{6}\ln|x - 1|$,
$\quad \frac{1}{4}\ln\left|\dfrac{x-1}{x+1}\right| - \frac{1}{2}\tan^{-1} x$, $\frac{1}{3}\tan^{-1} x - \frac{1}{6}\tan^{-1}\frac{1}{2}x$.

Chapter 12

8. $\pi/4a$.
9. $\frac{4}{3}|a^3/b|$.
10. $2\ln 2 - 1$.
12. The curves meet at $x = \pm 2$, and in $-2 \leqslant x \leqslant 2$, both functions are positive. The area is $\displaystyle\int_{-2}^{2} (12 - 2x^2)\,dx - \int_{-2}^{2} x^2\,dx = 32$.
13. $\frac{4}{3}$.

Answers

14. $2 \ln 2 - \frac{1}{2}$ (the function is negative in $0 < x < 1$).
15. Require area under $x^2 + (y + b)^2 = a^2$, $-\sqrt{(a^2 - b^2)} \leqslant x \leqslant \sqrt{(a^2 - b^2)}$.
    Answer: $a^2 \sin^{-1} \sqrt{(1 - b^2/a^2)} - b\sqrt{(a^2 - b^2)}$.
16. $(\frac{2}{3}\pi - \sqrt{3}/4)a^2$.
17. In $0 \leqslant t < l/v$, number $= \displaystyle\int_0^t Au^2 e^{-au} \, du$.

    When $t \geqslant l/v$, number $= \displaystyle\int_{t-l/v}^t Au^2 e^{-au} \, du$.
18. $dI/dx = -aI$. Intensity $= I_0 e^{-al}$.
20. 12.
21. $2\pi^2$.
22. $1 - \sqrt{3}/2$.
26. (i) $\displaystyle\int_a^b h(x) \, dx$ and $\displaystyle\int_A^B H(y) \, dy$ are the areas between the curve and the x- and y-axes, respectively; (ii) $H(y) = \sin^{-1} y$, $h(x) = \sin x$.
    Answer is $c \sin^{-1} c + \sqrt{(1 - c^2)}$; (iii) $H(y) = \dfrac{\sqrt{(1 - y^2)}}{y}$, $h(x) = \dfrac{1}{\sqrt{(1 + x^2)}}$.
    Answer is $\frac{1}{2}\sqrt{3} - \ln(2 + \sqrt{3})$.
27. $\frac{1}{2}c^2$.
29. $\frac{3}{2}c^2\pi$.
30. $\pi(a^2 + \frac{1}{2}b^2)$.
31. $\frac{1}{2}(a^2 + b^2)$.
32. $\frac{1}{4}a^2 e^{2\alpha}(e^{4\pi} - 1)$.
33. $r = a e^{\theta \cot \alpha}$. Area $= \dfrac{a^2}{2 \cot \alpha}(e^{2\pi \cot \alpha} - 1)$.
35. $2\pi a(c^2 - b^2)$.
36. $\frac{1}{6}\pi l^3$.
37. $\pi \displaystyle\int_0^1 x^2 \dfrac{1 - x}{1 + x} \, dx = \pi(2 \ln 2 - \frac{4}{3})$.
38. $\frac{4}{3}\pi ab^2$.
39. $\pi(\frac{2}{3}a^3 + \frac{1}{3}b^3 - a^2 b)$.
40. $2\pi a^3 \left[\sqrt{\left(1 - \dfrac{c^2}{a^2}\right)\left(\dfrac{1}{3}\dfrac{c^2}{a^2} + \dfrac{2}{3}\right)} - \dfrac{c}{a} \sin^{-1} \sqrt{\left(1 - \dfrac{c^2}{a^2}\right)} \right]$.
41. $\frac{32}{15}\pi a^3$.
42. $2\pi(\frac{1}{8}\pi - 3 \ln 2)$.
43. $2\pi^2 abc$.
45. $\pi a^2(\rho_0 l + \frac{1}{2}\varepsilon l^2)$.
46. 0.325π gm.
47. $10^5 + 2.10^2$ gm.
48. $\displaystyle\int_{r_0}^{r_1} \rho c(a + b \ln r) 2\pi r \, dr = \pi\rho c[(a - \frac{1}{2}b)(r_1^2 - r_0^2) + b(r_1^2 \ln r_1 - r_0^2 \ln r_0)]$.
49. $\displaystyle\int_{R_0}^{R_1} \rho c\left(A + \dfrac{B}{R}\right) 4\pi R^2 \, dR = 4\pi\rho c[\frac{1}{3}A(R_1^3 - R_0^3) + \frac{1}{2}B(R_1^2 - R_0^2)]$.
50. $\displaystyle\int_0^a A(a^2 - r^2) 2\pi r \, dr = \frac{1}{2}\pi A a^4$.

460 *Introductory Analysis*

51. Area of strip $\approx 2\pi a \sin\theta \,.\, a\delta\theta$; flow rate across strip $= \alpha a \cos\theta$/unit area, θ measured from vertical through centre. Total flow rate across strip $= 2\pi a \sin\theta \,.\, a\delta\theta \,.\, \alpha a \cos\theta = \delta F$, say, and $\lim_{\delta\theta\to 0} \delta F/\delta\theta = 2\pi a^3\alpha \sin\theta \cos\theta$.

Total flow $= 2\pi a^3\alpha \int_0^{\frac{1}{2}\pi} \sin\theta \cos\theta \,\mathrm{d}\theta = \pi a^3\alpha$.

52. At time t the point is $\int_0^t v(t)\,\mathrm{d}t = \frac{1}{2}\beta t^2$ behind the front, so it is below a point in the cloud discharging at $(A/\sqrt{2})\beta^{\frac{1}{2}}t\,\mathrm{e}^{-\alpha(\frac{1}{2}\beta t^2)}$ gal/hr, mile^2. In time δt it therefore receives a rainfall $\delta F = (A/\sqrt{2})\beta^{\frac{1}{2}}t\,\mathrm{e}^{-\frac{1}{2}\alpha\beta t^2}\,\delta t$ gal/mile^2.
Answer $= [A/(\sqrt{2\alpha\beta^{\frac{1}{2}}})](1 - \mathrm{e}^{-\frac{1}{2}\alpha\beta t^2})$ gal/mile^2.

66. (a) $0.859 \leqslant I \leqslant 1.719$; (b) $0.718 \leqslant I \leqslant 1.219$; (c) $0.693 \leqslant I \leqslant 1.886$; (d) $1.000 \leqslant I \leqslant 1.884$; (e) $1.000 \leqslant I \leqslant 1.221$.

67. $-1 \leqslant \sin\dfrac{1}{x} \leqslant 1$ gives $-\frac{1}{3}\varepsilon^3 \leqslant \int_0^\varepsilon x^2 \sin\dfrac{1}{x}\,\mathrm{d}x \leqslant \frac{1}{3}\varepsilon^3$.
$\varepsilon < 0.144$ would satisfy the condition.

68. Note $x^4 \leqslant x^2$ in $[0,1]$. We have $\sqrt{(4-x^2)} \leqslant \sqrt{(4-x^2+x^4)} \leqslant \sqrt{4}$.

69. (a) Form $\int_0^x \ldots \mathrm{d}x$ of the inequality; (b) $0.678 \leqslant I \leqslant 0.717$.

70. Consider the cancellation of positive and negative parts of the integrand. Better bounds from $\frac{1}{2} \leqslant (1 + x^2)^{-1} \leqslant 1$.

Chapter 13

4. Put $\dfrac{b-a}{n} = h$. Then $\sum_{i=1}^n [x + (i-1)h]^2 \leqslant A \leqslant \sum_{i=1}^n (x + ih)^2$. The two sums tend to the same limit.

5. Divide the interval $[1,m]$ into $n(m-1)$ intervals.

6. $\delta A \approx \frac{1}{2}f^2(\theta_{i-1})\,\delta\theta$, leading to $\int_{\theta_1}^{\theta_2} \frac{1}{2}f^2(\theta)\,\mathrm{d}\theta$.

7. Divide the wall into n horizontal narrow strips of width δx, area $a\delta x$. Force on a strip, $\delta F_i \approx a\delta x \,.\, Av^\alpha(x_{i-1})$. Total force $= \int_0^h aA(Bx^\beta)^\alpha\,\mathrm{d}x$,
$= \dfrac{aAB^\alpha}{\alpha\beta + 1}h^{\alpha\beta+1}$. Moment on a strip is $x_{i-1}\delta F_i$, leading to
$$aAB^\alpha \int_0^h x^{\alpha\beta+1}\,\mathrm{d}x = \dfrac{aAB^\alpha}{\alpha\beta + 2}h^{\alpha\beta+2}.$$

11. Divide the rod into n sections, each of length $\delta x = l/n$. The charge on each is $q\delta x/l$.

13. (b), (d), (g), (i) if $(a^2/b^2) > 1$, (k), (l).

21. The indefinite integrals appropriate to the separate intervals differ only by constants, and F must be continuous.

22. Jump in $f = \lim_{x\to a^+} f(x) - \lim_{x\to a^-} f(x)$. Jump in slope $= \lim_{x\to a^+} \dfrac{\mathrm{d}F}{\mathrm{d}x} - \lim_{x\to a^-} \dfrac{\mathrm{d}F}{\mathrm{d}x}$.

23. $\phi(x)$ is an indefinite integral.

26. $\int_a^{a+\frac{1}{2}h} f(u)\,\mathrm{d}u = F(a + \frac{1}{2}h) - F(a)$, etc., and $F(x)$ is differentiable at $x = a$.

APPENDIX OF TABLES

TABLE 1. SQUARE ROOTS AND CUBE ROOTS

x	$x^{\frac{1}{2}}$	$x^{\frac{1}{3}}$	x	$x^{\frac{1}{2}}$	$x^{\frac{1}{3}}$	x	$x^{\frac{1}{2}}$	$x^{\frac{1}{3}}$
1·0	1·000	1·000	4·4	2·098	1·639	7·7	2·775	1·975
1·1	1·049	1·032	4·5	2·121	1·651	7·8	2·793	1·983
1·2	1·095	1·063	4·6	2·145	1·663	7·9	2·811	1·992
1·3	1·140	1·091	4·7	2·168	1·675	8·0	2·828	2·000
1·4	1·183	1·119	4·8	2·191	1·687	8·1	2·846	2·008
1·5	1·225	1·145	4·9	2·214	1·698	8·2	2·864	2·017
1·6	1·265	1·170	5·0	2·236	1·710	8·3	2·881	2·025
1·7	1·304	1·193	5·1	2·258	1·721	8·4	2·898	2·033
1·8	1·342	1·216	5·2	2·280	1·732	8·5	2·915	2·041
1·9	1·378	1·239	5·3	2·302	1·744	8·6	2·933	2·049
2·0	1·414	1·260	5·4	2·324	1·754	8·7	2·950	2·057
2·1	1·449	1·281	5·5	2·345	1·765	8·8	2·966	2·065
2·2	1·483	1·301	5·6	2·366	1·776	8·9	2·983	2·072
2·3	1·517	1·320	5·7	2·387	1·786	9·0	3·000	2·080
2·4	1·549	1·339	5·8	2·408	1·797	9·1	3·017	2·088
2·5	1·581	1·357	5·9	2·429	1·807	9·2	3·033	2·095
2·6	1·612	1·375	6·0	2·449	1·817	9·3	3·050	2·103
2·7	1·643	1·392	6·1	2·470	1·827	9·4	3·066	2·110
2·8	1·673	1·409	6·2	2·490	1·837	9·5	3·082	2·118
2·9	1·703	1·426	6·3	2·510	1·847	9·6	3·098	2·125
3·0	1·732	1·442	6·4	2·530	1·857	9·7	3·114	2·133
3·1	1·761	1·458	6·5	2·550	1·866	9·8	3·130	2·140
3·2	1·789	1·474	6·6	2·569	1·876	9·9	3·146	2·147
3·3	1·817	1·489	6·7	2·588	1·885	10	3·162	2·154
3·4	1·844	1·504	6·8	2·608	1·895	20	4·472	2·714
3·5	1·871	1·518	6·9	2·627	1·904	30	5·477	3·107
3·6	1·897	1·533	7·0	2·646	1·913	40	6·325	3·420
3·7	1·924	1·547	7·1	2·665	1·922	50	7·071	3·684
3·8	1·949	1·560	7·2	2·683	1·931	60	7·746	3·915
3·9	1·975	1·574	7·3	2·702	1·940	70	8·367	4·121
4·0	2·000	1·587	7·4	2·720	1·949	80	8·944	4·309
4·1	2·025	1·601	7·5	2·739	1·957	90	9·487	4·481
4·2	2·049	1·613	7·6	2·757	1·966	100	10·000	4·642
4·3	2·074	1·626						

TABLE 2. TRIGONOMETRIC FUNCTIONS

Angle		Sine	Co-sine	Tan-gent	Angle		Sine	Co-sine	Tan-gent
Degree	Radian				Degree	Radian			
0°	0·000	0·000	1·000	0·000					
1°	0·017	0·017	1·000	0·017	46°	0·803	0·719	0·695	1·036
2°	0·035	0·035	0·999	0·035	47°	0·820	0·731	0·682	1·072
3°	0·052	0·052	0·999	0·052	48°	0·838	0·743	0·669	1·111
4°	0·070	0·070	0·998	0·070	49°	0·855	0·755	0·656	1·150
5°	0·087	0·087	0·996	0·087	50°	0·873	0·766	0·643	1·192
6°	0·105	0·105	0·995	0·105	51°	0·890	0·777	0·629	1·235
7°	0·122	0·122	0·993	0·123	52°	0·908	0·788	0·616	1·280
8°	0·140	0·139	0·990	0·141	53°	0·925	0·799	0·602	1·327
9°	0·157	0·156	0·988	0·158	54°	0·942	0·809	0·588	1·376
10°	0·175	0·174	0·985	0·176	55°	0·960	0·819	0·574	1·428
11°	0·192	0·191	0·982	0·194	56°	0·977	0·829	0·559	1·483
12°	0·209	0·208	0·978	0·213	57°	0·995	0·839	0·545	1·540
13°	0·227	0·225	0·974	0·231	58°	1·012	0·848	0·530	1·600
14°	0·244	0·242	0·970	0·249	59°	1·030	0·857	0·515	1·664
15°	0·262	0·259	0·966	0·268	60°	1·047	0·866	0·500	1·732
16°	0·279	0·276	0·961	0·287	61°	1·065	0·875	0·485	1·804
17°	0·297	0·292	0·956	0·306	62°	1·082	0·883	0·470	1·881
18°	0·314	0·309	0·951	0·325	63°	1·100	0·891	0·454	1·963
19°	0·332	0·326	0·946	0·344	64°	1·117	0·899	0·438	2·050
20°	0·349	0·342	0·940	0·364	65°	1·134	0·906	0·423	2·145
21°	0·367	0·358	0·934	0·384	66°	1·152	0·914	0·407	2·246
22°	0·384	0·375	0·927	0·404	67°	1·169	0·921	0·391	2·356
23°	0·401	0·391	0·921	0·425	68°	1·187	0·927	0·375	2·475
24°	0·419	0·407	0·914	0·445	69°	1·204	0·934	0·358	2·605
25°	0·436	0·423	0·906	0·466	70°	1·222	0·940	0·342	2·747
26°	0·454	0·438	0·899	0·488	71°	1·239	0·946	0·326	2·904
27°	0·471	0·454	0·891	0·510	72°	1·257	0·951	0·309	3·078
28°	0·489	0·470	0·883	0·532	73°	1·274	0·956	0·292	3·271
29°	0·506	0·485	0·875	0·554	74°	1·292	0·961	0·276	3·487
30°	0·524	0·500	0·866	0·577	75°	1·309	0·966	0·259	3·732
31°	0·541	0·515	0·857	0·601	76°	1·326	0·970	0·242	4·011
32°	0·559	0·530	0·848	0·625	77°	1·344	0·974	0·225	4·331
33°	0·576	0·545	0·839	0·649	78°	1·361	0·978	0·208	4·705
34°	0·593	0·559	0·829	0·675	79°	1·397	0·982	0·191	5·145
35°	0·611	0·574	0·819	0·700	80°	1·396	0·985	0·174	5·671
36°	0·628	0·588	0·809	0·727	81°	1·414	0·988	0·156	6·314
37°	0·646	0·602	0·799	0·754	82°	1·431	0·990	0·139	7·115
38°	0·663	0·616	0·788	0·781	83°	1·449	0·993	0·122	8·144
39°	0·681	0·629	0·777	0·810	84°	1·466	0·995	0·105	9·514
40°	0·698	0·643	0·766	0·839	85°	1·484	0·996	0·087	11·43
41°	0·716	0·658	0·755	0·869	86°	1·501	0·998	0·070	14·30
42°	0·733	0·669	0·743	0·900	87°	1·518	0·999	0·052	19·08
43°	0·751	0·682	0·731	0·933	88°	1·536	0·999	0·035	28·64
44°	0·768	0·695	0·719	0·966	89°	1·553	1·000	0·018	57·29
45°	0·785	0·707	0·707	1·000	90°	1·571	1·000	0·000	∞

Appendix of tables

TABLE 2—*Continued*

Angle		Sine	Cosine	Angle		Sine	Cosine
Radian	Degree			Radian	Degree		
0·00	0·00°	0·000	1·000				
0·02	1·15°	0·020	1·000	0·92	52·71°	0·796	0·606
0·04	2·29°	0·040	0·999	0·94	53·86°	0·808	0·590
0·06	3·44°	0·060	0·998	0·96	55·00°	0·819	0·574
0·08	4·58°	0·080	0·997	0·98	56·15°	0·830	0·557
0·10	5·73°	0·100	0·995	1·00	57·30°	0·841	0·540
0·12	6·88°	0·120	0·993	1·02	58·44°	0·852	0·523
0·14	8·02°	0·140	0·990	1·04	59·59°	0·862	0·506
0·16	9·17°	0·159	0·987	1·06	60·73°	0·872	0·489
0·18	10·31°	0·179	0·984	1·08	61·88°	0·882	0·471
0·20	11·46°	0·197	0·980	1·10	63·03°	0·891	0·454
0·22	12·61°	0·218	0·976	1·12	64·17°	0·900	0·436
0·24	13·75°	0·238	0·971	1·14	65·32°	0·909	0·418
0·26	14·90°	0·257	0·966	1·16	66·46°	0·917	0·399
0·28	16·04°	0·276	0·961	1·18	67·61°	0·925	0·381
0·30	17·19°	0·296	0·955	1·20	68·75°	0·932	0·362
0·32	18·33°	0·314	0·949	1·22	69·90°	0·939	0·344
0·34	19·48°	0·333	0·943	1·24	71·05°	0·946	0·325
0·36	20·63°	0·352	0·936	1·26	72·19°	0·952	0·306
0·38	21·77°	0·371	0·929	1·28	73·34°	0·958	0·287
0·40	22·92°	0·389	0·921	1·30	74·48°	0·964	0·268
0·42	24·06°	0·408	0·913	1·32	75·63°	0·969	0·248
0·44	25·21°	0·426	0·905	1·34	76·78°	0·973	0·229
0·46	26·36°	0·444	0·896	1·36	77·92°	0·978	0·209
0·48	27·50°	0·462	0·887	1·38	79·07°	0·982	0·190
0·50	28·65°	0·479	0·878	1·40	80·21°	0·985	0·170
0·52	29·79°	0·497	0·868	1·42	81·36°	0·989	0·150
0·54	30·94°	0·514	0·858	1·44	82·51°	0·991	0·130
0·56	32·09°	0·531	0·847	1·46	83·65°	0·994	0·111
0·58	33·23°	0·548	0·836	1·48	84·80°	0·996	0·091
0·60	34·38°	0·565	0·825	1·50	85·94°	0·998	0·071
0·62	35·52°	0·581	0·814	1·52	87·09°	0·999	0·051
0·64	36·67°	0·597	0·802	1·54	88·24°	1·000	0·031
0·66	37·82°	0·613	0·790	1·56	89·38°	1·000	0·011
0·68	38·96°	0·629	0·778	1·58	90·53°	1·000	− 0·009
0·70	40·11°	0·644	0·765	1·60	91·67°	1·000	− 0·029
0·72	41·25°	0·659	0·752	1·62	92·82°	0·999	− 0·049
0·74	42·40°	0·674	0·738	1·64	93·97°	0·998	− 0·069
0·76	43·54°	0·689	0·725	1·66	95·11°	0·996	− 0·089
0·78	44·69°	0·703	0·711	1·68	96·26°	0·994	− 0·109
0·80	45·84°	0·717	0·697	1·70	97·40°	0·992	− 0·129
0·82	46·98°	0·731	0·682	1·72	98·55°	0·989	− 0·149
0·84	48·13°	0·745	0·667	1·74	99·69°	0·986	− 0·168
0·86	49·27°	0·758	0·652	1·76	100·84°	0·982	− 0·188
0·88	50·42°	0·771	0·637	1·78	101·99°	0·978	− 0·208
0·90	51·27°	0·783	0·622	1·80	103·13°	0·974	− 0·227

$\frac{1}{2}\pi = 1\cdot5708$, $\pi = 3\cdot1416$, $\frac{3}{2}\pi = 4\cdot7124$, $2\pi = 6\cdot2832$

TABLE 3. NATURAL LOGARITHMS

	0	1	2	3	4	5	6	7	8	9
1·0	0·0000	0100	0198	0296	0392	0488	0583	0677	0770	0862
1·1	0·0953	1044	1133	1222	1310	1398	1484	1570	1655	1740
1·2	0·1823	1906	1989	2070	2151	2231	2311	2390	2469	2546
1·3	0·2624	2700	2776	2852	2927	3001	3075	3148	3221	3293
1·4	0·3365	3436	3507	3577	3646	3716	3784	3853	3920	3988
1·5	0·4055	4121	4187	4253	4318	4383	4447	4511	4574	4637
1·6	0·4700	4762	4824	4886	4947	5008	5068	5128	5188	5247
1·7	0·5306	5365	5423	5481	5539	5596	5653	5710	5766	5822
1·8	0·5878	5933	5988	6043	6098	6152	6206	6259	6313	6366
1·9	0·6419	6471	6523	6575	6627	6678	6729	6780	6831	6881
2·0	0·6931	6981	7031	7080	7129	7178	7227	7275	7324	7372
2·1	0·7419	7467	7514	7561	7608	7655	7701	7747	7793	7839
2·2	0·7885	7930	7975	8020	8065	8109	8154	8198	8242	8286
2·3	0·8329	8372	8416	8459	8502	8544	8587	8629	8671	8713
2·4	0·8755	8796	8838	8879	8920	8961	9002	9042	9083	9123
2·5	0·9163	9203	9243	9282	9322	9361	9400	9439	9478	9517
2·6	0·9555	9594	9632	9670	9708	9746	9783	9821	9858	9895
2·7	0·9933	9969	1·0006	1·0043	1·0080	1·0116	1·0152	1·0188	1·0225	1·0260
2·8	1·0296	0332	0367	0403	0438	0473	0508	0543	0578	0613
2·9	1·0647	0682	0716	0750	0784	0815	0852	0886	0919	0953
3·0	1·0986	1019	1053	1086	1119	1151	1184	1217	1249	1282
3·1	1·1314	1346	1378	1410	1442	1474	1506	1537	1569	1600
3·2	1·1632	1663	1694	1725	1756	1787	1817	1848	1878	1909
3·3	1·1939	1969	2000	2030	2060	2090	2119	2149	2179	2208
3·4	1·2238	2267	2296	2326	2355	2384	2413	2442	2470	2499
3·5	1·2528	2556	2585	2613	2641	2669	2698	2726	2754	2782
3·6	1·2809	2837	2865	2892	2920	2947	2975	3002	3029	3056
3·7	1·3083	3110	3137	3164	3191	3218	3244	3271	3297	3324
3·8	1·3350	3376	3403	3429	3455	3481	3507	3533	3558	3584
3·9	1·3610	3635	3661	3686	3712	3737	3762	3788	3813	3838
4·0	1·3863	3888	3913	3938	3962	3987	4012	4036	4061	4085
4·1	1·4110	4134	4159	4183	4207	4231	4255	4279	4303	4327
4·2	1·4351	4375	4398	4422	4446	4469	4493	4516	4540	4563
4·3	1·4586	4609	4633	4656	4679	4702	4725	4748	4770	4793
4·4	1·4816	4839	4861	4884	4907	4929	4951	4974	4996	5019
4·5	1·5041	5063	5085	5107	5129	5151	5173	5195	5217	5239
4·6	1·5261	5282	5304	5326	5347	5369	5390	5412	5433	5454
4·7	1·5476	5497	5518	5539	5560	5581	5602	5623	5644	5665
4·8	1·5686	5707	5728	5748	5769	5790	5810	5831	5851	5872
4·9	1·5892	5913	5933	5953	5974	5994	6014	6034	6054	6074
5·0	1·6094	6114	6134	6154	6174	6194	6214	6233	6253	6273
5·1	1·6292	6312	6332	6351	6371	6390	6409	6429	6448	6467
5·2	1·6487	6506	6525	6544	6563	6582	6601	6620	6639	6658
5·3	1·6677	6696	6715	6734	6752	6771	6790	6808	6827	6845
5·4	1·6864	6882	6901	6919	6938	6956	6974	6993	7011	7029

TABLE 3—*Continued*

	0	1	2	3	4	5	6	7	8	9
5·5	1·7047	7066	7084	7102	7120	7138	7156	7174	7192	7210
5·6	1·7228	7246	7263	7281	7299	7317	7334	7352	7370	7387
5·7	1·7405	7422	7440	7457	7475	7492	7509	7527	7544	7561
5·8	1·7579	7596	7613	7630	7647	7664	7681	7699	7716	7733
5·9	1·7750	7766	7783	7800	7817	7834	7851	7867	7884	7901
6·0	1·7918	7934	7951	7967	7984	8001	8017	8034	8050	8066
6·1	1·8083	8099	8116	8132	8148	8165	8181	8197	8213	8229
6·2	1·8245	8262	8278	8294	8310	8326	8342	8358	8374	8390
6·3	1·8405	8421	8437	8453	8469	8485	8500	8516	8532	8547
6·4	1·8563	8579	8594	8610	8625	8641	8656	8672	8687	8703
6·5	1·8718	8733	8749	8764	8779	8795	8810	8825	8840	8856
6·6	1·8871	8886	8901	8916	8931	8946	8961	8976	8991	9006
6·7	1·9021	9036	9051	9066	9081	9095	9110	9125	9140	9155
6·8	1·9169	9184	9199	9213	9228	9242	9257	9272	9286	9301
6·9	1·9315	9330	9344	9359	9373	9387	9402	9416	9430	9445
7·0	1·9459	9473	9488	9502	9516	9530	9544	9559	9573	9587
7·1	1·9601	9615	9629	9643	9657	9671	9685	9699	9713	9727
7·2	1·9741	9755	9769	9782	9796	9810	9824	9838	9851	9865
7·3	1·9879	9892	9906	9920	9933	9947	9961	9974	9988	2·0001
7·4	2·0015	0028	0042	0055	0069	0082	0096	0109	0122	0136
7·5	2·0149	0162	0176	0189	0202	0215	0229	0242	0255	0268
7·6	2·0281	0295	0308	0321	0334	0347	0360	0373	0386	0399
7·7	2·0412	0425	0438	0451	0464	0477	0490	0503	0516	0528
7·8	2·0541	0554	0567	0580	0592	0605	0618	0631	0643	0656
7·9	2·0669	0681	0694	0707	0719	0732	0744	0757	0769	0782
8·0	2·0794	0807	0819	0832	0844	0857	0869	0882	0894	0906
8·1	2·0919	0931	0943	0956	0968	0980	0992	1005	1017	1029
8·2	2·1041	1054	1066	1078	1090	1102	1114	1126	1138	1150
8·3	2·1163	1175	1187	1199	1211	1223	1235	1247	1258	1270
8·4	2·1282	1294	1306	1318	1330	1342	1353	1365	1377	1389
8·5	2·1401	1412	1424	1436	1448	1459	1471	1483	1494	1506
8·6	2·1518	1529	1541	1552	1564	1576	1587	1599	1610	1622
8·7	2·1633	1645	1656	1668	1679	1691	1702	1713	1725	1736
8·8	2·1748	1759	1770	1782	1793	1804	1815	1827	1838	1849
8·9	2·1861	1872	1883	1894	1905	1917	1928	1939	1950	1961
9·0	2·1972	1983	1994	2006	2017	2028	2039	2050	2061	2072
9·1	2·2083	2094	2105	2116	2127	2138	2148	2159	2170	2181
9·2	2·2192	2203	2214	2225	2235	2246	2257	2268	2279	2289
9·3	2·2300	2311	2322	2332	2343	2354	2364	2375	2386	2396
9·4	2·2407	2418	2428	2439	2450	2460	2471	2481	2492	2502
9·5	2·2513	2523	2534	2544	2555	2565	2576	2586	2597	2607
9·6	2·2618	2628	2638	2649	2659	2670	2680	2690	2701	2711
9·7	2·2721	2732	2742	2752	2762	2773	2783	2793	2803	2814
9·8	2·2824	2834	2844	2854	2865	2875	2885	2895	2905	2915
9·9	2·2925	2935	2946	2956	2966	2976	2986	2996	3006	3016
10·0	2·3026	3036	3046	3056	3066	3076	3086	3096	3106	3115

Table 4. Common logarithms

	0	1	2	3	4	5	6	7	8	9
10	0·0000	0043	0086	0128	0170	0212	0253	0294	0334	0374
11	0·0414	0453	0492	0531	0569	0607	0645	0682	0719	0755
12	0·0792	0828	0864	0899	0934	0969	1004	1038	1072	1106
13	0·1139	1173	1206	1239	1271	1303	1335	1367	1399	1430
14	0·1461	1492	1523	1553	1584	1614	1644	1673	1703	1732
15	0·1761	1790	1818	1847	1875	1903	1931	1959	1987	2014
16	0·2041	2068	2095	2122	2148	2175	2201	2227	2253	2279
17	0·2304	2330	2355	2380	2405	2430	2455	2480	2504	2529
18	0·2553	2577	2601	2625	2648	2672	2695	2718	2742	2765
19	0·2788	2810	2833	2856	2878	2900	2923	2945	2967	2989
20	0·3010	3032	3054	3075	3096	3118	3139	3160	3181	3201
21	0·3222	3243	3263	3284	3304	3324	3345	3365	3385	3404
22	0·3424	3444	3464	3483	3502	3522	3541	3560	3579	3598
23	0·3617	3636	3655	3674	3692	3711	3729	3747	3766	3784
24	0·3802	3820	3838	3856	3874	3892	3909	3927	3945	3962
25	0·3979	3997	4014	4031	4048	4065	4082	4099	4116	4133
26	0·4150	4166	4183	4200	4216	4232	4249	4265	4281	4298
27	0·4314	4330	4346	4362	4378	4393	4409	4425	4440	4456
28	0·4472	4487	4502	4518	4533	4548	4564	4579	4594	4609
29	0·4624	4639	4654	4669	4683	4398	4713	4728	4742	4757
30	0·4771	4786	4800	4814	4829	4843	4857	4871	4886	4900
31	0·4914	4928	4942	4955	4969	4983	4997	5011	5024	5038
32	0·5051	5065	5079	5092	5105	5119	5132	5145	5159	5172
33	0·5185	5198	5211	5224	5237	5250	5263	5276	5289	5302
34	0·5315	5328	5340	5353	5366	5378	5391	5403	5416	5428
35	0·5441	5453	5465	5478	5490	5502	5514	5527	5539	5551
36	0·5563	5575	5587	5599	5611	5623	5635	5647	5658	5670
37	0·5682	5694	5705	5717	5729	5740	5752	5763	5775	5786
38	0·5798	5809	5821	5832	5843	5855	5866	5877	5888	5899
39	0·5911	5922	5933	5944	5955	5966	5977	5988	5999	6010
40	0·6021	6031	6042	6053	6064	6075	6085	6096	6107	6117
41	0·6128	6138	6149	6160	6170	6180	6191	6201	6212	6222
42	0·6232	6243	6253	6263	6274	6284	6294	6304	6314	6325
43	0·6335	6345	6355	6365	6375	6385	6395	6405	6415	6425
44	0·6435	6444	6454	6464	6484	6474	6493	6503	6513	6522
45	0·6532	6542	6551	6561	6571	6580	6590	6599	6609	6618
46	0·6628	6637	6646	6656	6665	6675	6684	6693	6702	6712
47	0·6721	6730	6739	6749	6758	6767	6776	6785	6794	6803
48	0·6812	6821	6830	6839	6848	6857	6866	6875	6884	6893
49	0·6902	6911	6920	6928	6937	6946	6955	6964	6972	6981
50	0·6990	6998	7007	7016	7024	7033	7042	7050	7059	7067
51	0·7076	7084	7093	7101	7110	7118	7126	7135	7143	7152
52	0·7160	7168	7177	7185	7193	7202	7210	7218	7226	7235
53	0·7243	7251	7259	7267	7275	7284	7292	7300	7308	7316
54	0·7324	7332	7340	7348	7356	7364	7372	7380	7388	7396

Appendix of tables

TABLE 4—*Continued*

	0	1	2	3	4	5	6	7	8	9
55	0·7404	7412	7419	7427	7435	7443	7451	7459	7466	7474
56	0·7482	7490	7497	7505	7513	7520	7528	7536	7543	7551
57	0·7559	7566	7574	7582	7589	7597	7604	7612	7619	7627
58	0·7634	7642	7649	7657	7664	7672	7679	7686	7694	7701
59	0·7709	7716	7723	7731	7738	7745	7752	7760	7767	7774
60	0·7782	7789	7796	7803	7810	7818	7825	7832	7839	7846
61	0·7853	7860	7868	7875	7882	7889	7896	7903	7910	7917
62	0·7924	7931	7938	7945	7952	7959	7966	7973	7980	7987
63	0·7993	8000	8007	8014	8021	8028	8035	8041	8048	8055
64	0·8062	8069	8075	8082	8089	8096	8102	8109	8116	8122
65	0·8129	8136	8142	8149	8156	8162	8169	8176	8182	8189
66	0·8195	8202	8209	8215	8222	8228	8235	8241	8248	8254
67	0·8261	8267	8274	8280	8287	8293	8299	8306	8312	8319
68	0·8325	8331	8338	8344	8351	8357	8363	8370	8376	8382
69	0·8388	8395	8401	8407	8414	8420	8426	8432	8439	8445
70	0·8451	8457	8463	8470	8476	8482	8488	8494	8500	8506
71	0·8513	8519	8525	8531	8537	8543	8549	8555	8561	8567
72	0·8573	8579	8585	8591	8597	8603	8609	8615	8621	8627
73	0·8633	8639	8645	8651	8657	8663	8669	8675	8681	8686
74	0·8692	8698	8704	8710	8716	8722	8727	8733	8739	8745
75	0·8751	8756	8762	8768	8774	8779	8785	8791	8797	8802
76	0·8808	8814	8820	8825	8831	8837	8842	8848	8854	8859
77	0·8865	8871	8876	8882	8887	8893	8899	8904	8910	8915
78	0·8921	8927	8932	8938	8943	8949	8954	8960	8965	8971
79	0·8976	8982	8987	8993	8998	9004	9009	9015	9020	9025
80	0·9031	9036	9042	9047	9053	9058	9063	9069	9074	9079
81	0·9085	9090	9096	9101	9106	9112	9117	9122	9128	9133
82	0·9138	9143	9149	9154	9159	9165	9170	9175	9180	9186
83	0·9191	9196	9201	9206	9212	9217	9222	9227	9232	9238
84	0·9243	9248	9253	9258	9263	9269	9274	9279	9284	9289
85	0·9294	9299	9304	9309	9315	9320	9325	9330	9335	9340
86	0·9345	9350	9355	9360	9365	9370	9375	9380	9385	9390
87	0·9395	9400	9405	9410	9415	9420	9425	9430	9435	9440
88	0·9445	9450	9455	9460	9465	9469	9474	9479	9484	9489
89	0·9494	9499	9504	9509	9513	9518	9523	9528	9533	9538
90	0·9542	9547	9552	9557	9562	9566	9571	9576	9581	9586
91	0·9590	9595	9600	9605	9609	9614	9619	9624	9628	9633
92	0·9638	9643	9647	9652	9657	9661	9666	9671	9675	9680
93	0·9685	9689	9694	9699	9703	9708	9713	9717	9722	9727
94	0·9731	9736	9741	9745	9750	9754	9759	9763	9768	9773
95	0·9777	9782	9786	9791	9795	9800	9805	9809	9814	9818
96	0·9823	9827	9832	9836	9841	9845	9850	9854	9859	9863
97	0·9868	9872	9877	9881	9886	9890	9894	9899	9903	9908
98	0·9912	9917	9921	9926	9930	9934	9939	9943	9948	9952
99	0·9956	9961	9965	9969	9974	9978	9983	9987	9991	9996

TABLE 5. EXPONENTIAL FUNCTIONS

x	e^x	e^{-x}	sinh x	cosh x
0·00	1	1	0	1
0·01	1·0101	0·9900	0·0100	1·0001
0·02	1·0202	0·9802	0·0200	1·0002
0·03	1·0305	0·9704	0·0300	1·0005
0·04	1·0408	0·9608	0·0400	1·0008
0·05	1·0513	0·9512	0·0500	1·0013
0·06	1·0618	0·9418	0·0600	1·0018
0·07	1·0725	0·9324	0·0701	1·0025
0·08	1·0833	0·9231	0·0801	1·0032
0·09	1·0942	0·9139	0·0901	1·0041
0·1	1·1052	0·9048	0·1002	1·0050
0·2	1·2214	0·8187	0·2013	1·0201
0·3	1·3499	0·7408	0·3045	1·0453
0·4	1·4918	0·6703	0·4108	1·0811
0·5	1·6487	0·6065	0·5211	1·1276
0·6	1·8221	0·5488	0·6367	1·1855
0·7	2·0138	0·4966	0·7586	1·2552
0·8	2·2255	0·4493	0·8881	1·3374
0·9	2·4596	0·4066	1·0265	1·4331
1	2·7183	0·3679	1·1752	1·5431
2	7·3891	0·1353	3·6269	3·7622
3	20·086	0·0498	10·018	10·068
4	54·598	0·0183	27·290	27·308
5	148·41	0·0067	74·203	74·210
6	403·43	0·0025	201·71	201·72
7	1097	0·0009	548	548
8	2981	0·0003	1490	1490
9	8103	0·0001	4051	4051
10	22026	0·000045	11013	11013

Appendix of tables

TABLE 6. INVERSE FUNCTIONS

x	$\sin^{-1} x$	$\cos^{-1} x$	$\tanh^{-1} x$	x	$\sin^{-1} x$	$\cos^{-1} x$	$\tanh^{-1} x$
0·00	0·000	1·571	0·000	0·50	0·524	1·047	0·549
0·01	0·010	1·561	0·010	0·51	0·535	1·036	0·563
0·02	0·020	1·551	0·020	0·52	0·547	1·024	0·576
0·03	0·030	1·541	0·030	0·53	0·559	1·012	0·590
0·04	0·040	1·531	0·040	0·54	0·570	1·000	0·604
0·05	0·050	1·521	0·050	0·55	0·582	0·988	0·618
0·06	0·060	1·511	0·060	0·56	0·594	0·976	0·633
0·07	0·070	1·501	0·070	0·57	0·607	0·964	0·648
0·08	0·080	1·491	0·080	0·58	0·619	0·952	0·662
0·09	0·090	1·481	0·090	0·59	0·631	0·940	0·678
0·10	0·100	1·471	0·100	0·60	0·644	0·927	0·693
0·11	0·110	1·461	0·110	0·61	0·656	0·915	0·709
0·12	0·120	1·451	0·121	0·62	0·669	0·902	0·725
0·13	0·130	1·440	0·131	0·63	0·682	0·889	0·741
0·14	0·140	1·430	0·141	0·64	0·694	0·876	0·758
0·15	0·151	1·420	0·151	0·65	0·708	0·863	0·775
0·16	0·161	1·410	0·161	0·66	0·721	0·850	0·793
0·17	0·171	1·400	0·172	0·67	0·734	0·837	0·811
0·18	0·181	1·390	0·182	0·68	0·748	0·823	0·829
0·19	0·191	1·380	0·192	0·69	0·761	0·809	0·848
0·20	0·201	1·369	0·203	0·70	0·775	0·795	0·867
0·21	0·212	1·359	0·213	0·71	0·789	0·781	0·887
0·22	0·222	1·349	0·224	0·72	0·804	0·767	0·908
0·23	0·232	1·339	0·234	0·73	0·818	0·752	0·929
0·24	0·242	1·328	0·245	0·74	0·833	0·738	0·950
0·25	0·253	1·318	0·255	0·75	0·848	0·723	0·973
0·26	0·263	1·308	0·266	0·76	0·863	0·707	0·996
0·27	0·273	1·297	0·277	0·77	0·879	0·692	1·020
0·28	0·284	1·287	0·288	0·78	0·895	0·676	1·045
0·29	0·294	1·277	0·299	0·79	0·911	0·660	1·071
0·30	0·305	1·266	0·310	0·80	0·927	0·644	1·099
0·31	0·315	1·256	0·321	0·81	0·944	0·627	1·127
0·32	0·326	1·245	0·332	0·82	0·961	0·609	1·157
0·33	0·336	1·234	0·343	0·83	0·979	0·592	1·188
0·34	0·347	1·224	0·354	0·84	0·997	0·574	1·221
0·35	0·358	1·213	0·365	0·85	1·016	0·555	1·256
0·36	0·368	1·203	0·377	0·86	1·035	0·536	1·293
0·37	0·379	1·192	0·388	0·87	1·055	0·516	1·333
0·38	0·390	1·181	0·400	0·88	1·076	0·495	1·376
0·39	0·401	1·170	0·412	0·89	1·097	0·473	1·422
0·40	0·412	1·159	0·424	0·90	1·120	0·451	1·472
0·41	0·422	1·148	0·436	0·91	1·143	0·428	1·528
0·42	0·433	1·137	0·448	0·92	1·168	0·403	1·589
0·43	0·444	1·126	0·460	0·93	1·194	0·376	1·658
0·44	0·456	1·115	0·472	0·94	1·223	0·348	1·738
0·45	0·467	1·104	0·485	0·95	1·253	0·318	1·832
0·46	0·478	1·093	0·497	0·96	1·287	0·284	1·946
0·47	0·489	1·082	0·510	0·97	1·325	0·246	2·092
0·48	0·501	1·070	0·523	0·98	1·370	0·200	2·298
0·49	0·512	1·059	0·536	0·99	1·429	0·142	2·647
				1·00	$\pi/2$	0·000	∞

These functions are defined only in $-1 \leqslant x \leqslant 1$. For negative arguments note that $\sin^{-1} x$ and $\tanh^{-1} x$ are odd, and $\cos^{-1} x$ is even.

TABLE 6—Continued

x	$\tan^{-1} x$	$\cot^{-1} x$	$\sinh^{-1} x$	$\cosh^{-1} x$	x	$\tan^{-1} x$	$\cot^{-1} x$	$\sinh^{-1} x$	$\cosh^{-1} x$
0·00	0·000	1·571	0·000	—	5·00	1·373	0·197	2·312	2·292
0·10	0·100	1·471	0·100	—	5·10	1·377	0·194	2·332	2·313
0·20	0·197	1·373	0·199	—	5·20	1·381	0·190	2·351	2·332
0·30	0·291	1·279	0·296	—	5·30	1·384	0·186	2·370	2·352
0·40	0·381	1·190	0·390	—	5·40	1·388	0·183	2·388	2·371
0·50	0·464	1·107	0·481	—	5·50	1·391	0·180	2·406	2·390
0·60	0·540	1·030	0·569	—	5·60	1·394	0·177	2·424	2·408
0·70	0·611	0·960	0·653	—	5·70	1·397	0·174	2·441	2·426
0·80	0·675	0·896	0·733	—	5·80	1·400	0·171	2·458	2·443
0·90	0·733	0·838	0·809	—	5·90	1·403	0·168	2·475	2·461
1·00	0·785	0·785	0·881	0·000	6·00	1·406	0·165	2·492	2·478
1·10	0·833	0·738	0·950	0·444	6·10	1·408	0·162	2·508	2·495
1·20	0·876	0·695	1·016	0·622	6·20	1·411	0·160	2·524	2·511
1·30	0·915	0·656	1·078	0·756	6·30	1·413	0·157	2·540	2·527
1·40	0·951	0·620	1·138	0·867	6·40	1·416	0·155	2·555	2·543
1·50	0·983	0·588	1·195	0·962	6·50	1·418	0·153	2·571	2·559
1·60	1·012	0·559	1·249	1·047	6·60	1·420	0·150	2·586	2·574
1·70	1·039	0·532	1·301	1·123	6·70	1·423	0·148	2·601	2·590
1·80	1·064	0·507	1·350	1·193	6·80	1·425	0·146	2·615	2·605
1·90	1·086	0·484	1·398	1·257	6·90	1·427	0·144	2·630	2·619
2·00	1·107	0·464	1·444	1·317	7·00	1·429	0·142	2·644	2·634
2·10	1·126	0·444	1·487	1·373	7·10	1·431	0·140	2·658	2·648
2·20	1·144	0·427	1·530	1·425	7·20	1·433	0·138	2·672	2·662
2·30	1·161	0·410	1·570	1·475	7·30	1·435	0·136	2·686	2·676
2·40	1·176	0·395	1·609	1·522	7·40	1·436	0·134	2·699	2·690
2·50	1·190	0·381	1·647	1·567	7·50	1·438	0·133	2·712	2·704
2·60	1·204	0·367	1·684	1·609	7·60	1·440	0·131	2·726	2·717
2·70	1·216	0·355	1·719	1·650	7·70	1·442	0·129	2·739	2·730
2·80	1·228	0·343	1·753	1·689	7·80	1·443	0·128	2·751	2·743
2·90	1·239	0·332	1·786	1·727	7·90	1·445	0·126	2·764	2·756
3·00	1·249	0·322	1·818	1·763	8·00	1·446	0·124	2·776	2·769
3·10	1·259	0·312	1·850	1·797	8·10	1·448	0·123	2·789	2·781
3·20	1·268	0·303	1·880	1·831	8·20	1·449	0·121	2·801	2·794
3·30	1·277	0·294	1·909	1·863	8·30	1·451	0·120	2·813	2·806
3·40	1·285	0·286	1·938	1·895	8·40	1·452	0·118	2·825	2·818
3·50	1·292	0·278	1·966	1·925	8·50	1·454	0·117	2·837	2·830
3·60	1·300	0·271	1·993	1·954	8·60	1·455	0·116	2·848	2·842
3·70	1·307	0·264	2·019	1·983	8·70	1·456	0·114	2·860	2·853
3·80	1·313	0·257	2·045	2·010	8·80	1·458	0·113	2·871	2·865
3·90	1·320	0·251	2·070	2·037	8·90	1·459	0·112	2·882	2·876
4·00	1·326	0·245	2·095	2·063	9·00	1·460	0·111	2·893	2·887
4·10	1·332	0·239	2·119	2·089	9·10	1·461	0·109	2·904	2·898
4·20	1·337	0·234	2·142	2·114	9·20	1·463	0·108	2·915	2·909
4·30	1·342	0·228	2·165	2·138	9·30	1·464	0·107	2·926	2·920
4·40	1·347	0·223	2·187	2·162	9·40	1·465	0·106	2·937	2·931
4·50	1·352	0·219	2·209	2·185	9·50	1·466	0·105	2·947	2·942
4·60	1·357	0·214	2·231	2·207	9·60	1·467	0·104	2·958	2·952
4·70	1·361	0·210	2·252	2·229	9·70	1·468	0·103	2·968	2·963
4·80	1·365	0·205	2·272	2·251	9·80	1·469	0·102	2·978	2·973
4·90	1·369	0·201	2·293	2·272	9·90	1·470	0·101	2·988	2·983

$\cosh^{-1} x$ is defined for $x \geqslant 1$. $\tan^{-1} x$, $\cot^{-1} x$, $\sinh^{-1} x$ are defined in $-\infty < x < \infty$: for negative arguments note that these are odd functions.

Table 7. Constants

π	3·14159	e	2·71828
2π	6·28319	$\log_{10} e$	0·43429
$\tfrac{1}{2}\pi$	1·57080	ln 10	2·30258
$1/\pi$	0·31831	ln 2	0·69315
$2/\pi$	0·63662		
$1/2\pi$	0·15915		

Table 8. Trigonometric formulae

$$\sin^2 x + \cos^2 x = 1$$
$$\sec^2 x = 1 + \tan^2 x$$
$$\operatorname{cosec}^2 x = 1 + \cot^2 x$$
$$\sin^2 x = \tfrac{1}{2}(1 - \cos 2x)$$
$$\cos^2 x = \tfrac{1}{2}(1 + \cos 2x)$$
$$\sin 2x = 2 \sin x \cos x$$
$$\cos 2x = 2\cos^2 x - 1 = 1 - 2\sin^2 x$$
$$\sin(A \pm B) = \sin A \cos B \pm \cos A \sin B$$
$$\cos(A \pm B) = \cos A \cos B \mp \sin A \sin B$$
$$\tan(A \pm B) = (\tan A \pm \tan B)/(1 \mp \tan A \tan B)$$
$$\sin A \cos B = \tfrac{1}{2}[\sin(A+B) + \sin(A-B)]$$
$$\sin A \sin B = \tfrac{1}{2}[\cos(A-B) - \cos(A+B)]$$
$$\cos A \cos B = \tfrac{1}{2}[\cos(A-B) + \cos(A+B)]$$
$$\sin A + \sin B = 2 \sin \frac{A+B}{2} \cos \frac{A-B}{2}$$
$$\sin A - \sin B = 2 \cos \frac{A+B}{2} \sin \frac{A-B}{2}$$
$$\cos A + \cos B = 2 \cos \frac{A+B}{2} \cos \frac{A-B}{2}$$
$$\cos A - \cos B = -2 \sin \frac{A+B}{2} \sin \frac{A-B}{2}$$

TABLE 9. DERIVATIVES OF FUNCTIONS

Note. The derivatives are given for the values of x for which they exist: they do not necessarily exist for all x for which f is defined.

Function $f(x)$	Derivative $f'(x)$	Function $f(x)$	Derivative $f'(x)$
x^ν	$\nu x^{\nu-1}$	$\cot^{-1} x$	$-1/(1+x^2)$
$\sin x$	$\cos x$	$\sec^{-1} x$	$1/|x|(x^2-1)^{\frac{1}{2}}$
$\cos x$	$-\sin x$	$\mathrm{cosec}^{-1} x$	$-1/|x|(x^2-1)^{\frac{1}{2}}$
$\tan x$	$\sec^2 x$	$\sinh^{-1} x$	$1/(1+x^2)^{\frac{1}{2}}$
$\cot x$	$-\mathrm{cosec}^2 x$	$\cosh^{-1} x$	$1/(x^2-1)^{\frac{1}{2}}$
$\sec x$	$\sec x \tan x$	$\tanh^{-1} x$	$1/(1-x^2)$
$\mathrm{cosec}\, x$	$-\mathrm{cosec}\, x \cot x$	$\coth^{-1} x$	$1/(1-x^2)$
e^x	e^x	$\mathrm{sech}^{-1} x$	$-1/x(1-x^2)^{\frac{1}{2}}$
$\ln x$	$1/x$	$\mathrm{cosech}^{-1} x$	$-1/|x|(1+x^2)^{\frac{1}{2}}$
$\ln |x|$	$1/x$	$\ln [x+(x^2+1)^{\frac{1}{2}}]$	$1/(1+x^2)^{\frac{1}{2}}$
a^x	$a^x \ln a$	$\ln \dfrac{1+x}{1-x}$	$2/(1-x^2)$
$\sinh x$	$\cosh x$		
$\cosh x$	$\sinh x$	$\ln \left|\dfrac{x-1}{x+1}\right|$	$2/(x^2-1)$
$\tanh x$	$\mathrm{sech}^2 x$		
$\coth x$	$-\mathrm{cosech}^2 x$	$\ln [x+(x^2-1)^{\frac{1}{2}}]$	$1/(x^2-1)^{\frac{1}{2}}$
$\mathrm{sech}\, x$	$-\mathrm{sech}\, x \tanh x$	$\ln \left[\dfrac{1}{x}+\dfrac{(1+x^2)^{\frac{1}{2}}}{|x|}\right]$	$-1/|x|(1+x^2)^{\frac{1}{2}}$
$\mathrm{cosech}\, x$	$-\mathrm{cosech}\, x \coth x$		
$\sin^{-1} x$	$1/(1-x^2)^{\frac{1}{2}}$	$\ln \left[\dfrac{1+(1-x^2)^{\frac{1}{2}}}{|x|}\right]$	$-1/x(1-x^2)^{\frac{1}{2}}$
$\cos^{-1} x$	$-1/(1-x^2)^{\frac{1}{2}}$		
$\tan^{-1} x$	$1/(1+x^2)$		

Rules for derivatives:

$$\frac{\mathrm{d}}{\mathrm{d}x}[af(x)] = af'(x),$$

$$\frac{\mathrm{d}}{\mathrm{d}x}f(ax+b) = af'(ax+b),$$

$$\frac{\mathrm{d}}{\mathrm{d}x}(u_1 + u_2 + \ldots + u_n) = \frac{\mathrm{d}u_1}{\mathrm{d}x} + \frac{\mathrm{d}u_2}{\mathrm{d}x} + \ldots + \frac{\mathrm{d}u_n}{\mathrm{d}x},$$

$$\frac{\mathrm{d}}{\mathrm{d}x}(u_1 u_2 \ldots u_n) = \frac{\mathrm{d}u_1}{\mathrm{d}x}u_2 \ldots u_n + u_1 \frac{\mathrm{d}u_2}{\mathrm{d}x} \ldots u_n + \ldots + u_1 \ldots u_{n-1}\frac{\mathrm{d}u_n}{\mathrm{d}x},$$

$$\frac{\mathrm{d}}{\mathrm{d}x}\left(\frac{u}{v}\right) = \frac{v\dfrac{\mathrm{d}u}{\mathrm{d}x} - u\dfrac{\mathrm{d}v}{\mathrm{d}x}}{v^2}.$$

$$\frac{\mathrm{d}f(t)}{\mathrm{d}x} = \frac{\mathrm{d}f(t)}{\mathrm{d}t} \cdot \frac{\mathrm{d}t}{\mathrm{d}x},$$

$$\frac{\mathrm{d}^n}{\mathrm{d}x^n}(uv) = \frac{\mathrm{d}^n u}{\mathrm{d}x^n}v + n\frac{\mathrm{d}^{n-1}u}{\mathrm{d}x^{n-1}}\frac{\mathrm{d}v}{\mathrm{d}x} + \frac{n(n-1)}{2!}\frac{\mathrm{d}^{n-2}u}{\mathrm{d}x^{n-2}}\frac{\mathrm{d}^2 v}{\mathrm{d}x^2} + \ldots + u\frac{\mathrm{d}^n v}{\mathrm{d}x^n}$$

TABLE 10. INTEGRALS

$f(x)$	$\int f(x)\,\mathrm{d}x$
$x^\alpha,\ \alpha \neq -1$	$x^{\alpha+1}/(\alpha+1)$
x^{-1}	$\ln\|x\|$
$\ln x$	$x\ln x - x$
$\sin x$	$-\cos x$
$\cos x$	$\sin x$
$\sec^2 x$	$\tan x$
$\operatorname{cosec}^2 x$	$-\cot x$
$(1-x^2)^{-\frac{1}{2}}$	$\sin^{-1} x$ or $-\cos^{-1} x$
$(1-x^2)^{\frac{1}{2}}$	$\tfrac{1}{2}[x\sqrt{(1-x^2)} + \sin^{-1} x]$
$(1+x^2)^{-1}$	$\tan^{-1} x$ or $-\cot^{-1} x$
$(1+x^2)^{-\frac{1}{2}}$	$\sinh^{-1} x = \ln[x + \sqrt{(1+x^2)}]$
$(1+x^2)^{\frac{1}{2}}$	$\tfrac{1}{2}[x\sqrt{(1+x^2)} + \sinh^{-1} x]$
$(x^2-1)^{-1}$	$\tfrac{1}{2}\ln\|(x-1)/(x+1)\|$
$(x^2-1)^{-\frac{1}{2}}$	$\cosh^{-1} x$ for $x > 1$; $-\cosh^{-1}(-x)$ for $x < -1$
	or $\ln\|x + \sqrt{(x^2-1)}\|$ for $\|x\| > 1$
$(x^2-1)^{\frac{1}{2}}$	$\tfrac{1}{2}x\sqrt{(x^2-1)} - \tfrac{1}{2}\int (x^2-1)^{-\frac{1}{2}}\,\mathrm{d}x$
$\dfrac{1}{(x+a)(x+b)}$	$\dfrac{1}{a-b}\ln\left\|\dfrac{x+b}{x+a}\right\|$
$\dfrac{1}{x^2+px+q}$	$\begin{cases}\dfrac{2}{\sqrt{(4q-p^2)}}\tan^{-1}\dfrac{2x+p}{\sqrt{(4q-p^2)}},\quad 4q > p^2\\[6pt] \dfrac{1}{\alpha-\beta}\ln\left\|\dfrac{x-\alpha}{x-\beta}\right\|,\quad 4q < p^2\end{cases}$
	where α and β are the roots of $x^2 + px + q = 0$.
$x\sin x$	$\sin x - x\cos x$
$x\cos x$	$\cos x + x\sin x$
$x^2\sin x$	$2x\sin x - (x^2-2)\cos x$
$x^2\cos x$	$2x\cos x + (x^2-2)\sin x$
$\operatorname{cosec} x$	$\ln\|\tan\tfrac{1}{2}x\|$ or $-\tfrac{1}{2}\ln\left(\dfrac{1+\cos x}{1-\cos x}\right)$
	or $\ln\|\operatorname{cosec} x - \cot x\|$
$\sec x$	$\ln\|\tan(\tfrac{1}{2}x + \tfrac{1}{4}\pi)\|$ or $\tfrac{1}{2}\ln\left(\dfrac{1+\sin x}{1-\sin x}\right)$
	or $\ln\|\sec x + \tan x\|$
$\sin mx \cos nx$	$-\dfrac{\cos(m-n)x}{2(m-n)} - \dfrac{\cos(m+n)x}{2(m+n)}$
$\tan x$	$-\ln\|\cos x\|$
$\cot x$	$\ln\|\sin x\|$
$\sin^{-1} x$	$x\sin^{-1} x + \sqrt{(1-x^2)}$
$\cos^{-1} x$	$x\cos^{-1} x - \sqrt{(1-x^2)}$
$\tan^{-1} x$	$x\tan^{-1} x - \tfrac{1}{2}\ln(1+x^2)$
$\cot^{-1} x$	$x\cot^{-1} x + \tfrac{1}{2}\ln(1+x^2)$
$\sinh^{-1} x$	$x\sinh^{-1} x - \sqrt{(1+x^2)}$
$\cosh^{-1} x$	$x\cosh^{-1} x - \sqrt{(x^2-1)}$
$\tanh^{-1} x$	$x\tanh^{-1} x + \tfrac{1}{2}\ln(1-x^2)$
$\coth^{-1} x$	$x\coth^{-1} x + \tfrac{1}{2}\ln(x^2-1)$
$x^n \mathrm{e}^x$	$\mathrm{e}^x(x^n - nx^{n-1} + n(n-1)x^{n-2} - \ldots + (-1)^n n!)$
$\mathrm{e}^{ax}\sin bx$	$\dfrac{\mathrm{e}^{ax}}{a^2+b^2}(a\sin bx - b\cos bx)$
$\mathrm{e}^{ax}\cos bx$	$\dfrac{\mathrm{e}^{ax}}{a^2+b^2}(a\cos bx + b\sin bx)$

Index

Abscissa 1
Absolute value 74
Acceleration 336
Aitken's δ^2-method 237
Algebraic function 80
Algorithm, defined 110
  for bisection method 110
Angle, between curves 54, 146
  positive 9, 28
Anti-derivative 333
Approximation, numerical 124–25
  to derivative 125, 167
  to integral 400–407
Arch, masonry 38
Area, differential equation for 364, 366
  inequality for 364
  integral representation of 362
  in polar coordinates 381
  signed 377
Area analogy 376, 412
Argument of function 65
Arithmetical progression 247
Asymptote 70, 71, 72, 180
  of hyperbola 51
Average, moving 439
Axes, left-handed 3
  right-handed 3, 7
  rotation of, 26–28, 55
Axial symmetry 385
Axis 1
  major 45
  minor 45
  polar 28

Bacteria, growth of 290, 296
Bending moment 167, 174
Bernoulli numbers 292
Binomial expansion 280
Biological growth 291

Bisection method 109–112
Bomb, motion of 38
Braking distance 74
Branch 319
  principal 320
Bridge, suspension 38

Car cylinder, rebore 190
Cardioid 384
Catenary 309
Centre, of circle 31
  of conic 57
  of ellipse 45
  of hyperbola 51
Centroid 20
Chain rule 146–151
Charge 145
  force of 423
Chemical reaction 290
Chord, approximating curve 17, 402
Circle 31, 58
  centre of 31
  normal to 41
  tangent to 40
  through 3 points 32
Circles, coaxial 34
Coefficients of polynomial 78
Comet, motion of 39
Cone, depth in 120, 149
  volume of 387
Conic 31–63
  section 58
  central 58
Conics, confocal 53
Constant 64
Contact, point of 39
Continuity 101
  at end-point 102
  on interval 102
  piecewise 427
  of polynomial 105

Convergence, absolute 260
  conditional 260
  interval of 269
  of sequence 218
  of series 246
Coordinate axes 1
Coordinates 1
  polar, 28–30, 34
    in integration 381
Cosine 7
Cube root 137, 463
Current 145, 167
Curve, concave downward 175
  concave upward 175
  of second degree 54–63
Curves, angle between 54, 146

Decimal places 238
Decimal, recurring 248, 249
Density 394
  heat 394
  inequality for 395–396
Derivative, absence of 138
  of constant 130
  defined 118
  of implicit function 143, 160
  of inverse function 325, 330
  logarithmic 301
  nth 158
  numerical approximation to 125, 130, 167, 190, 210
  of polynomial 130
  positive 128
  of power series 274
  of product 138
  of quotient 140
  rules for 474
  second 158
  of sum 132
  term-by-term 274
  third 158
  of trigonometric function 151–154
  vanishing of 186
Derivatives, table of 474
Differential coefficient 118
  equation 186, 332
  for area 365
Directrix, of ellipse 49
  of hyperbola 53
  of parabola 34

Discontinuity 73, 103
  infinite 103
  integral over 436
  removable 103
Distance between points 3–6, 29
Domain 65
Drug concentration 291

Eccentricity, of ellipse 45
  of hyperbola 51
Electricity, cost of 74
Ellipse 44–49, 58, 60, 64
  area of 367, 371
  normal to 48
  in polar coordinates 49
  tangent to 46, 49
Equation, differential 332
Error analysis 187–190
  round-off 209, 238–244
  truncation 209
Errors, relative 155
  small 154–157
Exponent of number 239
Exponential function 285–292, 470
  exponent of 287
Extrapolation 212

Fermat's principle 172
Floating point 239
  exponent of 239
  fractional part of 239
Flow-chart 112, 137, 207
  use in law 114
Focus, of ellipse 44, 49
  of hyperbola 50
  of parabola 34
Focus-directrix property 58
Force, work done by 423
Fraction, partial 80–83, 360–361
Function 64–74
  algebraic 80
  argument of 65
  composite 146, 160
  continuous 101–109
    extremes of 107
    on interval 102
  differentiable 118
  discontinuous 103, 104
    on interval 105

Function (*contd.*)
  double-valued 71, 72
  even 67, 72
  exponential 285–292, 470
  function of 146
  Heaviside unit 96
  hyperbolic 309–315
  implicit 143, 160
  increasing 128
  inverse 318–331, 471–472
    branch of 319
  multiple-valued 66
  not-differentiable 138
  odd 67
  piecewise continuous 427
  rational 80, 87
  single-valued 65
  step 73, 416
  transcendental 80
Fundamental theorem of calculus 435

Gas, adiabatic law 156
  behaviour of 156
G.P.O., charges of 73, 173
Geometric series 247
Glass, drinking 174
Graph plotter 54, 63
Gravity, error in 156
  motion under 24, 38, 129, 134, 337

Heat content 420
  density 394
Heaviside unit function 96
Highway Code 74
Hyperbola 49–54, 58, 61
  asymptotes of 51
  branches of 51
  normal to 52
  rectangular 51
  tangent to 52, 53
Hyperbolic functions 309–315, 470
  inverse 328–331, 471–472

Income tax 174
Indeterminate forms 192–200
Induction 89
Inequality sign, defined 46

Infinite series 245–284
  absolutely convergent 260
    operations with 264
  alternating 258, 261
    error in 259
  Cauchy's test 257
  comparison test 251
  computation of 281
  conditionally convergent 260
  convergent 246
    operations with 263
  divergent 246
  power 267–276
  ratio test 254
Infinity, negative 97
  positive 97
Inflection, point of 177, 180
Information theory 305
Initial condition 336
Integral as inverse of derivative,
    area analogy for 376, 412
  definite integral derived from 369
  definition 333
  factorable 351
  function of upper limit 380
  indefinite 333
  inequalities for 410
  notation for 338
  numerical estimation 400
  power series 408–410
  properties of 375, 381
  variable of integration in 370
Integral, defined as sum,
  definition 426
  derivative of 433
  existence of 426
  function of upper limit 431, 433
  fundamental theorem 435
  indefinite 435
  intermediate value theorem 434
  limiting process 416, 424
  properties of 428
Integrals, table of 475
Integrand 338
Integration,
  arbitrary constant 334
  factorable integrand 354
  functions of $ax + b$ 344
  partial fractions 360
  rotational functions 356

Intermediate value theorem 434, 439
Interpolation 210–212
  Lagrange's polynomial 211
  linear 17, 189
Inverse function 318–331
Isosceles triangle 5
Iterative method 110, 137, 227–238
  accuracy of 242–244
  order of 234–236
  for roots of equation 228, 232–234, 299, 307

Lagrange interpolation coefficients 211
Latus rectum, of ellipse 48
  of hyperbola 52
  of parabola 35
Leibnitz's theorem 315
Lemniscate 384
Lens, properties of 157
L'Hôpital's rule 192, 303
Light, Fermat's principle 172
  in inhomogeneous medium 420
  intensity of 373
  through lens 157
  reflection of 44, 49, 171
  refraction of 172
  strength of 174
Limaçon 384
Limit 84–95
  infinite 99
  left 97
  one-sided 95–99
  right 96
  rules for 87
  for trigonometric expression 93
  uniqueness of 95
Line, division in given ratio 18
  equation of 20–25, 64
    in polar coordinates 30
  perpendicular distance from 22, 24
  projection of 23
  slope of 11–20
Lines, angle between 15
  intersecting 58
  parallel 24, 60
  perpendicular 15, 24, 63
Liquid, viscous 399
Locus, defined 33
Logarithm, calculation of 298
  common 305, 468–469
    characteristic of 306
    mantissa of 306
  general 304
    base of 304
  Napierian 293
  natural 293–299, 466–467
    base of 304
    series for 295
Logarithmic derivative 301

Maclaurin's theorem 203, 212
Mantissa 306
Masonry arch 38
Mass of non-uniform beam 396, 423
Maximum, local 162
Mean, arithmetic 76
  geometric 76
Mean-value theorem 184–187
  generalized 191
Military supply 174
Minimum, local 162
Modulus 74
Motion in parabola 38

Newton's method 135–138, 154, 210, 227, 234, 292
  order of 235, 236
Noble, B. 244
Norm of partition 425
Normal to curve 39–44, 54
  ellipse 48
  hyperbola 52
  parabola 43
Numerical approximation 124–125
  to derivative 125, 130, 167, 190, 210
  for integral 400–407
  in iteration 242–244

Ordinate 1
Origin 1
  change of 25–26, 54

Parabola 34–39, 57, 59, 65
  area under 373
  degenerate 57
  normal to 43
  tangent to 42
  vertex of 35

Parallelogram 19
Parcels, charges for 73
Partial fraction 80–83
   in integration 360–361
Partition 425
   norm of 425
Perpendicular distance from line 22, 24
Perpendicular lines 15, 24, 63
Piecewise continuous function 427
Polar axis 28
   coordinates 28–30, 34
Pole 28
Polynomial 77–83
   coefficients of 78
   continuity of 105
   of degree n 78
   derivative of 130
   division of 78
   identical vanishing of 79, 183
Power of a number 299
Power series 267–276
   integral of 408
Pressure 156, 423
Principal branch of inverse function 320
Probability distribution 290
Projection 23

Quadrant 2

Radar navigation system 53
Radian 9, 91, 464–465
Radius of circle 31
Rainfall 399
Range 65
Rate of change 116–123
   of area 365
   average 117
   of circle 120, 150
   of cube 119
   of sphere 119, 150
   of square 119
   of volume 121
Rational function 80, 87, 356
Reaction, chemical 166, 290
Rectangle 19
Recursion formula 214
Resistance 157
Riemann sum 424–425

Rolle's theorem 181–183
Root of equation, bisection method 109
   iterative method 228, 232
   Newton's method 135–138, 154, 292
   by Rolle's theorem 182
Round-off error 209, 238–244

Sector, area of 381
Sequence 213–244
   bounded 220, 226
   convergent 218
   defined 214
   divergent 218, 220
   limit of 218, 219
Sequences, addition of 221
   division of 224
   multiplication of 222
Series, alternating 258, 261
   arithmetic 247
   exponential 287
   geometric 247
   harmonic 250
      generalized 252
   infinite 245–284
   logarithmic 295
   power 267–276
      continuity of 273
      convergence of 269
      derivative of 274
      operations with 273
   Taylor 277–282
Significant figures 238
Simpson's rule 406
Sine 7, 464
   hyperbolic 309, 470
Slide rule 305
   log-log 305
Slope 11
   infinite 13
Snell's laws 172
Solar system 39, 48
Speed 117, 121
   average 117, 186
   under gravity 24, 129, 134, 151
Sphere, volume of 387
Square root 135, 463
Stationary point 163
Step function 416

Summation notation 282–284
Symmetry, about line 35, 67
 about origin 67

Tangent of angle 7, 17
Tangent to curve 39–44, 54, 145
 length of 42
 rule for conics 44, 63
Taylor expansion 201
Cauchy's remainder 202, 206
Lagrange's remainder 201, 209
 series 277–282
Taylor's theorem 200–205
 numerical applications of 205–210
Telephone calls, cost of 74
Temperature, Centrigrade 24
 on equator 186
 Fahrenheit 24
Thinking distance 74
Torus 389
Traffic flow 373
Transcendental function 80
Trapezium rule 402
Triangle, centroid of 20
 change in 154, 155
 formulae for 10
 isosceles 5
Trigonometric formulae 7–11
 error in 157
 limit of 93
 numerical values 17, 20
Trigonometric function, derivative of 151–154
 interpolation of 190
 inverse 321–328, 471–472
 values of 328
 series for 307–308
 Taylor expansion of 208
Truncation error 209

Variable 64
 dependent 65
 independent 65
 of integration 338
Velocity 336, 419
Vertex of parabola 35
Viscous flow 399
Volume, equation for 387, 392
 inequality for 386
 of solid of revolution 385

Washing machines 170
Whispering gallery 49
Work done 423

x-axis 1

y-axis 1